READING TO BARBADOS AND BACK

READING TO BARBADOS AND BACK

Echoes of British History: a Narrative of the Tudor Family of Haynes of Reading

Stewart Johnson

Book Guild Publishing
Sussex, England

First published in Great Britain in 2011 by
The Book Guild Ltd
Pavilion View
19 New Road
Brighton, BN1 1UF

Copyright © Stewart Johnson 2011

The right of Stewart Johnson to be identified as the author of
this work has been asserted by him in accordance with the
Copyright, Designs and Patents Act 1988.

All rights reserved. No part of this publication may be reproduced, transmitted, or stored in a retrieval system, in any form or by any means, without permission in writing from the publisher, nor be otherwise circulated in any form of binding or cover other than that in which it is published and without a similar condition being imposed on the subsequent purchaser.

Typesetting in Garamond by
YHT Ltd, London

Printed and bound in Great Britain by
CPI Group (UK) Ltd, Croydon, CR0 4YY

A catalogue record for this book is available from
The British Library.

ISBN 978 1 84624 611 1

'The evil that men do lives after them; the good is oft interred with their bones.' – William Shakespeare: *Julius Caesar*.

'It is good to go back to the rock whence we were digged and sometimes historically instructive . . . If it is good to read history why should one exclude living in one's own past, especially if in many respects it is more agreeable than the present?' – Edmund Sidney Pollock Haynes: *The Lawyer, a Conversation Piece*, Eyre & Spottiswood, 1951, *pp. 53-4.*

This book has been written for Trilby, Bart, Meriel, Tallulah and Monty, some of whose forebears appear in these pages.

Contents

Acknowledgements   xiii

Introduction   xv

Part I   Tudor and Stuart England and the Commonwealth   1

1   The Family of Haynes of Reading in the Early Sixteenth Century   3

Richard Haynes of Whitley - Thomasine Folkes - Whitley - Conduit Close - Hayneses in the Royal Household - Churchwarden of Reading St Giles - Reading martyrs - Exemplification of Arms - Children

2   Who Was Thomasine?   19

Visitation of London 1568 - Foxley or Folkes? - Daughter of More - Nicholas More and John Folkes - Burials in St Giles - Maternal descent from John Foxley of Berkshire and Sussex

3   Children of Richard and Thomasine Haynes in Berkshire: Richard and Christian Haynes of Wargrave and Thomas Haynes of Whitley   31

Richard Haynes - House at Hinton - Marriage to Anne Ford - Death - Will - Marriage of widow

Christian Haynes - Marriage to Thomas Gunnell - Children - Richard's bequests - Wider Gunnell family

vii

Thomas Haynes - Inheritance of Whitley - Will and bequests - Marriage to Agnes Wheatley - Churchwarden of Reading St Giles

4   John Haynes Esq of Marylebone, Sergeant of the Acatry [1512-1608]   39

Royal Household - Yeoman of the Store - Purveyor - Underkeeper of Marylebone Park - Sergeant of the Acatry - Tottenham Court - Sayes Court - Creslow Pastures - Responsibilities - Fishmonger - A Sergeant's Lot - Major Perks - St John's Wood - Justice of the Peace - Fall from Grace - Retirement - Family - Agnes and Joan - Son William - Will and Bequests - The Gunnells

5   Christopher Haynes, Customer and Taverner of Arundel [c.1527-1586]   74

Childhood in Reading - Marriage to Elizabeth Colbroke - Burgess of Arundel - Mayor of Arundel - Customer - Serious Allegations - William More of Loseley, Vice-admiral for Sussex - Properties at Ford and Climping - Smuggling and piracy - The Falcon - Purveyor of the Sea-fish - In the Wake of Throckmorton - William Shelley - Escape of Paget and Arundell to France - Examinations - Taverner - Unanswered Questions - Family and Circle - Christopher Haynes of Billingshurst

6   Nicholas Haynes of Hackney, Purveyor of Grain [1529-1594]   109

Reading - Marriage to Agnes Enowes - Harbinger - Purveyor - Norfolk Wheat Transactions - Arms and the Man - Family Life - Children - Earls Colne - Hackney - Death from Plague

7   William Haynes of London, Purveyor of Sea-fish [c.1532-1598/9]   130

CONTENTS

Fishmonger - Perks - The Stocks Market - Marriage and Family - Rye - Dissent and Disorder - Replacement - Other Williams - Last Years

8 Birth, Marriage and Death in the Haynes Family   154

Birth - Marriage - Death - Causes of Death

9 In the Queen's Service   161

Domus Providencie - Past Hayneses in royal service - Hurst - The diet: allowances to household staff - Entertainment at Court and elsewhere - Dispersal of the family

10 Comparisons of Coats of Arms of Possibly Related Families   172

Arms with three Crescents - Bezants and Annulets

11 The Next Generation   186

Haynes of Reading and Hackney: Children of Nicholas Haynes

   Thomasine Haynes; children by William Dartnall - Joan Haynes - Richard Haynes - Alice Haynes - Robert Haynes, son or grandson?

Gunnells of Wargrave: Children of Christian Haynes

   Sara Gunnell - Moses [Moyses] Gunnell - Daniel Gunnell - Christopher Gunnell - William Gunnell - Note on William Gunnell junior and Ralph Gunnell - Other Children of the Gunnells of Wargrave

12 Children of Richard Haynes of Hackney [1600–1686]   216

Plague victims: John, Ann and Elizabeth - Richard putative émigré or figment? - William - Susanna, father's heiress - Edward Hopkinson - Nicholas Meade - Thomas Ballard - Susanna's children

Part II	**Barbados**	225
13	Settlement on Barbados [c.1635-1680]	227

Settlement - Origins - Known facts - Acquisition of land - The Earliest Hayneses - William and Robert Haynes - Elizabeth Haynes [-c.1674] - Edmund Haynes [- bef Oct 1685] - John Haynes [?1639-1678] - William Haynes [-1679] - Other early Barbadian Hayneses - The Dutch Connection

14	Children of John and Anne Haynes of St John, Barbados [c.1660-c.1730]	254

Robert Haynes [c.1666-1696] - John Haynes of New York [-1689] - A Victim of Leisler's Rebellion - Elizabeth Haynes - Anne Haynes

15	The Family Multiplies - the Plantocracy [1700-1800]	272

Richard Haynes [1694-1739] - Captain Robert Haynes [1720-1753] - Edmund Haynes [c.1739-1811] - Major-General Richard Downes Haynes [1746-1793]

16	Lieutenant-General Robert Haynes [1769-1851]; End of an Era	289

School - Stepmother - America - Marriage - Newcastle - Family - Expansion - Approaching Abolition - England - Back to Planting - Insurrection - Speaker - Resignation - Ann Lovell

Part III	**Back in England**	341
17	Emancipation and the Move to England [1835-1851]	343

Cutting the ties - Distribution of wealth - Bath - Death of Ann Thomasine - Marriage to Anna Caitcheon - Reading - Major Robert Haynes Lovell - Ann Cave Haynes Lovell - Will - Obituaries

CONTENTS

18 The General's Siblings [1772-1860]   356

Richard Downes Haynes [1772-1793] - Captain Henry Haynes RN [1776-1838] - Ann Elcock Haynes [1777-1857] - Dorothy Haynes [1779-aft. 1850] - Edmund Haynes [1781-1846]

19 The Diarist's Family [1792-c.1931]   364

Richard Haynes [1792-1859] - Robert Haynes [1795-1873] - George Barrow Haynes [1796-1825] - Thomasine Haynes [1799-1880] - William Clark Haynes [1800-1831] - Henry Husbands Haynes [1801-1852]

Epilogue - A Family of High Respectability   399

Index   405

Acknowledgements

What little has been achieved could not have been done without the assistance of many individuals and organisations. I acknowledge my debt of gratitude to the late Ivor Johnston of Helensburgh who started me on this quest with the loan of E.M.W. Cracknell's: '*The Barbadian Diary of Gen. Robert Haynes 1787-1836*', copies of correspondence with his Barbadian family and his mother, Brenda's: '*Memories and Confessions of a Grandmother*', privately published in 1949, which documented her Haynes forebears. I also received much helpful encouragement from the late Major Peter Hutchinson, another descendant of the General, who was an enthusiastic archivist of the Hutchinson descendants. I am also grateful to Sue and John Hutchinson for a copy of the notes made during their visit to the Barbados Department of Archives in 2003. They and more recently my daughter, Meriel Johnson, took many photographs of the monumental inscriptions of St John's Church and some of the plantation houses. Over the years I have enjoyed much electronic correspondence with the general's wider living family in Barbados, Australasia and the United States and I am particularly grateful to Grant Houston of Lake City, Colorado, for providing copies of his photographs and some quotations from *The Barbadian*, to Ernest M Wiltshire, Friends of the Barbados Archives, Ottawa, for information held by him on the Haynes and Reece families, and to Angela Lloyd-Roberts for her notes on the family of William Clarke Haynes.

I am pleased to acknowledge the assistance of the archivists, team leaders and staff of the Barbados Department of Archives, Berkshire Record Office, British Library, East Sussex Record Office, Essex Record Office, Hackney Archives Department, London Metropolitan Archives and Conservation Workshop, Society of Genealogists, Surrey History Centre (especially for permission to quote from the Loseley Manuscripts), West Sussex Record Office and The National Archives, in supplying much of the manuscript material and granting permission to reproduce extracts of wills, deeds, letters and other manuscripts. The permission of The Berkshire Archaeological Society to reproduce the illustration of the Old Conduit, Whitley, of Norfolk Record Society for permission to quote extracts from *The Papers of Nathaniel Bacon of Stiffkey*, and of the East Sussex Record Office for permission to use the illustration and transcription of William Haynes's letter are also gratefully acknowledged; Christopher Haynes's signature is reproduced by courtesy of the County Archivist, West Sussex Record Office. The Cowdray and Wiston Archives, accessed at the West Sussex Record Office, were kindly made available by Lord Cowdray and the Goring Family respectively, and their use is gratefully acknowledged. I am also indebted to Professor Alan Macfarlane for his generous permission to use extracts from the database of his monumental Earls Colne Project and to the Archivist, Senate House Library, University of London, for permission to publish extracts of Robert Haynes's correspondence from the Papers of the Newton Family.

Transcription of Tudor and Stuart English and translations from Latin were initially time-consuming but latterly overwhelming. To the rescue with these and research tasks at Hackney, Kew and elsewhere, came Richard Samways of Portland, Dorset, whose skill and input was as invaluable as it was impressive. To him I owe a huge debt of gratitude. I am also indebted to Christopher Whittick for his guidance on interpreting much of the material on Christopher Haynes.

Finally and above all, indulgence of this kind is not without cost: time lost while away from family, the generosity of family and others trying to appear interested while understandably stifling those yawns, endless time and interest of my dearest wife Trilby in selflessly urging me to complete this review of what the two of us managed to unearth. I do so hope that the account in these pages will go some small way towards compensating her not only for her love, encouragement, helpful criticism, and understanding, which made this book possible, but also for those lost hours when we might otherwise have been together.

Stewart Johnson, March 2011

Introduction

It was in the summer of 1974 when the late Ivor Johnston of Helensburgh, on learning of our planned holiday to Barbados, mentioned his distant connections with that island and of a visit he made there in the 1930s. When we next met, he brought with him his copy of E.M.W. Cracknell's: *The Barbadian Diary of Gen. Robert Haynes 1787-1836*, privately published in 1934. Ivor was a direct descendant of the general and had an infectious interest in his family about which he had so many historical anecdotes. On reading the '*Diary*', which Ivor kindly left behind along with his mother's correspondence with distant cousins on Barbados, I was hooked. It was the first time I had come close to a family history spanning three centuries ostensibly passed on by word of mouth – yet the more I reflected on it, the more I began to question it. Such questioning, never entirely quiescent, inevitably gave way to the important things of everyday life, but it was always the intention to return to the subject in an attempt to unravel the truth. This proved possible only after retirement in 2000 and this book is the product of research and further reflection since then.

General Robert Haynes, born on a sugar plantation in Barbados, had a strong belief in the importance of his family and of his own destiny from an early age. He had been weaned on stories of his ancestors and how they came to be in the West Indies and was clearly proud of the fact that he could trace his forebears to the early years of settlement of the island. He believed that he was descended from a Berkshire family living in the times of Henry VII and Henry VIII, whose coat of arms he would continue to use, and around his eighteenth birthday he penned the following brief outline of his origins:

'Sept 1787. At the house of Mrs Sarah Bradford, High Street, Bridgetown, Barbados.

✓ Myself, Robt Haynes born Septr 27, 1769 at the old home, Newcastle, Parish of St John's, Barbadoes, a descendant of Richard Haynes, Royalist, forced to emigrate to Barbadoes in the time of Cromwell. Here their ancestors have held landed property which has descended from father to son ever since.

When the first of our family emigrated to Barbadoes I cannot ascertain, but they were here at an early period after the settlement of the Island, being with those who, defying the Arch-Tyrant, Cromwell, did openly proclaim His Majesty, King Charles II as King of England and this Country.

Our family is one of high Respectability and ancient lineage, originally being natives of Berkshire and sometime owners of Haines Hill, Berks, nr Reading, England.

In 1500 lived Richard Haynes, who married Thomasine, dau: and Heiress of John Foxley Esqre, thereby quartering the Arms of Foxley with his own. These lived at Foxley Grange nr Reading. Of their six sons, four are known to have held office in the household of Queen Elizabeth, one being the Groom of her Bedchamber, and another, Nicholas, Purveyor of Her Majesty's Grain and Malt at Hackney. Here he died, and was buried Janry, 1593.

This Nicholas had a son – Richard – whose uncles devised to their nephew Lands in the Parish of Hurst, Berks, and one – John – his Seal of Office with his Arms upon it'.[1]

Robert Haynes inherited Newcastle Plantation on his father's death and was at once elected to the House of Assembly as a representative of St John's parish, which he served with distinction through a particularly turbulent period. He also became a Lieutenant-General of Militia. In 1825, he was unanimously elected Speaker of the House of Assembly, an office he held for four years. He retired from government at a critical time for Barbados and remained to see the island through the emancipation process with which he was not entirely sympathetic. It was at this point that he and many of his family moved to England; he to the Reading of his forebears, and where on 18 April 1851, he died.

Several points in the above extracts from the 'diary' suggested that the story might have been uncritically accepted, embellished, or modified through the generations of its telling. For example Thomasine being the daughter of John Foxley of Berkshire was difficult to reconcile with the fact that no evidence for such person existed; I discovered that the last male member of the Bray family of that name, Thomas, had died in 1436 – and there were very few Foxleys reportedly living in Berkshire around 1500. The general's wording is

INTRODUCTION

Lieutenant-General Robert Haynes, Speaker of the House of Assembly.
(© National Portrait Gallery, London)

such that it appears to have been taken from the 'Exemplification of Arms' of Nicholas Haynes of 1578, subsequently published in the Visitation of Berkshire of 1665-6.[2] Secondly, Foxley Grange, although located a short distance from the site of the fourteenth-century Foxley manor in Bray parish, was most unlikely to have been constructed as early as 1500. Furthermore, the house with Elizabethan origins known as Haines Hill was first known by that name in the 1590s when in the possession of the Windebank family. What was the evidence for the general's assertion that his family were *sometime owners* of it? I thus began to question the veracity of other statements concerning the Haynes's remote lineage. Was the general really descended from Richard and Thomasine Haynes, through their son, Nicholas of Hackney? Were there four sons who worked in the Royal Household? Was Richard Haynes the son of Nicholas the one to whom his uncles devised their lands? Did he or his son migrate to Barbados? Was the romantic notion plausible of his open support for King Charles II being the reason for Richard's migration? Was it true that the same plantation was passed from father to son in an unbroken line ever since?

There were also questions in relation to English, later British, political events. How did the family cope with the Dissolution, repeated changes of religion, the Jesuit threat to Queen Elizabeth and trans-Atlantic travel? What was the impact on the family in Barbados of the colonisation of the Americas, slavery, rebellions, the American war of independence, the wars with France, slave uprisings and ultimately the abolition of slavery?

This book sets out to piece together the surviving fragmentary evidence left by this family *'of high respectability'* that Robert Haynes passionately believed to be his own and the main descent of the family to the general. The reader will be struck, as was the writer, by how the story of this family parallels the events of the nation, in which on not a few occasions it was directly involved. Part I, 'Tudor and Stuart England and the Commonwealth', traces the general's putative forebears in England of the sixteenth and seventeenth centuries, beginning with Richard Haynes and his wife Thomasine in Reading, and each of their children and grandchildren. The likely identity of Thomasine is revealed for the first time as are the details of her three sons in the Royal Household, including their perks and daily work and of a fourth, peripherally involved in the Throckmorton conspiracy. The coats of arms are scrutinised in order to identify possibly related families. Finally, the dangers to life and limb from disease and from politics that led people to uproot themselves and their families for distant climes are all discussed.

Part II deals with the general's ancestors in Barbados. The very earliest Haynes family on the island is reconstructed, but a leap of faith is nevertheless required to link it with the family back in England. The family is traced as it emerges as part of the sugar plantocracy and how it coped with political and natural cataclysms, slavery and emancipation; ending as a family at its zenith, but nervous of its future. Part III, 'Back in England' covers the consequences for the family of the abolition of slavery including the return of the general, with several of his siblings and children, to England, the remainder staying behind to run their plantations.

The subject material includes newly discovered information enabling the assertions and conjectures of the general and others to be challenged; yet the task is far from complete. It is very much work in progress which has concentrated on the early English end of the story. Much remains to be done in piecing together the story of the early Barbadian Hayneses, which should therefore be regarded as preliminary, awaiting confirmation or refutation. Furthermore, Chapter 19 contains only the briefest of sketches of the General's grandchildren, which might otherwise have been included as an

INTRODUCTION

Appendix, its main purpose being to enable living descendants more easily to link to their Haynes forebears. The author is neither historian nor writer, merely a delver into archives, the results of which have enabled him to string together the traces left by this family. There will undoubtedly be errors of omission, judgment and interpretation, but it is hoped that there will be enough new material to stimulate others not only to fill in the gaps, but also to place the material that deserves to be so into its appropriate historical context.

Notes and references

1. E.M.W. Cracknell, ed. (1934), *The Barbadian Diary of Gen. Robert Haynes 1787-1836*, The Azania Press, Medstead, Hampshire, UK.
2. Visitation of Berkshire in 1665-6, Harleian Society Vol. LVI, p 144, 1907.

Part I
Tudor and Stuart England and the Commonwealth

1

The Family of Haynes of Reading in the Early Sixteenth Century

Richard Haynes of Whitley – Thomasine Folkes – Whitley – Conduit Close – Hayneses in the Royal Household – Churchwarden of Reading St Giles – Reading martyrs – Exemplification of Arms – Children

Haynes was a common name in the fifteenth and sixteenth centuries, occurring time and time again in the Close and Patent Rolls. There are many variant spellings of the name including Haines, Heynes, Heines, Haynnes, Haynys, Hawnes, Hawenys, all with or without the 's' or 'es' and occasionally without the 'h'; none is useful for distinguishing between family groups in the sixteenth century and the occurrence of two or more variants was not uncommon in a single document. It appears to have been the locative name of Haynes or Hawnes in Bedfordshire. By the sixteenth century many Haynes families had become armigerous in the heraldic sense, their most common blazon comprising three crescents. Their sheer numbers render genealogical search both difficult and unreliable before 1500, which is therefore where this story begins.

Richard Haynes and Thomasine Folkes alias Foxley

Richard Haynes lived with his wife, Thomasine, in the Parish of St Giles, Reading, Berkshire, where Richard held lands of the Abbot of Reading, in the hamlet of Whitley.[1] This was in the early sixteenth century, the first decade of which saw the births of their first children. With the Dissolution, the Abbey lands were sequestered and in 1548, Whitley Manor and the Abbot's deer park were granted to Edward, Duke of Somerset. But after his attainder four years later, the manor was leased by Queen Mary to Sir Francis Englefield, from whom it was sequestered in 1564 and the reversion granted by Queen Elizabeth to Sir Francis Knollys [Knolles] her Treasurer.[2] The Haynes family continued to occupy their Whitley home through all these changes in their superior landlord. Knollys himself lived in the Abbey House at Reading and by the time the Haynes family had become his tenant, three of Richard and Thomasine's sons were coincidentally already working under him in the Royal Household.

We know that the Haynes family lived in Whitley because of what was mentioned in the Charter granted to Reading by Queen Elizabeth in 1559. Paragraph 57 deals with the granting to the Mayor and Burgesses of the revenues of the Englefield estates and several other messuages, tenements, lands, etc. to be held by fealty only. One of these properties appears: *'And all that our close or pasture with the appurtenances lying at the end of Syvyorstrete [Silver Street] called the Conduit Close containing by estimation two acres now or late in the tenure or occupation of Richard Haynes at will or his Assigns'*.[3]

Conduit Close near the old turnpike at the southern end of modern Whitley Street and approximately bounded today on the east by Highgrave Street, was the location of the spring head of the water supply to Reading Abbey and was part of Whitley Manor. In Richard's time, the conduit was a brick structure about 9ft long and 5ft 6ins wide and rising 4ft above ground level, situated in conduit field. Access to the water was by a brick archway on the west side about 3ft wide with steps down to the water level.[4] There were two internal arches 7ft wide from east to west covering a cistern of water of 14ft 6ins in length from north to south. When the conduit was cleaned out in the nineteenth century, the height from the cistern floor to the head of the arch was about 7ft. The remarkable fact is that Reading Abbey was supplied with fresh, cold, soft spring water conveyed through about a mile of lead piping which actually passed under the river. The head of water high up in Whitley would have probably operated a fountain and filled troughs and other receptacles in the Abbey. During the years following the Dissolution, many lengths of the lead conduit were dug up and sold. The water from the conduit was the main supply of the neighbourhood and would have been the Haynes family's source of fresh water situated as it was on their land. For several centuries the water was reputed to be of medicinal benefit for inflammatory eye complaints. In the 1730s the conduit was situated in the garden of the Fountain ale-house which stood on the left hand just above Whitley turnpike. A drawing of the conduit was made by T. T. Cumming in 1908 (Figure 1.1).[5]

On Richard's death the tenancy passed to his eldest surviving son, Thomas, who in his will was styled '*of Whitley in the p'ishe of St Gyles in Redinge, gent*'.[6] There is one confirmatory piece of information to be gleaned from Thomas's will proving that the Whitley location was the same as the one mentioned in the Reading Charter, for he left to '*all my tenants in the Cundyte (Conduit) Close xijd a peese*'.

Figure 1.1 The Whitley Conduit – a drawing by T. T. Cumming in 1908
(Copyright Berkshire Archaeological Society)

The family's home in Whitley was probably the typical hall house of the period with the hall in the centre and a two-storey, gabled block at each end. It could have been built of brick, stone or timber with plaster infill, rose-red bricks made from Reading clay being the most likely. Whitley farmhouse can be excluded from consideration as it was in the possession of Stephen Cawet and his family in 1542 and subsequently it belonged to his wife, Margaret and son, Thomas.[7] If it was the Whitley manor house, it might have had a gatehouse. Rooms at one end of the hall would have included a dairy, buttery and cellar. Although cooking in the early sixteenth century tended to be done at the hall fire, the Haynes' house may have had a separate kitchen. There would also have been a brewing house. The family accommodation was at the opposite end to the kitchen and brewing house and definitely included a parlour, which could double up as a dining room, when guests were entertained, a bed-chamber for visitors and a room for Thomasine and the children. There would be sleeping accommodation over both wings reached by separate staircases. Water would be from an outside pump: there was a particularly plentiful water supply from the conduit with its spring, and the manor house might well have had its own supply piped directly off it to the pump, although the same spring water could easily have been tapped into for more than a mile around the conduit. Outside would have been a range of barns, animal houses, chicken runs and an orchard. It is possible that they, or more probably their servants, farmed the surrounding fields, known as Spittlefields and Crownfields, and obtained their fuel from the coppice, known as the Grove, later Gaunder's Coppice or High Grove, the southern part of which joined the end of Conduit Close, although apart from the reference to Richard Haynes owning the two acres of Conduit Close with a tenement and garden, the full extent of the Richard's demesne remains unknown. Richard's son, Thomas, who took over the family home after his parents, appears to have sub-let properties in Conduit Close, so presumably the Haynes family house was a little distance away; Whitley hamlet itself covered 902 acres, much of it taken up by the park. At the time of the Dissolution, the manor was valued at the annual rent of £26 18s 4d and grazing in the park at £3. Another pasture in Whitley known as Catelsgrove was let at £4 13s 4d. Whitley also contained extensive pleasure grounds and fish ponds.[8]

We have some idea of the kind of furnishings they possessed from the description in Thomas Haynes's will. At the end of his life, Thomas was using the parlour at Whitley as a bed-chamber (presumably he could no longer climb the stairs) where he had his joined bedstead with coloured hangings (*'paynted cloth hereunto belonginge'*). There was a joined table standing in the hall with wooden stools, a cupboard, a press and a wooden case in which glassware was kept. This last served partly as a settle which was also hung with *'paynted clothes'*. The fact that there were wall hangings suggests that, by 1582, the central hearth and fire of the original hall, the smoke from which would have originally left by a central opening in the roof, had been replaced by a wall hearth and chimney. Windows may well have been glazed, and if so they would have been of small panes, and shuttered. Bed furnishings consisted of feather beds, bolsters, pillows and white coverlets. The sheets would have been of white linen and the blankets of wool, woven in Reading. There were assorted brass pans of different sizes including a frying pan and other kitchen utensils. Other possessions as was usual for the minor gentry, included table silver, a great service book, jewellery and assorted firearms.

Richard and Thomasine would have employed a number of live-in servants or those living in their cottages in Conduit Close. Help was essential in both the house and on the land. Women servants tended to help with the housework; the men's work was out of doors. Thomasine would have supervised or done the cooking herself, all on an open fire (needing a plentiful supply of wood). The dairy and brewhouse would also have been her responsibility. We have yet to ascertain Richard's occupation. If he was the manorial tenant, then he would have had husbandmen to help with the farmwork, but he would have had to work hard himself alongside his men. There was a Richard Haynes who was a fruiterer to King Henry VIII. Around 1520, he was recorded as being the first to plant a cherry orchard in Tenham, Kent.[9]

Thomas Haynes, who eventually inherited his parents' house, left Robert Prosser '*all such workinge tolles (tools) as he hath of myne*'. He left small sums of money *to 'everie servant in my house'*, and mentioned another servant, Margaret Somerfyld, by name, as she received only half of that left to each of the others. All his tenants in Conduit Close were remembered. Clearly this was a substantial household; the use of the phrase '*everie servant*' suggests four or more in addition to Margaret, which for a childless couple was a large number, doubtless reflecting the size of the house.

In a deeply religious age, the day for the household would begin with prayers attended by Richard and Thomasine, their children and servants. As a person of importance in the Parish Church, Richard would have been totally committed to his daily devotion. Richard Haynes was a churchwarden of Reading St Giles, from 1520 to 1522 and was responsible for keeping the account book for those years which has survived.[10] He was first elected on Good Friday 1520 to replace John Horthorne and he served for two years. The appointment was duly entered on 4 October 1521 in the Records of the Borough of Reading.[11] He was responsible for receiving the church's income from a variety of sources, especially tithes and rents of about 30 tenements and lands owned by the church, seat fees, burial fees, etc. He would also have supervised the payment of the organist – the organ had been installed two years earlier – and others in the church's employ as well as the bills for repairs to the church.[12] Among his many duties would have been that of winding the clock also installed in 1518. On Good Friday 1522 appears the following entry in the St Giles account book: *'At this accompte hath been dysmyssed Richard* Hayne[13] *ed [and] chosen William A Dene'*.[14] The account book also has entries mentioning Richard's son, Nicholas.

The Vicar of St Giles during Richard Haynes's period as churchwarden was John Eynon (vicar 1520–1533), who was executed with Hugh Faringdon, the last Abbot of Reading under King Henry VIII for refusing to acknowledge the supremacy of the King as the Head of the Church of England. Eynon, the Abbot and John Rugg, a former prebendary of Chichester, who was living in retirement at Reading Abbey, were 'tried' in Reading before a jury of local gentry, who declared them guilty, despite their being allowed no defence. Indeed it is entirely possible that Richard Haynes might have served on this jury. They were hanged, drawn and quartered outside the abbey gate on 14 November 1539, an event which might well have been attended by the Haynes family and other prominent townspeople.[15] Eynon, Rugg and Faringdon were beatified by Pope Leo XIII in 1895.

The borough records also include a list of soldiers sent from Reading towards Scotland on 12 September 1542, as back-up support for King Henry's action against James V at Solway Firth, and among these we find that Richard Haynes provided 'a harnes', i.e. plate armour and accoutrements, for '*Soldear John Hopton*'.[16] This armour would have incurred significant costs on the part of the donor, who must have been a person of substance. This large outlay may be in part explained by the fact that John Hopton was an important person locally, a person of that name having been the Mayor of Reading in 1515 and 1519.[17] Most of the well-known families from the surrounding communities held property in Reading, and a tenement and garden owned by Richard Haynes not far from the Dolphyn tavern, then occupied by Richard Mathews, was assessed in the Reading Charter of 1560 at one shilling rent payable to the corporation.[18] John Hopton's rent was two shillings.

All this confirms Richard Haynes and his family being centred on Reading and the church of St Giles, the poor of which were to be remembered in the wills of his eldest two sons, and their Reading domicile is evidenced by two other documents. The later of the two, entitled: *'Haynes of Reading. Exemplification of Armes and Creast'*, was the result of an application in 1578 to the Clarenceux King of Arms, Robert Cooke, by Richard's son, Nicholas, to have the combined Arms of Haynes and Foxley confirmed on him. It is published along with the records of the Visitation of Berkshire of 1665.[19] In it are mentioned Nicholas's father, Richard Haynes *'of Redinge'* and his mother, Thomasen, daughter and coheir of John Foxley, gentleman of Berkshire:

> ' ... And beinge required of Nicholas Haynes of Hackney in Midlesex fowrth son of Richard Haynes of Redinge in the County of Barkeshire gentilman and of Thomasen his wife daughter and coheire of John ffoxley of ... gentilman in ye County of Barkshire to make serche in ye Registers and Recordes of my Office for the auntient Armes and Creast belonging to that name and famyly whereof he is desended. Wheareuppon I have made serche accordingely and doe fynde that he maye lawfully beare as his Auncestors heretofore have borne thes Armes and Creast heareafter followeing That is to say ye first for Haynes Argent three Cressants unde azure and gules, the second for ffoxley gules twoo bares humite (cut short) silver and so quarterlly and to his Creast uppon the healme on a wreathe silver and gules a Heron volany the boddye in proper couller wynges si[l]ver Legged and beaked goulde houldinge up one of his ffeete Mantelled gules doubled silver as more playnly is depicted in ye Margent ... '

The *'quartering of arms'* referred to in this document appears to have been used by Nicholas's descendants ever since. It was only applicable to the male offspring of Richard and Thomasine. Nicholas is given as the fourth son and the differencing mark[20] referred to for the fourth son is a martlet, the tiny legless bird which appears twice on the Arms illustrated in the *'exemplification'*. The order of male issue by the Clarenceux King of Arms, was in respect of surviving heirs for, as we shall see, Nicholas was in fact the fifth-born son, one brother having already died.

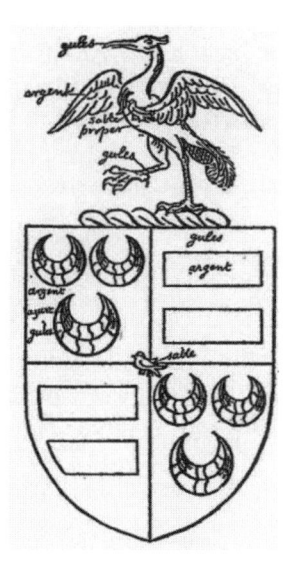

Figure 1.2 Arms and crest confirmed on Nicholas Haynes in 1578

The second and earlier document was the Visitation of London of 1568,[21] the record of an official visit by a Herald undertaken to examine pedigrees and determine the right to bear coats of arms by those then residing in the City of London, which at that time included John Haynes and his brother, William. This gave Richard and Thomasine's family in some detail:

> 'Richard Haynes of Redinge in com. Berk. gent. maried Thomasin daughter of John Folkes (Folx) and by her had yssue Thomas Haynes first Sonne that maried Agnes daughter of John Whelly and John Haynes seconde Sonne now sargeant of the Catre to our sovereyne lady Quene Elizabeth which John Haynes maried Agnes daughter of Thomson, Christopher third Sonne maried Elizabeth daughter of John Colbroke, Nycholas Haynes forth sonne maried Anne daughter of Rychard Enowes and had yssue Richard Thomasin Johan Alis, William Haynes fifte Sonne maried Ellen daughter of Harman Bonne of Wolvorhampton, Christian maried to thomas Gounde (?) of barkshere'.

Some of the following points of interest that emerge from this Visitation of London will be discussed in the next chapter. First, mention is made of a daughter, Christian, and a first son Thomas. Second, we are informed that all siblings were married, and to whom. Third, Nicholas and his wife Anne were the only ones mentioned to have had children, two of whom took their grandparents' forenames. Fourth, only the children still alive were recorded in this pedigree – Richard had died two years earlier. Fifth, Thomasine is not the daughter of John Foxley, but of John Folkes. Sixth, a number of transcription errors appear in the Visitation record. Last, and most important, there is a footnote containing the statement:

> 'F.1 [F.1. 311] has the pedigree in tabular form, beginning 'John folx mar d of Moor (and had issue) Thomasin married Richard Haynes of Redynge,' ... There is a shield (repeated twice) tricked: Argent, two bars vert between nine martlets gules'.

In other words, there is an illustration (trick) of the Arms of Thomasine's mother's family, Moor or More, in the original manuscript compiled by the Herald in 1568, an important clue as to Thomasine's ancestry, as the arms were those of More and atte More of Berkshire and Hampshire.

The churchwardens' account book for the Parish of St Giles, Reading, has a receipt for payment by John Folkes for a church seat for himself and his wife entered in the year 1521-2, when Richard Haynes was one of the churchwardens.[22] Eight years later[23] [1529-30], a receipt was recorded: *'for the grave of John ffolkes vis viijd'*. Four years later still[24] [1533-4], the entry read: *'It[em] for the grave of Mawd ffolkes vjs viiid'*. The evidence for these being the parents of Thomasine Haynes is presented in the next chapter. The fact that Thomasine's parents and grandparents lived in the parish of Reading St Giles or neighbouring parishes suggests that it was here that she met and married her husband.

Children of Richard and Thomasine Haynes

It is not known how many children Richard and Thomasine had in total. We do know, however, of their seven children, six sons and a daughter, who reached adult life and married (Figure 1.3). The ages of only two of them, John and Nicholas, are documented, from which it appears that John, the third child was born in 1512 and Nicholas, the sixth, in 1529. Thus, Thomasine's pregnancies must have been between about 1509/10 and 1531/2, about 22 years between the first and last birth. The average spacing between births in the sixteenth century was 22 months;[25] thus, Thomasine could be expected to have had up to 12 children. It is therefore possible that we may still have to account for up to five more children.

Three, possibly four, of Richard and Thomasine's sons came to be employed by the Royal Household, suggesting a history of family service to the sovereign, the commonest means of entry into royal employment at that time, and it is of interest that a number of Richard Hayneses had been in royal employ in the fifteenth century.[26,27] Literacy and numeracy were prerequisites for employment in the Royal Household, and the boys must have received a sound education. As a group, the Tudor gentry like the Hayneses were becoming increasingly keen on enrolling their children in both the petty schools [infants' schools] between the ages of 5 and 7 and grammar or monastic schools thereafter to the age of 14, at which time they would have emerged numerate, literate and fluent in spoken and written Latin. It is possible that the Haynes boys attended the grammar school in Reading linked to Reading Abbey, which had a famous teaching staff providing a liberal education.[28] Education tended to progress from grammar school to university, but there is no evidence that the Haynes boys received higher education, and this would have depended on parental financial circumstances.

If a suitable grammar school was not available, it was the parental aspiration of gentlemen for their sons to enter service in the house of an aristocrat where they would be educated along with the children of that household and learn all the attributes of a gentleman. They were known as 'henchmen' and were taught by the same teachers who taught the sons of the house. The boy in service and the sons of the house would live as brothers and could form lifelong friendships. The learning of behavioural codes such as deportment; demeanour; manners (especially at table); courtesy and consideration for others; superiority and inferiority – their place in the house; ability to interact appropriately with people at all levels of society; parental respect and duty; nepotism – forwarding family interests, was an important part of education. It was the acquisition of these social attributes which could best be acquired from service in a great house and which were desired by gentlemen for their sons.

Schooling began early, usually by 6 o'clock, the time the rest of the household would be on the move. The syllabus, although lacking structure, included elements of grammar (prose and poetry);

handwriting; French and classical languages; mathematics; sport, especially equestrian activities [hunting and jousting] and athletics; music [recorder, harp and lute] and dancing. Latin was still commonly used and many of the available books were in that language. Accomplishment in Latin was essential for university entry, the law, business and many other activities, and the Haynes boys would have spent much of their schooling translating English sentences into Latin and vice versa from classical literature. Religious instruction was not a subject of schooling, but daily religious observance was obligatory. All the Haynes children would have been brought up in the Catholic faith. Discipline, while strict, was generally fair; insolence or insubordination shown to adults was not tolerated, and beatings were the norm for those who transgressed.

Depending on where and how they were educated would determine whether their academic education continued, either at university or while still in service. Grammar school education lent itself to being followed by an apprenticeship to one of the trades. Such further training and apprenticeships would last at least seven years during which time it was generally forbidden to marry.[29] It is known that John and William were fishmongers and that their positions at court required them to to be members of that 'Mystery' or of another City Company. However, there were three ways to join such a guild: by apprenticeship; by patrimony – sons of members of guilds could claim their freedom without having to be apprentices; or by redemption by which the wealthy paid for their guild membership in lieu of seven years' toil. So John and William at least may have served an apprenticeship, particularly as they were not the eldest of the brothers, unless they entered the guild by redemption. It is not known why they chose this trade, but there was at least one Haynes, Thomas, who had been a wealthy London fishmonger in the 1440s, who might have been a forebear.[30] The lengths of indenture to their masters may have varied, but were unlikely to be much less than seven years, at the end of which time they would have been about 24 years old, the age limit set by the Statute of Artificers for completion of an indenture. It can, therefore, be appreciated that from the age of seven the Haynes boys would have seen comparatively little of their parents, yet as will be evident, their family ties were to remain strong throughout their lives, and each child emerged as a self-assured individual, well-equipped both socially and technically for his future employment.

Nicholas had no need to become a guildsman as, like Richard, Thomas and Christopher, he neither lived nor worked in the City of London. Indeed, he was still in Reading in 1545 as in the churchwarden's account book of St Giles, Reading, for that year, a note is made of payment of wages to Nicholas who would then have been 16 years old.[31]

Wills have so far been identified for four of the six Haynes brothers: Richard (not mentioned in the London Visitation), Thomas, John and Christopher, and for the husband of Christian, Thomas Gunnell. All of these wills refer to other Haynes siblings. There is also an administration record for Nicholas, who died intestate at the height of a plague epidemic. Only the will of William has yet to be discovered. However, the instructions to William under the terms of Christopher's will were that, on William's death, all properties in Arundel formerly owned by Christopher, were to be left to their nephew Richard Haynes of Hackney.[32] That this did in fact occur suggests that William did not die intestate, although a copy of Christopher's will may have sufficed for Richard [of Hackney] to claim the Arundel estate under a letter of administration. It is of interest that the initial religious preambles to the brothers' wills indicate that none of them had adhered to the Catholic faith by the time they died.

The following order of birth of the sons given in the Visitation of London has been accepted in this book: namely, Thomas, John, Christopher, Nicholas and William. However, the wills of both Thomas and Christopher put William before Nicholas in order of legatees. Furthermore, the date of Richard's marriage suggests a birth date of the mid-1520s for him, despite the fact that he was given his father's name.

Only two of the siblings, Christian and Nicholas, are known to have had children who outlived them. Those of Christian were not mentioned in the Visitation of London. The known sons of John and William predeceased their parents, although there is the possibility that William left grandchildren.

Richard Haynes was still living in 1542, when he provided the armour for John Hopton. He was clearly dead, as was his son Richard, when at an inquisition taken in Reading before William Dunninge, esquire, Sheriff of Berkshire, on 4 June 1571, enquiring into the whereabouts of Richard Haynes and two others, the twelve-man jury returned that as of the previous 10 February, Richard had

THE FAMILY OF HAYNES OF READING IN THE EARLY SIXTEENTH CENTURY

Richard Haynes
m. (c 1507)
Thomasine Folkes

Richard
(d. 1566)
m. (1549)
Anne Ford
(Anne m. (1567)
William Rockall)

Thomas
(d. 1582)
m. Agnes Wheatley
(d. 1589)
d.s.p

John
(esquire;
b. 1512; d. 1608)
m1. Agnes Thomson
m2. Joan
d.s.p.s

William

Christian
(d. 1573)
m. (1551)
Thomas Gunnell
(d. 1586)

Dorytie
Moses
Richard
Daniel
Christopher
Susanna
Christian
Sara
William
Thomas
Richard
William

Christopher
(d. 1586)
m. Elizabeth Colbroke
(living 1586)
d.s.p

Nicholas
(b. 1529; d. 1594)
m. Agnes Enowes
(d. 1605)

Thomasine
Joan/Jane
Richard
Alice
Robert (?)
Christian (?)
Nicholas (?)

William
(d. c 1598)
m. Ellen Bonne
(living 1586)
d.s.p.s

William

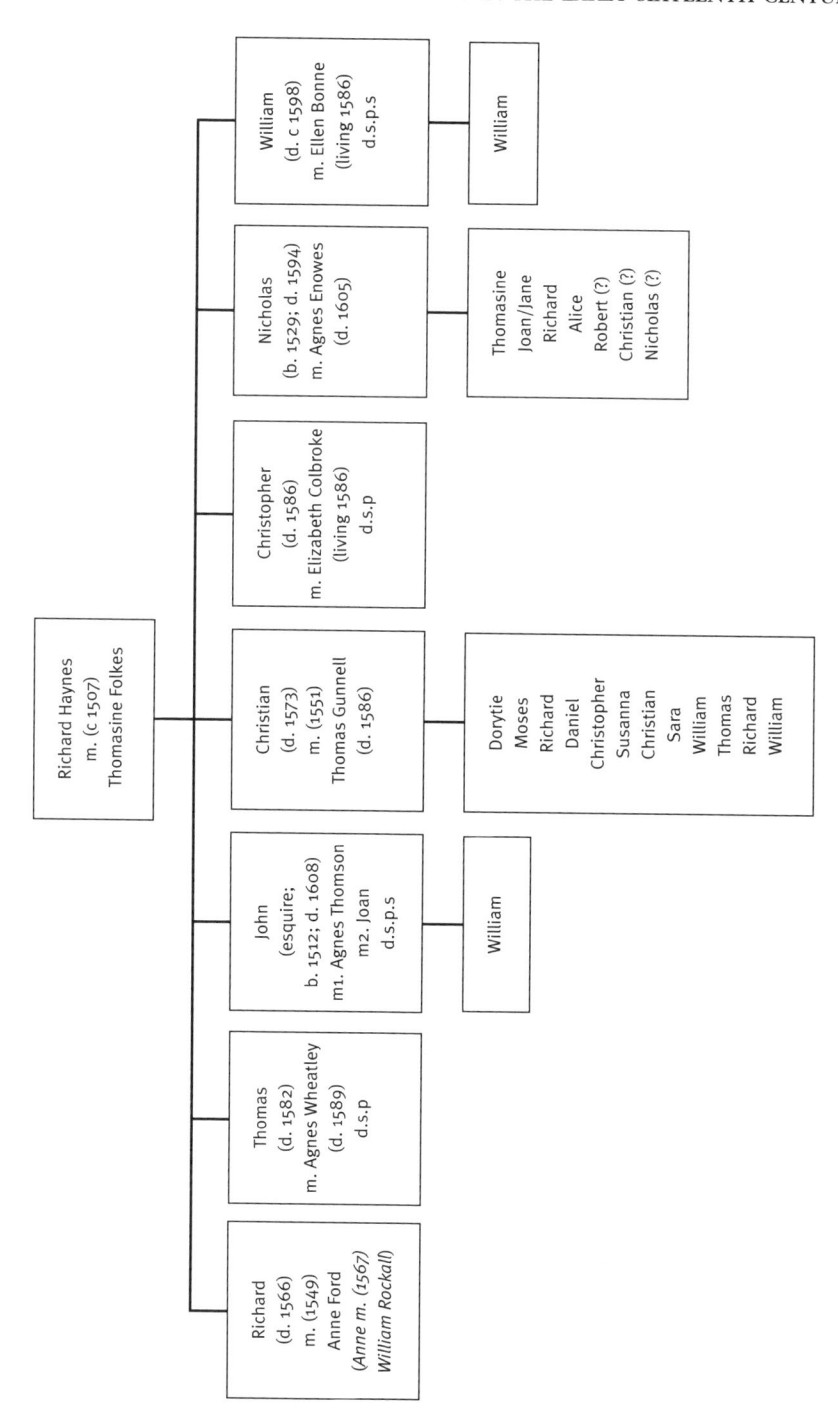

Figure 1.3 The family of Richard Haynes and Thomasine Folkes of Reading

no goods, chattels, lands, or tenements in Berkshire, which could be appraised, valued or taken into the hands of the Queen as required by the writ setting up the inquisition.[33] The reason for this inquisition remains a mystery.

Notes and references

1. At that time, according to Dugdale, the manor of Whitley with tithes was worth £26 18s. 4d, customary rents £34 9s 0½d and assize rents, £1 16s 1d.
2. Sir Francis was granted the Manor of Caversham and other estates in Reading, Oxfordshire and Berkshire, by King Edward VI in 1552 [Cal Pat Rolls, 28 Apr 1552 EdwVI, vol. 4, 1550-1553, pp. 344-5, m. 24] and when, two years later, Sir Francis was appointed Treasurer of the Royal Household, he was granted the lease of Abbey House, the old abbey buildings at Reading which had been converted to a Royal Palace after the Dissolution, but which Queen Elizabeth rarely had occasion to use. At about the same time on 20 July 1565, Sir Francis appears to have surrendered his Caversham estates [Cal Pat Rolls, 28 Apr 1552 EdwVI, vol. 4, 1550-1553, pp. 344-5, m. 24].
3. *Reading Charters, Acts and Orders, 1253-1911*, edited by C. Fleetwood Pritchard, Wyman & Son, Reading & London, pp. 43, 47. We are not informed of by what obligations (fealty) Richard Haynes was bound to his lord.
4. Edward Margrett, 1908, 'The Old Conduit at Whitley, Reading', *Berks, Bucks & Oxon Archaeological Journal*: vol. 14: 52-58; the precise location in 1908 was given as the west side of Highgrove Street 50 yards north from Christchurch Road.
5. Charles Coates, 1802, *The History and Antiquities of Reading*, J. Nichols & Son, pp. 279-280.
6. Berkshire Record Office: D/A1/76/205; Will of Thomas Haynes of Whitley, Parish of St Giles, Reading, 1582. Text reproduced by permission of the Berkshire Record Office.
7. The Will of Stephen Cawet of Reading, Berkshire, 7 February 1542, proved 10 April 1543, PCC: Spert; PROB 11/29.
8. J. Doran 1835, *The History and Antiquities of the Town and Borough of Reading in Berkshire with some notices of the most considerable places in the same county*, Reading: Samuel Reader, pp. 41, 88, 151.
9. A. Stephens, 1809: *Public Characters*, Vol. 10, p. 538.
10. *The Church-Wardens' Account Book for the Parish of St Giles, Reading*, Part I (1518-1546), transcribed from the Manuscript by W. L. Nash, p.17.
11. Records of the Borough of Reading Vol 1, ed. George William Palmer, p.144.
12. J. Doran, 1835, *The History and Antiquities of the Town and Borough of*

THE FAMILY OF HAYNES OF READING IN THE EARLY SIXTEENTH CENTURY

Reading in Berkshire with some notices of the most considerable places in the same county, Samuel Reader, Reading, p. 151.
13. It is worth noting in passing that from the fifteenth century onwards, the locative name Haynes (i.e. of Haynes, e.g. in Bedfordshire) was sometimes spelt with a double n, or as Hayne, Hayn, Hain[n][e][s], Han[n]e[s], Han[n]ys, or as Heyn[e][s], Hein[e][s] and even lacking an aitch. The name could appear in two or more forms in a single document, as in Richard Haynes's will, and has survived as most of them.
14. *The Church-Wardens' Account Book for the Parish of St Giles, Reading*, Part I (1518-1546), transcribed from the Manuscript by W. L. Nash, p.17.
15. The following year, 1540, saw the appointment of John More as vicar of St Giles, who may have been related to Thomasine Haynes on her mother's side; see Chapter 1. A Nicholas More had been the vicar of St Giles from 1497 to 1499.
16. 'Records of the Borough of Reading' Vol. 1, ed. George William Palmer, pp.180-182. By *'a harnes'* was meant the plate armour of the soldier.
17. Doran, 1835, p. 58.
18. Reading Charters, Acts and Orders, 1253-1911, edited by C. Fleetwood Pritchard, Wyman & Son, Reading & London, pp. 43, 47.
19. Visitation of Berkshire in 1665-6, Harleian Society Vol. LVI. In this Visitation, Rich: Haynes was summoned to appear under the Reading Hundred, Parish of St Laurence (Reading; p. 33, second Visitation). This summons was struck through, the most likely reason being Richard's domicile in the Reading Hundred, Parish of St Giles.
20. Differencing marks or cadences, are small heraldic charges added to indicate a specific position in the family: a crescent for a second son, a mullet [star] for a third, a martlet [legless beakless bird] for a fourth, an annulet [ring] for a fifth, a fleur de lys for a sixth, rose for a seventh, etc.
21. Visitation of London 1568, ed. Sophia W. Rawlins. Harleian Society, London, 1963.
22. *The Church-Wardens' Account Book for the Parish of St Giles, Reading*, Part I (1518-1546), transcribed from the Manuscript by W. L. Nash, p. 16.
23. Ibid. p. 39.
24. Ibid. p. 44.
25. P. Glennie, 2004, 'Life and Death in Elizabethan Cheshunt', ch. 5, p. 79, in *Hertfordshire in History*, ed. Doris Jones-Baker, Univ. Herts Press.
26. On 23 May 1414 there was a grant to four of the King's servants, Richard Haynes, William Malbon, Robert Couper and Thomas Porter; Cal Pat Rolls Hen V, vol. 1, pp. 198, 288.
27. Sir Harry Nicolas, ed. 1837, *Proceedings and Ordinances of the Privy Council of England*, p. xxxiv.
28. John Long was appointed Master in 1503 and remained until his death in 1530, when Abbot Hugh Cook Faringdon appointed Leonard Coxe, the Master of Reading School until its closure with the Suppression of the Abbey in 1539. Coxe was reappointed Master when the school reopened in 1541; J. Oakes and M. Parsons, 2005, *Reading School, the First 800 Years*. DSM Peterborough for Reading School, p. 7.

29. Alison Sim, 2006, *Masters and Servants in Tudor England*, ch. 1, p. 29, Stroud: Sutton Publishing.
30. Cal Pat Rolls 1444, 8 July, p. 218, m. 25.
31. *The Church-Wardens' Account Book for the Parish of St Giles, Reading*, Part I (1518–1546), transcribed from the Manuscript by W. L. Nash..
32. Will of Christopher Haynes of Arundel, 28 November 1586 PCC, PROB 11/69.
33. TNA: E 199/37/24. The other two that were the subject of the inquisition were John Hancock and Robert Botyler. Hancock was said to be living in Chevenesse and Botyler at Woodham Walter, Essex.

2

Who Was Thomasine?

Visitation of London 1568 - Foxley or Folkes? - Daughter of More - Nicholas More and John Folkes - Burials in St Giles - Maternal descent from John Foxley of Berkshire and Sussex

Richard Haynes and his family were known to the heralds as '*Haynes of Reading*', and all the evidence points to Whitley in the parish of Reading St Giles being their place of domicile. There can be no doubt of the existence of Richard's wife, Thomasine, whose life spanned the closing decade of the fifteenth and early decades of the sixteenth centuries: at least four generations of her female successors took her Christian name at baptism. She is nearly always referred to as Thomasine Foxley, doubtless as a result of the statement that she was the daughter and coheir of John Foxley of Berkshire made by the Clarenceux King of Arms in the exemplification of the coat of arms confirmed on her sons Nicholas and William in 1578.[1] However, examination of contemporary documents makes a Foxley parentage improbable, but a more remote Foxley kinship remains plausible, likely even, as will appear.

Visitation of London 1568[2]

The Heralds' Visitations were undertaken in the sixteenth and seventeenth centuries to determine the right to bear coats of arms after King Edward VI had become concerned about their unauthorised use. The collection of pedigrees formed much of their work, and their original records are archived at the College of Arms, many of which have been published either in facsimile or contemporary print. One such is the record of the family of Haynes of Reading. However, the recording in manuscript form was not without its problems, as both errors and inaccuracies occurred. Accuracy of a

Visitation report would require accuracy on the part of the writer, of the recollection of the person providing the information and, if not the original, at subsequent transcription. It also demanded the integrity of the heralds, who were frequently accused of fabricating coats of arms for payment – cash for honours being by no means only a recent phenomenon!

In the published 1568 Visitation of London return, there are at least three names in the single paragraph recording the Haynes pedigree that were either wrong or inaccurate. For example, Thomas Haynes's wife, Agnes, is said to be the daughter of John Whelly. His name was actually Wheatley, as is clear from Agnes's will, and one can hazard a guess that the Herald's spelling was Whetly, with the letter 't' wrongly transcribed as an 'l' in the publication. Christian married Thomas Gunnell, often spelt Gonnell at the time, but the Herald's record was unclear and the transcription was Gounde with a question mark by the name, indicating difficulty in reading the letters. In fact on scrutiny of the original document, the name was clearly Gonnell. Nicholas Haynes's wife was given as Anne Enowes, but, as we shall see, she was definitely Agnes Enowes. In this case, however, the Herald had probably correctly noted the information given, as Anne was the common abbreviation of Annes or Annis, reflecting the anglicised French pronunciation of Agnes. It was presumably John, a London resident, who had provided the Herald with the information on the family in 1568, rather than Nicholas, who did so ten years later when he applied for new heraldic arms.

There is, however, one major inconsistency between the account in the 1568 Visitation of London and the subsequent 'Exemplification of Arms'. In the Visitation, Thomasine's surname is not Foxley but Folkes, and again, in a footnote, as Folx. These were not transcription errors. Despite the minor inaccuracies in the published version of the Visitation of London, most of the facts could be independently corroborated, so there is no *prima facie* reason to suspect that Folkes would not also be correct, and for three reasons: first, the name was spelt phonetically as Folx; second, unlike *Whelly* or *Gounde*, Folkes was a well-known name in its own right in the Reading area; and third, in a rectangular box by the name Folx were the words *'3 leane sheaves'*, suggestive of heraldic arms. We cannot be certain which of the Haynes brothers gave his maternal grandfather's name as Folkes to the Herald at the time of the Visitation of London – it was probably John – but within ten years we know that in Nicholas's application for new arms, his grandfather's name appears to have become John *Foxley* of Berkshire, but with no location in Berkshire whereby John Foxley's existence might be verified. In the light of the survival of very little evidence for a John Foxley of Berkshire, but with abundant evidence for the existence of a John Folkes around 1500 in Reading, one could be forgiven for questioning Nicholas's veracity and also that of his brother William, who undoubtedly cooperated in seeking this confirmation.

Foxley or Folkes?

As mentioned already, there are no grounds for believing that Folkes was a misspelling by the herald of the earlier form of the name, Foxle, as the name John Folkes was then a well-documented name in the Reading area, whereas in comparison, no evidence exists for a John Foxley *of Berkshire* at that time. Folkes, Faulkes, Faux, Fawkes, Fokes and numerous variants were interchangeable with Fox or Foxes. It is therefore possible that over the generations a branch of the family dropped the 'ley' and assumed the name Fox or a variant. Certainly the Foxley family associated themselves with the 'fox'-part of their name, which appears in Sir John de Foxley's [1378] memorial brass in Bray church as his helm. However, by this time the two names were already recognised as different, and Foxley had no distinct latinised form, whereas Folkes did – Fulconis. The justice Sir Robert Fokes, who witnessed a deed at Stratfield on 18 April 1273, has been identified with the royal justice, Sir Robert Fulconis or Fulco [d. 1287], who had interests in the Reading area.[3] The mention of the armorial *'3 leane sheaves'* by the name Folx in the original heraldic document strongly suggests that it was not an abbreviation of Foxley. It is worth recalling that, in 1787, General Robert Haynes, the diarist, never claimed that Thomasine was descended from a particular person beyond her father, John Foxley.[4] In this he was probably quoting and believing in the facts as set out in the 'exemplification'. This assumption was made by Cracknell, the editor of the diary, based on the diarist's statement that Thomasine's family *'lived at Foxley Grange nr Reading'*; and based on the fact that Foxley Grange was at Bray, part of Foxley Manor, which was named by Sir John de Foxley [c.1258–1324], together with the fact that a modified version of the 'Bray' Foxley arms was used in the quartering by the herald, it was a fair assumption. However, there is no evidence that Foxley Grange even existed in 1500. We can however be reasonably certain that the 'grange' was one of the few very old timber-framed buildings surviving at the time the Haynes diary was written (1787) that bore the Foxley name.

By the time Thomasine married Richard Haynes in the early sixteenth century, the eminent Berkshire and Hampshire Foxley family had already died out in the male line with Thomas in 1436. This was not uncommon, as knightly and noble families were prone to extinction in what was an era of high natural mortality, a quarter dying out in the male line every 25 years.[5] However, even though clear records of continuous succession do not exist in one branch, wherever the late medieval Foxleys were recorded, whether in Berkshire, Norfolk, Northamptonshire, Sussex, or Wiltshire, the name survived here and there into the sixteenth century. Some consideration as to how this might have come about would seem in order. First, one of the known branches believed to have died out, hadn't; it was merely 'lost', i.e. unrecorded; second, descent was through a collateral branch whose male line ended with John Foxley, Thomasine's father; third, the John Foxley named in the 'exemplification' was in fact not of Berkshire but of elsewhere; fourth, descent was through a female line with the Foxley name adopted later by a male descendant as an alias, as in the case of the Northamptonshire Foxleys who were actually called Chete;[6] fifth, descent was through a female line, with no male issue adopting the Foxley name until the 'Exemplification of Arms and Crest' in 1578.

Are any of these explanations plausible as far as Thomasine's father is concerned? For the first two alternatives someone called Foxley, preferably John Foxley, is needed in the Reading area around 1500. Only one Foxley has been found in Reading during Thomasine's life-time. This was Richard Foxley, who was buried within St Lawrence Church, Reading, in 1530-1, as recorded in the church's Obituary.[7] Nothing is known of him or his family and there is certainly no record of a John.

Outside Berkshire, however, is a different matter. The Archdiaconal Court of Beaconsfield, at its sitting at Amersham on 26 March 1496, heard a case against Isabella Saunder brought at the request of a John Foxley, the identity of whom is unknown,[8] but possibly of Northamptonshire. The Foxleys of Highworth, Wiltshire, must also be considered, as a John Foxle was recorded in Purton, in that county, in 1539. There was also a Thomas Foxley, a farmer, who died in Bagendon, Gloucestershire, about 15 miles from Highworth, in 1525, who may have been of the Highworth family. In his will, dated 1525, he mentioned his wife, Alice, their son, Thomas and daughter, Jane.[9] His father, John Foxley, had been a churchwarden of Bagendon church.[10] The possibility that Thomasine was the daughter of John Foxley of Bagendon, and sister of Thomas (d. 1525), cannot easily be dismissed, as Bagendon and nearby Daglingworth and Duntisborne, formed the seat of a Haynes (Haines) family, whose Arms still to be seen in Daglingworth Church, were the same as those of Richard Haynes of Reading, suggesting common ancestry. The Hayneses of Daglingworth were major sheep farmers and wool producers, and Richard Haynes, were he not the first son, might have moved away to become involved in the woollen cloth industry then centred on Reading. But this is pure speculation.

So there was the odd John Foxley in neighbouring counties around 1500, but if one of these Foxleys was of the Bray/Bramshill family, from whom did he descend? Were there collateral male branches of the family through which Thomasine might have claimed descent? This certainly remains a possibility as there is evidence for adult male siblings of, for example, Sir John de Foxley [d. 1378] and John Foxley [d. 1419] his son; but all traces of their families, if any, are lost. There were also separate lines of male Foxleys from the Sussex branch known to have outlived the last Sir John [d. 1378]. Later Foxleys are also to be found in Kent, Norfolk, Wiltshire and Yorkshire, and such Foxleys in Hull and Northamptonshire used the same coat of arms as the Foxleys of Berkshire (gules two bars argent). However, although the survival of a John Foxley cannot be entirely discounted, in view of the well-documented history of the Bray Foxleys, as a possibility it must be considered unlikely.

The third option that Thomasine might have been the daughter of a John Foxley outside Berkshire such as one of the John Foxleys of Northamptonshire although possible is also unlikely. There is no mention of Thomasine in the will of John Foxley of Blakesley, dated 29 December 1535,[11] in which all other female members of his family were remembered, although of course, she might have died before then, her last known child being born about 1531/2. Thomasine would certainly have been born too early to have been the daughter of John Foxley junior of Northamptonshire. In any case, these Foxleys were actually called Chete alias Foxley. Could Thomasine's father, John Folkes, who married the *'daughter of Moor'* have assumed the name Foxley as an alias if she were an heraldic heiress of Foxley? The answer must again be 'yes' in theory, but in the surviving documents relating to this John Folkes no such alias was ever used. Furthermore, his forebears were also known as Folkes for at least three generations. So there remains only the consideration of the adoption of the Foxley arms by the Haynes brothers through maternal descent.

Maternal descent?

The Visitation of London possessed a marginal note of some importance informing that the original Haynes pedigree was *'in tabular form beginning 'John Folx mar d of Moor (and had issue) Thomasine married Richard Haynes of Redynge'... There is a shield (repeated twice) tricked: Argent, two bars vert between nine martlets gules.'* Thus through this we learn that the family of Thomasine's mother, was originally that of atte More, now More or Moore, a well-known family of Berkshire and Hampshire.[12] Of interest in this regard was the observation that these arms of More had many features in common with those of Ailleward.[13] One of the sisters of John de Foxley (d.1378]) is known from his will[14] to have married into the Ailleward family and had a daughter Joan, and this coincidence required an examination of the intriguing possibility that Thomasine might possibly have been able to claim descent from Thomas de Foxley [d.1360], Sir John's father, via the More/Ailleward connection.

There is one certain link between the Foxleys and Mores. On the death in 1419 of John Foxley of Rumboldswyke, Sussex (a son of Sir John de Foxley of Bray by Joan Martyn), two coheirs were mentioned. One was the deceased's brother Thomas Foxley, the other was John More.[15] This strongly favours John More being related to John and Thomas Foxley by marriage or common descent and the possibility cannot be excluded that it was through this John More, perhaps married to a sister, half-sister or, cousin of these Foxley brothers, that Thomasine's descent from the Foxleys of Bray was justified by Nicholas and William Haynes, all male Foxleys having died out (Thomas Foxley being the last to die in 1436). It may be relevant that Foxley manor in Bray, the seat of Thomas Foxley, was situated near the manor of Mores in Bray to which a John More had succeeded by 1396.[16] He died leaving a widow, Joan.[17] If this Joan was born Joan Ailleward,[18] the niece of Sir John de Foxley (d.1378), and Thomasine was descended from John More's heir, then this would explain her Foxley heritage. It would also explain the closeness of the coats of arms of the More and Ailleward descendants.[19] This possible descent is presented in Figure 2.1. After John Foxley's death in 1419, the protection of the interests of his infant daughter, Alice, was the subject of a number of notices. Alice's inheritance was initially committed to the care of a William Walton, with a William Ailleward of Berkshire acting as guarantor.[20] This last William must have been a close kinsman of the dead John, perhaps his cousin, the brother of Joan Ailleward (Figure 2.1). The intervention of William Ailleward would certainly make sense if it was on behalf of Joan (Ailleward) More, who had been recently widowed.

Among the more interesting documents mentioning John Folkes in the Reading area around the time of Thomasine's early life are those in which he had dealings with Nicholas More of Barkham and Coleman's More. For example, there are deeds relating to transfers [enfeoffments] of lands, tenements, meadows, woods, pasture, rents, reversions and services with all their appurtenances and easements in Barkham, Wokingham, Arborfield and Sindlesham owned by Nicholas More of Colmans More to John Folkes, William More [presumably a kinsman of Nicholas] and others on 28 September 1510.[21] After Nicholas More's death in 1525, some of the lands were transferred back from John Folkes and the other feoffees, to Nicholas, the son and heir of Nicholas More, who was a fuller in the Reading cloth-making industry,[22] one of the witnesses being John More. Such significant land transfers in trust between John Folkes and Nicholas More and his son suggest that the two families were related and as John Folkes the father of Thomasine, married a *'daughter of Moor'*, these transactions are consistent with the suggestion that she might have been the daughter of Nicholas More the elder or one of his kinsmen (Figure 2.1). The choice of Nicholas as a Christian name by Thomasine Haynes for one of her sons also resonates with that of her putative maternal grandfather.[23]

Earlier references indicate where John Folkes might have lived and the identity of his father. Thus, on 20 September 1501 there was a deed of gift drawn up between *'John Folkes son and heir of John Folkes'* and John Lathbury of a house called 'Gyboncokes' with a

READING TO BARBADOS AND BACK

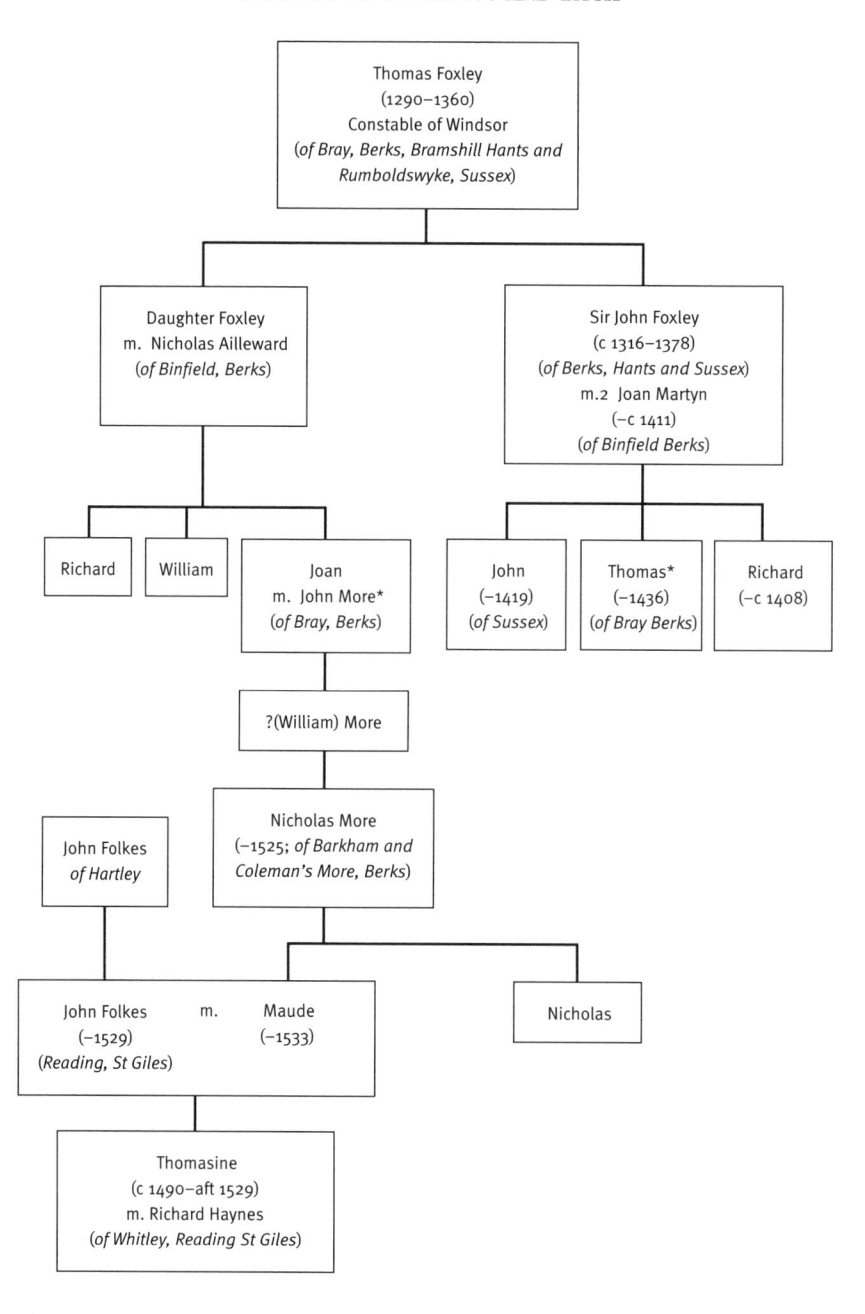

```
                    Thomas Foxley
                     (1290–1360)
                  Constable of Windsor
           (of Bray, Berks, Bramshill Hants and
                   Rumboldswyke, Sussex)
                          │
          ┌───────────────┴───────────────┐
   Daughter Foxley                  Sir John Foxley
   m. Nicholas Ailleward              (c 1316–1378)
    (of Binfield, Berks)          (of Berks, Hants and Sussex)
                                    m.2 Joan Martyn
                                       (–c 1411)
                                    (of Binfield Berks)
```

- Daughter Foxley's children: Richard, William, Joan m. John More* (of Bray, Berks)
- Sir John Foxley's children: John (–1419) (of Sussex), Thomas* (–1436) (of Bray Berks), Richard (–c 1408)

Joan m. John More* →
?(William) More →
Nicholas More (–1525; of Barkham and Coleman's More, Berks)

John Folkes of Hartley

Nicholas More's children: John Folkes (–1529) (Reading, St Giles) m. Maude (–1533); Nicholas

Thomasine (c 1490–aft 1529) m. Richard Haynes (of Whitley, Reading St Giles)

*Coheirs of John Foxley (d. 1419).

Figure 2.1 Possible maternal descent of Thomasine Folkes from Thomas Foxley scutifer, of Bray, Bramshill & Rumboldswyke, Constable of Windsor Castle

garden and animal enclosure, the building adjoining on the east the road leading from Lambwood Hill, Grazeley in Shinfield Parish, towards Hartley.[24] The deed was witnessed by William Wodecoke, Richard Gerard and Robert More and others.[25] The other name mentioned in the deed was John Folkes's attorney, his '*beloved in Christ, Thomas Jenyns*', who is also mentioned the following year (20 August 1502) in a quitclaim by him and Sir William Norreys of the same land to John Folkes the son, that they had previously had as a gift and feoffment from his father.[26] The above transactions are of added interest in view of the fact that one of the witnesses was called More, and also of the proximity of these Folkes and More families to Whitley and the Hayneses whose home was literally on the St Giles/Shinfield parish boundary. Although we have not undertaken a systematic search, earlier Folkeses in the area are not difficult to find. As far back as 1352 a William Folk of Hartley (Hertleigh) was involved in a legal action over property in Whitley, Reading.[27]

Not only were the Folkes and More families living near to Whitley, the home of the Hayneses, in the Parish of St Giles, Reading, but the churchwardens' account book for the years 1518 to 1546 gives the possible identity of John Folkes' wife as Maude. As mentioned in the previous chapter, the receipts for the costs of the burials of Maude[28] [1533-4] and John[29] [1529-30] were also recorded. Both John and Maude Folkes were eventually buried together within St Giles, as the cost of their graves (6s 8d each), was the same as that paid by others who were specified as having been buried inside the church.

In summary, Thomasine's father was, until proved otherwise, John Folkes son of John Folkes of Hartley and Shinfield. There is no evidence for him being a member of the Foxley family. The balance of evidence suggests that Thomasine's Foxley heritage was at best remote and on her mother's side. Coats of arms with titles that lay dormant could be revived in favour of distant heirs. Even though the line had died out, as in the case of the Foxleys in 1436, the coat could be modified by the herald and reassigned.[30] This is of interest as the 'Foxley' quarters of the arms confirmed on Nicholas and William were not identical with those of the original Foxley family; the two bars argent being modified significantly to two silver bars 'humite' (or 'hummetty') meaning 'cut short' or 'couped'. The only remaining obstacle to accepting these arms as being a reassignment by the herald of the coat of a distant forebear is the statement of

Thomasine being the *daughter* of John Foxley, but in sixteenth-century usage *daughter* could also simply mean 'female descendant' however distant. We are therefore faced with the conclusion that there was probably no such individual as a John Foxley of Berkshire, the alleged father of Thomasine, in the late fifteenth century; rather he was the more remote ancestor described as John Foxley, gentleman, of whom Thomasine was the descendant of a coheir. The only one of that lineage, name and status, who also named two coheirs, was John Foxley esquire of Berkshire and Sussex who died in 1419.

Notes and references

1. 'Visitation of Berkshire in 1665-6', *Harleian Society*, London, Vol. LVI.
2. 'Visitation of London 1568', ed. Sophia W. Rawlins. *Harleian Society*, London, 1963.
3. B. R. Kemp, 1987, 'Reading Abbey Cartularies', vol. II, p.327, charters 1237, 880-3 and 1029-30. This was probably the Robert Fulco who bequeathed land in New Street adjoining Reading Abbey to the Grey Friars in 1288 [Victoria, County History of Berkshire, vol. ii, p. 90].
4. E.M.W. Cracknell, ed. (1934), *The Barbadian Diary of Gen. Robert Haynes 1787-1836*, The Azania Press, Medstead, Hampshire, UK.
5. M. Prestwich 1980, *The Three Edwards; War and State in England 1272-1377*, Weidenfeld & Nicholson, London.
6. Cal Pat Rolls 16 Hen VII 18 January 1501, pp. 224-5, m. 21(6).
7. C. Kerry, (1883). 'A History of the Municipal Church of St Lawrence Reading', p. 189.
8. 'The Courts of the Archdeaconry of Buckingham, 1483-1523', *Buckinghamshire Record Society* 19, 240 pp. 169-170.
9. Gloucestershire Record Office: Will of Thomas Foxley, Bagendon 1525.
10. G. E. Rees, 1932, *The History of Bagendon*, Cheltenham: Thomas Hailing Ltd, Oxford Press, pp. 63, 154, 173.
11. Northamptonshire Record Office: Will of John Foxley of Blakesley, 1535.
12. M. Burrows, 1886, *The Family of Brocas of Beaurepaire and Roche Court*, London: Longmans Green & Co.
13. 'Visitations of Surrey', Harl. Soc. Vol. 43, pp. 43-4, 1899; 'Visitation of Wiltshire 1623', pp. 226-7, 1954.
14. W. H. Gunner and Albert Way, 'The Will of Sir John de Foxle, of Apuldrefeld, Kent; AD 1378', *Archeological Journal* 15, 1858, 267-277.
15. Cal Close Rolls 1 June 1425, 1422-1429, m.1, pp. 179-180. John More's wife, Joan, was still living in 1433 [C. Kerry, 1861, *The History and Antiquities of the Hundred of Bray in the County of Berkshire*, London: Savill and Edwards, p. 123].

16. Court Rolls (Gen. Ser.) portf. 153., No. 63 quoted in VCH Berks, vol. iii, p. 105.
17. Court Rolls (Gen. Ser.) portf. 153., No. 64 quoted in VCH Berks, vol. iii, p. 105. John More of Bray died before 1430 when his heir William was in possession of Mores. Richard More held at his death in 1495 a messuage, 85 acres of land and £4 rent in Bray, which he bequeathed to his widow Katherine for life with reversion to John More his son and heir [Chan Inq PM (ser. 2), xi, 108; VCH Berks, vol. iii, p. 105].
18. W. H. Gunner and Albert Way, 'The Will of Sir John de Foxle, of Apuldrefeld, Kent; AD 1378', *Archeological Journal* 15, 1858, 267-277.
19. 'Visitations of Surrey', Harl. Soc. Vol. 43, pp. 43-4, 1899; 'Visitation of Wiltshire 1623', pp. 226-7, 1954.
20. Cal Fine Rolls IV, 1413-22, 7 HenV, m. 27, p. 286.
21. TNA: E 210/6653.
22. TNA: E 210/6624.
23. There is also an interesting parallel in the names of the descendants of William atte More of Wyeford, Pamber [d.1481] father of Henry whose wife was called Christian. Henry [d. 1496] had a son, Nicholas. This Nicholas More [d. 1496] who was married to Jane Druez, had two daughters, Jane [b. 1495; d.1563] and Christian [b 1496; d. bef. 1524], who could have been Thomasine's contemporaries. All of these individuals would have been descended from those with the same coat of arms as Thomasine's mother. A number of other entries in the St Giles account book may also be of relevance, such as the frequent references to a Xpine (Xrine), Cristine, Cristyne or Christian More who was first a Sexton of St Giles and later a paid member of the choir. She appears to have lived in accommodation owned by the church and would presumably have been related to John More, who in 1540 was appointed vicar of St Giles. This may be of significance in that Thomasine named her only daughter, Christian – possibly after a family member.
24. A later deed places the Gyboncokes in Hartley Battle (Berkshire Record Office: D/EX217/4), which was the location of the manor of Moor Place (see Parishes: Shinfield; VCH Berks, 1923, Vol. 3, pp. 261-267).
25. Berkshire Record Office: D/EX217/2; Moor Place was partly situated on land held by the Woodcock family of Battle Abbey (Parishes: Shinfield; VCH Berks, 1923, Vol. 3, pp. 261-267).
26. Berkshire Record Office: D/EX217/3; The Norreys family held lands in Whitley in the sixteenth century; Fine Rolls Berkshire and Buckinghamshire 44 Eliz. pt. i, no. 44; Chan. Inq. p.m. (Ser. 2), cccxiv, 127: 'The Borough of Reading: Manors', *A History of the County of Berkshire*: Vol. iii (1923), pp. 364-367. John Folks sen., might be identified with John Fulks, who in the reign of Henry VII was one of the woodwards of the forest of Windsor; he was woodward of the abbot of Abingdon, of his wood called Hirstenhalderst [Hurst] – Whistley/Hurst were held by the Abbacy of Abingdon until its Dissolution [*A History of the County of Berkshire*; 1923: Vol. ii].
27. B. R. Kemp, *Reading Abbey Cartularies*, 1987; see Charters 1265 and

1266, pp. 338-9. See also: Berkshire Record Office: D/EE/T1/1/11. It is of interest that a Master Fulk was the King's gardener at Windsor in 1277 (Pat Rolls 5EdwI mm. 11, 6, 5; *A History of the County of Berkshire*; 1923: Vol. ii, p. 344).

28. *The Church-Wardens' Account Book for the Parish of St Giles*, Reading, Part I (1518-1546), transcribed by W. L. Nash, p. 44.
29. Ibid. p. 39.
30. N. Denholm-Young, 1969, *The Country Gentry in the Fourteenth Century with Special Reference to the Heraldic Rolls of Arms*, OUP, p. xii.

3

Children of Richard and Thomasine Haynes in Berkshire: Richard and Christian Haynes of Wargrave and Thomas Haynes of Whitley

Richard Haynes *- House at Hinton - Marriage to Anne Ford - Death - Will - Marriage of widow*
Christian Haynes *- Marriage to Thomas Gunnell - Children - Richard's bequests - Wider Gunnell family*
Thomas Haynes *- Inheritance of Whitley - Will and bequests - Marriage to Agnes Wheatley - Churchwarden of Reading St Giles*

Richard Haynes [c.1508–1566]

Richard Haynes, son of Richard and Thomasine Haynes, and assumed to be the first of six brothers for no other reason than he took his father's name, held a house and land in Hinton, to the north of Hurst, Berkshire.[1] He wasn't the only Haynes mentioned in Hurst around that time, for in 1528 a William Hayne, husbandman, held lands and tenements there in tail subject to certain covenants for which William deposited a bond for £20. Part of the land was actually leased by a man called Pether[2] for the next seven years. The relationship of William Hayne to Richard, if any, is unknown, but it is possible that Richard's house was the one that lent the Haynes name to subsequent dwellings on this site, for by the end of the sixteenth century is mentioned a house by the name of Haynes Hill, the spelling of which eventually changed to Haines Hill. As far as is known, the house in Hinton mentioned in Richard's will is the only evidence to support this connection, first made by Lieutenant-General Robert

Haynes of Barbados in his 'diary'.[3] Richard Haynes married Anne Ford, a member of a well-known local farming family of Ford House, Wargrave,[4] at Wargrave Church on 5 February 1549. The Fords had widespread landholdings in Wargrave parish including some adjoining the A'Bears and Gunnells at Hare Hatch and the three families would eventually intermarry. Richard, who styled himself 'yeoman', probably farmed the lands brought to him by his wife, although his actual occupation is unknown. General Robert Haynes believed that four Haynes brothers worked in the Queen's Household and although we haven't established that Richard was one of them, on 3 February 1565 a Richard Haynes was granted the office of a *'Queen's Gunner'* in the Tower of London, on the death of a previous incumbent, Robert Stonestrete. His wages were set at 6d a day.[5] This may have been the same Richard Haynes, master gunner, who in 1546, was mentioned lodging near the river in Estsmythefeelde.[6]

Richard and Anne lived in Wargrave, but they retained their house in Hinton three miles away which may have been his parents' old house or that of the aforementioned William Hayne. Hinton was also only a short ride from Waltham St Lawrence and Binfield where Thomas Cressell and his wife Margaret held lands. It was on 14 September 1558 that Richard, with William Cressell, Thomas Smyth and John Grove were witnesses to the will of Thomas Cressell.[7] Richard's own will was dated 11 March 1566;[8] he died and was buried on 2 December 1566 and the will was proved twelve days later by William Grove, yeoman of Wargrave, proctor for Anne his widow. He was described as *'Richard Haynnes of the parish of waregrove in the countie of Barks yoman'*. Wargrave was clearly his place of domicile as judged by the fact that he left ten shillings to the poor there and also to the poor of Shiplake which is the village directly opposite Wargrave on the other side of the Thames. He desired to be buried in Wargrave Church at the end of his pew. He left nothing to the poor of Hurst or nearby Ruscombe, whose church, St James's was the nearest to Hinton, which again reinforces the view that Hinton was not his main home. It is not known whether or not Richard predeceased his parents, but the fact that he was not in possession of the family home at Whitley suggests either that he predeceased his father if he was the eldest son, or alternatively, if he died after his parents – which is more likely – he could not have been their eldest son. As mentioned in Chapter 1, the date of Richard's marriage is more consistent with a birthdate of the mid-1520s for him, even though he has been assumed to have been the first-born. The relevant item in Richard's will is:

> 'I give and bequeth unto my Brother Thomas Haynes fouretie shillinges and vi silver spoones and after his decesse the same spones to remeyne to my Brother Nicholas children Also I give to the saide Thomas my howse with the appurtenances sett lyinge and being in Henton [Hinton] and after the decease of the saide Thomas the same to remeyne unto Richard Hayn the sonne of Nicholas Hayne my brother and if he dye withoute yssue the same to remeyne to his Daughter' [Note the three different spellings of his surname in this extract].

Richard, who after 17 years of marriage had no living issue at the time of his death, made Anne his executrix and residuary legatee. Apart from property, Richard's will mentioned leases, horses, silverware and his residual estate; he left a total of £18 14s 10d in monetary bequests.

Anne continued to live in their Wargrave house, and a little over five months after Richard's burial she married William Rockall (Rockell; Rokholl; Rock[h]all) in Wargrave church on 12 May 1567. William Rockall, along with William Ford (probably Anne's brother), would be one the overseers and witnesses of the will of Richard's brother-in-law Thomas Gunnell in 1580.[9] She had no children by her second husband, which suggests that she, rather than her husbands, might have been sterile. Anne Rockall died in January 1603 a year or so after her husband[10,11] and was buried on the eleventh of that month in Wargrave churchyard.

Christian Haynes [c.1525–1573]

Christian was clearly close to her brothers and it was her son William, who was ultimately destined to inherit the Haynes' Berkshire estates. We have estimated that she was born around 1525, probably between Christopher and Nicholas [below]. On the 8 November 1551, she (Christian Heines) was married to Thomas Gunnell (the name was written 'Gonnell' until the seventeenth century), yeoman of Wargrave, where her brother Richard had already married and settled, possibly near Hare Hatch where both the Gunnells and Fords held lands. Thomas Gunnell was a man of some substance, and the couple had at least six sons (Moses [Moyses], Richard, Danyell, Christofer, William, Thomas) and four daughters, Dorytie, Susanna, Christian and Sara. Dorytie [1552] was the first-born but died after only four weeks. Two years later their first son, Moses was born [1554]. The picture is somewhat confused by the existence of another Thomas Gunnell, the son of Thomas Gunnell of Hare Hatch, who in his will named his grandchildren Richard, William, John and Abraham, each of whom were to receive 20d when they reached 14 years of age.[12] This condition relating to age serves to distinguish at least John and Abraham from their kinsmen of the same name (the sons of William Gunnell, Christian's brother-in-law), who were already beyond their fourteenth birthday.

Christian and her children were remembered in her brother Richard's will: *'I give unto my sister Christian Gunnell fouretie shillinges also unto the children of the said Christiane foure poundes to be divided equallie amonges them ...'*[13] At least four of the first six, born before their uncle Richard's death, were still living by March 1566, as the £4 was kept for them by their father Thomas, who eventually left 20 shillings to each of the four children born during their uncle's life and surviving at the time their father drew up his own will in 1580: '... *all wyche money was delyvered into my handes for theire behofe by theire uncle Rycharde Haynes'.*[14]

Some of the Gunnell children also received bequests from their other Haynes uncles, Thomas Haynes leaving them 50s in 1582 to be shared equally. This suggests that there were only five of Christian's children surviving in 1582, each of whom would receive 10s. Christian predeceased her husband, being buried in Wargrave churchyard on 13 February 1573. The baptisms of the possible children of Christian and Thomas are given in the following table:

Table 3.1 Baptisms of possible Children of Thomas Gunnell [Gunel; Gunnel; Gonnell; Gonell; Gunhill; Gonnhill etc.] and Christian Haynes, based on the assumption that it was Christian and not her daughter of the same name who was buried in 1573

Date of Baptism	Name	Son [s] or Daughter [d]	Parent recorded	Comments
28 Oct 1552	Dorytie	d	Thomas	Bur 23 Nov 1552
7 Jan 1554	Moses	s	Thomas	Mentioned in father's will
27 Mar 1556	Richard	s	Thomas	
15 Mar 1559	Daniel	s	Thomas	Mentioned in father's will

CHILDREN OF RICHARD AND THOMASINE HAYNES IN BERKSHIRE

Unk*	Christopher	s	–	Mentioned in father's will as born bef 1566
20 Jun 1562	Susanna	d	Thomas	Bur 13 Feb 1573
24 Aug 1564	Christian	d	Thomas	
17 Feb 1566	Sara	d	Thomas	Mentioned in father's will; alive [widow] in Jan 1606.
5 Sep 1568	William	s	Thomas	Mentioned in father's will; inherited John Haynes's estate
31 Aug 1569	Thomas	s	Thomas	
2 Dec 1571	Richard	s	Thomas	**
24 Jan 1573	William	s	Thomas	**Witness of John Haynes's will; 'William Gunnell, jun'.

*Between Mar 1559 and June 1562 is the most likely place given the regular interval between baptisms and the average interval between Elizabethan births.[15]
** Possible son of Thomas of Hare Hatch.

Thomas Gunnell had two brothers, William and John, whom he mentioned in his will.[16] A Thomas and a William Gunnell (William Gonnell and Thomas Gonall) were witnesses of the will of William A'Beare of Hare Hatch, Wargrave, dated 7 February 1553. They also, with others, prepared the inventory of A'Beare's effects after the will was proved.[17] They may have been the sons of Thomas, William, or John Gunnell of Wargrave rather than of William Gunnell of Hare Hatch, all of whom were mentioned in the Lay Subsidies for Wargrave in 1524/5.[18]

Thomas Haynes [c.1510–1583]

Thomas Haynes styled himself gentleman in his will[19] dated 2 February 1582: *'I Thomas Haynes of Whitley in the p[ar]ishe of St Gyles in Redinge in the Countie of Barks gent'*. Given his parents' suggested pedigrees, his father would also have been styled 'gentleman', and the fact that his brother Richard, was styled *'yeoman'* at the time of his death in 1666 suggests that Richard had either not yet received his father's estates, was of yeoman rank in his occupation, or was not actually the eldest son. Thomas who appears to have inherited the family estate had a great number of possessions, the bulk of which he left sequentially to his brothers John, William and Nicholas followed by Nicholas's heirs:

'Itm I give to my brother John Haynes my best gold ringe with all the lands and tenements that I have in this Realme of England during his naturall lyfe and after his lyfe I give and bequeath all the sayd lands tenements and all other promises unto my brother William Haynes to have and holde the same duringe his naturall lyfe and after the decease of the sayde Willim I give and bequeath all my promises unto my brother Nicholas Haynes and to his Heyres lawfullie begotten of his bodie for and provided alwayes and it is my will that my sayd brother John Haynes nor William Haynes shallnot sell lett nor laye to morgage or my wyfe, do or putt a waye the sayd lands tenements possessions and hereditaments or anie parte or p[ar]cell thereof without the speciall licence consent and agrement of the sayd Nicholas and his heyres'.[19]

Thomas also made bequests to his nieces Jane, Alice [Alles] and nephew Richard, all Nicholas's children. Nicholas's other daughter named in the Visitation of London, Thomasine, was not mentioned, as she had died in the interim. His prized possession was *'my best gold ringe'*, which he bequeathed to his brother John. This may have been the *'great goulde seal'* that John subsequently bequeathed to his nephew Richard Haynes on his death in 1608[20]. All Thomas's five brothers, including the deceased Richard, and his sister, Christian, were mentioned by name in Thomas's will, as well as the three children of Nicholas.

Thomas married Agnes (Annis) Wheatley. It is possible that they met through his Royal Household connections, as a William Wheatley was Yeoman Purveyor in the Scullery responsible for Bowls and Trays.[21] When she died in 1590, Agnes bequeathed all of her effects to the Wheatleys (including Solomon, son of William Wheatley) and others who did not bear the Haynes name.[22] A probate inventory appended to Agnes's will valued her possessions at £22 13s 1d.

Thomas Haynes was a churchwarden of St Giles in Reading for the years 1559 and 1560.[23] In 1560 the decision was taken to remove from the chancel the four ancient statues of St Giles, St Christopher, St Mary and St John along with the altar dedicated to St John and also the high altar. Thomas duly recorded the cost of 4d for pulling down the statues and 2s 8d for destroying the altars and carting away the rubbish.[24] His occupation is unknown, but his wealth is suggestive of membership of the mercantile elite. It is possible, in view of his brother Christopher's alleged involvement with the Reading clothiers [below], that Thomas was involved in the cloth-making trade, this being the most important industry in Reading in Elizabethan times.[25] However, according to C. Reg Haines,[26] Thomas was a Yeoman of the Guard to Queen Elizabeth. If this is true, he would be the fourth of the brothers to have served in the Royal Household, although Christopher and Richard too may have even better claims. But for none of them is the evidence revealed to date really compelling.

Thomas's means had been boosted by the earlier bequest from his brother that enabled him to live in fashionable Whitley. He left at least two houses and lands in Berkshire and tenanted property in Reading along with some silverware, jewellery, quality furniture, firearms and £8 12s in monetary bequests. Added to this was a residual estate valued in excess of £20, positioning him at the upper end of the Tudor wealth scale. He died in February 1583 leaving no living issue. He was buried in St Giles' churchyard on thirteenth of that month.[27]

Notes and references

1. Then part of Wiltshire; H. Farrar, 1984, *The Book of Hurst*, Barracuda Books Ltd, Buckingham.
2. TNA: E210/913. If the man Pether was Bastian Pether, he turns up 20 years later occupying land in Hartley, Shinfield, previously owned by John Folkes.
3. E.M.W. Cracknell, ed. (1934), *The Barbadian Diary of Gen. Robert Haynes 1787-1836*, The Azania Press, Medstead, Hampshire, UK
4. Ernest B. Pope, 1929, *History of Wargrave*, Hitchin, Carling.
5. About £9 a year: Cal Pat Rolls Eliz 1 Vol. II, 1563-6, p. 254, (Para 1226); HMSO 1960.
6. Letters and Papers Foreign and Domestic Vol XXI, 38 Henry VIII (i) 1546-7, Part II, No. 147, p. 65.
7. The will of Thomas Cressell of Binfield (Byndfeld), PCC Welles, dated 14 September 1558, proved 7 December 1558; PROB 11/42A. John Grove was probably a kinsman of the William Grove who would in due time assist Richard's widow as proctor.
8. Will of Richard Haynnes 14 December, 1566, TNA: PCC 33 Crymes & Morrison, PROB 11/48.
9. Will of Thomas Gunnell or Gonnell, Yeoman of Wargrave, 1582, TNA: PCC Windsor, PROB 11/69.
10. She was Widow Rockoll in the Wargrave Lay Subsidy of 1602; TNA: E179/75/312.
11. William Rockoll was last mentioned in the Lay Subsidy of 1601/2; TNA: E179/75/303. In her will in 1589, Agnes, widow of Thomas Haynes, made

bequests to the wife of Richard Rockell and her two children, Joan and Robert.
12. Will of Thomas Gunnell of Hare Hatch, February 1579.
13. Will of Richard Haynnes 14 December, 1566, TNA: PCC 33 Crymes & Morrison, PROB 11/48.
14. Will of Thomas Gunnell or Gonnell, Yeoman of Wargrave, 1582, TNA: PCC Windsor, PROB 11/69.
15. P. Glennie: 2004, 'Life and Death in Elizabethan Cheshunt', Ch. 5, p. 72, in *Hertfordshire in History,* Doris Jones-Baker, ed., University of Hertfordshire Press.
16. Will of Thomas Gunnell or Gonnell, Yeoman of Wargrave, 1582, TNA: PCC Windsor, PROB 11/69.
17. The other witnesses were: Richard Spensar, Ralf Newbery, Richard Stretten and Thomas Dowglas. Probate was granted at Reading on 15 April 1553, at which time an inventory of his effects was made by *'Ralf Neubery, Thomas Barnard, William Gonell and Thomas Gonell'.*
18. TNA: E179/73/135; E179/73/132; E179/73/141.
19. Berkshire Record Office: D/A1/76/205; Will of Thomas Haynes of Whitley, Parish of St Giles, Reading, 1582. Text reproduced by permission of the Berkshire Record Office.
20. Will of John Haynes of the Parish of Marybone, 1 Jan 1605/6, TNA: PROB 11/111.
21. Allegra Woodworth, 1946, 'Purveyance for the Royal Household in the Reign of Queen Elizabeth', *American Philosophical Society,* New Ser. 35, p. 71; List of Purveyors 1586, Lansdowne 46, No 84.
22. Berkshire Record Office: D/A1/77/60, Will of Agnes Haynes, Parish of St Giles, Reading 1589.
23. Leslie Harman, 1946, *The Parish of St Giles-in-Reading*, p.93, published by the Vicar and Parochial Church Council, St Giles.
24. J. Doran, 1835, *The History and Antiquities of the Town and Borough of Reading in Berkshire*, Samuel Reader, Reading, p. 153.
25. Daphne Phillips, *The Story of Reading*, 3rd Edn, 1999, Countryside Books, Newbury, pp. 32–34. *'Reading tawny'* was a heraldic yellow coloured cloth used for the Earl of Oxford's livery in the sixteenth century.
26. C. Haines, Reg. 1901, *Notes & Queries,* 9th Series VII, March 2.
27. Parish Registers of Reading St Giles; Berkshire Record Office: D/P96/1/1.

4

John Haynes Esquire of Marylebone, Sergeant of the Acatry [1512-1608]

Royal Household - Yeoman of the Store - Purveyor - Underkeeper of Marylebone Park - Sergeant of the Acatry - Tottenham Court - Sayes Court - Creslow Pastures - Responsibilities - Fishmonger - A Sergeant's Lot - Major Perks - St John's Wood - Justice of the Peace - Fall from Grace - Retirement - Family - Agnes and Joan - Son William - Will and Bequests - The Gunnells

John Haynes was the third son of Richard and Thomasine. He is one of only two of their seven children for whom it is possible to determine his year of birth, for in a deposition dated 1601 he was said to be '*aged fowerscore and nyne yeares or thereabout*', making his birth year approximately 1512.[1]

John spent much of his working life employed in the Royal Household but of his early life next to nothing is known. The earliest record of one who might have been our man was to John Haynes of London, who with many others was granted a licence to sell fuel previously regarded as substandard, in London and the surrounding counties. This appears to have come about because of the shortage of fuel available for the coming winter. The next mention is of John Haynes, merchant of the City of London, who had somehow infringed the law, as in 1542 there is a record of him being bound over to the King in his own recognisance of £100, by the Privy Council at its meeting at Greenwich on 2 October, on condition that he should attend the Council daily to answer interrogatories.[2] The interrogatories and John's responses have not come to light and the nature of the offence was not disclosed, although the amount of his recognisance suggests it was serious; the Privy Council generally only took such pains over the examination of individuals believed to have information relating to plots or treasons of one form or another. He was probably giving evidence as a witness as there is no record of his trial or sentence.

In January 1547, the month King Henry died, a John Haynes was recorded as being granted a tenement with shops, cellars and solars in the parish of St Gabriel Fenchurch in Langbourn Ward, London, owned by John Banyster, esquire, who obtained a licence costing 12s paid to the Hanaper for this purpose.[3] In October, 1547 Haynes was accused of encroaching eastward beyond the frame of his building to make room for a jakes (privy; the encroachment, which was substantial, was probably made by his predecessor).[4] Two years later a large number of properties in London were granted to William Warde, gentlemen, which were previously owned by the Cluniac Abbey of St John of Pontefract, Yorkshire, 'dissolved' by royal command in 1539. These included two messuages in *'Byllyngesgate Strete within the parish of St Mary Antill [St Mary-at-Hill] ... lately leased to John Haynes there ...'.*[5] This location suggests that the John Haynes in question was involved in the fish business at this time and makes him more than likely to be our man, then aged 37.

It is not possible to be certain when he became a servant in the Royal Household, although it might have been at the time of his Billingsgate Street lease. There is certainly a reference to a person called Haynes listed in the Household of Queen Mary. It was customary during the reigns of Queens Mary and Elizabeth for the various members of the Household to present the monarch with New Year's gifts, of which an exact inventory was enrolled and signed by the sovereign before witnesses. Several such rolls have survived and in one from the reign of Philip and Mary, for 1556-7, is listed a gift from *'- Haynes, a table, with a picture of Christ and his Mother'*.[6] A few lines below this entry, in the Queen's hand, is *'Marye the quene'*. We cannot say that this was John Haynes son of Richard and Thomasine, but it could have been, for John would have been 44 in 1556 and he would have needed to have been employed within the Royal Household for many years to achieve the promotion to Sergeant twelve years later.

When, in 1559, John Haynes was employed as *'yeoman of the store'*,[7] he was licensed to export from London 100 tons of beer free of duty.[8] This was almost certainly because too much beer had been obtained under royal warrants of purveyance and the excess was undoubtedly being sold abroad at a profit. This happened twice in

Elizabeth's reign. In 1564 he was recorded as the *'Queen's Servant'* and within two years he had become one of the Queen's purveyors, which was probably one and the same job. By March 1567 he had become the underkeeper of Marylebone Park [*Parke of Marybone*] reporting to the keeper, Edward Carye esquire,[9] and he held on to this position and the house in Marylebone that went with it, after he had been promoted to sergeant of the acatry in 1568.[10]

The acatry[11] was the largest and most important division within the Royal Household, responsible for the purchase of meat, fish and salt.[12] The sergeant at its head usually had reporting to him some six separate purveyors for beef, oxen and sheep, and both fresh fish and sea-fish. In this position John was not only responsible for the acatry's purveyors but occasionally helped them personally with the purchase of cod, ling and stockfish, a job he doubtless knew well from his time as yeoman purveyor and earlier at Billingsgate. As sergeant of the acatry, he reported to a succession of Lords Steward, who were notoriously ineffectual, until William Cecil, Lord Burghley, stepped in to fill the vacuum and effectively managed most of the Household's affairs during Queen Elizabeth's reign. John was now at the top of his department's hierarchy which ran from sergeant through yeoman and groom to page. His day-to-day work involved supervising the selection of an adequate supply of cattle, sheep and fish, and being an expert judge of their quality and of arranging for their holding and storage. He would also have supervised the killing, skinning and dressing of the animals and the scaling of the fish prior to storage. Meat was kept 'on the hoof' for as long as possible, which sometimes necessitated the hire of pasture.

Tottenham Court, Sayes Court and Creslow Pastures

There were three major areas of land set aside for pasture for the acatry: at Tottenham, northwest of London, of which John himself leased part; Sayes Court, Deptford, in Kent and Creslow Pastures in Buckinghamshire.[13] They were used for holding the herds purchased by the purveyors, and also to a lesser extent as breeding units. The one at Tottenham Court, part of Marylebone Park, also housed and provided pasture for John's private herd, which Allegra Woodworth, an expert on Tudor purveyance, suggested was an unapproved perk of the sergeant, and of any courtier whom he wished to favour.[14] The

Elizabethan manor of Sayes Court, Deptford, was situated near the Thames about three miles south-east of London. It was adjacent to the Royal Naval Dockyard, separated from it only by a long wall. On another side was the royal slaughterhouse which was situated beside the Ravensbourne at Harold Wharf. Its function was to supply Greenwich Palace with meat from cattle grazed at Sayes Court. It measured 160 feet from east to west and was 50 feet wide with a wharf and a pond at the west end. The date of its foundation is unknown, but was around 1550. John would have been responsible for this abattoir during his time as sergeant. The Browne family, Clerks of the Green Cloth, in the late sixteenth and early seventeenth century, took over Sayes Court and its slaughterhouse after John's departure. Their buildings at Sayes Court included 34 bays of ox-stalls of which eight were reserved for the Queen's cattle. There was also a building known as the *'Doghouse'*. These structures were *'somewhat decayed'* in 1608 and demolished by 1649. The size of the Sayes Court pastures is unclear, but was in excess of 210 acres.[15] Only a remnant survives today, a small park still bearing the name, Sayes Court. The three-gabled oak-framed manor house of Sayes Court had 18 rooms on two storeys. It was surrounded by orchards and gardens and reached via a tree-lined avenue from Deptford through its two gates.[16] It was under John's keepership and must have been a useful residence when the Queen was at Greenwich.

Creslow Pastures, six and a half miles miles from Aylesbury on the Whitchurch road, was the name given to the manor and lands at Creslow in the sixteenth century which came into the possession of the crown at the time of the Dissolution, after which time it was used as pasture for the sheep and cattle of the Royal Household under the keepership of successive sergeants of the acatry. The extensive pastures covering more than 850 acres sloped down to Creslow Brook. In John's time the manor consisted mainly of an early fourteenth-century house comprising a hall with a wing at the north end containing the kitchen and associated rooms, and a chamber block to the south. There were two cellars called *'the crypt'* and *'the dungeon'*. The crypt, hewn out of the underlying limestone was a vaulted cellar under the eastern end; it originally had a passageway to the dungeon. There was a tower at the south-west with an octagonal stair turret at its north-west angle. There was also a wing built on the east end of the south wall of the solar block.[17]. The large window in the south wall of the solar was of fourteenth-century origin. The whole building with its gables and three-storey tower and buttress must have been an imposing house in the sixteenth century as it remains today. The house adjoined the parish church, but services had ceased by the time John Haynes was keeper. In 1596 the manor was still known as *'Hayne Houses ground and pasture'*,[18] presumably after John, who occupied it until 1592 when James Quarles, chief clerk of the kitchen, sergeant of the poultry and victualler of the Queen's ships, was appointed keeper.[19] Quarles was too busy to look after the manor and the pasture and the buildings soon fell into disrepair.[20]

From the foregoing can be gained some idea of the considerable perquisites granted to the sergeant of the acatry and his staff. However, John lived mainly in a lodge, built in Marylebone Park for one of the previous keepers, either Thomas Savage or Andrew Wedon. Savage had been granted life-keepership at 4d a day with permission to keep 12 cows, 1 bull, 4 geldings and 2 foals. After Wedon's death in 1560 the keepership was held by Savage only, for whom John might have worked. In recognition of his increased responsibility, the keeper was granted pasture for more of his own animals and a further 30 loads of hay in winter and 20 loads of brushwood for heating fuel. It is therefore to be expected that when John took over his *'underkeepership'* role, after Savage's death around 1564, he would have retained the permission to keep livestock along the lines of his predecessor – he certainly seems to have assumed as much. However, there is evidence that a condition of his tenure of the lands at Tottenham Court, was *'after pasture'*, and that only after the mowing and provision of the first *'50 loads of hay to be delivered at the Mews for and towards her Majesty's provision there,'*[21] could he then graze his own animals. William Necton, feodary (an officer of the Court of Wards) of the county of Middlesex, in his survey notes made while he was producing a plan of Tottenham Court in 1591, mentioned that the *'chief house, the stable and two barns and a little close called Pond Close with the orchard and the two closes called Murrell's'* were used for feeding the queen's cattle *'at the discretion of Her Majesty's officers'*.[22] Murrell's closes occupied a site on the north side of present-day Euston Road, extending eastwards from Tottenham Court to the Duke of Bedford's Road, which went northwards from the rear of Bedford House, Bloomsbury Square.[23] It seems from this that towards the end of the sixteenth century the custom had evolved whereby the decision whether or not to pasture the Queen's cattle was at the discretion of John Haynes and his herdsmen and that Murrell's had become more of a holding station pending slaughter than primarily a farm for the Queen's herds.

It is quite clear that John behaved in every way as autonomous park keeper despite the fact that titular keepership (at 8d a day) was held in succession by the Carye family (right up to the Commonwealth), for it took more than 20 years to emerge that John's own cattle were taking precedence over those of the Queen, a factor in his eventual replacement. It was John's responsibility to keep the fences around the pasture stockproof and the hay barns in good order. The royal pastures would definitely have been self-sufficient in hay, but for the overgrazing by John's cattle and those of others not only at Tottenham but particularly at Creslow and Sayes Court. He clearly couldn't fulfil his *'after pasture'* contract on the royal pastures and hay had to be bought in at the Queen's expense in reality to feed great numbers of cattle owned by others rather than the Queen, to the extent that the Queen's animals *'cam leane to the Corte and yor Matie [Majesty] charged to buye haie ...'*.[24]

Marylebone Park was a private royal deer-hunting park of some 554 acres created by the Queen's father King Henry VIII, who in 1538 had compulsorily appropriated the constituent properties of Tyburn Manor, with its orchards, pastures, farm and mill; lands of the Prior of St John of Jerusalem; lands of Eton College; lands of the Prebend of Rugemere; those held of St Paul's Cathedral, Tottenham Court, St Giles church, and several individuals. It consisted mainly of woodland and pasture of which about three-quarters was emparked. Queen Elizabeth took a great interest in the Park, as did King Edward VI and Queen Mary before her, and spent a great deal of money on improving the fences, water-courses and buildings. During 1567, an immense amount of work was undertaken in and around the Park which took about 20 months to complete.

Of interest to us are the records relating to John Haynes from the receipts and payments for the Park works recorded by Edward Carye, keeper, and John Fothergill, woodward for Middlesex.[25] On 20 March 1567 John is recorded as handing over to Edward Carye sums of money due for the essential repair to his lodge in the Park, namely: 7 days' work by the carpenter [8s 6d]; 10 days' work by the labourer [7s 6d]; 8 days' work by the tiler [8s] plus costs for 300 nails [2s], 50 floor boards [2s 6d], lathe nails [6d]; 100 lathes [12d]; 200

tiles [9d]; 37 ruff tiles [2s 8d]; tile pins; [2d]; lime [5s 4d]; 200 bricks [2s 2d]; carriage of load of sand [8d]. The whole repairs to the lodge building came to 49s 8d with an additional 32s 8d for post, rail and paling fencing. On the 31 December following, Robert Carter was paid 44s 6d for glazing John's lodge *['...for glassynge at John Heyns lodge according to a byll of the particularities thereof']*. So John appears to have had a major refurbishment of his house at the the time he became sergeant. He appears also to have charged for work done by his own staff, for on 31 December 1567 he was paid £9 6s 7d for the felling of 40 trees in the park by his servant. He was also reimbursed 5s 8d for the carriage of posts and rails to the river and planks and posts to the bridge for which he himself had paid.

Responsibilities

John had total responsibility for supervising the work of his staff, everyone from the yeomen to the herdsmen. Thus he had two grooms; a yeoman purveyor of veal, who looked after not only the veal but also stirks (yearling heffers and bullocks) and pork in all its forms; a yeoman purveyor of beef; a yeoman purveyor of mutton; a yeoman purveyor of sea-fish (his brother William); one of freshwater-fish; another of the salt stores. In addition there were the herdsmen who looked after the animals on the hoof at the holding pastures, whose job it was to report any animals stolen or dead from disease.[26] His clerk kept the accounts of the office.

Placement by an existing household sergeant of a member of his own family, as a token of royal favour, was not uncommon, and it is likely that John would have had a big say in William's appointment, as purveyor of sea-fish carried the rank of yeoman in the acatry. John appears to have carried on influencing the appointment of acatry staff even in 1590 when he himself was under a cloud.[27] An appointment of a herdsman at Sayes Court, for example, was made behind the back of Sir Francis Knollys while he was away from Court. The Queen had commanded Knollys to make this appointment and Knollys was incandescent and immediately fired off a terse complaint to Lord Burghley.

The Lord Steward (Burghley) was responsible for that part of the Royal Household responsible for the practical running of the Court, named by Edward IV the *Domus Providencie*, as well as accounting and household management. He was aided by the Board of Green Cloth, which consisted of the treasurer of the Household, the comptroller and various clerks.[28,29] John would have had to submit daily accounts to the clerk of the green cloth, but it was he who had responsibility for keeping the financial records of the acatry. There was very little explanation required as to why certain purchases had been made or money spent, and it would have been easy for money or goods to have been misappropriated without detection.[30]

There were some purveyors who were appointed in the place of previous purveyors, not to a salary or wage, but to enjoy *'rights, profits, privileges and advantages'*, and these were clearly considerable.[31] John's rank of sergeant carried an annual salary of £11 8s 1½d plus a notional 1s 4d a day (rising to 1s 9½d in 1576) board wages, with an additional allowance of £10 a year for delivering his cattle residues at specified places, and many other sources of income. For example, he would also have received around £6 a year and associated benefits for his position as underkeeper of Marylebone Park. Thus his potential monetary income totalled more than £40 a year. In addition, his recognised perquisites included the heads, midriffs, hearts, tripes and feet of the oxen.[32] These and property perquisites compensated for his absolute lack of further preferment in the Household, for sergeant was regarded as a job for life and he would have had no chance of promotion to the Board of Green Cloth.

John's good communication with the kitchen was vital as his job was to fulfil its daily requirements of salt and fresh meat, salt and fresh fish and salt. But as sergeant he was by no means confined in responsibility to the acatry. He would be able to petition the Queen directly and being in her presence would not have been taken as unusual. He needed to travel a great deal especially in those counties that provided the Acatry with the bulk of its supplies, or to and from the pastures, and it was possibly in this context that in 1587 he was employed in *'carrying letters for the Court'*.[33] We learn from his will[34] that his work for Queen Elizabeth had been recognised in the form of a grant, by letters patent, of property in Old Fish Street, London.

Fishmonger

His promotion to sergeant would have created the vacancy for a yeoman purveyor, which may have been the one filled by his brother William. In 1567/8[35] John Haynes, described as a *'merchant'*, was purchasing fish from the masters of foreign vessels arriving at the Port of London. Thus, on 13 October 1567 he bought 5 lasts of fish from the master of the *Mychell* of Amsterdam, Henrik Johnson, for £30. Likewise from Peter Garetson master of the *Merman* of Amsterdam, John purchased 5 lasts of fish for £30 on 23 October 1567. On 25 November 1567 he bought 720 lings and 3 lasts of fish for £42 from Christian Cornelis, master of the *George* of Amsterdam, and on 16 August 1568, 18 barrels of fish for £9 from Sebrant Cornelis, master of the *George* of Purmerend, North Holland. These were enormous quantities of fish – a last was the equivalent of about 12 barrels – so the bulk of these purchases must have been destined for the Royal Household. They may have been John's last such purchases as he was already sergeant. Doubtless some of the fish found its way 'round the corner' to Old Fish Street where John is known to have run a retail fish business in the premises provided by the Queen. He could also legitimately augment his income by the sale of the Royal Household's leftovers that he was allowed to purchase.[36] In November 1569 John and William Haynes of London, fishmongers, were left a quarter of wheat between them in the will of William Wrighte, clerk, of Little Laver, Essex. Other London fishmongers were also beneficiaries and may have known Wrighte when he held the cure of St Michael Cornhill from 1554 to 1562.[37]

John Haynes's shop, like that of his brother William, was said to carry the finest stock, which rightly or wrongly was widely believed to be due to their use of the royal purveyors' right of pre-emption in purchasing (having first pick of the best fish at a fixed low price before it could be put on general sale) in order to feather their own nests. Apart from the advantage of buying fish for less than the market price, they might have been tempted to purchase more fish for the household than was necessary and either siphon off the excess or 'legitimately' buy any leftovers from the Green Cloth and then sell it at a higher price. Although in May 1582, in an attempt to stamp out such practices, Burghley issued a rule that no officer should buy any provisions for their own use, in order to obviate suspicion that they were doing so for personal gain (*'of p'vat lucre*'),[38] royal purveyors could also have delayed payment of large sums, as was the custom of the Green Cloth. The suspicion was that they did all of these things.[39] On 18 April 1579, John Haynes, esquire, received a commission of provision as sergeant of the acatry, to take up six lasts of salmon for the provision of the Queen's Household.[40]

John's fish business, a shop and warehouse in Old Fish Street, London, given to him by Queen Elizabeth, may have been occupied by earlier royal purveyors. Robert the purveyor of sea-fish to Henry VIII and John's predecessor but one, lived, died and, in 1541, was buried in the adjoining church of St Nicholas Cole Abbey. The property is possibly that confiscated by Henry VIII from the London Minoresses (Poor Clares) at the time of the Dissolution, records of which can be traced back to the fourteenth century. In January 1547 it was held by Agnes Blockeswiche.[41] William Haynes also had accommodation and/or a shop in this parish (St Nicholas), as in 1582 he paid 50 shillings in the London Subsidy for Bread Street Ward. There was another John Haynes in the same parish, not obviously related to the sergeant. He was a clothworker (merchant tailor) and had a partner called Thomas Kowdale (Cowdall) another merchant tailor, who were jointly assessed at 15s in the 1582 Subsidy.[42]

Shops of the day were mainly warehouses, but that of a fishmonger would possibly have included stone slabs, stalls or similar structures for display. Henry III had prohibited the landing of fish from fishing boats anywhere in London except Queenhithe, so that he could be sure of augmenting the customs receipts. For three centuries, therefore, the fish traders had concentrated themselves in the Billingsgate area and the old fish market extended from Bread Street in the east to Old Change in the West, and from the northern ends of those streets to Old Fish Street in the south,[43] where the major fish stalls were located. This area made up a considerable part of Bread Street Ward. The immediate vicinity of Old Fish Street was famous for fish dinners, especially on Fridays, and was a favourite haunt of Londoners, including courtiers. Popular venues of John's time were the four taverns in Old Fish Street: The Swan on the southeast corner with Bread Street; The Bores Head; The Will Somers, named after the famous Court Jester (d. 1560); and The Dolphin. It has been held that The Will Somers was named by a local – Robert Hunsing the royal purveyor of sea-fish has been considered a possibility – who gained influence at court from the friendly offices of the jester, who had the ear of Henry VIII; but the tavern could equally have been named by John Haynes.

John's shop received a supply of Thames water from the conduit situated by the churches of St Mary Magdalen and St Nicholas Cole Abbey near Old Fish Street. Although we don't know exactly where his fish premises were located, there was a row of small houses in the middle of Old Fish Street, known as Middle Row, all of them occupied by fishmongers, which began life in earlier times as moveable boards or stalls for displaying the fish on market days, and then were licensed to set up sheds. Eventually these grew into houses of three or four stories, some of which were only 8ft by 18ft at ground level. It is possible that John's property was one of these, as we know from his will that it comprised a shop, warehouse, plus three other rooms. Many rich and famous fishmongers held these properties, including William Newport, who became Sheriff of London in 1575 and Bernard Randolph, a wealthy benefactor, who in 1583 gave the Fishmongers' Company £900 towards the cost of the stone and lead cistern next to St Nicholas Cole Abbey *'for the ease and conveniency of the fishmongers and other inhabitants in and about Old Fish Street'.*[44] John Haynes's nephew, Moses Gunnell, was also a fishmonger and lived with his wife and family in the same parish as John, with whom he may have worked. Moses' sister Sara Gunnell married Thomas Bigg another fishmonger of St Nicholas' Parish and became a favourite niece of John (below).

All of the properties in Old Fish Street were destroyed in the Great Fire of 1666, mainly because of their proximity to each other. Davenant, writing before the Great Fire, humorously observed: '... *the goodly landskip of Old Fish street! ... where the garrets, perhaps not for want of architecture, but through abundance of amity, are so made that opposite neighbours may shake hands without stirring from home'.*[45] This architecture no doubt explained the buildings' susceptibility to the fire when it came.

No fish could be sold in the City of London unless by fishmongers – freemen of the Mystery of Fishmongers – and then only in three markets: Bridge Street, Old Fish Street and the Stocks. In this way large quantities of the best fish would be available in the same place, to the particular convenience of the royal purveyors as well as others.

In the Visitation of London, 1568 John Haynes was described as being *'now sargeant of the Catre to our sovereyne lady Quene*

Elizabeth', the 'now' perhaps indicating his recent promotion from yeoman. His predecessor was John Dunning, who continued to live in Rye as a jurat and customer,[46] and where, in 1567, he was elected mayor for the first of four occasions. This election may explain Dunning's departure from the household, although he continued to be involved with fish buying along with William Haynes.[47]

Mention has already been made that surplus items in the acatry, including fish, meat, animal skins and hides, may have been considered the property of John and his senior officers or, at least, they would have had first refusal on their purchase. They would then have been sold to supplement respective incomes. John received £10 a year for delivering the remains of the acatry's cattle to pre-assigned places.[48] It is of interest in this regard, that the three who were witnesses of John's will, Richard Deacon, Edward Clarke, and William Gunnell, junior, were all skinners,[49] William Gunnell, Junior, being the younger brother or cousin of William Gunnell, John's nephew, executor and principal beneficiary, who worked for his uncle, as his 'servant'. It is possible that he helped to look after John's cattle and horses at Marylebone.

A Sergeant's lot

The troubles John experienced through the activities of his yeoman purveyors sometimes had to be dealt with by higher authority. In the Spring of 1574 the Privy Council dispatched a letter to the Lord Mayor of London, Dr Thomas Wilson, Master of the Court of Requests and Mr Peter Osburne, instructing them to examine the facts behind a '*supplicacion by John Heynes, Sergeaunt of her Majesties Catery, and others*' against William Hunt who had unlawfully molested the purveyors for bringing into the country fish in barrels for which they had a valid licence under Letters Patent.[50] This was probably an action by Hunt against John Dunning and/or William Haynes who in 1964 had been granted a licence to bring fish into the country in barrels or casks.[51] In 1578 the Privy Council required Robert Fynnett, the mayor of Dover, to attend them in person when he had completed his mayoralty, to answer to certain misdemeanours committed by him against John Heynes, '*a servaunt of Mr Comptroler's*[52] *in causing him to open his maister's letters perforce*' [i.e. for allegedly forcing John Haynes to open letters addressed to the controller]. This was possibly associated with the problems that his brother William Haynes was encountering with the fishermen of Rye, and was a serious offence, if true. In March the previous year, a Sergeant Haynes and two others, Thomas Stanley and Rafe Isake, described as *'Constables in Holbourne'*, were summoned to appear before the Privy Council as soon as possible. The reason was not given[53] athough the fact that a sergeant of the Household had to attend with the constables, suggests that it had to do with an irregularity arising from purveyance or its administration.

Problems arose when it emerged that two of John Haynes's purveyors on the meat side, Richard Besbitche followed by his son Mathew, had not paid some of their suppliers for more than ten years. On 17 February 1580 the Queen issued a proclamation addressed to all such suppliers *'to whom either Richard Besbitche or Mathew his son, now purveyor in his place, owe money for stock or other provisions taken by them or their deputies'*.[54] The proclamation required all suppliers who hadn't been paid by the Besbitches over the preceeding ten years to take their bills to John Haynes *'Sergeant of the Account'*, within forty days for settlement. Mathew Besbitche was ordered to recompense all such suppliers. If the suppliers felt that they were being inadequately recompensed, they were given permission lawfully to refuse to provide their animals.[55]

Major perks

Like all members of the Household, John Haynes received lucrative benefits of income from property leases granted by the Crown, and there were perks of a different kind in the form of licences to export goods. Thus in November 1564, as the *'Queen's Servant'*, he was granted the lease for 21 years of the Rectory of Kirkdale, Yorkshire, once the Monastery of Newburgh (Newborowe; now Newburgh Priory), and also the mansion house of Sturminster Newton, Dorset.[56] The yearly rents were respectively £26 8s and 5s. There were repair and other conditions attached to the leases for which a fine of £51 was paid at the Exchequer,[57] but there were clearly significant profits to be made. On 11 October 1566, as one of the Queen's purveyors, he was granted a lease by the Exchequer of the Manor of Raynehurst in Chalke, Kent, previously owned by Sir Thomas Wyatt,[58] who in

1541 had transferred the manor to the King in exchange for other properties.[59] His brother Nicholas was also granted a lease for part of the Raynehurst estate. The properties were in a parlous state of decay because of past flooding, and John is reported as being prepared to undertake the cost of the necessary repairs and maintenance.[60] On 10 May 1582, with five years of the lease still to run, the Raynehurst parcels of land, excluding the woods and quarries, were leased to Reynold Knatchbull.[61]

For his services as underkeeper of Marylebone Park and as sergeant of the acatry, John was granted several leases in the Marylebone area and was granted a 30-year lease of a cottage in Hampstead, being part of the Manor of Chalcottes and Wyldes in the parishes of Hendon, Finchley and Hampstead, with 1½ acres of pasture, an orchard of 1 acre and another 2 acres of pasture, at £3 6s 8d a year. He sold this last lease on to Alexander Glover on 13 December 1577. Glover, who had been one of the clerks belonging to the Receipt of the Exchequer, was appointed to *'the office of heard [herd] under Her Majesty'* with the custody of her house and grounds at Totnam Courte *'used for the keeping and pasturing of her Majesty's provision of cattle and sheep for Her Highness's Household'*. In this role he might have reported to John as sergeant of the acatry. Glover, with his wife Blanche, daughter of Richard Loftis, mercer, of St Magdalen, Milk Street, London, who owned the adjoining Manor of Chalcottes and Wyldes in Hampstead, greatly extended and improved the cottage bought from John to the tune of £60.[62] Interesting light is shed on this restoration by William Necton, who in 1591 noted that the Queen's cook, Daniel Clarke was living at Tottenham Court, in *'a very slender building of timber and brick',* which had earlier been a much larger building, but had been cannibalised in order to provide materials *'to amend some part of the house now standing which hath been repaired of late by Alexander Glover'.*[63] It appears that Glover had demolished part of the main house complex then in his care, in order to refurbish his own. Glover died around 1590 and was succeeded as *'heard'* by Clarke, who claimed Glover's cottage as part of the Queen's grant to him. This was strongly contested by Blanche Glover who claimed the cottage belonged to her.[64]

Two closes of land adjoining Tottenham Court [Tottenhall] Manor were also demised to John Haynes by the Earl of Arundel and the Earl of Leicester, as the Crown, through the Lord Steward, held a 99-year lease from 1560 on the Manor from the St Paul's Prebend, Thomas Watts, at £56 a year,[65] and the Lord Steward [Leicester] and Officers of the Green Cloth then sublet the manorial lands to their staff. The affinity of John's brother Christopher with the Earl of Arundel and John himself with the Earl of Leicester may have influenced this grant. It is suspected that, after his appointment as Lord Steward of the Household, Leicester kept his personal stock animals at Tottenham for his own domestic consumption through some private arrangement with John Haynes.[66] It has also been noted by Alan Haynes[67] that one of John Haynes's purveyors was John Dudley, and the Earl was known to have favoured his kinsmen.

The survey made on 6 April 1591 of the manor of Tottenhall [Tottenham Court] by William Necton[68] showed four closes of pasture land to be held by Sergeant Haynes. One of these closes, situated in Kentish Town, was only one mile from London (and John's fish business there), the second adjoining Tottenham Court extending to 4 acres in the Parish of St Pancras, were the ones demised by the two earls.[69]

There was an attempt on 20 January 1601 by one John Parker, in an action against John Vincent and John Abshawe [a former tenant of the manor house of Tottenhall], to claim a fifth close consisting of a 16-acre meadow, as part of the Prebendal Manor of Tottenhall [Tottenham Court]. This was because of its situation, being sandwiched between two parcels of the demesne lands of the manor and always having been considered part of it. The evidence taken by Commission in the Court of Exchequer was heard by the Commissioners [George Wrightington, Esq., John Knight, Henry Addy and Edward Fawcett, gent] sitting in the Maidenhead Inn in Dyott Street, Parish of St Giles in the Fields [now the site of New Oxford Street]. The close in question was referred to at that time as Arthur's Close in the *'parish of Pancresse'* after Arthur Rainscroft and his son who had been the owners in 1583. It also went under the name of *'Carres Close'* because it had previously also been in the occupation of Robert Carr.[70]

All those with information related to this land were called upon to answer interrogatories. This included John Haynes *'of Marybone in the county of Middlesex, esquire'*, from which we learn that he was then aged 89. John was able to provide conclusive proof based on his intimate knowledge of the copyhold, customary freehold and demesne crown leases of Tottenham Court, including his own, that the Dean and Chapter still maintained ownership and that the close was not part of the Prebendary of St Paul's Cathedral[71] then leased to the Crown.[72] He knew that Arthur Rainscroft the elder, who lived in Friday Street, London, had occupied the close and enjoyed the profits of it for many years, but he had never heard that the title held by the Dean and Chapter of '*Pawles*' had been transferred to Her Majesty or otherwise, until John Parker had brought this action. This was because his own predecessor on this site, Henry Bludder [or Bludde], yeoman of the counting house,[73] did not hold Arthur's Close under his sublease from the Crown, nor did John Haynes himself by the leases granted to him after the death of Bludder, take any profits of this land or '*pretend tytle in Her Majesty's right*',[74] i.e. it was not part of the demesne land leased to the Crown. He denied any knowledge of Alexander Glover or any other keeper of Her Majesty's grounds of Tottenham Court pasturing cattle on Arthur's Close with the permission of Arthur Rainscroft.[75] John was then asked if the plaintiff in the action, John Parker, or anyone on his behalf, had offered to bribe him to agree that the close was part of the demesne of the Manor of Tottenham Court. This he denied, but acknowledged that Parker had once asked him whether the close belonged to the Dean and Chapter of St Paul's, and Haynes said that it did not, on the grounds that Henry Bludder who had occupied part of the demesne lands and pastures of Tottenham Court, leased by the Dean and Chapter to the late Earl of Leicester and others, did not occupy or take profits of Arthur's Close, as John Haynes himself hadn't, when he leased the same demesne lands after Henry Bludder's death.[76] This was the true situation, Haynes asserted; Arthur's Close was not crown land.[77]

Finally John was granted, by virtue of the Queen's warrant of 23 July 1582, a 21-year lease on two further parcels of woodland in St Marylebone, formerly of the Hospital of St John of Jerusalem, for a yearly rent of 67s 3d.[78] This land, which covered an eighth of the whole wood now known as St John's Wood, had been acquired by King Henry VIII in 1538 which at that time included 14 acres of woodland and 12 acres of pasture.[79] The woodland had previously been leased to Edmund Downinge.[80]

Property leases in Elizabethan times generally carried the responsibility of the lessee to repair or replace structural defects whether in stone, brick, tiles, timber, glass, drains or lathe and plasterwork. Although John seems to have been able to claim for Robert Carter's reglazing of his Marylebone lodge, most repairs had usually to be funded by the tenant. For those who lived in cities, cesspools, filled via downpipes from the unsealed upstairs privy, had to be emptied manually at regular intervals by *'night soilmen'* and were then strewn with juniper to ward off the smell – all at considerable expense.

Dangers of Marylebone

As part of his working life at Marylebone, John and his assistants had to defend the vast grounds from poachers and other felons and there are regular cases recorded at the Middlesex Sessions of the trials of men apprehended by the keeper for taking deer.[81] One man, Thomas Tyanson, was bound over to appear at the next General Session of the Peace, to answer *'for takinge of henges from the gate of Marybone Parke'*.[82] John also appears to have acted as one of the Marylebone magistrates[83] as there are records of recognisances being taken before him (John Haynes, Esq. J.P.) in February 1588/9. William Cecil (Lord Burghley) seems to have been in possession of Tottenham Court Manor House or another residence at the time of a robbery there in 1577.[84] John himself experienced a burglary when London spinster, Elizabeth Atkins, broke into his house and stole two spits, a grid-iron and a pair of bellows worth 6s in all. She admitted the theft for which she was found guilty of petty larceny.[85] It is of interest that the thief chose to steal household ferrous utensils, as such a burglary as John experienced was often for items of silver, which were extremely common in Elizabethan households as is evident from the wills of several of the Haynes brothers. Elizabeth's sentence is not recorded but petty larceny generally carried a sentence of forfeiture of property to the Crown and whipping.

Despite his job as deputy keeper and the frequent arrest of felons in which he was involved, John managed to keep clear of the frequent affrays that were recorded in the Marylebone and Paddington areas – at least he doesn't appear to have suffered serious injury. Duelling to resolve quarrels, not only between gentlemen but between men from all social strata, was common and, unless it resulted in a fatality, would seldom come to the attention of the magistrates. Deaths caused by blades of all kinds, but especially by the rapier and dagger, were also common occurrences as is evident from the Middlesex Records, and deaths from gun-shot wounds were beginning to appear. The same records include many deaths of muggers and thieves, such as that of Robert Radclyffe in Longe [Lustie] Lane, *'Maribone Parke'*, killed in self-defence by Charles Wrenne of Gray's Inn.[86] The quiet roads around the perimeter of Marylebone Park were favoured haunts of highwaymen out to steal horses from those on their journeys northwards from London.[87]

As well as his work as a Justice of the Peace in Marylebone, John was occasionally required to serve on inquests or associated commissions in the City of London. Thus he was appointed with William Tooke, Auditor General of the Court of Wards and Liveries and escheator of the county of Middlesex, John Daye, doctor of laws, Christopher Reve, Robert Lydesey, William Danyell, esquire and William Necton, gentleman (feodary of the Court of Wards and Liveries) of the county of Middlesex, to serve on the inquisition following the death of Bartholomew Brookesbie, late scrivener of London,[88] who had died on 10 August 1579[89] seised of several properties in the Vintry and elsewhere in the City of London. Brookesbie had been a close neighbour of John in Bread Street Ward. He was buried in the choir of St Mary Magdalen Church, Old Fish Street, on 13 August 1579 under the altar. Some of the others serving on the inquisition were also locals.

That John had earlier been held in a position of high regard is evident from his assignment in October 1578, along with his brother, William, and Thomas Smythe, one of the customers of London,[90,91] in which they were entrusted with the responsibility of the receipt and safekeeping of considerable sums of money collected by the mayors and sheriffs of all counties raised from a nationwide appeal for the construction of a harbour at Hastings.[92] The three not only had to keep an account of the money, but also records of anyone that refused to make a contribution. They then had to deliver the money and bills to Anthony, Viscount Montague, KG,[93] William, Lord Cobham, warden of the Cinque Ports and Constable of Dover Castle and Thomas, Lord Buckhurst, Lord Lieutenant of Sussex, Richard Calverly and John Jefferey, two jurats of Hastings, and William Relfe of Ore, gentleman, or any three of them, of whom one of the right honorables aforesaid, or his deputy, was to be one. Calverley, Jefferey and Relfe were surveyors of the project. The estimated cost of the Hastings project was about £4000, so this represented considerable trust being placed in the Haynes brothers. The following March (29

March 1579) the three men were ordered to *'paie unto such persones as shall bring anye moneye towards the repairinge of the haven' iiijd in the pound'*.[94]

As so often happened when such a large amount of money was at stake, it *'was quickly converted into private purses, and the public good neglected.'* There was no suggestion that either of the Haynes brothers was involved in this misappropriation; indeed it was around this time that John became an esquire. The principal defaulter appears to have been one of the surveyors, John Jefferey, since on 22 July 1581 Sir William Brooke and Lord Cobham issued a warrant to the bailiff and jurats of Hastings to arrest him *'for certain contempts'*, but in a return made on 7 August it was stated that he was not to be found in the town.[95]

John Haynes remained a trusted member of the household in 1591 when, on 10 December, at the age of about 80, he was given three commissions of provision, to take 500 oxen and 5000 sheep and other necessities from the counties for Her Majesty's Household.[96]

Fall from Grace

John's fall from grace may have come about as the result of the inquiry following the Queen's command for a *'Commission for Household Causes'* to be set up by the Lords of the Privy Council. The Commissioners wrote to Justices of the Peace in all counties requiring them to seek out and document evidence on oath of *'intolerable abuses of pourveyers for the howsholde in takinge and convertinge to their own uses, great quantities and numbers of p'vicions wth manye other disorders ...'*. The evidence poured in and old laws concerning purveyors were enforced and new laws protecting the seller and all aspects of the purveyance process introduced[97] and a Bill was put before Parliament in February 1589,[98] many provisions of which the Treasurer of the Household [Knollys] opposed as unworkable. Penalties for transgression by purveyors were severe; the infringment of some of the new purveyance laws carried the penalty of death by hanging[99] and some purveyors were tried in the counties in which their abuses had occurred and were executed,[100] while others were imprisoned and died in prison. Some were removed from office yet shown *'tomuche favo[ur]'*, others managed to keep their jobs. Sir Francis Knollys in a letter to the

Queen indicated that if *'anye purveyor have bene complained upon'*, he would be *'sent into the countrye to be tried for his lyeffe'*.[101]

Four of John's purveyors, William Beachamp, John Dudley, Richard Owen and Mathew Bestbyche, were among those guilty of misconduct serious enough to necessitate their discharge.[102,103,104] However, some, like Richard Owen the purveyor of sheep and oxen, felt they had been unjustly dismissed and argued their corner. Owen petitioned Burghley for an enquiry into his removal, seeking to be restored to his post.[105] John was still in office in October 1594, for in an action against Edward Page esquire and his son William Page, gent., in the Court of Requests, he still appears as *'Seriannt of her Majesty's Caterie'*[106] and he is similarly described in November 1596. However, there can be no question that John was among those replaced in the reforms of mid-1590s, in his case, by William Lancaster; the reason apparently being given that he *'did greetlie deceave yor Ma[jes]tie as is p[ro]ved'*.[107] Although Woodworth summed up John as a man without scruples, who put self before duty,[108] his retention of the London and Marylebone leases after his departure from the acatry argues against him being held culpable of a really serious misdemeanour. It might have been simply the case that he was far too old to retain the job and, being long past retirement age, may even have lied about his age. Even a cursory review of the household records shows that the problem people were by no means limited to the acatry – the poultry and bakehouse appeared to have even worse offenders. If he was dismissed in 1596/7 for a serious misdemeanour, John appears to have been one of the lucky ones. He couldn't himself have been guilty of the abuses of his purveyors, even though he might have benefited from their misdeeds, but the most likely reason for his replacement was his use of the royal pastures for his own animals at the Queen's expense, which could reasonably have been interpreted as deception on his part. Added to this, age was probably telling and he would have been unable to manage such a large organisation as the acatry as competently as once he did. Perhaps his long service saved him from serious consequences; he would have been about eighty-three by this time and ready to relinquish his position.

The financial cost of his 'deception' came to light some time after 13 November 1596 when he was still sergeant,[109] probably in the early months of 1597. This can be gauged from the report to Lord

Burghley dated 1 July 1597 of the yearly savings resulting from the reorganisation of the acatry following the departure of John, William and other staff, as quantified by Master Trown, one of Lord Burghley's clerks: greater amounts of tallow in the slaughterhouse, £600; sale of wool, £100; cost of mutton, £100; replacing William Haynes with William Angell led to fewer cod and ling being purchased and left over, £200; there was also £200 saved by replacing the yeoman of the slaughterhouse, removing illicitly grazing cattle from the royal pastures at Creslow and Sayes Court and fattening up only the Queen's herds and incidentally producing a surplus of hay,[110] making a total annual saving in the running expenses of the acatry of £1200.[111] The problems at Marylebone, where John was personally in control appeared to be minor compared with events at Creslow and Sayes Court and in view of his great age John may not have been entirely aware of the extent of his staff's misdemeanours, although he was undoubtedly the one held responsible. He was clearly not alone. Savings were similarly made in the scullery, spicery, privy, bakehouse, cellar and stable and the total gain to the household budget during the three years after the reforms amounted to £9,451.10s.

Retirement

He retired to Marylebone, where in the first and third subsidies for Marybone in the Hundred of Osulston, Middlesex, 1597, he was assessed on goods valued at £8 on both occasions,[112] on which he paid 21s 4d on 16 December 1598[113] and 21s 8d on 20 January 1600[114]. He still had his woods and fields in Marylebone and grazed his animals there. He also retained his business interests in Old Fish Street, where, before retirement on 22 November 1594, he was valued after the rate of £20 in lands and fee on which he paid £4;[115] on 24 November 1595 he was valued in lands and fee at £10 on which he paid 26s 8d;[116] on 13 November 1596 he was assessed at £20 in land and £3 for goods on which he paid, as *'Seriaunt of be Majesties Catrie'*, 26s 8d.[117] In 1597, after his retirement, he was assessed at £60, but with no *'Queen's exoneration'*.[118] From around the time of his departure from the Royal Household, John's fish business was probably being run by his eldest nephew, Moses Gunnell, who lived in the Parish of St Nicholas Cole Abbey with his wife Katherine and their six children until the family was devastated by the plague in 1603.

However, even retirement was not without incident, for John's woodland in Marylebone was destined to become the cause of some anxiety. In 1602 John himself, then in his ninety-first year, became the defendant in a case brought by John Tavernor, Surveyor of Her Majesty's Woods South of the Trent. Acting on information received from Richard Page of the Parish of St Giles in the Fields, Tavernor accused John of felling trees in St John's Wood, and creating waste and spoil.[119] This may have been a vexatious case instigated by Page in revenge for the successful suit John had brought earlier against Edward Page and his son, William,[120] who had claimed repayment of a debt by John, the larger part of which had already been paid. The wood had been granted to John by the Queen in July 1582, although the interrogatories prepared for John[121] suggest that he had actually held them as tenant for only 17 years, i.e. since 1584. It comprised 50 acres out of the 420-acre wood. The four sworn deponents (William Gregory, 52; William Syster, 36; Henry Epton, 60; Francis Fynch, 43), all husbandmen from the Marylebone area, claimed intimate knowledge of St John's Wood and of the 50 acres tenanted by John. They agreed that the wood was mainly of small trees and coppice with few large oaks or great timber trees. They had some difficulty in estimating by how much the trees had grown over the 17 years of John's tenancy and deposed that, to their knowledge, the defendant had not cut down large trees, only coppice. Most of the trees they estimated at no older than about 40 years, which certainly fits in with the fact that the area wasn't all woodland when Henry VIII had commandeered it. Any wood felling that had been undertaken in John's 50 acres was by skilled woodsmen, one of whom was named. They had only removed timber in accordance with statute. There appeared to be no waste or spoil in John's woodland, which indicates that he had been falsely accused. Two deponents mentioned seven or eight small trees being felled by one Fossett, in his own part of St John's Wood, expressly for the purpose of renovating the house owned by the Queen at Marylebone, which Fossett then occupied. There was clearly no support gleaned from these depositions for the accusation that John Haynes had committed any illegal act such as the felling of timber trees or causing waste in his woodland.

Personal and family life

John had a house in Queenhithe in the Parish of St Nicholas Cole abbey and St Mary Somerset. He appears in the Lay Subsidy Rolls for London, 1576 and 1582, where as *'s[er]ieant Cater'* and *'Sergeaunt cator'*, on each occasion he paid five per cent tax on goods valued at £60,[122] less the amount he paid for being in the Queen's service of £1 6s 8d. The level of this assessment suggests that his house would have been a substantial building and John's main London residence, which he presumably shared with his wife Agnes Thomson during her lifetime, when not at the Marylebone Park lodge, Creslow Manor or Sayes Court. We learn of Agnes only from the 1568 Visitation of London.[123] She is presumed to have been the daughter of William Thomson who lived within the fishmongers' community. Her baptism took place on 29 August 1547 at Saint Mary Magdalene Church, Old Fish Street, London, making her some 35 years younger than her husband. John had at least one child, for the baptism of a son, William, was registered at St Nicholas Cole Abbey on 11 July 1575 (Figure 4.1), when John was 63 years old:[124]

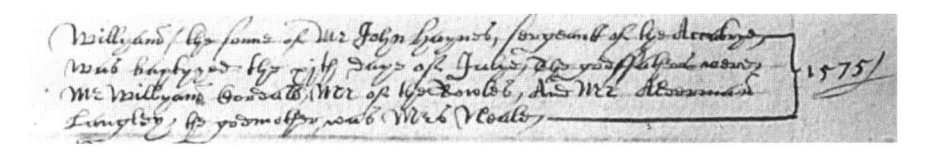

Figure 4.1 Registration of the baptism of William, Son of John Haynes, Sergeant of the Acatry

This entry was uncommonly detailed in so far as two godfathers and a godmother were also mentioned, and reflects not only the status of John Haynes as sergeant of the acatry, but also in his choice of godparents, the circle in which he moved. The *'Mr Willyam Cordall, Mr of the Rowles'* was the eminent Sir William Cordell of Long Melford, Suffolk, Master of the Rolls of the Queen's High Court of Chancery; *'Mr Alderman Langley'* was Sir John Langley, goldsmith, Alderman of London and Lord Mayor the year after William's baptism. William is not mentioned in the wills of either of his godfathers (1581 and 1577 respectively) or in his father's will and may have died in infancy, although there is no record of his burial at St

Nicholas. The identity of Mrs Neale, William's godmother remains to be established, although a Margery Neale became the second wife of Walter Bigge, fishmonger, at St Nicholas Cole Abbey on 13 December 1584, eventually becoming related by marriage to John's niece Sara Gunnell, who married Thomas Bigge there in 1590. We are not given the name of the mother of William, though it is likely to have been Agnes, who would have been only 28 years old in 1575. One can speculate that Agnes may have died soon after William's birth, as John had no further children.

It is known that John remarried, as in his later years his wife was called Joan. She died in 1587, the registration of her burial that year at St Nicholas Cole Abbey recording: *'Mrs Joan Haynes ye wife of Mr Seargeant Haynes was buryed ye 29th of December in ye vault'*, where John would eventually join her.[125]

Nothing is known of John's family life after the birth of his son, William. He was clearly in touch with brothers Nicholas and William during their lifetimes by virtue of their work for the Household. Of his contact with Christopher we know of only one letter, or more precisely verbal accounts of it. The letter was written in December 1583 after the arrest of Sir Francis Thockmorton for his role in the Catholic plot to kill the queen, when two of his sympathisers had made their escape to France from the Sussex coast. John had received some alarming news from within the Household that Francis Walsingham, the Queen's spymaster, suspected that Christopher had been involved in assisting in this escape and that he was to be interrogated. John appears to have written to warn Christopher that he should expect to be *'thoroughly examined'* and to have his explanations ready. He apparently added that if the story was true and Christopher was found to have been implicated in treachery, he could expect no help from his brothers, who would disown him. This last threat was probably John's own insurance in case his letter came into Walsingham's hands. Christopher appears to have panicked and discussed John's letter with his indiscreet nephew, William Dartnall, before destroying it. The letter and its supposed contents are discussed in the next chapter. Christopher did not forget his brother's helpful warning and, in his will three years later, he left John a valuable nest of silver-gilt bowls and a grey gelding. Indeed it was while visiting John that Christopher fell ill, completed his will, died and was buried in John's parish church.

John still appears to have had a good relationship with his nephew, Richard Haynes, as late as 1594, when at the time of his action against Edward and William Page in the Court of Requests, Richard served as his solicitor.[126]

John Haynes's will was dated 1 January 1605 [i.e. 1606] at Marylebone; he was in his ninety-fourth year. His nephew William Gunnell, eldest surviving son of his sister Christian, was John's principal beneficiary and inherited most of John's property, including lands, tenements, rents, etc. in Reading, Hinton and Twyford and elsewhere in Berkshire and Wiltshire. [127]

William Gunnell is described as John's *'loving kynnesman and faithfull servannte'*. Elsewhere in the Will he is described as John's cosen (not nephew), but also as *'my sisters sonne'*, leaving no doubt as to the relationship. He left a large gold ring seal with his arms engraved on it to his nephew Richard, son of his brother, Nicholas. This was possibly the *'best gold ring'* which his brother Thomas had left him in 1582, and, although the diarist claimed John's ring to be his Seal of Office,[128] it is more likely to have been engraved with the Haynes' arms,[129] even if used to seal his official correspondence. Of interest is the request for information on a seal in the possession of the queriant, *'Spalatro'*, in Notes and Queries for 17 November 1860.[130] It bore the Haynes of Reading Arms (argent, three crescents barry wavy of six azure and argent), with a mullet as a differencing mark. It was surmounted by an esquire's helm and the crest was *'a stork, heron, or crane rising'*. The motto was *velis et remis* (literally 'with sails and oars'; the motto means full speed ahead or making an all-out effort). The mullet (star) indicates that these were the arms of the third son, and it is tempting to speculate that this might have been the seal (or an impression of it) passed on by John Haynes to his nephew. A respondent identified the Arms as those of Nicholas Haynes that were quartered with those of Foxley in 1578,[131] but ignored the fact that those of Nicholas bore a martlet as his differencing mark (fourth son). Because of the mullet, the 'Spalatro' seal could not have belonged to Nicholas or William.

Thomas Haynes had also left John all his lands and tenements in Berkshire and Wiltshire to be passed on in turn to their brother William, then after William's death to their brother Nicholas, then to Nicholas's heirs. Interestingly under the terms of Thomas's will[132] the lands should have been passed on to Nicholas's son Richard, but John, the last of the brothers to survive, possibly as a result of the outcome of a court action involving himself and Richard[133] and in the knowledge that Richard had come into Christopher's considerable estate, decided in favour of the children of his only sister who had worked for him in his later years. To what extent the Gunnell family influenced the testator in deciding in their favour is not known, but must be considered a possibility. Several lived either with John, or in the environs of Old Fish Street, and could easily have influenced the decisions of a vulnerable old man of uncertain testamentary capacity. Clearly suspicions were voiced, for the will was contested by John's great nephew, Jasper Dartnall, through his proctor, the notary public Richard Stubbs, possibly on this last point, for it was not finally determined in favour of William Gunnell until 1 June 1608, by Judge John Bennett, who on advice, found that John Haynes was '*of sound memory when he made, ordained and established his said testament containing his last* will'.[134] There is no record of Richard Haynes, hitherto John's heir-apparent, contesting the will, although it seems that a number of challenges had occurred for on 1 June 1608, it was stated that '*the case was pending, undecided for some time, and is still pending*' and mentions William Gunnell's '*allegation*' presented '*at another time in this matter given, exhibited, and brought before* us'.[135] However, by the time the date was fixed for judgment, Richard's family had been wiped out by the plague, and the will was left to be contested only by his sister's son, Jasper Dartnall.

John also left a substantial bequest to William Gunnell's sister Sara Bigge, then a widow with a teenage son, Alexander, of whom there was no mention in John's will. Sara Bigge was left a gold ring with the words: '*Parj Jugo dulcis tractus*'[136] engraved on it. Sara was also left John's three rooms in Old Fish Street at the time rented to Thomas Parre, from the income of which she had to pay her brother William Gunnell, a pound a year. John was clearly concerned for his widowed niece's welfare: as well as £5 in money and other legacies to Sara, he required William to look to her maintenance.[137] It is possible that Sara lived with her uncle John after the death of her husband, who came from the same parish, and hence left her the Old Fish Street accommodation in case she wished to live there.

The extended Gunnell family came in for further bequests. Thus, John left Thomasine Gunnell, William's daughter, a gold ring mounted with pointed diamonds to have on her eighteenth birthday or on her marriage. In the event of her premature death the ring was to go to her brother, John. Presumably this ring had belonged to John's late wife. Provision was made for William's son in the event that William predeceased his wife, Elizabeth, in that the shop and warehouse in Old Fish Street were in tail to his son, John Gunnell *'in as large and ample manner as the same were granted to me by L[ett]res patente from the late Quene Elizabethe'*.

Being the last survivor of the six brothers, John had very few monetary bequests to make: £8 13s 4d in all. His five properties, goods, chattels, residual money and plate mostly went to William Gunnell. The three people who witnessed his signature were all skinners, including *'William Gunnell Junior'*.

John outlived all his siblings and died around 12 January 1608; he was in his ninety-sixth year. Probate was granted 14 January 1608, the date of his funeral. He was buried alongside his wife, Joan, who had predeceased her husband in 1587, in the vault (crypt) of St Nicholas Cole Abbey. They appeared to have left no living issue; John's son, William, had presumably predeceased them. The entry in the parish register reads: *Mr John Haynes Esquire Sergeant of ye Queens Acatry was buryed in ye vault ye 14th of January 1607'*.[138]

Notes and references

1. Survey of London, 1938 v.19, p. 17, London County Council; TNA: E134/43 Eliz/Hil4.
2. Cal Pat Rolls 1550-1553, vol. 4, 6 Edw VI, p. 349 m. 36: Acts of the Privy Council, 1542-7, p. 40. The members of the Privy Council at this meeting were: *'Tharchebisshop of Canterbury; the Lorde Chawncellour off Englond; Therle of Hertford; The Lorde Admirall; the Bisshop of Winchester; Sir John Gage, Comptrollour; Sir Thomas Wriothesley, Secretary; Sir Rafe Sadlair, Secretary'*.
3. TNA: C66/805 m.49; Cal Pat Rolls 1547 1 Edw VI, part VII, p. 217 m. 49. The Hanaper was an office of the English court of chancery in which writs relating to the business of the public, and the returns to them, were kept in a wicker basket or hanaper (hamper). St Gabriel's Church was not rebuilt after its destruction in the Great Fire, the parish being united with St Margaret Pattens.
4. John Haynes's encroachment: Misc. MSS Box 91 [C]: 1547-1549 (nos 206-266; [214]), London viewers and their certificates, 1508-1558: Certificates of the sworn viewers of the City of London (1989), pp. 85-104. URL: http://www.british-history.ac.uk/report.aspx?compid=36059. Date accessed: 6 July 2010.
5. Cal Pat Rolls 3 Edw VI, Part IV, pp. 324-5 m.18.
6. Unfortunately the donor's forename is omitted; it could not have been

Simon Haynes, the well-known cleric, Dean of Exeter, etc., who died in 1552.
7. It is possible that this was the *'Freshstore'* an alternative for *'Sea Fish'*, of which William was subsequently also the yeoman purveyor.
8. Cal Pat Rolls 1E1, 1 Nov 1559: Vol. 1 1558-60, p.43 m.24. See also J.R. Tanner, 1922, *Tudor constitutional Documents A.D. 1485-1603 with historical commentary*, p. 599.
9. Carye was a groom of the Privy Chamber (British Library: Lansdowne 29, No.68 f.161v, 1580; 59, No.22 f.43v, 1589) and master and treasurer of the King's Jewels and plate; of Great St Bartholomew's, London. He was subsequently knighted.
10. 'Visitation of London 1568', ed. Sophia W. Rawlins, *Harleian Society*, London, 1963.
11. Achatry, Acatery, Chaterie, Catre, Cater, Catery, or Kathery, from Fr. *acheter*, to buy. It was finally abolished in the household reforms of January 1779.
12. Allegra Woodworth, 1946, 'Purveyance for the Royal Household in the Reign of Queen Elizabeth', *American Philosophical Society*, New Ser. 35, p. 7.
13. Ibid. p. 62.
14. Ibid. p. 62.
15. In 1658 the lands of the Manor of Sayes Court, Deptford Strand alias West Greenwich, Kent, comprised a field called Broomefield (57a); a parcel of marsh called Broade Marsh alias Mould Meade (43a.3r.20p.); 26 acres land abutting Mould Mead east and south; a parcel of pasture called Pott Mead alias Crabtree Mead (36a.25p); 1 parcel of meadow (34a.1r.) abutting on the Thames on the north, being part of Great Crane Meadow which adjoins the other part of the said meadow on the east; another parcel of meadow called Neales Marsh (8a.1r.8p): Cambridge Records Office: CON 3/6/2/11. In March 1588, Richard Miller, Keeper of the Queen's grounds at Deptford, and 'Serjeant Haynes, Overseer of the Queen's Meadows', were ordered to board and fill up with earth 60 roods of the manor wall, near the water meadows, by the following Michaelmas, or forfeit 20s for every rood then undone ['Addenda, Queen Elizabeth - Volume 30: March 1588', Calendar of State Papers Domestic: Elizabeth, Addenda 1580-1625 (1872), pp. 243-248].
16. Guy de la Bédoyère, ed., *The Diary of John Evelyn*, 2004, pp. 65-6.
17. 'Parishes, Creslow', *A History of the County of Buckingham*, Vol. 3, 1925, pp.335-8.
18. Cal Pat Rolls 38 Eliz vol. 4, m. 37.
19. Cal Pat Rolls 38 Eliz vol. 4, m. 37.
20. Cal SP Dom 1591-4, p. 240; Cal Pat Rolls 38 Eliz vol. 4, m. 37.
21. E. B. Chancellor, 1930, *London's Old Latin Quarter: Being an Account of Tottenham Court Road and its Immediate Surroundings*, J. Cape, pp. 41-2. Savage was also granted the position of Yeoman of the Crown for life at 6d a day from 1546; Cal Pat Rolls 1549-1551, vol. 3 1929, p. 187.

22. E. B. Chancellor, 1930, *London's Old Latin Quarter: Being an Account of Tottenham Court Road and its Immediate Surroundings*, J. Cape, pp. 41-2.
23. Ibid. p. 42.
24. British Library: Lansdowne 83, No. 53 ff.142-143v, 1597.
25. TNA: E101/544/29.
26. R. Folkestone Williams, 1660, *Domestic memoirs of the Royal Family and of the Court of England*, Vol. 2, London: Hurst and Blackett, pp. 133-4. J. Nichols, 1823, *The Progresses and Public Processions of Queen Elizabeth,* London: John Nichols and Son, pp. 43, 46. The cost of the 127 sheep stolen or dead from illness reported by the Keeper of the Pastures was £31 15s in 1576.
27. British Library: Lansdowne 64, No.53 f.127, 1590.
28. I. W. Archer, S. Adams et al., 2003, *Religion Politics and Society in Sixteenth Century England*, Camden Fifth Series, Vol. 22, p. 7.
29. J. L. Laynesmith, 2004, *The Last Medieval Queens: English Queenship 1445-1503*, OUP, p. 225.
30. Allegra Woodworth, 1946, 'Purveyance for the Royal Household in the Reign of Queen Elizabeth', *American Philosophical Society*, New Ser. 35, pp. 9, 16, 30. Petty graft and dishonesty were tolerated by the Household Treasurer, Sir Francis Knowles and the Comptroller, Sir James Croft; p.9.
31. Amanda Bevan, 2002, *Tracing your ancestors in the Public Record Office*, 6th edn, p.332.
32. R. Folkestone Williams, 1660, *Domestic memoirs of the Royal Family and of the Court of England*, Vol. 2, London: Hurst and Blackett, pp. 133-4. The notional 1s 9½d daily allowance set out in the Queen's book in 1576 was in reality almost doubled (3s 3d) that year; J. Nichols, 1823, *The Progresses and Public Processions of Queen Elizabeth,* London: John Nichols and Son, pp. 43, 46.
33. E.M.W. Cracknell, ed. (1934), *The Barbadian Diary of Gen. Robert Haynes 1787-1836*, The Azania Press, Medstead, Hampshire, UK, p. 6.
34. Will of John Haynes of Marylebone, 1 January 1605/6: Windebanck Quire Nos 1-55: PROB 11/111; probate granted 12 January 1608.
35. London Port Book, numbers 34, 57 (f. 16b), 169 and 730 (f.234). E. 190/4/2; ['A Calendar', from: 'London Port Book, 1567-8', Nos. 1-99 (Sept-Nov, 1567), *The port and trade of early Elizabethan London: documents* (1972), pp. 1-28].
36. Allegra Woodworth, 1946, 'Purveyance for the Royal Household in the Reign of Queen Elizabeth', *American Philosophical Society*, New Ser. 35, pp. 9, 16.
37. Essex Record Office: D/AER 11A/54; Will of William Wrighte, clerk, parson of Little Laver, 18 August 1569. Wrighte, once a monk of Hatfield Regis (Hatfield Broad Oak), may have held Colchester St Leonards and Little Laver; see James E. Oxley, 1965, *The Reformation in Essex to the Death of Mary*, Manchester University Press, p. 114.
38. British Library: Lansdowne 34, No. 23 ff.48-49v, 1582.
39. Allegra Woodworth, 1946, 'Purveyance for the Royal Household in the

Reign of Queen Elizabeth', *American Philosophical Society*, New Ser. 35, pp. 9, 16.
40. 'Calendar of the Manuscripts of the Most Hon. the Marquis of Salisbury', Preserved at Hatfield House, Hertfordshire. Vol. 13:12C-1597.
41. Will of John Haynes of Marylebone, 1 January 1605/6: Windebanck Quire Nos 1-55: PROB 11/111; probate granted 12 January 1608. The property granted by patent to John Haynes in Old Fish Street can be identified with that confiscated by Henry VIII from the London Minoresses (Poor Clares) at the time of the Dissolution. Records, almost certainly of this property, can be traced to 1395. It may have been originally held by Henry le Waleys, the Lord Mayor of London who died in 1302; see J. Röhrkasten, 2004, *The Mendicant Houses of Medieval London 1221-1539*, Li Verlag, Münster, p. 255. In January 1547 it was in the tenure or occupation of Agnes Blockeswiche ('Letters and Papers, Foreign and Domestic, of the reign of Henry VIII', Vol. 21, Part 2, London PRO, p. 415).
42. R. G. Lang, ed., 1993, 'Two Tudor subsidy rolls of the city of London, Breadstreet Warde,' *London Record Soc.* 29, pp.159-164. The Cowdalls and his fellow clothworker, John Haynes (d. Sept 1593), had large families to be found in the parish register of St Nicholas Cole Abbey (London Metropolitan Archives: St Nic CA). Adding to our difficulty, this John Haynes had a son John, who precedeased his father (February 1592).
43. John Stow, 1842, *Survey of London*, Whitaker and Co., p. 129.
44. W. Herbert, 1836, *The History of the Twelve Great Livery Companies of London*, pp. 16-17.
45. In H. B. H. Beaufoy, 1855, *Descriptive catalogue of the London traders, tavern, and coffee-house tokens current in the seventeenth century*, p. 98.
46. Dunning was customer of the Port of Chichester, responsible for customs matters from Hythe to Chichester, from the early Elizabethan period until his death in 1579; G. Mayhew, 1987, *Tudor Rye*, Falmer, University of Sussex, p. 103.
47. Cal Pat Rolls Eliz. Vol. VI, 1103, m. 37, 1572-5. Green usually referred to fresh fish that had been heavily salted while wet; stockfish was air-dried fish. Both processes served to preserve the fish from decay during its slow journey from the coast.
48. R. Folkestone Williams, 1660, *Domestic memoirs of the Royal Family and of the Court of England*, Vol. 2, London: Hurst and Blackett, pp. 133-4.
49. Will of John Haynes of Marylebone, 1 January 1605/6: Windebanck Quire Nos 1-55: PROB 11/111; probate granted 12 January 1608.
50. Acts of the Privy Council of England, 1574, p. 233.
51. Cal State Papers, Domestic 1547-1580, Vol. XXXV, p. 247, No 36.
52. Acts of the Privy Council 1578, p. 320 HMSO; PC2/12.
53. Acts of the Privy Council, XXth of Marche, 1577/8, p.190, 1895; TNA: PC2/12. The prices agreed between a royal purveyor and owners had to be in the presence of a constable; the purveyor would then give a copy of a receipt to the constable and before he left, he would give the constable

(or petty constable) a statement of all his purchases to be approved and signed. See: Allegra Woodworth, 1946, 'Purveyance for the Royal Household in the Reign of Queen Elizabeth', *American Philosophical Society*, New Ser. 35, p. 32.

54. Robert Jones was Richard Besbitche's deputy (Allegra Woodworth, 1946).
55. *Bulletin of the Institute of Historical Resesearch*, 1986, p. 104.
56. The Dorset house was originally part of the Manor of Sturminster Newton, once a possession of the Monastery of Glaston, Somerset.
57. Cal Pat Rolls Vol. III EI 1563-6, p. 341, m. 15, 28 November 1564.
58. Sir Thomas Wyatt, poet and diplomat, had earlier been accused of high treason by Bishop Edmond Bonner in the reign of Henry VIII (1541), tried and acquitted. He, Bonner and a clergyman, Dr Simon Haynes, Dean of Exeter (1537-1552), Master of Queen's College, Cambridge, and later to be one of the King's chaplains, were sent by Henry VIII on an embassy to Emperor Charles V (see John Bruce, 1850, 'Recovery of the Lost Accusation of Sir Thomas Wyatt, the Poet, by Bishop Bonner', *The Gentleman's Magazine*, June 1850, pp. 563-570). I have speculated over this connection of Wyatt, Haynes and Bonner. I Haynes related to John and his brothers? Was Bonner linked to Ellen Bonne or Bonner of Wolverhampton, the wife of William Haynes? Simon Haynes, a Cambridge graduate, was said to be *'a remarkable man'* (Edward A Freeman, 1887, Exeter. Longmans Green & Co. p.19). He had been Dean of Windsor in 1535 and later participated in the compilation of the first English liturgy (see Derek Plumb, 'The social and economic status of the later Lollards', in M. Spufford, ed., 1995, *The World of Rural Dissenters, 1520-1725*. Ch. 2, p. 122.
59. 'Parishes: Chalk', Th*e History and Topographical Survey of the County of Kent*, Vol. 3 (1797), pp. 457-71.
60. Cal Pat Rolls, vol. III, 8EI, 2579, m. 42 (p. 465).
61. Cal Pat Rolls EI vol. IX (HMSO, 1986), 1897, mm. 36-7 (p. 258).
62. *Survey of London*, 1938, v.19, London County Council, pp. 14-17. See also TNA: E134/43Eliz/Hil4 and St Pancras Notes and Queries, 1901, 179 and 185, pp. 145-151. Miller mentions a memorial to the daughter of A. Glover of Tottenhall Court over the vestry of St Pancras Church. She died in 1588 (Frederick Miller, 2008, *St Pancras*, p. 16), soon after her father's acquisition of the lease. Arthur Rainscroft, innkeeper, lived at the Bell, Friday Street, London. In his will, proved in 1583, he left to his son Arthur Rainscroft a meadow containing 16 acres near Tottenham Court.
63. E. B. Chancellor, 1930, *London's Old Latin Quarter: Being an Account of Tottenham Court Road and its Immediate Surroundings*, Boston and New York: Houghton Mifflin Company, p. 42.
64. *Survey of London*, 1938, v.19, London County Council, pp. 14-17. See also TNA: E134/43Eliz/Hil4 and St Pancras Notes and Queries, 1901, 179 and 185, pp. 145-151.
65. Allegra Woodworth, 1946, p. 62.
66. *Survey of London*, 1938, v.19, London County Council, p. 17.

67. Alan Haynes, 1987, *The White Bear; the Elizabethan Earl of Leicester*, London: Peter Owen, pp. 144-5.
68. Ann Saunders, 1969, *Regent's Park*, David and Charles, p. 31; TNA: E101/544/29.
69. E. B. Chancellor, 1930, *London's Old Latin Quarter: Being an Account of Tottenham Court Road and its Immediate Surroundings*, J. Cape, pp. 41-2.
70. *Survey of London*, 1938, v.19, London County Council, pp. 14-17. See also TNA: E134/43Eliz/Hil4 and St Pancras Notes and Queries, 1901, 179 & 185, pp. 145-151.
71. *Survey of London*, 1938, v.19, London County Council, p. 17.
72. *Survey of London*, 1938, v.19, London County Council, pp. 14-17. See also TNA: E134/43Eliz/Hil4 and St Pancras Notes and Queries, 1901, 179 and 185, pp. 145-151.
73. In 1554 Henry Bludder *'our beloved servant'* was granted by the King and Queen on the advice of the Treasurer, a field called Courtfelde and the adjoining two crofts of land and pasture amounting to 40 acres and Kynge Meade a 10-acre meadow and adjoining 7 acres of land known as Dayes Pytt, all in the parish of Aylesforthe, Kent. These lands had been confiscated from Sir Thomas Wyat, attainted and convicted of High Treason; TNA: C66/895 m.22.
74. *Survey of London*, 1938, v.19, London County Council, pp. 14-17. See also TNA: E134/43Eliz/Hil4 and St Pancras Notes and Queries, 1901, 179 and 185, pp. 145-151.
75. Ibid.
76. Ibid.
77. In 1605, the widow of Robert, Earl of Leicester, claimed inherited rights to the grounds. Board of Greencloth Records I, f. 91d, quoted by Allegra Woodworth, 1946, p. 62.
78. Cal Pat Rolls Eliz vol. V, 3331.
79. TNA: SC1/23/65.
80. Cal Pat Rolls Eliz vol. V, 3331.
81. Ann Saunders, 1969, *Regent's Park*, David and Charles, p. 31.
82. J. C. Jeaffreson, 1886, Middlesex County Records (Old Series), Vol. 1, Greater London Council, 1972, pp. 165-6.
83. Ibid. p. 212.
84. E. B. Chancellor, 1930, *London's Old Latin Quarter: Being an Account of Tottenham Court Road and its Immediate Surroundings*, J. Cape, pp. 41-2.
85. J. C. Jeaffreson, 1886, Middlesex County Records (Old Series), Vol. 1, Greater London Council, 1972, p. 144.
86. Ibid. pp. xliii-xliv; p. 182; pp.183-4.
87. Ibid. p. 258.
88. TNA: C66/1246 m.9d.
89. 'Inquisitions: 1580', *Abstracts of Inquisitiones Post Mortem for the City of London*, Part 3 (1908), pp. 19-32; Chan. Inq. p. m., 22 Eliz., part 2, No.

26. This Inquisition was taken before Sir William Cordell, Master of the Rolls and godfather of John's son William in 1575.
90. A Thomas Smythe had been *'clerk of the Kytchinge in the Queen's householde'* in 1559/60: London Metropolitan Archives: HMD/X/041-2.
91. Smythe was relied on to return £1500 a quarter to the exchequer in 1588: British Library, Lansdowne 58, ff.109-110v, 1588. He appeared before the Middlesex Sessions [Jeaffreson, 1886, p. 163 dated 20 Aug 27 Eliz; 1584] with John Haynes, yeoman, accused of not attending church for the previous year. Thomas Smythe was Clerk of the Green Cloth during 1575 [Cal Pat Rolls 1572-5, p. 550 (3233)]. He was found at an inquest postmortem in January 1585 to have killed Richard Awdyence during a sword fight [Jeaffreson, 1886, p. 156, 12 Jan 27 Eliz)].
92. Cal Pat Rolls vol. VII, 2888 m.24 (pp. 238-9).
93. Anthony Browne of Cowdray Park, Sussex, who was raised to the peerage in 1554 as Viscount Montague (Cal Pat Rolls, 1554-5, p. 88). The letters patent are entered in B.M., Add. MS. 31952, f. 48v.
94. Acts of the Privy Council, 1578-80, p. 91.
95. Sussex Archaeological Collections 1862, p. 85.
96. Cal Dom Eliz Vol ccxl, p. 136.
97. British Library: Lansdowne 56, No. 21 ff.64-65v, 1588.
98. British Library: Lansdowne 58, No. 50 ff.111-112, 1589.
99. British Library: Lansdowne 56, No. 30 ff.85-87, 1588.
100. At the sessions at Newgate held in January 1889, Nicholls one of the Queen's purveyors was sentenced to be hanged for infringing the law prohibiting the taking, buying, or commandeering carriage for the same, not contained in his commission; or for taking more than was delivered to the Household, or for taking goods for which payment was not made: E. Coke, 1797, *The Second Part of the Laws of England: Containing the Exposition of many Ancient and other Statutes*, Vol. II, p. 545, London: E & R Brooke.
101. British Library: Lansdowne 73, No. 34 f.117-117v, 1593.
102. Allegra Woodworth, 1946, p. 65.
103. British Library: Lansdowne 46, No. 84, 1586; No. 87 ff.197-183, 1586; No.91 f.91, 1586: Lansdowne 58, No. 56, 1588.
104. 'Calendar of the Manuscripts of the Most Hon. the Marquis of Salisbury', preserved at Hatfield House, Hertfordshire, Vol. 14, entry 185: 1596-7.
105. Ibid.
106. TNA: REQ1/18/73; REQ1/18/133-4; REQ1/18/159. An action had been brought by William Page in the Court of Common Law against John Haynes for £20 for which sum John, William and Nicholas Haynes had signed a bond in favour of Edward Page. The actual debt was 20 marks (£13 6s 8d). In October 1694 John countered this action in the Court of Requests as Nicholas Haynes, by that time deceased, had already repaid £8 6s 8d to Edward Page, part of the 20 marks, leaving only £5 still owing. John Haynes paid the residual £5 into the court which then ordered the Pages to cease their action in common law under pain of a £100 injunction, until all the defendants had appeared before the court to argue their

case for the full £20. The Pages agreed this settlement and, on 8 November, the court passed the £5 to Richard Haynes, John's nephew and solicitor, for settlement. Richard's receipt of this sum was minuted by the court on 15 November 1594. He was still Sergeant of the Acatry on 13 November 1596 (TNA: E115/195/106) as, according to his taxation record of that date he was *'resyaunt and abydinge here at the Courte in the time of taxacion and for the most part of the year before'*.

107. British Library: Lansdowne 83, No. 53 ff.142-143v, 1597.
108. Allegra Woodworth, 1946, p. 65.
109. TNA: REQ1/18/73; REQ1/18/133-4; REQ1/18/159.
110. British Library: Lansdowne 83, No. 53 ff.142-143v, 1597.
111. As early as 1582, the high price of fish, meat and salt had been noted and a suggested saving of £1000 mooted (British Library: Lansdowne 34, No. 34).
112. TNA: E179/142/234; E179/142/ 239.
113. TNA: E115/182/108.
114. TNA: E115/204/25.
115. TNA: E115/199/112.
116. TNA: E115/202/28.
117. TNA: E115/195/106. On 15 January 1591 records show he was assessed in the Parish of St Nicholas Cole Abbey, £20 in lands and fee and 60s for goods (TNA: E115/200/8), the same in lands as on 20 August 1586 (TNA: E115/197/17).
118. R. G. Lang, ed., 1993, 'Two Tudor subsidy rolls for the city of London, Quenehithe Warde', *London Record Soc.*, 29, pp. 274-78. See also the London Lay Subsidy Roll for 1576; TNA: E179/145/252f in which John was also assessed on goods (including wages) of £60. In the second subsidy granted in 1597 for Queenhithe Ward [assessment date 1 October 1599] John was still assessed at £60 [St Nicholas, St Mary Somerset and St Peter's parishes].
119. TNA: E133/10/1514.
120. TNA: REQ1/18/73; REQ1/18/133-4; REQ1/18/159.
121. These interrogatories were written questions submitted to John from his adversary to obtain answers in writing and signed under oath. They were a discovery device used to enable the individual to learn the facts that could be the basis for, or support of, the accusation with which John has been served.
122. See note 118.
123. 'Visitation of London 1568', ed. Sophia W. Rawlins, *Harleian Society*, London, 1963.
124. London Metropolitan Archives: Parish Records of St Nicholas Cole Abbey (St Nic CA).
125. Ibid.
126. TNA: REQ1/18/73; REQ1/18/133-4; REQ1/18/159.
127. Will of John Haynes of Marylebone, 1 January 1605/6: Windebanck Quire Nos 1-55: PROB 11/111; probate granted 12 January 1608.

128. E.M.W. Cracknell, ed. (1934), *The Barbadian Diary of Gen. Robert Haynes 1787-1836*, The Azania Press, Medstead, Hampshire, UK, p. 13.
129. We know that his brother William was by 1580 using a seal on official correspondence of about 2cm diameter depicting the arms of Haynes quartered with Foxley and the heron crest, confirmed in 1578 [East Sussex Record Office: Rye MS. 47/25, part II, no. 184]. This seal may have been passed on to John after William's death; alternatively John might have had his own version of these arms.
130. *Notes & Queries*, 17 November 1860, 2nd Series, Vol. X, p. 387.
131. J.G.N., *Notes & Queries*, 2nd Series, Vol. XI, 12 January 1861, p. 38.
132. Berkshire Record Office: D/A1/76/205; Will of Thomas Haynes of Whitley, Parish of St Giles, Reading, 1582. Text reproduced by permission of the Berkshire Record Office.
133. TNA: C4/27(2)/40.
134. Sentence of the Will of John Haynes, 1 June 1608 [The sentence was a final judgment about a disputed will given at the conclusion of any litigation]; Windebanck Quire Numbers: 56-114; TNA: PROB 11/112.
135. Ibid.
136. Commonly used words suggesting mutual affection and harmony: lit., '*sweet the load when the yoke is evenly harnessed*'; a very common refrain in Elizabethan madrigals and poetry. The words were later used (on 28 March 1607) by Sir Francis Bacon in the Commons' continuing debate on the Union with Scotland; Jnl House Commons 1, 1802.
137. Will of John Haynes of Marylebone, 1 January 1605/6: Windebanck Quire Nos 1-55: PROB 11/111; probate granted 12 January 1608.
138. I.e. 1608. London Metropolitan Archives: Parish Records of St Nicholas Cole Abbey (St Nic CA).

5

Christopher Haynes, Customer and Taverner of Arundel [c.1527-1586]

Childhood in Reading - Marriage to Elizabeth Colbroke - Burgess of Arundel - Mayor of Arundel - Customer - Serious Allegations - William More of Loseley, Vice-admiral for Sussex - Properties at Ford and Climping - Smuggling and piracy - The Falcon - Purveyor of the Seafish - In the Wake of Throckmorton - William Shelley - Escape of Paget and Arundell to France - Examinations - Taverner - Unanswered Questions - Family and Circle - Christopher Haynes of Billingshurst

Christopher, the fourth son of Richard and Thomasine Haynes, was in some respects the most colourful of all the brothers. Next to nothing is known of his early life although it can be assumed he was born in the family home at Whitley, Reading, possibly in the mid-1520s. As all the brothers were clearly literate, it is likely that they were educated first in the petty school from the ages of five to seven and then at the grammar school in Reading, under its internationally distinguished Head Master, Leonard Coxe. Reading School was a foundation of the Abbey, noted for its teaching of the liberal arts, which, after the Dissolution in 1539, continued under the control of Reading Corporation with the active support of the King, who guaranteed the Master's stipend. Education tended to progress from grammar school to university at the age of 14, but there is no evidence that Christopher or his brothers received a higher education, possibly because of the disruptive impact of the Dissolution on school life. Nevertheless, at least four of the brothers subsequently entered positions for which literacy and fluency in spoken and written English with some Latin were pre-requisites.

The prosperity of Reading was built on the manufacture of woollen cloth which had developed there because of its fast-flowing rivers ideal for fulling and its accessibility to the wool farmers of the Cotswolds and nearby counties. It is possible that the Haynes family was involved in cloth-making, but of the six brothers, only Thomas, the eldest, continued to live in the family home at Whitley after their parents had died. Richard had moved down the road to farm at Wargrave and his other brothers had sought their futures away from Berkshire by working in the Royal Household. Christopher eventually left Reading for Sussex and the cloth-exporting port of Arundel where he spent his last 30 years.

Meanwhile there is reason to believe that his first move from Reading was to nearby north Hampshire, as a Christopher Haynes, gent, was mentioned in the will of William More of Sherfield-on-Loddon in 1548.[1] What makes it a distinct possibility that this Christopher was the one from Reading, is that his maternal grandmother was of the same family of More[2] as William, which would make the two men second cousins. However, Christopher, who was both a beneficiary and a witness, was described by William as his *'loving friend'* rather than his kinsman. Although this may appear to leave their true relationship unresolved, the use of the term 'friend' also referred to kin in the late medieval period. If he was indeed Christopher of Reading, this description of his close relationship and the fact that he was also left a cow by William, suggests that Christopher may have been living in the vicinity of Sherfield-on-Loddon at the time of William's death, or at least a frequent visitor from Reading, barely half a day's ride away.

Christopher moved to Sussex around the mid-1550s, by now styled gentleman, he married Elizabeth Colbroke daughter of John Colbroke,[3] a member of the Midhurst family of that name, and he and his wife made their home in Arundel. The marriage may have been childless for they had no children surviving at the time of Christopher's death and there is no evidence that they ever had any. By 1558, Christopher was called by the escheator to serve with others of like status as juror of inquisitions post mortem,[4] which indicates that he had by then risen to prominence in the community. Indeed this service occurred not long after his election to the Arundel council.

Burgess

Christopher soon became active in local civic affairs, being made a burgess of Arundel at the meeting of the council on 16 January 1557, from when his name appears in most entries in the Arundel Borough Minute Book up until October 1585.[5,6] At one of his earliest meetings of the burgesses held in the Town House at the corner of High Street and Tarrant Street, he seems to have been rather too outspoken for his own good, as in an entry dated 27 June 1558,[7] he is recorded as being ejected from the council for being rude, disrespectful and *'saying shameful words'* to the mayor, Robert Styles, a note being made that he was subsequently reinstated on 6 September, that year. He was himself mayor of Arundel in 1562, and 1578.[8] Under Bridgewarden's Accounts for 13 August 1578, when he was mayor, there is a memorandum concerning the lease of a shop by Christopher Haynes, which he held for the rest of his life. His signature appears at the top of those attending listed on the minutes of this meeting, as was customary for the mayor. The mayor of Arundel was also the bridgewarden, a post which carried the responsibility during his tenure for repairing or even rebuilding the bridge, with timber generally given to the council for that purpose by the Earl of Arundel. The Bridgewarden's Account Book records the income received by the bridgewarden, which was mainly derived from the rent on a shop or storehouse on the quay and an annuity arising from the Crown House.

Being a burgess incurred the significant expense of an entry fee, £6 13s 4d in January 1560 rising to £10 in May 1568, but it carried with it one particular advantage: right to the valuable brook lands known as the *'burgesses' brooks'*, about 100 acres of meadowland between the castle and the river. Only the 12 burgesses[9] were allowed to lease the brooks for fixed numbers of cattle and horses and Christopher would undoubtedly have exercised this privilege for most of his time in Arundel. The burgesses also held a fund, known as the *'burgess chest'*,[10] which may have been to enable the widows of burgesses to enjoy their late husbands' perquisites. Christopher would also have been allowed to use the borough seal if he could justify such use. Because there were so few burgesses, the title of burgess became a valuable commodity which could be sold. There was however one important condition attached; a burgess had to live within the confines of Arundel, which Christopher did. In the first year that Christopher was mayor, the Earl of Arundel nominated two burgesses, and as their election was by the mayor and existing burgesses, it may have signified a growing closeness between Christopher and the Earl. The Earl's nominee was admitted as a burgess in August 1562 without paying the fee.[11]

As a burgess, Christopher would have had a large say in the running of Arundel. The burgesses controlled trading and associated licensing, markets, weights and measures, the quality of ale, tolls, the port, care of the poor, schools, building control, and repairs to the bridge and some public buildings such as the church. They were also responsible for law and order within the town, in which they were assisted by bailiffs, constables and reeves. Christopher's last entry in the Borough Minute Book was on 4 October 1585 when he witnessed the installation of the new burgesses for the last time.[12] It must have been a poignant moment as he signed the Minute Book on this occasion [Figure 5.1] along with his old friends including John Humfrey, the Arundel coroner and one-time customer, and Edmund Sheppard, who had been with him on the borough council for some 28 years. Why he ceased to be a burgess is not known: he would have been about 60 years old and may have been unwell, as he was to live for only another 13 months.

Figure 5.1 Signature of Christopher Haynes taken from the Arundel Borough Minute Book, 1585 (by courtesy of the County Archivist, West Sussex Record Office); the script is much freer than that of his brothers.

There is a court record dated 18 February 1566[13] of an action of Christopher Haynes against Ralph Westdean to do with a legal infringement over a pledge of 10s made by Thomas Colbroke [Haynes's brother-in-law] to William Trevet.[14] Christopher through his attorney, William Smitton, filed a declaration, unfortunately no longer preserved, but Westdean failed to turn up at court on the appointed day and was fined 3d. The case was adjourned to the next court on 25 February when the defendant again did not show up and this time was fined 6d. On 4 March 1566 both Christopher Haynes and Ralph Westdean appeared at the court, Westdean asking through his attorney for a copy of Haynes's declaration and an extra day to negotiate with Christopher, which was granted. Application was then made to the court a week later for a licence to settle the action. Most of the cases of this court seem to have been concerned with minor debts, or unjust withholding or occupancy of property and it seems likely that this case was similar.

Customer

Christopher Haynes held the customs franchise at Arundel, which included the port of Littlehampton. He seems to have purchased the right of custom-collection from the Crown (Her Majesty's Warrant) on all manner of goods passing through the port except wine.[15] As customer of Arundel, he would have been entitled to repay himself from receipts and to keep any profits[16] and also receive the usual 'portage' of £1 for every £100 delivered to the Queen's Cofferer.[17] Christopher worked with his brother-in-law, Thomas Colbroke, who was the Searcher, the officer of the custom house appointed to search ships, baggage or goods for dutiable or contraband articles. The third member of the customs' team was Thomas Brydger who, following in Christopher's footsteps and doubtless with his support, became one of the Arundel burgesses.[18] Brydger, a former mariner from Felpham, was the controller, the person who kept a separate set of accounts or counter-rolls;[19] he was also said to be related to the Colbrokes. Contemporaries would have seen no necessary conflict of interest in three closely-related individuals working in the same government enterprise, but their kinship would in time be the cause of concern, rumour and innuendo that the three were lining their pockets at the country's expense. That Christopher Haynes and Thomas Colbroke were related by marriage proves beyond question (if proof were needed) that the Christopher Haynes of Arundel was definitely the son of Richard and Thomasine Haynes of Reading, as it confirms his marriage to Elizabeth Colbroke recorded under the *'Haynes of Reading'* entry in the Visitation of London of 1568.

Serious allegations

On 15 August 1572 serious allegations were penned by an unidentified person, who from the detailed information given, was clearly an informant with considerable local knowledge. This document is in the form of a letter held in the Loseley Manuscripts, the records of the More and More Molyneux family of Loseley Park, Surrey. This location gives some insight as to the likely recipient for in 1559, William More of Loseley had been appointed to the office of vice-admiral for Sussex.[20] In order to fulfil this role, More had, in December 1569, appointed William Morgan of Chilworth Manor in Surrey as his deputy vice-admiral,[21] to one of whom the letter in question had presumably been addressed.

The allegations referred to were that Christopher Haynes, customer of Arundel, aided and abetted by Colbroke and Brydger, had increased the customs receipts for their own profit and before the interests of Queen or country. It was also alleged that Christopher Haynes had persuaded the clothiers of Reading to bring their cloth to Arundel, where he had assisted in its sale. The sale of cloth was strictly regulated in Elizabethan times by being offered for sale in cloth markets where the so-called 'hallage' dues were levied on all woollen cloth. Duties were payable by the seller on each piece or parcel of cloth sold, and also sometimes on the weekend storage of unsold cloth.[22] The accusation implied that such duty was being withheld by Christopher. It is unlikely that Christopher Haynes had been doing anything strictly unlawful, as even the person making the allegations recognised that Christopher had purchased the right of custom-collection in Arundel[23] (except for the customs on wine which Owen Bynyon had bought) or so it was said! The informant insinuated that the customs had been given by the Queen to someone else from whom Christopher had then acquired them and that he did not really have the right to them.

It was further alleged that Christopher Haynes had a vessel laden with cloth or similar merchandise in France, which was at that time making its return journey to England.[24] Such suspect goods had in the past been secreted in a house or warehouse held by Haynes at Ford or Climping, villages nearer the coast to the south of Arundel. They had been seen many times being secretly transported to and from this house usually by packhorses. The inference was that Christopher Haynes was either a smuggler bent on withholding duty from the Queen or, at best, importing foreign goods without a licence.

Little could be done, the writer of the allegations continued, because of the fact that all three against whom the allegations of wrongdoing were made, were '*lynked togithere so in kindred*', that if anyone of the town were to be openly a witness, he would lose his position or the lease on his property [owned by the Earl] or even his right to remain in Arundel. Nevertheless, he urged that all of the allegations could be investigated without the need to reveal the identity of the informer.[25]

Such allegations had evidently been brought to Christopher's attention before, as the writer also mentions what can only have been Christopher's earlier responses: that those peddling these rumours were people with malice to the Earl [of Arundel] or to the borough of Arundel itself. They said the Earl should not transport his timber overseas as he was licensed to do. Christopher Haynes maintained that he had a valid licence '*on the seas*', and as a licensed merchant was permitted to sail from England. He didn't hold back from mentioning that he had the Earl's ear and if necessary would use his powerful influence at Court through his brothers who worked in the Queen's Household.

Although the allegations appear to have come to nothing, there are reasons to suppose that at least some of them were well-founded. There are several reports to the Lord Treasurer alleging that customers of different ports had diverted receipts to their own advantage.[26] At least two of the Arundel men, customer and searcher, were definitely related by marriage. Christopher did indeed have a vessel and he is said to have taken part in voyages for '*taking and spoiling such things as he could meet withall upon the seas*', after which he conveniently locked his plunder away in the Queen's custom house at Arundel.[27] If true, he was not alone: in 1569 James Milles, the customer of Rye kept most of the spoils which had been seized upon the arrest of Captain Chichester who was brought to that town. Christopher may have been one of those whom Elizabeth, through the Earl of Lincoln, Lord Admiral of England, had sent to cleanse the seas of pirates who were harassing the fishing fleet and trading ships.[28] In view of his position as customer, it is possible that he behaved legally, or at least through a smokescreen of legality, as under a patent granted by Queen Mary, the Lord Admiral, through the vice-admiral for Sussex and the customers, had the right to the goods and chattels seized from pirates and traitors, which right was not withdrawn until 8 June 1577.[29] It would therefore have been perfectly legal for Christopher to have used his boat in order to intercept contraband or pirated goods. Thus, the actions of some of the customers were often difficult to distinguish from those of the pirates and provided significant opportunities for both 'private enterprise' and malicious denunciation.

Christopher's vessel was a 40-ton hoy called the *Falcon*, which had been duly notified to the vice-admiral of Sussex on 25 January 1569 as berthed at Arundel.[30] It was a single-masted vessel twice the size of a fishing boat, rigged as a sloop, and such hoys were commonly used to carry passengers and freight on short coastal journeys. One of 40 tons burden, the heaviest reported in the returns, would have been capable of crossing the Channel. Christopher presumably sailed it with his custom house men (Brygder had been a mariner) or hired a crew when necessary. Christopher's vessel was the only one listed under Arundel for that year, although another hoy called the *Dragon* was usually berthed there; next to the entry were noted Richard Bettes [or Battes], shipmaster, and the mariners: Thomas Stanner, John Corde and Thomas Tredcrafte, as *'able to serve'*, who may have been his crew. Christopher would have definitely needed a master for a boat of the *Falcon's* size.

On 25 April 1574, the Bishop of Chichester informed the Lord Treasurer of the considerable harm being done by the *'rovers upon the sea towards this coaste'*, stealing corn and foodstuffs with the active assistance of the locals *['whereof a great sorte be thoughte to lyve by these rovers']*, and by piracy on the seas.[31] As Arundel market was described as among the chief corn markets in Sussex,[32] and grain was one of the town's biggest exports being destined mainly for London and various south coast ports,[33] the pirates were waiting to intercept it as it left the Arun. So much grain seemed to be misappropriated in this way, that prices had rocketed to 4s 4d a bushel and the populace was becoming restive.[34] One is left wondering about Christopher's role in all this. He was almost certainly correct in claiming that the town had the benefit from the transport and delivery of *'wheate as the Quene hath lycensed them to doe'*.[35] A letter has survived dated 24 August 1579 from the Privy Council to the Sheriff and Justices of Peace in Sussex ordering them *'to suffer Christofer Heynes, Purveyor of the seafishe, to transporte v^c [500] quarters of wheate'*.[36] The likelihood is that such a volume of grain could have been carried from Sussex to London only by sea, and Christopher was given the job of organising it, possibly even transporting it himself. The intriguing aspect of this reference is to Christopher being a purveyor of sea-fish and one wonders whether their Lordships had intended the letter to read 'Christofer Heynes, *brother of William Heynes,* Purveyor of seafishe,' etc., or whether Christopher was [by 1579] indeed the elusive fourth servant of the Royal Household. The latter is probably unlikely unless he was temporarily deputising for his brother William, for in January 1584, his occupation was noted as taverner of Arundel.[37] However, the transport of wheat by sea did seem from time to time to be licensed to the purveyors of sea-fish, for William Angell, who took over this position from William Haynes, and William Massam, contracted with the Council in March 1597 to deliver 2500 quarters of wheat, from abroad, for the Queen's use.[38]

The allegation that Christopher had persuaded the clothiers of Reading to bring their cloth to sell in Arundel also has the ring of truth about it. Reading was Christopher Haynes's home town, then one of the most important white cloth-making centres.[39] His brother, Thomas, may have been involved in the cloth industry there; his widow, Agnes Haynes made several bequests of uncoloured broadcloth in her will of 1589.[40] If Christopher's wife was one of the Midhurst Colbrokes, at least one of her kinsmen, Richard, was a clothier.[41] Thus, Christopher would have had a trusted and confidential supplier of the merchandise based in Reading and may have had a 'family' outlet in Sussex. Furthermore, Arundel was a port from which cloth was exported to Normandy and Ireland in the sixteenth century[42] and Christopher also had the means of shipping it in the *Falcon* and doubtless the benefit of any custom receipts on its export.

Christopher Haynes is known from his own will to have had lands at Ford,[43] which may have been brought to him in marriage, and to have had a house in Climping, to the south of Arundel, half way to the sea. This may give a clue as to how his deeds had been detected, for the single vantage point from which such activities could have been observed at either Ford or Climping was from the 40-foot-high Norman tower of St Mary's Church Climping situated between the two villages, with its commanding views of both sea and river stretching up to Arundel, its original purpose being a look-out post for the approaches to Arundel Castle. Could Christopher have been shopped by information being passed to the informer by the St Mary's incumbent, Richard Mery or a churchwarden?

Why the allegations against Christopher Haynes came to nothing one can only surmise. They would certainly have been investigated, so the customer's version of events must have prevailed possibly with some assistance from his brothers at Court. What is known, however, is that on 4 October 1572, William More the Sussex vice-admiral, had replaced his deputy, William Morgan, with William Lussher of Thakeham (near Billingshurst)[44] for the Rapes of Arundel, Bramber, Lewes, Pevensey and Hastings, but not the Rape of Chichester. The timing of this appointment, within two months of the allegations against Christopher, suggests that the two events might not have been unrelated and that pressure might have been put on More to remove Morgan. Lussher eventually moved to Arundel, where he became a burgess and a close and *'trusty'* friend of Christopher Haynes.[45] It wasn't long before Lussher and his servants were seen helping themselves to goods of dubious provenance recovered from the sea[46] and he also reported in his official capacity that the truly enormous quantities of goods *'taken up owt of the sea within the Lord Admiralls jurisdiction'*, including almost 33 tuns of oil and 4½ tuns of olives, had been dispersed among so many recipients as to be irrecoverable for either the Admiral or their owners.[47] These goods may have been the spoils taken off Dover from a Flemish hoy bound for Antwerp[48] that was attacked by 50-60 men in a pirate ship. The goods were indeed subsequently found in small parcels all over Sussex and as far away as Alton, Hampshire. William Lussher was said to have received a bag containing three bushels of pepper from a constable.[49]

In the wake of Throckmorton

This was not the only mention of *'suspicious packs'* being delivered to one of Christopher Haynes's houses. According to Cracknell, *'it was admitted in the examination before Lord Buckhurst that very suspicious looking packs were conveyed by Simon Smyth, Secretary to the Earl [of Northumberland of Petworth House], to Christopher Haynes' house in Arundel'*.[50] This referred to the aftermath of the conspiracy for which Francis Throckmorton was arrested in November 1583 and subsequently found guilty for his part in plotting a Spanish invasion of England combined with a Catholic rebellion at home. A staunch Roman Catholic, Francis and his brother Thomas had moved abroad and became involved with fellow Catholic expatriates in the Netherlands, including Sir Francis Englefield and others, with whom Francis discussed the possibility of the government being overthrown by a foreign invasion. In the early 1580s Throckmorton returned to England in order to serve both Rome and the imprisoned Queen Mary of Scotland. In the spring of 1583, secretary of state Walsingham had become aware through his contact in the French embassy that Throckmorton and Henry, Lord Howard, had become the Scottish Queen's chief agents. In November 1583 Throckmorton was arrested at his London house where among his papers was a list of names of Catholic noblemen and gentry including those of the Earl of Northumberland and William Shelley, alongside details of sites and harbours in Sussex suitable for the landing of a foreign invasion force. He was subsequently found guilty of treason, on evidence obtained after he had been racked. He was executed at Tyburn on 10 July 1584.

There had been a number of plans on the part of the Spanish and French to invade England with the intention of removing Queen Elizabeth and replacing her with Queen Mary, and in 1583, Charles Paget, the Catholic fourth son of William, Baron Paget, had volunteered to travel to England from his self-imposed exile in France under an assumed name, with the purpose of enlisting support of several persons for one such invasion, including Henry Percy of Petworth, Earl of Northumberland, and William Shelley of Michelgrove in the Parish of Clapham, Sussex, and to learn what assistance for the invasion might be obtained from English Catholics. On about 8 September 1583, Charles Paget secretly returned to England through Arundel where he was taken first to the home of William Davies a farmer and a servant of William Shelley, who lived at Patching three miles from Arundel and half a mile down the valley from Shelley at Michelgrove House. Paget subsequently met both Northumberland and Shelley at Petworth where he stayed for about a week in Conigar Lodge in the Park. Thomas, Lord Paget, was invited by the Earl to meet his brother Charles at Petworth where he also met William Shelley and thereby became unintentionally implicated in the conspiracy. Thomas Paget was also a Catholic who had been imprisoned in 1581 for assisting the Jesuit Edmund Campion on his return to England, but he had not been a party to the present plotting.

Charles Paget returned to Davies's house the following week and had a further meeting with Shelley at Patching Copse on the 16 September. It was from Davies's house that John Haler, a mariner with a large vessel moored at Arundel, had collected Charles Paget, together with Anthony Snap another of Shelley's servants, and an unnamed stranger, for the return journey on Wednesday 25 September at night time.[51]

Charles Arundell and Henry Howard, uncle of the Earl of Arundel, were cousins, nephews of Queen Catherine Howard. Both were courtiers, who had in 1581 been wrongly imprisoned for 'plotting', based on the false testimony of the Earl of Oxford.[52] The pair fought long and hard to clear their names, and on examination, Howard was able to prove his innocence and Arundell was released without charge after more than ten months in custody; they were subsequently placed under house arrest, Arundell at Sutton, ten miles from Petworth in West Sussex. As with all members of recusant families they had also been under suspicion for their Catholicism, and when news emerged of Throckmorton's imprisonment, Arundell, fearing he would be implicated in the plot hatched so close to his house, decided enough was enough, and with Lord Paget determined to flee to France. Charles Arundell remained at Sutton and Lord Paget and his manservant lodged with William Davies at Patching while William Shelley made arrangements for their escape.

The Petworth meetings and subsequent events came to the attention of Francis Walsingham; the Privy Council, which had for some time been aware of what was being planned in France, had summoned the French ambassador and required him to inform King Henri III, to which the King immediately ordered a halt. Despite this outcome, as evidence was revealed by Throckmorton's torture, Walsingham ordered the arrest and interrogation of the Earl of Northumberland and William Shelley (who had failed to make his own escape) and the examination of all those involved in the circumstances leading to the flight of Arundell and Paget, among whom was numbered Christopher Haynes.

In Arundel, Christopher Haynes had a cargo that needed shipping to Dieppe that month. This apparently innocent cargo was to lead to him to being implicated and questioned over the events culminating in the flight of Arundell and Paget. What appears to have happened is that in November 1583, Haynes had ordered two large packs, supposedly of sawn timber, from a supplier in London called Isham on behalf of one Lawrence Adams in Dieppe. This was actually a repeat order as he had filled a similar one two months previously. Arrangements were made for a carrier to pick up the packs at Southwark and to take them to Petworth, a usual carrier stopover. Haynes then needed a local carrier to transport the packs over the last seven miles to Arundel and so he sought the help of Simon Smyth, a servant of the Earl of Northumberland, in arranging for their transport from Petworth to his house in Arundel. Haynes, who by this time no longer had his own ship, then commissioned John Haler, who was the co-owner with John Wood of a hoy called the *Dragon* moored in Arundel haven, to ship the timber to Dieppe. Haler had also handled the previous load. What Christopher Haynes did not know (or at least denied knowing) was that his earlier cargo had accompanied Charles Paget on his return to France in September. When on 17 December 1583[53] he was questioned on this specific point, Christopher Haynes said he knew of '*no such personages that repared out of France and was in the house of William Davyes*'. Haynes explained that he kept a tabling (i.e gambling) house[54,55] and tavern in Arundel, frequented by merchants, but he knew of none that went to Davies's house. He claimed ignorance of the whole business. Haler's testimony supported that of Haynes in that his passengers went from the haven straight to and from Davies's house and didn't stop off in Arundel town.

However, the heavy packs delivered to him from Petworth that November were wrongly suspected of containing the possessions of Lord Paget and Charles Arundell. Thus, if his testimony is to be believed, Christopher was inadvertently caught up in the conspiracy to assist in the escape of the two fugitives. As will become clear, he became implicated for other reasons as well.

During the questioning on 28 December 1583[56] by Henry Goringe and Richard Lewkenor (Sergeant-at-Law), of Simon Smyth who, as mentioned already, was in the service of the Earl of Northumberland at Petworth and of John Ramsden, a carrier, it transpired that Smyth had hired Ramsden and Richard Thwaites about the end of November to carry two packs to the house of Christopher Haynes in Arundel. Their delivery to Arundel had been at the specific request of Christopher Haynes to Smyth personally. These packs had earlier been brought from the White Hart Inn, Southwark,[57] by another carrier called Downer, to Smyth's house in Petworth. Although Smyth had been at the White Hart, Southwark, when the packs had been received there by Downer, he couldn't recall who had brought them, and assumed it had been a merchant or his apprentice. The carrier Ramsden, the next to be examined,[58] confirmed that Simon Smyth had hired him to carry an enormous package, that looked like a wool pack, marked with ink at both ends, to Christopher Haynes in Arundel, from where it was to be sent overseas. Smyth had told him that it belonged to a merchant who was staying with Haynes. Ramsden complained that the pack was so heavy that it had *'almoste spoiled his horse in carrying of itt betweene Petworth and Arundell'*. Thwaites had carried a similar pack to Christopher Haynes's house. Neither carrier knew what they contained. It puzzled Smyth that, although he had agreed a price of 2s with Haynes for the carriers to make the seven-mile journey from Petworth to Arundel, when Ramsden and Thwaites arrived at their destination, Christopher Haynes paid them 4s 6d. The obvious explanation was that Haynes probably felt obliged to pay extra because of the detrimental effect of the weight on Ramsden's horse.

After Throckmorton's arrest, the word of the escape of the two fugitives, Paget and Arundell, had spread like wildfire. The two had been taken across the Channel by a local mariner called Thomas Clynsall in a vessel owned by Thomas Banckes, another Catholic. For some reason Christopher Haynes felt he needed to discuss what he knew of the escape with William Dartnall of Drungewick, Wisborough Green, his nephew by marriage to Thomasine, daughter of his brother Nicholas, doubtless in the belief that he could rely on his family's discretion. During the course of this discussion Haynes apparently expressed his opinion that William Shelley of Michelgrove who had organised the escape and had been arrested, would have been better to have spent whatever was necessary to hush up his involvement (a sum of £500 being mentioned; based on average earnings, more than £1 million today), rather than have the matter of his hiring of Thomas Clynsall to take Paget and Arundell to France being made known to the authorities, for according to Haynes, its revelation would create reverberations the likes of which had not been experienced for a century – *'it wode repe up suche a matter as had not bene repte up thys hundred yeres'*. Christopher Haynes had further told his nephew that his brothers who worked in the Royal Household, had got wind of (and were greatly concerned over) his name causing ripples, and the eldest, John, had written to Christopher the week before to say that if he were a traitor, he would be on his own and could expect no help from them; furthermore they would disown him *'for they wold no more take hym for theyr brother'*. Unfortunately for Haynes, Dartnall, while visiting Thomas Pellet, a yeoman farmer at North Stoke two miles outside Arundel[59] and a servant of Shelley, had seemingly passed on these details of Christopher's conversation. As with Shelley's other servants, Pellet was interrogated and in his examination on 9 December disclosed what Dartnall had told him. This was transmitted to Walsingham and undoubtedly led to Christopher being questioned the following week.

Christopher Haynes more or less corroborated Pellet's statements in his own evidence. Although he knew Thomas Pellet well, as a farmer living only a couple of miles away, he said he hadn't met him since Pellet had been employed by Shelley at Michelgrove and he hadn't been to Pellet's house at North Stoke for the past two years. Haynes believed that the remarks he had made of Shelley were to the comptroller at the custom house and also to William Dartnall, but *'he used no spiche of ripping up of matters'*. What is quite telling about this evidence is Christopher's likely action had he been in the same situation – he would have simply bribed his way out of it.

A curious event took place on 9 December 1583. Edward Caryll, Esquire, of Shipley, was examined at his own request by Thomas Lewkenor, at Angmering, because on the previous Saturday whilst in Arundel, the mayor had told him that Christopher Haynes had spread the rumour that Paget and Arundell had lodged in his (Caryll's) house at Shipley.[60] Caryll was incensed and so clearly concerned at being wronged by Haynes that he at once made for Lewkenor to put the record straight[61] by offering himself up for examination on the matter. He took along his servant John Mychell, yeoman, a member of a well-known Shipley land-owning family, to be examined as well, as Mychell had been left in charge of his house during his absence. In brief, Caryll knew nothing of Paget and Arundell going to France apart from what he had heard *'by the common voice yn London'*. He was only on nodding acquaintance with the two fugitives, whom he had passed by in the street a couple of times in London. He had no idea where they had stayed in Sussex or whence they had embarked, apart from the fact that rumour had it that they had left from Sussex. He had held no discussions on the matter with anyone, certainly not his servants.

John Mychell of Shipley was examined after Caryll had completed his evidence in his master's presence. He had no idea why he was being questioned. Asked about his master's movements, he said that Caryll had come home the previous Thursday from London, where he had been for almost a year, and that on Saturday he went to Arundel. He was then asked to name all those who had resided in Edward Caryll's house in which he himself had been living for the last few weeks and if any strangers or guests had been staying there,[62] to which he answered no-one except two servants of the Earl of Arundel, who had come to collect money due them. Also Francis Hobbes and Ralph Raynebowe, servants, Edward Caryll's kinsman, John Caryll of Warneham, and four others named Sares of Abinger, someone called Surrey, a yeoman; Richard Berde and Giles Vale, servants of William Palmer, had stayed over on last Friday night. As to whether he knew Lord Paget, he believed he had known him some twelve years previously before his enoblement, but probably wouldn't recognise him now; he had seen Charles Arundell in Angmering Park recently, shooting deer, but had never spoken to him. He had heard of the fugitives' departure from Giles More, his fellow servant, clerk to Edward Caryll's ironworks. His master hadn't said a word on the matter. When asked if he had heard any credible reports of where Paget and Arundell had lodged or where they had embarked, he said that he had heard in Horsham market the previous Saturday that they *'were shypped aboute Felffham'*, but he knew of nobody who had assisted in their departure.

William Shelley, it subsequently transpired, had indeed arranged for a barque, probably owned by Thomas Banckes, to be made ready for sailing with Lord Paget and Charles Arundell from the tiny coastal village of Ferring just to the west of Worthing. For this purpose Shelley had hired the services of Thomas Clynsall, a master mariner for £30.[63] Clynsall, like Shelley, then lived at Clapham, a village adjacent to Patching where William Davies, Shelley's servant, lived. Clynsall stated that it was Shelley and not the Earl of Northumberland (as accused by the Government), who had repeatedly asked of the readiness of the barque[64] and Shelley had arranged for the passengers to be at William Davies's house (at Patching) by 6 o'clock in the morning on the day of departure. This meeting place was undoubtedly chosen because it was just a short distance to the north of Ferring. One additional curious fact is that William Shelley's recusant brother, John, also lived at nearby Clapham, yet he appears to have been completely oblivious of the events unfolding on his doorstep and doesn't appear to have been examined.

It seems highly likely from Christopher Haynes's evidence that he himself introduced Thomas Clynsall to Shelley.[65] He had been to see Shelley at Michelgrove, on Sunday three weeks before, accompanied by Clynsall, whom Haynes had invited along only the day before. Haynes had clearly wanted him there, and despite what emerged at his inquisition, the choice of Clynsall could have been no accident, as he was the one mariner who lived near to Michelgrove, Davies's house and Ferring. The reason Haynes gave for his visit to William Shelley was one *'of good will'* as Shelley had just returned home,[66] and because Haynes leased land from him.[67] While there, he had been prevailed upon to keep company with Mr John Shelley, who was visiting from Clapham down the road. This presumably gave William Shelley the opportunity to negotiate with the mariner in private.[68] It turns out that Thomas Clynsall after his release from imprisonment in the Marshalsea for his role in the fugitives' escape,[69] became a tenant of Christopher Haynes in Arundel.[70]

The fugitives and their armed escort were observed riding towards the sea at Ferring by Thomas Barnard a resident there. It was Sunday evening about an hour after nightfall, when Barnard saw eight men on horseback in the main highway. One of them rode ahead with his sword drawn, six others followed in pairs and the eighth made up the rear probably also with his sword drawn, but Barnard didn't get a good view of them. Neither the men nor their horses returned the same way.[71]

Haynes's introduction of the mariner to Shelley begs the question as to whether he was also involved with the barque used by Clynsall, although it seems that he (Haynes) was twice interrogated only about his dealings with the mariner, John Haler, whose ship, the *Dragon*, based at Arundel, was commonly used for trips between England and France. It was Haler who had brought Charles Paget from France on 8 September and returned him on Wednesday 25 September,[72] without Paget's identity being disclosed. Haynes said that he had used Haler on two occasions only for the purpose of transporting packs of sawn timber to Dieppe.[73] The first of these probably accompanied Charles Paget on his return trip (on 25 September) the second, also from Isham, comprised the two large packs of boards carried to him, via Petworth, for the last seven miles by Ramsden and Thwaites at the request of Simon Smyth acting on

Christopher Haynes's behalf. These then were the *'suspicious-looking packs'* that had inadvertently been caught up in the conspiracy and mistakenly believed to contain the possessions of Lord Paget and Charles Arundell. For whatever reason, Haynes was clearly both defensive and evasive in playing down his knowledge of Haler, whom he must have known well as he had certified Haler and his vessel to the vice-admiral in 1561.[74] Furthermore, Haler had been paying dues for his wharfage for more than 20 years to the Arundel bridgewarden, a role undertaken by the mayor, and in 1562 when Christopher Haynes himself was mayor and bridgewarden, he is recorded as having received wharfage payment from John Haler.[75] Haler's home was at Wisborough Green, where Haynes's nephew, William Dartnall, lived: there can be no doubt that Christopher Haynes was well acquainted with the mariner. Haynes was adamant that he knew none of the personages involved or by what means they had been carried overseas from Arundel, and he had never been privy to any such information. He could not have been telling the truth. In 1583 Haler's hoy was the only vessel of its size based at Arundel haven, and it had been so for more than 20 years. Christopher Haynes had been the customer of Arundel for much of that time as well as assistant to John Humfrey the previous customer. If anyone would know about shipping movements and capability of ships to cross the Channel, it was he. He claimed to have no idea of anyone that had left Arundel without a licence to do so over the past year, although he was aware that Robotham, the Earl of Northumberland's servant, had recently travelled to France and back on behalf of the Earl,[76] a journey he had previously made a number of times.[77]

Haynes professed in evidence that he first heard of Lord Paget and Charles Arundell's departure for France in a letter from Mr Yonge (John Yonge, customer of the Port of Chichester whose jurisdiction included Rye and Arundel[78]) and it dawned on him that it had to be by the barque sailed by Clynsall. It was only then that he discussed the matter with the controller and William Dartnall *'that Mr Shelley was therby under'* (i.e. 'sunk'; done for). He agreed that he had received a letter from his brother, the sergeant (John Haynes), which confirmed what Mr Yonge had told him, that Paget and Arundell were going over to France at Arundel, which presumably he must know about, and advising him (Christopher) to make sure he could answer for his involvement in the matter, as he was bound to be thoroughly interrogated. Christopher Haynes, in response to the question on the whereabouts of his brother's letter, said that he had burnt it. He thought he had told both the searcher and the controller of the letter from Yonge, but couldn't remember whether he had told Dartnall anything, but he might have done. One intriguing item in the report of Christopher's examination was the mention of his right hand being lame, suggesting that he might have had a stroke and was unable to sign the document.

The report of Christopher Haynes's examination was important enough to be required to be sent to the secretary of state, Sir Francis Walsingham. It was evidently considered that Christopher might be implicated in the departure of the fugitives and *'would be undone'*. However, in his report, his examiner said that Christopher Haynes had blamed the searcher of Arundel for being negligent in his office, and a tippler,[79] and it seems that no further action was taken against him.

Unanswered questions

Were Christopher Haynes's responses to questioning entirely straight? Apart from the carriers who appear to have innocently entered the scene at a most unfortunate time, he knew all of the individuals questioned that December intimately, including Shelley. Whatever the motivation behind Christopher's assistance to Shelley, religion was not it, as judging from the preamble to his will, he was certainly no Catholic. It would not be out of keeping with the rest of his endeavours if his interest turned out to be solely pecuniary. Those who knew him well might have expected him of all people to be one who would keep his mouth shut, but clearly as the plot unravelled and the consequences became clearer, his instinct for self-preservation came to the fore.

Were the heavy packs delivered to Haynes in Arundel really packs of timber, or was this explanation a lie to throw the examiners off the scent? Certainly Francis Walsingham suspected they contained Lord Paget's possessions, furthermore he had documented that John Haynes knew that his brother was privy to the fugitives' plan. This would seem to make sense; why else would the Earl of Northumberland's servant, Smyth, be at Southwark to ensure that the packs were delivered to the carrier? This can only have been at the earl's behest and it would be truly surprising if he had sanctioned Smyth's presence at the White Hart for the purpose of assisting

Haynes with a timber purchase. In any case, why was the timber wrapped like a woolpack? Smyth had told the carrier Ramsden that the packs belonged to a merchant who was staying with Haynes, yet Haynes inferred that it was for an Englishman in Dieppe. If Christopher Haynes's evidence was untrue, the possibility remains that he was acting on behalf of the earl but was ignorant of the packs' true contents, having been told that they contained timber. Alternatively, if he knew or even suspected that the packs contained the fugitives' possessions, he must have concocted his plausible story to deceive the examiners. Whatever the explanation of these inconsistencies, Haynes never implicated the Earl of Northumberland, apart from identifying the earl's servant Robotham as a frequent traveller to and from France.

If the second lot of packs had contained the possessions of Lord Paget, it would also cast doubt on the earlier shipment that had accompanied Charles Paget on his return to France. Might this also have been organised by Christopher Haynes on behalf of Northumberland or the Pagets?

Was Edward Caryll's part as innocent as he made out? What was the nature of his business in Arundel that Saturday, two days after returning from a year's absence? Did he speak to Haynes? Why did Haynes implicate Caryll by telling the mayor that the fugitives had stayed at Caryll's house at Shipley, a long way from that of Davies in Patching? Was the location a mistake? Was it Caryll's London home in which Paget and Arundell had stayed? Caryll knew Shelley very well; their families were after all joined in marriage and both followed the Catholic religion and Caryll was known to be a recusant and presented as such in 1580.[80] Furthermore, Caryll was steward of the estates of Philip Howard, Earl of Arundel, also implicated in the Throckmorton affair, in whose will he was named as executor. It must also be of significance that after his examination Edward Caryll was imprisoned in the Tower.[81] Although he had trained as a lawyer, having entered the Inner Temple, in some way he managed to avoid taking the Oath of Supremacy.[82]

There is every possibility that Caryll's servant, John Mychell, was not as innocent as was made out before Lewkenor either. Who were the four men Mychell hardly knew: Sares, Surrey, Berde and Vale, who had stayed at Caryll's house the day after Caryll's return?[83] On Sunday the 8 December 1583, at 3 o'clock in the afternoon, a man called Braye unexpectedly arrived at the house of John Tawke

[Taulke], gentleman, with a young man called Anthony Crompton [or Crumpton], supposedly a priest, who was wanted for questioning along with Shelley. It seems that Crompton was hoping to escape to France as well. Tawke, a Catholic, who had previously been imprisoned in the Fleet for his beliefs, was prevailed upon to house Crompton for two or three days. But on the Tuesday (10 December) while in Chichester, Tawke had heard of Paget and Arundell's flight and of Shelley's involvement and arrest and became anxious. That same evening at about 8 or 9 o'clock, one of Shelley's servants called Fowler arrived at Tawke's house from Michelgrove on a white unshod mare without bridle or saddle that he had borrowed from his friend, Brydger (possibly Christopher Haynes's searcher). Fowler said that Braye had suggested he should come over to keep Tawke company. Fearing it would attract attention, Tawke sent the mare packing back to Shelley's, before asking Fowler about Shelley's whereabouts on the previous Saturday. Tawke was given to understand that Shelley was with Tawke's neighbour, William Davies. Tawke became really anxious lest he be inadvertently implicated in the trouble and managed to persuade Mychell another near neighbour, to let Crompton stay with him, which was where, on Thursday the 12 December, Crompton was apprehended. John Tawke, who was examined immediately by Lewkenor,[84] agreed to set down his detailed confession in his own hand on 15 December 1583[85] of which the foregoing is a summary. Why did Mychell say that he had heard that the fugitives had departed from Felpham, much further to the west? Was it to deflect Lewkenor's attention from the true departure point?

It is an interesting coincidence that it was on that same Saturday as Haynes's reported conversation with the mayor, that he had arranged for Clynsall, the only mariner to live near Davies and within an hour's walk of Ferring, to join him on a visit to Shelley the following day. Why did Christopher Haynes choose this mariner when there were so many in Arundel and Littlehampton? The obvious answer is that he must have known what was going on. Clynsall was imprisoned for the part he played and his subsequent release from the Marshalsea was upon recognisance. Who provided the surety for him? Why did Clynsall become a tenant of Christopher Haynes on his release[86] – was it because of an obligation on Christopher's part?

Clearly something had set the alarm bells ringing in London that had prompted the Haynes brothers to warn Christopher to get his story straight and to let him know that, if he turned out to be a traitor, to expect no support from them. There is a tantalising document listing three matters on which Christopher Haynes was to be charged. The first was that he had been accused '*by his brother to be privy to the going over of Ld Paget and Arundell*', the second (with Banckes) was that he had '*made [Thomas] Banckes set his hand to a blanke* [paper] *for the discharge of a cocket;* the third was that he was *of the counsel with Climsall in the conveying of Ld Paget and Arundell*'.[87] The first, though possible, could not be proved as Christopher had destroyed part of the evidence – the letter from his brother, John, who appears to have received information on Christopher's involvement from Yonge, possibly via Yonge's father-in-law who worked for Lord Burghley. The fact that Christopher had encouraged Banckes to sign the blank Arundel Council's cocket, the custom house warrant that would allow Clynsall to set sail, might if true, have implicated him to a similar extent as Shelley. He was after all it should be recalled, a member of the Council as well as the Arundel customer, so in theory could have issued the cocket without anyone else being involved – and falsification of customs documents including *'false cocketting'* turns out to have been the *modus operandi* of the crooked Chichester customer and his men.[88] However, it appears that the matter of the cocket was glossed over by his examiner as an example of the searcher's negligence.

Was it Banckes' boat that had been used, as Christopher had apparently suggested to Dartnall? Probably so! Thomas Banckes, a member of a staunch Catholic family, was under suspicion for his part in the disappearance of Paget and Arundell. His brother, Richard Banckes wrote to him from exile in Paris on 1 September 1583,[89] wishing him and his parents good health, then railing against the heretics opposing the '*Catholique and Romayne religion the true apostolicall and Christian religian*', whom he prayed would '*breake theare nekes headlong and to be blaste from the face of the yearth and the name of them not to be knowne, which God in his infinit mearcy graunt*'. Such was the strength of feeling harboured by the Catholics against their persecutors. He finished by urging his brother to keep this letter to himself, but the fact that it ended up in the state papers suggests that Walsingham had come by it.

Philip Howard, Earl of Arundel, was also suspected of plotting with Throckmorton, Charles Paget and Lord Paget. He was initially detained and then released, only to be re-apprehended on trying to leave the country from Sussex in April 1585. He was imprisoned in the Tower of London with Henry Percy, Earl of Northumberland, of Petworth. Howard was eventually convicted of treason in 1589, but died in his cell in 1595, Queen Elizabeth having commuted his death sentence. Percy apparently committed suicide and at a subsequent meeting of the Star Chamber he was again linked to Don Bernardino de Mendoza (the Spanish Ambassador), Throckmorton and Paget. In the evidence it appears that William Shelley had confessed to providing the ship at Percy's request that took Paget to France, although this evidence was seemingly extracted under the threat of duress (the '*racke*').

Shelley was tried by a special commission in Westminster Hall where on 12 February 1586 he admitted his guilt. He was condemned to death for high treason, but the sentence was not carried out as it was '*respited at her Majesty's good pleasure*', the Queen having earlier received a petition from Shelley's uncle Richard for her '*gracious suspending of his execucion*' and a plea for a pardon for his nephew.[90] Shelley himself acknowledged '*suche speciall grace in this my greate dystresse*' in a letter to Walsingham dated 10 November 1584.[91] William Shelley was still in the Tower in July 1588, but was subsequently transferred to the Gatehouse Prison, Westminster. After his attainder, Shelley's estates were confiscated and his immediate family reduced to penury. Shelley was a member of a well-known family of recusants considered dangerous by the state, and it seems likely that the temporary suspension of his death sentence was granted in order to control the troublesome activities of the rest of the family. His wife Jane prayed for her husband to die in his bed, and even seven years after his trial, suffered extreme anxiety over the possibility that William might yet be executed. He died in prison on 15 April 1597.[92]

Charles Arundell was pronounced guilty of high treason in 1585 and died in exile in Paris two years later. Thomas, Lord Paget was attainted *in absentia* and his lands confiscated, the income from which (almost £2300) was put towards the maintenance costs of Queen Mary of Scotland. He died at Brussels in 1589. Charles Paget was able to return to England with the accession of King James. His attainder was reversed and he recovered his lands. He died in 1612.

By June 1577 laws were enacted for the preservation of the Queen against Jesuits and Catholic priests plotting against her on the basis of the Bull of Pope Pius V of 1570. They included dismissal penalties and imprisonment against port officers who permitted other than those on pre-approved special government business, fishermen and merchants, to leave without licence of the Queen.[93] These laws impacted heavily on Christopher Haynes as customer of Arundel, as they did on all port officials including the mayors, bailiffs, jurats, customers and contollers who were called upon to identify and report in writing all vessels over sixteen tons' burden belonging to each port or creek, their owners, burden, crew, dates of departure, nature of their cargo, whether custom duty on it had been paid or not, destination and anticipated return date. Vessels not putting to sea had also to be likewise accounted for. Vessels of lesser burden, if acting 'suspiciously', were also to be reported. Mariners, masters and pilots from every port had to be listed by name along with their marital status. Vessels were henceforth prohibited, under pain of forfeiture, from putting to sea without official permission. If vessel owners disregarded this law, and performed acts of piracy or depredation, the port authorities would be held to account. Any ship arriving at a port was not permitted to land any cargo or receive any assistance or victualling until it was cleared of being involved in piracy. Those under suspicion of piracy or roving including Scotsmen sailing under letters of marque, were to be arrested, examined and the outcome reported to the vice-admiral. Those caught purchasing goods from such vessels were to be imprisoned.

In the returns for Lansing and Shoreham following these restrictions is listed a mariner, John Yonge.[94] It is now clear that in his later years Christopher worked under a customer of Chichester of this name, who was involved in the illegal export of arms and other goods using his own ships; diversion of grain overseas from its English destinations; charging merchants an additional handling fee; understating the amounts to be shipped and keeping the remainder, and omitting to record customs duty he had collected which he would then share with his searcher.[95] Yonge had always appeared to have been a 'bad lot'. For example, when Yonge was customer of Rye, he was accused by Thomas Fenner of great offences done in the course of his office, including cheating at dice, which led to the ruin of many unwary.[96] However, Yonge had influence at Court, where his father-in-law worked; he too would eventually fall from grace.[97] It therefore seems to have been a remarkable lack of judgment on William More's part when on 2 November 1585, he appointed John Yonge, by then customer of Chichester an office which included other Sussex ports, as his deputy vice-admiral of Sussex.[98]

Family and circle

Christopher was predeceased by three of his siblings. The first was his brother Richard, who died at his home in Wargrave, Berkshire, where he and his wife had married and spent their lives together. Richard's home was not far from his sister, Christian Gunnell. He was buried at the end of his pew inside the village church on 2 December 1566. Christopher received a bequest from Richard of 20 shillings. In February 1572, Christopher learned the sad news of the death of his only sister, Christian, ten weeks after the birth of her tenth child, Richard, in Wargrave. She was buried on 13 February in Wargrave churchyard. Christopher would eventually remember Christian's eldest son, Moses Gunnell, in his own will. Christopher Haynes was next mentioned in the will of his brother, Thomas, who died in February 1583, in which he was left a case of pistols and a caliver with powder flask and horn touchbox,[99] which he received after probate had been granted two months later in April. Perhaps Thomas was aware that, of all his brothers, Christopher would have more need of them. The ownership of calivers or other handguns was an indication of membership of the mercantile and trading elite of the times.[100]

Elizabethan gentlemen would consider both a rapier and a dagger to be essential items of attire as 'life-preservers'. Thieves, swindlers, rufflers [con-men], beggars, hookers (thieves who stole clothes by hooking them with a long pole), horse-thieves and prostitutes constituted a widespread vagabond problem, which the traveller ignored at his peril. Their numbers greatly increased during Elizabeth's reign, and all of the Haynes brothers must have had frequent encounters while making their journeys. Horseback was the commonest means of travel, as the carriers' carts were unsprung and unsuitable for passengers on long journeys. However, for added security, the Haynes brothers probably rode in the company of carriers, who travelled throughout the country delivering goods. In the case of Arundel the carrier stop at Petworth was a well-known destination from London, whence carriage to the coastal towns would have been by local carriers as was the abovementioned delivery of Christopher's packs from London. Carrier routes were dotted along their whole length by inns for stopovers.

We have only glimpses of Christopher's domestic life. In 1586 he and his wife, Elizabeth, lived with their manservant Richard Wool-ridge, or Wooldridge, who had arrived in Arundel from Stafford at the age of 18, in 1580. Their house adjoined The White Horse in the High Street, which was owned by wealthy burgess Francis Garton and let to Francis Cradle, a confessed receiver of pirated goods.[101] As with most properties in the town, Christopher's house was held of the Earl of Arundel. It is of interest that his fellow burgess, Garton, had obtained a licence for a tavern in his mansion house in Arundel; perhaps Christopher did likewise. Turning his house into a tavern or inn would have been an excellent business venture. The tavern was where much commerce was transacted, letters left for collection and, as we know in Christopher's case, where goods could be delivered and dispatched. It was also the haunt of the less desirable, the gamblers (usually betting on throws of dice) and prostitutes – many seaport taverns having rooms for hire for pleasure. The Haynes and Cradle buildings must have been very close to the location of the George Inn on the east side of the High Street in the late sixteenth century.

Woolridge's service was eventually generously rewarded by a bequest of £10 and the rent-free life tenancy of one of Christopher's houses in Arundel, which in 1586 was occupied by John Sibourne. Woolridge lived in Arundel for 15 years (between about 1580 and 1595), presumably in the house provided by Christopher.[102] His son, Henry, was baptised in Arundel on 21 November 1591. He and his family subsequently moved, first to Climping and then to Bosham.

In his detailed will,[103] Christopher made several bequests to family, including a nest of silver-gilt bowls to his brother, John, who was also left Christopher's grey gelding. He left £5 each to his nephew, Moyses (Moses Gunnell, eldest son of his sister Christian) and his niece, Alice Haynes, daughter of his brother Nicholas, and £10 to Richard, Nicholas's son. In total his monetary bequests amounted to more than £67. He left the wife of his brother, William, a set of ten silver-gilt cups. To his wife ('*nowe my wife*'), Elizabeth, he left household goods to the value of £20, an annuity of £20 and a choice of house for life, with reversion to William, his brother. All his properties in Arundel (five houses and a shop), were left to his brother William for the term of his life, in tail to his nephew, Richard Haynes. On Richard's death in 1634, all of Christopher's properties were subsequently passed to Richard's daughter, Susanna.[104] Christopher's total wealth at the time of his death was well in excess of £100, making him the most successful of all the brothers, especially as he had inherited practically nothing from the others.

The phrase, *'nowe my wife'*, used by Christopher for Elizabeth, suggests that she was not his first wife.[105] One of his initial bequests was to *'Jasper Barker my wifes kinsman five poundes in money'*. Interestingly, there is the record of a marriage of one, Christopher Haynes, to Mary Barker at Billingshurst, Sussex, on 12 May 1582.[106] This was near to Wisborough Green where the niece of Christopher of Arundel, Thomasine Dartnall, lived. She had named her second son, Jasper, possibly after Jasper Barker; her uncle Christopher was the godfather of her first son, William. All this suggests that there may have been a distant link between the two Christophers and Billingshurst. Clearly the Christopher Haynes of Billingshurst was not the same person as Christopher of Arundel, as the latter died in 1586 and the former went on to have his children after this date.

In the autumn of 1586 Christopher was suddenly taken ill while visiting his brothers in London. On 26 November there was a scramble to get his will in order. It was compiled by a secretary with the help of his brothers and then *'read distinctly to the said Christofer Haynes and by him subscribed sealed and delivered as hys deede and for his last will and testament'*. Christopher died within a day of completing his will and was buried later that day (27 November 1586) in the choir of the Church of St Nicholas Cole Abbey, the parish church of John Haynes. The entry in the parish register reads: *'Mr Christopher Haynes of Arundell was buryed ye xxviith of November in ye quire, 1586'*.[107] Probate was granted to his brother and executor, William Haynes, the following day. The will was challenged by Christopher's wife Elizabeth and his brother Nicholas and nephew Richard. However, the three challengers failed to appear in court despite being summoned several times and the sentence,[108] dated 27 February 1587, declared the will valid and in favour of William Haynes as the lawful executor. In the normal course of events, as was seen in the case of Thomas's will, the main beneficiary was the eldest male relative unless the property was in tail. As his elder brother, John, had inherited most of the Haynes family estate, Christopher decided in favour of William.

The possible identity of Christopher Haynes of Billingshurst

The relationship if any, of the two Christopher Hayneses awaits clarification, but it is safe to say that they were not father and son. Indeed Christopher of Billingshurst, whom we may call 'young Christopher', is said to have been the son of Robert Haynes, clerk, of nearby Wiggenholt who died in 1574.[109] Young Christopher is mentioned in the will of William Tredcrofte in 1593[110] who left Christopher's four children £10 between them. Tredcrofte appointed a Robert Barker as one of his overseers. C. P. Haines points out that the name Haynes spelt with an 's' was very uncommon in Sussex prior to 1600[111] and one wonders whether Christopher's father, Robert, might have been a kinsman of the Reading Haynes brothers, a cousin possibly, and therefore related to Thomasine Dartnall of Wisborough Green.[112] Young Christopher moved from his parental home at Wiggenholt to Billingshurst, where in 1582 he married Mary Barker, possibly the daughter of Jasper Barker [or Booker], and sister of Robert Barker, overseer of the Tredcrofte will. Jasper Barker was somehow also related to the wife of Christopher of Arundel.[113] Christopher and Mary's firstborn, a son, Christopher, was baptised on 3 March 1587 almost five years after their marriage. They went on to have three more children: sons John and Thomas and a daughter Joan.

Notes and references

1. Will of William More of Sherfield upon Loddon, PCC Populwell, proved 4 July 1549, TNA: PROB 11/32.
2. William More was the son of Thomas More and Margaret Cottesmore of Lancelevey, Sherfield on Loddon. Also mentioned in the same will was Richard Perkins or Parkyns, the testator's cousin, son of Thomas Perkins and Dorothea More, sister of Thomas More (A. Mary Sharp, 1892, *The History of Upton Court of the Parish of Upton in the County of Berkshire and of the Perkins Family compiled from Ancient Records*', London: Ellcot Stock, p. 201). This could have been the same William More

who with Christopher's grandfather John Folkes was enfeoffed land by Nicholas More of Colemans More in 1510 (TNA: E 210/6653).
3. 'Visitation of London 1568', ed. Sophia W. Rawlins. *Harleian Society*, London, 1963.
4. 'Post Mortem Inquisitions 1-25 Elizabeth', Sussex Rec. Soc. III, p. 155.
5. West Sussex Record Office (WSRO subsequently): Arundel Borough Archives A1 f.15 Minute Book 1539-1835.
6. WSRO, Arundel Borough Archives A1 f.30 Minute Book 1539-1835.
7. Ibid. f.16 Minute Book 1539-1835.
8. Ibid. f.19 Minute Book 1539-1835; WSRO Arundel Borough Archives F2/1 f.3 Bridgewarden's Accounts.
9. The corporation comprised a mayor, twelve burgesses, a steward, and other officers. The mayor was chosen annually at the court leet of the lord of the manor, and was a justice of the peace within the borough.
10. From: *'Arundel', A History of the County of Sussex*, Vol. 5, Part 1: 'Arundel Rape: south-western part, including Arundel' (1997), pp. 10-101.
11. WSRO, Arundel Borough Archives A1 f.19 Minute Book 1539-1835; WSRO Arundel Borough Archives F2/1 f.3 Bridgewarden's Accounts.
12. Ibid. f.30 Minute Book 1539-1835.
13. WSRO: Chichester City Archives W1a (m.5 dorso).
14. Possibly William Trevett a mariner and freemason, listed by John Shelley in his return of ships and mariners to the Vice-Admiral; Surrey History Centre: LM/488/9.
15. Surrey History Centre: LM/1936. This letter was eventually received by William More of Loseley, possibly through his deputy for Sussex, William Morgan, to whom it was probably originally sent. Brydger, the controller, was Thomas Brydger of Arundel, Sussex. Brydger and Colbroke were said to have been cousins. In 1561 Brydger had been certified by Philip Wyther, mayor, Haynes, John Humfrey, customer, and Richard Shepherd, controller, as a mariner from Felpham in a return to the Vice-Admirall for Sussex [Surrey History Centre: LM/488/4]. Thomas Colbroke, Elizabeth Haynes's brother, was most probably the person of that name who was bailiff's sergeant to the Mayor of Chichester in the 1570s.
16. The system was reintroduced by King James I [See: P. Erlanger, 1967, *The Age of Courts and Kings, Manners and Morals 1558-1715*, Weidenfeld & Nicolson, London, p.159]. Customers otherwise would receive an annual fee; the Customer of Chichester received £6 13s 4d and a reward of £21.
17. Sussex Archaeological Collections vol. IX, 1857, p. 107.
18. Thomas Brydger, one of the Arundel Burgesses; WSRO: Arundel Borough Archives A1 f16 Minute Book 1539-1835; WSRO: Arundel Borough Archives F2/1 f.3, Bridgewarden's Accounts; Brydger was aged 36 in 1576 by which date he had lived in Arundel for 14 years. He was born in 1540 in Easebourne near Midhurst; WSRO: Ep I/11/3. The Bridgers were an old armigerous Sussex family. See *Notes & Queries*, 1850, p. 78.

19. This was to check a treasurer or person in charge of accounts; controllers received a smaller annual fee than customers but a much greater reward.
20. Surrey History Centre: LM/1630.
21. Ibid. LM/1774/1.
22. D.W. Jones, 1972, 'The Hallage Receipts of the London Cloth Markets 1562-c.1720', *The Economic History Review*, New Series 25, p. 567.
23. Surrey History Centre: LM/1936.
24. Ibid.
25. Ibid.
26. E.g. British Library: Lansdowne 39 No. 28, 41 No. 22.
27. Neville Williams, 1962, *Captains Outrageous: Seven Centuries of Piracy*, Macmillan, p. 67. We have come across no evidence for Christopher's active involvement in piracy as opposed to receiving goods of questionable provenance. Even if it was true, he wouldn't have been the only member of Arundel Council to be so tarnished. Within six months of Christopher's death, William Barford was dismissed, accused of being a *'notorious person lyving by pyracies and other mysdemeanors'*; WSRO: Arundel Borough Archives A1 f.25 Minute Book 1539-1835.
28. See e.g. letter from Francis Walsingham to the Earl of Lincoln CP Vol. 160, No. 128; HMC Vol. 2 No. 457, p.150.
29. Surrey History Centre: LM/1787/1.
30. Ibid. LM/1795/2.
31. TNA: SP12/95, 82.
32. Cal. Pat. Rolls 1566-9, p. 169.
33. *'Arundel', A History of the County of Sussex*, Vol. 5 Part 1: 'Arundel Rape: south-western part, including Arundel' (1997), pp. 10-101.
34. TNA: SP12/95, 82.
35. Surrey History Centre: LM/1936.
36. Acts of the Privy Council 1578-80, HMSO, 1895, p. 250; PC2/12.
37. TNA: SP 12/167/59.
38. 'Calendar of the Manuscripts of the Most Hon. the Marquis of Salisbury', Preserved at Hatfield House, Hertfordshire, Vol. 7, entry 205: 1597.
39. Daphne Phillips, *The Story of Reading*, 3rd edn, 1999, Countryside Books, Newbury, pp. 32-4.
40. Berkshire Record Office: D/A1/77/60; Will of Agnes Haynes, Parish of St Giles, Reading, 1589.
41. WSRO: Cowdray Mss 4735/10.
42. *'Arundel', A History of the County of Sussex*, Vol. 5 Part 1: 'Arundel Rape: south-western part, including Arundel' (1997), pp. 10-101.
43. TNA: Will of Christopher Haynes Gentleman of Arundel, 20 November 1586; Windsor: PROB 11/69.
44. Surrey History Centre: LM/1774/2.
45. TNA: Will of Christopher Haynes Gentleman of Arundel, 20 November 1586; Windsor: PROB 11/69.
46. Ibid. LM/1977/2.
47. Ibid. LM/1798.
48. Ibid. LM/1977/4.

49. Ibid. LM/1977/3.
50. Lord Lieutenant of Sussex; E.M.W. Cracknell, ed. (1934), *The Barbadian Diary of Gen. Robert Haynes 1787-1836*, The Azania Press, Medstead, Hampshire, UK, p. 6. This was wrong: the examinations of Smyth and the carrier, John Ramsden, were before Henry Goring and Richard Lewkenor (SP12/164/66-67), not Lord Buckhurst.
51. TNA: SP12/164/45. Charles Paget, who was undoubtedly a Catholic conspirator, subsequently involved with the Babington Plot, was also acting as one of Walsingham's agents, passing him information, although he never had Walsingham's trust. Anthony Snap was a known recusant.
52. A. H. Nelson, 2003, *Monstrous Adversary: the Life of Edward de Vere, 17th Earl of Oxford*, Liverpool UP, pp 273-5.
53. TNA: SP12/164/33.
54. TNA: SP12/164, f 54; Cal State Papers Domestic 1581-90, Vol. II, Dec 17, 1583, No 33, ed. Robt Lemon, Longman 1865, in which 'Haines' has been wrongly transcribed as 'Harris'.
55. In 1580 Christopher Haynes occupied a tenement owned by the Earl of Arundel situated on the High Street abutting to the north, *'Le White Horse'* owned by Francis Garton but occupied by Francis Cradle, with the *'little park'* of Arundel to the east; WSRO: Lavington/153; the Little Park, north-west of the castle was at that time a partly wooded deer park of around 26 acres. Abutting 'The White Horse' on its north side was the house of John Fenne another burgess. In January 1578, Garton had also acquired a licence to keep a tavern or wine cellar in his mansion house in Arundel (WSRO: Lavington/152), which specified the prices that could be charged. Thus French Burgundy (from Gascony, Guyenne and La Rochelle) bought for £11 or less a tun [c. 252 gallons] could be sold at not more than 16d a gallon [c. 50% profit], sack malmsey and sweet wines bought for £8 or less a butt or pipe [c. 130 gallons] could be sold for no more than 2s a gallon [c. 60% profit], i.e. both about 1p a pint in today's money; muscatel on the other hand had no restrictions placed on the sale price. Such taverns in Arundel served the needs of the visiting sailors. The Haynes and Cradle houses approximated to where the George Inn was located in the 16th century and the latter may have started as a small domestic tippling house. Garton, originally of Billingshurst, was mayor of Arundel in 1583 and 1585-7. Francis Cradle confessed to receiving pirated salt and fish (Cal State Papers Domestic [1856] Vol. 1, 12 April 1578, p. 588). The 'white horse of Arundel' was the sinister supporter of the arms of the Earl of Arundel, and became a pseudonym of the Earl himself.
56. TNA: SP12/164/66-67.
57. An inn at the sign of the White Hart was established in the Borough High Street, Southwark, on the road from London Bridge. It was where carriers rented rooms overnight, and plays were staged in its courtyard from the 1570s to 1590s (it was mentioned by Shakespeare in *Henry VI*). It was also a haunt of thieves and pickpockets who robbed the unwary carriers. It became famous as a coaching inn and with the advent of rail travel was finally demolished in 1889. Carriers from Arundel, Billingshurst, Rye, and

other Sussex towns also arrived at the Queen's Head in Southwark on Wednesdays and Thursdays.
58. TNA: SP12/164/67.
59. TNA: SP12/164/59.
60. Caryll's house was Bentons Place, Shipley, six miles south-west of Horsham.
61. TNA: SP12/164/23.
62. TNA: SP12/164/23. The relationship of Mychell to Caryll may not have been the traditional master–servant one. John Mychell was from a well-known landowning family of Shipley and usually styled *'gent'*; WSRO: Wiston Mss 4383/4384.
63. TNA: SP12/167/59.
64. See the account in L. Hicks, 1964, *An Elizabethan Problem: Some Aspects of the Careers of Two Exile Adventurers*, p. 33, London, Burns & Oates.
65. TNA: SP12/167/59.
66. Shelley held property in Queenhithe Ward, St Michael's Parish, London, as he appears there in the 1582 subsidy, so he may well have known Christopher's elder brother, John, who resided in the same Ward. By coincidence, Francis Throckmorton is the first to be named in the subsidy as a resident of Queenhithe Ward [St Peter's Parish], which may be how the two conspirators first became acquainted.
67. Shelley owned the property leased by Christopher Haynes in Climping, a village to the south of Arundel known for its association with smuggling, which property subsequently came into the hands of Shelley's nephew Sir John Caryll and Richard Gawyn, details of which are recorded in a deed of feoffment dated 10 May 1605 (WSRO: SAS/BA/467). Sir John Caryll or Carrell, the son of Edward Caryll, was married to Mary Cotton, niece of William Shelley through his sister, Mary, who was the wife of Sir George Cotton (A. T. Everitt, *Notes & Queries*, 10th S. IV, 15 July 1905, pp. 56-57). He died in 1613 (Sussex Record Society 70). The land in the tenure of Christopher Haynes at Climping passed down several generations of the Gawyn [Gawen] family. It appears to have comprised a messuage and farmhouse in Climping together with several closes of land belonging to the messuage. There was also a 6-acre wood adjoining a parcel of ground called Sixteen Acres, the wood being part of the land known as 'the Hydes'; WSRO: SAS/BA/470.
68. TNA: SP12/164/33.
69. Acts of the Privy Council, vol. 10, 1895, p. 227.
70. Thomas Clynsall lived at Clapham, Sussex, a tiny village adjacent to Patching. After his release from prison, Thomas Clynsall went to live in one of Christopher Haynes's houses. The Clynsall family eventually settled in Barfham near Angmering, where on 4 February 1632/3, Edmund Climsoll married Joan Bennett. After Joan's death in 1644, her probate inventory was valued at £369 15s 2d (mostly in agricultural produce including 340 sheep), indicating a family that had risen to the top of the wealth scale; WSRO: Ep I/29/006/042.
71. TNA: SP12/164/23.

72. Wednesday before Michaelmas; TNA: SP12 164/45.
73. TNA: SP12/164 f 54; Haler returned with *'ix small fardelles of a yard long a pece'* for Isham, which Haynes was told comprised cards and writing paper.
74. Surrey History Centre: LM/488/4; Arundel Borough Archives, F2/1, Bridgewarden's Accounts f.2v.
75. Ibid.
76. TNA: SP12/164/33.
77. On his return from France in September 1583, Robotham was given the job of caring for Charles Paget; in L. Hicks, 1964, *An Elizabethan Problem: Some Aspects of the Careers of Two Exile-Adventurers,* London: Burns & Oates, p. 25.
78. British Library: Lansdowne 49, No 12 f.38, 1586; TNA: E133/10/1629; E133/10/1630; E134/36Eliz/Hil20. In 1586 Thomas Fenner provided evidence that John Young, customer of Chichester, had kept ships for the purpose of illegally exporting all manner of goods including ordnance. He provided evidence of Young's falsification of accounts, including the omission of receipts, which the customer and searcher then split between them. In one year prior to 1586 Fenner accused Young of passing grain overseas while claiming it was travelling between English ports, and of the false cocketting thereof; in one year £313 in duty had been withheld from the customs accounts. He was accused of sharing his ill-gotten gains with his searcher. Young was subsequently arraigned for concealment of customs in Rye and Chichester and falsification of entries in books. In particular, he was accused of collecting custom payments on cloth and then omitting such payments from the receipt books, presumably in order to pocket the money. In one year by his actions the Queen lost £800 of the customs due to her. When there was plague at Rye, all the merchants sent their goods to Lewes, where Young and his servants falsified entries of cloth, etc. shipped overseas. A man called Reve of Queenborough, Kent had a note of packs sent from London to Rye. When he saw the original books at Rye, he found that several packs had been omitted and brought a case against Young. Young with great audacity took money out of the Queen's customs to pay him off to drop the case.
79. TNA: SP12/167/59.
80. M. C. Questier, *Catholicism and Community in Early Modern England: Politics, Aristocratic Patronage and Religion, c. 1550-1640*, Cambridge, pp. 51.
81. Ibid. pp. 51-2. Mychell was mentioned in Caryll's will; TNA: PROB 11/115, fo. 467r.
82. Sussex Archaeological Collections, 1986, p. 195.
83. TNA: SP12/164/23.
84. TNA: SP12/164/21.
85. TNA: SP12/30v; SP12 164/30r.
86. TNA: Will of Christopher Haynes Gentleman of Arundel, 20 November 1586; Windsor: PROB 11/69.
87. TNA: SP12/167/110v.

88. British Library: Lansdowne 49, No12 f.38, 1586; TNA: E133/10/1629; E133/10/1630; E134/36Eliz/Hil20. See also note 78. The Arundel customs technically came under the jurisdiction of the customer of the Port of Chichester; Graham Mayhew (1987), *Tudor Rye*, Falmer, University of Sussex, p.103.
89. TNA: SP12/203/31.
90. British Library: Lansdowne 51, No10 f.20. The death sentence was respited, i.e. temporarily suspended during the Queen's pleasure, but not commuted.
91. TNA: SP12/175/5.
92. TNA: SP12/195/32; L. Hicks, 1964, *An Elizabethan Problem: Some Aspects of the Careers of Two Exile-Adventurers,* London: Burns & Oates, p. 48; although Pollard states that Shelley was executed on 12 February 1586 (A. P. Pollard, 1905, *The History of England from the Accession of Edward VI to the Death of Elizabeth (1547-1603)*, Longmans Green & Co., New York, p. 386), this was the date he was sentenced by Justice Anderson [see A.W. Brian Simpson, 1995, *Leading Cases in the Common Law*, Chapter 2: 'Politics and Law in Elizabethan England: Shelley's Case (1581)', Oxford University Press pp. 13-44]. Cal Dom Eliz 1591-1594, vol 244, 17 Feb 1593, 316-318: W. Betham, 1801, *The baronetage of England or the History of the English baronets etc* London: William Miller, 66-76.
93. Surrey History Centre, LM/1795/1.
94. Ibid. LM/488/5.
95. British Library: Lansdowne 49, No12 f.38, 1586; TNA: E133/10/1629; E133/10/1630; E134/36Eliz/Hil20. See also note 78.
96. Cal Dom Eliz 1581-1590, vol. 150, 23 Sep 1581.
97. It appears that Young's father-in-law, was eventually dismissed by Burghley ['Calendar of Manuscripts of the Most Honorable Marquess of Salisbury', Preserved at Hatfield House, Hertfordshire. 1902, vol. 9 p. 371]; he was removed on 13 October 1599.
98. Surrey History Centre: LM/1631/2.
99. Christopher was left a case of dagges [pistols] and a caliver with powder flask and horn touchbox. The caliver was a larger, trigger-operated matchlock gun – a more advanced form of arquebus, like a musket but lighter and fired a smaller ball. It did not need to be fired from a rest, but was not particularly accurate at more than 40 yards.
100. Mayhew, Graham (1987), *Tudor Rye*, Falmer, University of Sussex, p. 187.
101. See note 55.
102. WSRO: Ep I/11/8. Richard Woolridge then moved to Climping for two more years, before moving to Bosham. IGI: Henrie Woldridge baptised 21 November 1591. It may be of interest that a marriage licence is recorded by the Archdeaconry of Chichester for 13 July 1590 of Henry Woolridge and Elena Haynes (or Heyns), widow of Arundel. 'Calendar of Sussex Marriage Licenses recorded in the Consistory Court of the Bishop of Chichester for the Archdeaconry of Chichester, June 1575 to December 1730', Edwin H. W. Dunkin, 1909, *Sussex Record Society*, vol. IX, p. 15.

103. TNA: Will of Christopher Haynes Gentleman of Arundel, 20 November 1586; Windsor: PROB 11/69.
104. Court of Wards and Liveries, 24 February 1638 [1639]; TNA: WARD 7/89/51 and WARD 7/90/21.
105. When we have come across this phrase elsewhere, e.g. in the will of Thomas Ballard, the testator had always been married more than once.
106. IGI.
107. London Metropolitan Archives: Parish Records of St Nicholas Cole Abbey (St Nic CA).
108. Sentence on the Will of Christopher Haynes of Arundel: TNA: Spencer Quire Numbers, 1-40, 22 February 1587, PROB 11/70.
109. C.P. Haines, *A complete memoir of Richard Haines (1633-1685); a forgotten Sussex worthy, with a full account of his ancestry and posterity*, p. 7.
110. TNA: PROB 11/83, PCC, Dixy Quire Numbers: 1-44; Will of William Tredcrofte Yeoman of Billingshurst, Sussex, dated 10 September 1593: *'Item I will and bequeath unto Christopher Haynes fower children Xli (£10) of lawfull money of Englond to be devided equallie amongst them at the age or ages of xxitie yeres'* ... *'I ordaine and make John Ovey Leonard Richbell Robert Barker myne overseers of this my last will'*. Proved 19 February 1594. A William Tredcrofte is listed as a mariner [Surrey History Centre: LM/488/9]; Thomas Tredcrafte was a mariner at Arundel [Surrey History Centre: LM/1795/2].
111. C.P. Haines, *A complete memoir of Richard Haines (1633-1685); a forgotten Sussex worthy, with a full account of his ancestry and posterity*, p. 7.
112. Wealden timber and wood were among the chief exports mentioned in the record of royal customs collected at Arundel port between the later fifteenth and the mid-seventeenth centuries. It was loaded on vessels at Pallingham Quay in Wisborough Green and transported down the Arun to Arundel. From: 'Arundel', A History of the County of Sussex: Volume 5 Part 1: Arundel Rape: south-western part, including Arundel (1997), pp. 10-101. Drungewick is today in the parish of Loxwood.
113. TNA: Will of Christopher Haynes Gentleman of Arundel, 20 November 1586; Windsor: PROB 11/69.

6

Nicholas Haynes of Hackney, Purveyor of Grain [1529-1594]

Reading - Marriage to Agnes Enowes - Harbinger - Purveyor - Norfolk Wheat Transactions - Arms and the Man - Family Life - Children - Earls Colne - Hackney - Death from Plague

Nicholas was born in Reading in 1529,[1] where he spent his early years, and as a boy or teenager seems to have served in St Giles Church, where his father had been one-time churchwarden, as in the churchwardens' account book for 1545-6, Nicholas Hayne was paid *'ijs wages for the iij qre'*.[2] In what capacity the 15 year-old earned this money we are not told. It may have been for his services as a teacher of one or more boy choristers who had to learn reading and singing, for in the previous year a clerk called Whitborne was paid xijs by the St Giles' churchwardens for the year as remuneration for teaching two choristers *['to teche ij children for the quere']*. We can be fairly certain from entries in the wills of both Richard and Thomas Haynes that St Giles was the regular place of worship for the family.

Nicholas had met and married Agnes Enowes around 1560 and they lived much of their lives in Hackney, Middlesex, where their known children were baptised. Like at least two of his brothers, Nicholas entered into service within the Royal Household, but in a different department. The entry recording the burial of Nicholas Haynes at Hackney Church, describes him as *'purveyour for her mate [Majesty's] grayne'*,[3] a position which reported to the sergeant of the bakehouse.

His career in the Royal Household began when he was about 35 years of age, as Nicholas's liability for taxation in the Royal Household dates from around 1565 before which his residence for taxation purposes was in the hundred of Ossulston (Hackney).[4] He began as one of the small team of royal harbingers which included his friend from Hackney, Hyppolite Lynnet. They served as grooms of the Household between progresses, but as soon as the Queen's itinerary had been decided, the harbingers had the responsibility for riding along the route, in order to identify and reserve suitable accommodation, stabling, animal fodder (mainly oats and hay), bedding straw and associated provisions. Much time would have been taken up in finding beds of suitable quality for the different 'degrees', from Earls to gentlemen. When Cracknell mentioned that Nicholas was *'one of the harbingers sent to make herbage at Henley, Reading'*,[5] this was not as it reads today: it meant that he was to obtain accommodation in that town – the word 'herbage' being derived from the French *herberge* (hostelry; c.f. *'auberge'*); the old English word for harbinger being 'herbergere'.[5]

The logistics facing the harbingers before and during a progress were immense. Once the Queen had decided on the route that her progress would take, the Lord Chamberlain would draw up a comprehensive itinerary. This would then be passed to the harbingers. If the Queen's own houses were to be used, these would be checked by the harbingers and made ready. They would then source the additional accommodation necessary and allocate rooms according to status. More often, the Queen would stay at private estates and castles at their owners' expense and the harbingers would advise the hosts on how to allocate the accommodation and then decide who had to be lodged elsewhere usually in inns of nearby towns. They would then draw up the accommodation list and prepare dockets for each occupant. Harbingers probably looked forward to organising the accommodation in private houses, for it was customary for them to receive generous tips from the nobility and gentry whose houses were chosen – and several such houses would be used during a single progress. The amounts they received could be substantial. They doubtless received similar rewards from the inn-keepers in towns for choosing their accommodation and livery.[6] This was an age when petty corruption and extortion were so commonplace that it would have been difficult to determine whether the average harbinger drew the line at accepting tips only for services rendered, or crossed the line to promise services only if tips were forthcoming.

Between four and six hundred carts with draught horses were required for the Queen's furniture, equipment, catering and other supplies for the whole Court. Most of the transport was then commandeered from farmers and others by the royal cart-takers at frequent stopping points on the actual journey. Cart-taking was part of the royal prerogative, which also allowed the commandeering of horses, mares and draught animals[7] and these also had to be fed and watered. Simple mathematics confirms there must have been in excess of a thousand draught horses in addition to those used by the Queen and her retinue. Towards the end of Nicholas's time as harbinger, the royal stables housed around 270 riding horses and Court members would have ridden their own animals. The royal stables had a generous allowance of about 16 quarts (half a bushel; about 17½ litres) grain per horse per day, so if this applied during a progress, 1000 horses would consume some 500 bushels or 4000 gallons a day, 28,000 gallons a week for the duration of the progress. The organising of feed on this scale was a considerable logistic achievement.

The harbingers would also note the condition of the roads that had been chosen and report back to the surveyor who would organise any necessary repairs or surfacing by the local parish or borough.

Nicholas was a yeoman harbinger throughout the seventies and worked on the lengthy progresses to Kent, Gloucesterhire, Norfolk, Somerset (Bristol), Staffordshire, Wiltshire and Worcestershire. Their work was not without its frustrations as the Queen was prone to changing her plans. On three occasions after Nicholas' appointment, the harbingers made arrangements for the Queen's progress to Leicester and had identified accommodation, supplies and transport which had then been purchased, only to find the arrangements cancelled at the last minute.[8]

Right from the start Nicholas appears to have enjoyed the rank of yeoman and within a year or so he was attracting the benefits of his office including a daily allowance of 6d. On 9 May 1566 in the Patent Roll for Kent appears: *'The Queens Majestys pleasure signified by Mr Secretary is that Clement Norrys, Hippolit Lynnet, and Nicholas Haynes, III of Her Majestys harbingers, shall have a lease in reversion'*. This lease was subsequently enrolled on 5 July, when the three '*Yeoman Harbingers*', were jointly granted a 31 years' lease on properties, all part of the manor of Raynehurst in Chalke, near Gravesend, Kent, by letters patent for their service, *'their good and true services upon us heretofore done and bestowed'*.[9] The properties, all of which were subleased for variable unexpired periods, carried yearly rents totalling £41. One was a tenement called Filborowe [Filborough (or Felborough) Farmhouse] in Eschalke [East

Chalk], a farm known as Clamelane [Clamlane] in Westchalke [West Chalk], a tenement also called Clamelane in Westchalke and an unnamed tenement in Westchalke. All of these parcels of land had been earlier in the possession of Sir Henry Wyatt, then his son Sir Thomas Wyatt of Allington Castle, who granted the manors of Raynehurst and Tymberwood with all his other lands in the parish of Chalk to the King in exchange for other properties.[10] Nicholas was to be joined in Raynehurst, Chalk, by his elder brother, John, the following October (11 October 1566), so between them the four men must have held most of the manor. In May 1582 the Raynehurst properties of both Nicholas and John were relinquished in favour of Sir Amyas Powlett [Paulet][11] and Reynold Knatchbull respectively.[12] However, on 21 January 1587 Hippolite Lynnet and Nicholas Haynes successfully petitioned the Queen for another lease in reversion of lands of the manor of Raynehurst, valued at £32.[13]

On 3 April 1584 mention is made that Nicholas Haynes, Queen's harbinger ['*one of lez herbengers*'], had requested the Queen that a 21-year lease of the rectory of Owston, Yorkshire, be granted to William Adam.[14]

Nicholas was a harbinger for more than 20 years when out of the blue he was appointed yeoman purveyor of wheat. Throughout the dangerous eighties the Queen made no progresses until the defeat of Spain in 1588 and Nicholas's change of post may have been related to this. The files of the Exchequer contain records of the tax status of royal employees as they were liable for taxation in the Royal Household rather than in their actual places of residence. This generally meant a reduced rate for all subsidies, and the documents show that of Nicholas being based on the rate of £5 in fee. However, whenever a payment was due, Nicholas had to produce a certificate of residence signed by the Lord Treasurer, comptroller and cofferer identifying his name, position within the Household and tax status. Several of these subsidy documents have survived and in them we can see that Nicholas's change of title from harbinger to purveyor took place some time between 15 October 1586 and 30 September 1587.[15]

On Tuesday, 4 May 1591 the Queen embarked on one of her short progresses from Greenwich. Her first stop was Hackney where she remained for six days. Nicholas would almost certainly have been present, and this time he could have enjoyed the spectacle with his family, no doubt thankful that he would no longer have had to sort out the accommodation. On Monday 10 May the Queen left for Theobalds and then took in Theydon Bois, Havering, Luxborough and Leytonstone, before returning to Greenwich on 21 May.[16]

Purveyor of grain

It was therefore not until fairly late in his working life at the age of 58 that Nicholas became a purveyor of wheat, one of four yeoman purveyors of the great bakehouse for the ordinary household. His wages were about £5 a year plus 4d to 6d a day for board wages.[17] This took place in 1587[18] on the death or departure of three long-serving purveyors, including Bennett Byshley and Richard David. Byshley was replaced by Thomas Langley and David by one Boade.[19] Nicholas began by helping Erasmus Skidmore in Sussex and within a year he was recorded in an account of wheat to be furnished by composition,[20] as working in Sussex and Norfolk, two of eleven counties from which wheat for the household was purchased – Berkshire, Essex, Hampshire, Hertfordshire and Kent being the biggest providers.[21] Nicholas reported to Edward Jewkes, sergeant of the bakehouse, who had succeeded Thomas Fisher[22] and who was still heading that department in King James' reign. There were also two additional yeomen, the *'purveyors of the queen's mouth'*, who purchased wheat only of the finest quality for the Privy Bakehouse to make bread for the Queen's table. In Nicholas's time these were Hercules Turner and George Gates.[23] The primary function of the bakehouse was to make bread, and wheat was therefore the main commodity it purchased; oats and barley were also bought. Wheat was needed in huge quantities, and Nicholas and the other purveyors had to visit the county centres to which the grain was delivered by the farmers in order to assess quality and quantity, and to arrange for its transport. In 1577 over the year, 3450 quarters were purchased[24] at 6s 8d a quarter [total cost £1150]. The grain was kept in the garners (*'Garnars for the Q[ueens] p[ro]vis[io]nes'*; granaries; Edward Jewkes had been the garnitor in the 1570s) including one at Whitehall where it was delivered by boat[25] and at the '*Mewse*' where 250 quarters were delivered across country (Hertfordshire and Middlesex wheat only).

In arable areas the grain purveyors were as unpopular as the fish purveyors in the fishing ports and also suspected of corruption. This was because they delayed payment for about a year and sometimes didn't pay at all on the pretext that the grain had been spoiled by mould or vermin.[26] According to a letter to the Privy Council dated 8 June 1586 from the Justices of Hertfordshire,[27] the grain purveyor in that county had used a bushel measure so large that he gained a bushel of oats [about 12.5 per cent] more than he was paying for in every quarter. Furthermore, when the farmers were eventually paid, the purveyor withheld 6d out of every pound and was about to increase it to 12d [5 per cent], otherwise he told them they would have to seek their payment directly from the Court, a physical impossibility for almost all of them. Added to this were considerable fluctuations in grain prices which compounded the suppliers' problems, as the producer when paid late, was parting with grain worth much more than the cheap price agreed earlier with the purveyor.

Nicholas Haynes became the purveyor of wheat in Norfolk and some of his signed dockets for his transactions in the north and west of that county have survived. In Norfolk the chief constables and others in receipt of money in relation to a public service had to account in person for all receipts including those for wheat, oats or wax for the Queen's provision, at a date specified at the *'signe of the Crowne in St Stephens Street'*, Norwich: those relating to wheat sales to Nicholas Haynes, would be among them.

Nicholas's purveyance work in Norfolk coincided with the government's attempts to stamp out corruption among purveyors and to reform its buying procedures. As mentioned in Chapter 4, in 1588 the Queen had commanded the setting up of a 'Commission for Household Causes' to take evidence on oath of *'intolerable abuses of pourveyers, for the howsholde in takinge and convertinge to their own uses, great quantities and numbers of p'vicions wth manye other disorders ...'*. Parliament passed the Bill in February 1589, and the infringment of some of the new purveyance laws carried the death penalty. Some purveyors were actually tried in the counties in which their abuses had occurred and were executed;[28] others were imprisoned.

The Commissioners appointed were Christopher Hatton, Chancellor, William Burghley, John Fortescue and Thomas Buckhurst, who on 29 July 1591 had written to the JPs in Norfolk, but having received no response, sent a reminder eight months later. The recipients had to swear in four or six *'honest substantiall men of everie parishe'* to certify the provisions provided to the household purveyors over the preceding two years, giving the '*purveyors names that took the same & from whom, at what tyme in the year, & for what price, & what remaineth due for anie provicions taken*'.[29] They had also been asked to report on misdemeanours and abuses committed at fairs, markets and elsewhere by any royal purveyor.[30] The names were also demanded of those who had earlier failed to respond – clearly a threat meant to elicit a response – which it certainly did. Certificates poured in within a month and those that have survived provide a good deal of information on Nicholas's purchases and payments, or rather his delay in payment. Thus, the Chief Constables of the Hundred of Brothercross certified that on 15 May 1587 3qrs, 5 bushels, 1½ pecks (8 quarts = 1 peck; 4 pecks = 1 bushel; 8 bushels = 1 quarter) of wheat had been delivered to Nicholas Haynes at Cley at 36s a quarter, '*wherof he payd unto us but the sum of £4 8s 0d*' i.e. 24 shillings a quarter. The remainder was '*in his handes d[e]we to the Hundreth as apereth by his docket the sum [of] 44 shillinges*'.[31] The docket signed by Nicholas in acknowledgement of this wheat was attached to the constables' certificate along with a note recording the payment on 17 July 1590 of £4 8s (i.e. a rate of 24s a quarter), leaving 44s owed by the purveyor. It would thus appear that Nicholas was only prepared to pay the market price at the time of payment, rather than at the time of purchase. This last date was Nicholas's next visit to Norfolk when at Cley he purchased 3 qrs, 1 peck of wheat at 20s a quarter, the price being 4s less a quarter than the then current market price.[32] The purveyor paid for this in full (£3. 0s. 7½d), but not for the 5½ qrs, 3 bushels and ½ peck he purchased at Cley four days later at the same discounted price, for which payment remained due at the time the return was made to the Commissioners. The Chief Constables of the Hundred of Holt certified that, on 17 July 1590, 6 qrs, 5½ bushels of wheat had also been delivered to Nicholas Haynes at 20s a quarter, at a similar reduction on the market price to that delivered at Cley. No payment was made on that occasion either, being still outstanding almost two years later.[33] On 13 March 1592 the Chief Constables of the Hundred of South Erpingham certified that Nicholas Haynes, purveyor of wheat to the Queen's Household had, on 17 July 1590, taken up 10 qrs, 6 bushels of wheat at 20s, for which payment remained due.[34] In North Erpingham that day 7 qrs, 3 bushels of wheat were bought by Nicholas Haynes for 16s a quarter. There were two other purchases made by Nicholas on 17 July 1590:

one was in the Gallowe Hundred on 17 July 1590 when 8 qrs, 4½ bushels were purchased at 20s instead of the market price there of 22s a quarter; the second was in the Hundred of Smithdon, when 9½ qrs were bought at 19s a quarter, an even bigger discount on the market price of 24s a quarter.[35] Again, payment had not been made by the time of the return to the Commissioners on 31 March 1592. These returns show that of the amount due by Nicholas Haynes to the farmers of this part of north Norfolk for wheat purchased between 25 May 1587 and 17 July 1590 was £32 6s 6d, some 81 per cent of the total discounted price, and this was still owing in March 1592.

The above practice of Nicholas Haynes accords with that of the Green Cloth. It was always short of money, and this led not only to delays in payment, but also to the practice among some purveyors of lending the Household money at interest while delaying payment still further. It has to be said that the delays in paying the Norfolk suppliers are on the long side even for a Court strapped for money. The prices paid per quarter for some of the above-mentioned wheat, particularly that Nicholas purchased in 1587, were very high. These prices reflected the poor harvest of 1586 following which prices increased steeply and not because Nicholas was purchasing grain of the finest quality destined for the Privy Bakehouse to bake bread for 'the Queen's mouth'.

It was the Court's policy to pay less than market price for all provisions, and Nicholas Haynes's practice in this regard was entirely normal. The purchasing policy was well documented and the signed receipts for purchasing and payment were essential aspects of this. Nicholas's paperwork would appear to withstand scrutiny in that it tallied with that provided by the Chief Constables and appended to their reports. Indeed the lack of urgency of the Norfolk sheriff and JPs in responding to the Commissioners' requests suggests that, although the suppliers did not like being paid less for their produce than its true market value, in the case of grain its quality and market prices were so variable, that the discount required by the purveyors was 'lost' in the market variance, and the discount was less in Norfolk than in some other counties such as Hertfordshire.[36] Payment in arrears, although undesirable, could be acceptable to long-term suppliers after they had survived the delay of the first payment. In any case the JPs had found that the Household burden could be spread by assessing each farmer for a number of bushels, rather than levying a monetary tax. In this way the load was spread so than no individual had to carry a disproportionate share of the provision. In general, therefore the counties were satisfied with their wheat composition agreements. The JPs had to deliver the grain to specified places in the county, as noted above for Norfolk, and the purveyor then had to organise its carriage. There is no indication that the Commissioners found Nicholas Haynes to be guilty of any abuse or misdemeanour; indeed on 1 December 1591 he was again given a commission of provision to take up wheat for Her Majesty's Household[37] and he remained in office until his death. In this respect at least he was to differ from his brothers.

Arms and the man

As mentioned in the Introduction, Nicholas had the quartered arms of Haynes and Foxley confirmed on him by the Clarenceux King of Arms in 1578.[38] [Figure 1.2, p.10] This clearly came about as the result of his application to the herald and Nicholas would have been required to provide information in support, although such information was not often onerous or verifiable, an assertion that the arms on a seal or piece of silver allegedly owned by a forebear would generally suffice. His brother William made an identical application at the same time.

In the arms granted to Nicholas, there are two small sable-coloured martlets (birds), one in the centre of the quartered shield, the other on the heron's breast. These were the differencing marks for Nicholas as fourth son. The Haynes crescents, the two upper being smaller than the lower, are two-toned red and blue. According to Clarke [1824] the arms of Haynes of Reading were gules, three crescents paly wavy argent and azure,[39] i.e. the tones of the crescents were silver and blue. This was therefore a deliberate differentiation from the original on the part of the Herald, analogous to the shortened bars instead of the original full-width bars in the Foxley quarters.

Exemplification of arms must never be accepted without question. For some years towards the end of Elizabeth's reign, the heralds were suspected of selling coats of arms. Sir Thomas Smith in *De Republica Anglorum* (1551, published 1583) sneered at the individual who managed to '*be taken for a gentleman*' by obtaining arms from the heralds *'for money'*.[40] Robert Greene also showed contempt against those whose: *'own conceit was the Herald to blazon their descent from an old house, whose great grandfathers would have been glad of a new cottage to hide their heads in'*.[41] After Elizabeth's death, this practice became so widespread that bogus heralds set themselves up to cash in on the business.[42] King James created more than 800 new knights for payment, rather than based on services in battle, and he introduced baronetcies for £1095 each. In 1577, the year before Nicholas's confirmation of arms, William Harrison (quoting Thomas Smith almost 30 years earlier) pointed out that when a man such as a lawyer, doctor, university graduate, or other who could live without manual work, was able to *'bear the port, charge, and countenance of a gentleman, he shall for money have a coat and arms bestowed upon him by the heralds (who in the charter of the same do of custom pretend antiquity and service, and many gay things thereunto, being made so good cheap), be called master (which is a title that men give to esquires and gentlemen) and reputed for a gentleman ever after ....'*.[43]

However, there was no obvious reason why Nicholas or his brother would wish to acquire these new arms so late in their lives for their own personal benefit; certainly not fabricated ones. The original Haynes arms of Nicholas Haynes are to be found illustrated in the Hackney Archives;[44] they were argent three crescents barry wavy azure and gules. In the 'Exemplification of Arms' for Nicholas, the description was: argent three crescents unde (wavy as *in paly wavy*) azure and gules; yet in the illustration,[45] the Haynes quarter was presented as: argent three crescents azure, each charged with two barrulets wavy gules, much closer to the original depicted for Nicholas. The quartering for William is argent three crescents paly wavy azure and gules.[46]

Nicholas and William were entitled to bear the arms of Haynes and it is unlikely that the new arms quartered with those of a family whose last male, and illegitimate, descendant, Thomas Foxley, had died 142 years earlier, would be considered more prestigious in the eyes of their peers in the Royal Household or those with whom they did business. So what was it that drove the brothers to differentiate their arms in this way? Was it simply to distinguish themselves from others in the Household called Haynes? This seems unlikely as their job titles would have served this purpose. If it was because of a perceived advantage to their descendants – the most plausible explanation – then as William's only recorded son, William, had died three years earlier, he must have had other children, or grandchildren, living in 1578. Without any obvious advantage, it is less likely that this was simply a case of 'cash for honours'.

It was established in Chapter 2 that Nicholas's mother, Thomasine, was born Thomasine Folkes, daughter of John Folkes who was unrelated to the Foxley family. Her mother however was Maude More, who could probably claim distant descent from the Bray Foxleys. If we assume that such a claim was the basis for the herald's decision to confirm the new arms on Nicholas and William, then the herald must have used his words economically; '*daughter*' of John Foxley simply meaning a *female descendant* of that individual. As mentioned in Chapter 2 (note 30), coats of arms with titles that lay dormant could be revived in favour of a distant heir. Even though the line had died out, as in the case of the Foxleys which ended with the death of Thomas in 1436, the coat could be modified by the herald and reassigned. Certainly the 'Foxley' quarters of the arms confirmed on Nicholas and William were such deliberate modifications of the original Foxley arms, which is consistent with them being reassigned.

Family life

Nicholas Haynes married Agnes, daughter of Richard Enowes of Earls Colne, Essex, then a servant of the Earl of Oxford, around 1555–60. Agnes went by the name Ann, being the abbreviation for Annis, the anglicised form of Agnes pronounced in the French way, '*Anyaise*'. In the sixteenth century Agnes and Ann were interchangeable. Four children of Nicholas and Ann are known from documentary evidence, Thomasine, Joan, Richard and Alice, but the poor state of preservation of the Hackney parish records prevents the confirmation of others including Robert, Christian and Nicholas, for whom the evidence, to be discussed later, is circumstantial. It is therefore impossible to say how many children Nicholas and Ann actually had.

Ann Enowes was born around 1540 in Earls Colne, Essex. Her mother's Christian name is not known although she was a daughter of John and Alice Copyn.[47] Members of the Copyn family can be traced back at least three more generations in Earls Colne and environs. Ann's mother seems to have been a colourful character. As

Richard Enowes' first wife, she was before the manorial court as '*a common baker and brewer within this lordship*' who had '*broken the assize*' (i.e. the regulations governing weights, measures, quality and prices of food and drink); in 1534 and 1537 she was fined 2d for being a '*common tipler of bread and beer within this lordship*'.[48] Richard Enowes led an equally colourful life.[49] The Enowes' home in Earls Colne from 1544 was a messuage with gardens and orchards called Littmans and Giffords. Littmans, a timber-framed building dating from around 1500, still exists as numbers 94 and 96 on the south side of the High Street; it was listed Grade II in June 1962.

The Enowes (Enos; Ennew), a family of Dutch ancestry, seems to have had their centre at Coggeshall; some were clothiers or in related trades (fullers, tailors, etc.) in the sixteenth and seventeenth centuries. With the influx of Dutch immigrants into north-eastern Essex, mainly refugees from the Spanish occupation of the Low Countries, considerable numbers were employed in spinning and weaving cloth for export all over Europe. Richard Enowes was initially in the service of John de Vere, 16th Earl of Oxford, until about 1560, but later was involved in brewing and farming.[50] In December 1548 he was described as '*Richard Enowes, Yeoman*' and he was still so designated in his will of 25 January 1589, written about a week before his death.[51] He appears in a huge number of notices, some of them involving the more 'delicate' aspects of the life of the earl.

Ann was identified in the Visitation of London 1568 as Anne Enowes[52] along with four of their children: '*Richard, Thomasin, Johan, Alis*'. Her children Richard, Jane [Joan] and Alice were all mentioned in the will of their uncle Thomas Haynes (1582). Thomasin occurs only in the Visitation of London, 1568. Possible fifth and sixth children, John and Robert, were born much later than the others. Nicholas Haynes is recorded as father at their baptisms not in Hackney but at St Clement Danes, Westminster on 9 March 1577 and 12 April 1579 respectively. However, these children, if legitimate sons of Nicholas Haynes of Hackney, appear to have been unknown to (or ignored by) their uncle Thomas, as they weren't mentioned in the latter's will in 1582, or in those of Christopher and John. It is possible that Nicholas the father of John and Robert was the son of Nicholas Haynes of Hackney, making John and Robert the latter's grandsons. There is therefore considerable room for doubt that John and Robert were the legitimate children of Nicholas and Ann Haynes of Hackney. Another person who could have been of Nicholas's family is Christian Haynes, possibly named after Nicholas's only sister. She married Rowland Coytemore of Wapping, a mariner, at nearby St Dunstan's, Stepney on 13 January 1591, so it can be assumed that she was born round about 1570. She died before 1595, possibly in the 1593 plague epidemic, as her husband married a widow, Dorothy Harris, at St Mary's Whitechapel, on 28 March 1595.[53]

Hackney

Nicholas and Ann lived in Hackney for much of their married life. Their children were born there. It seems to have been the fashionable place to be. Nobles and courtiers frequented Hackney in Tudor and early Stuart times, and monarchs visited the parish. Margaret, dowager Countess of Lennox, daughter of Margaret Tudor, Queen of Scotland (wife of James IV), lived in Hackney where she died suddenly on 9 March 1578, following an agreeable supper entertaining the Earl of Leicester. The Earls of Warwick and Oxford, Sir William Brooke, Lord Cobham, the Queen's Spanish Ambassador and Edward, Lord Zouche were among the peers who held seats in Hackney. The last created a physic-garden there. Samuel Pepys admired girls at the fashionable schools and Daniel Defoe mentioned the opulence, said to surpass that of any village in the kingdom. In short, Hackney was a healthy, accessible, desirable retreat, and it may be that Nicholas felt the need for his aristocratic background and the pressure to seek confirmation of his family's illustrious past. However, Hackney attracted the less desirable elements, from which Nicholas was not immune, for on 10 June 1576 one Agnes Woorte or Wort,[54] formerly of London, spinster, was recorded as being pardoned for the *'theft of clothing belonging to Nicholas Haynes of Hacney, co Middx'*. This was a fortunate outcome for her, as such theft was a capital offence, especially for someone with a previous conviction, which she had. Two years earlier, being over 14 years old and having no lawful means of livelihood, she with four others had been found guilty of vagrancy at the Middlesex Sessions and sentenced to be severely flogged and burnt (i.e. branded) on the right ear. [55]

On 26 May 1585 an entry in the Court Rolls for the Manor of Kingshold, Hackney recorded that *'Henry Offley [Offeley], Esq., has lic[ence] to demise to Nicholas Haynes, gent.*'[56] To what this referred remains to be discovered, but Henry Offley (son of Sir Thomas Offley, Merchant Tailor and former Mayor of London; d. 1582) possessed Hasting's Mead, otherwise known as Jerusalem Close (containing an estimated 16 acres, with appurtenances formerly in the possession of the Hospital of St John of Jerusalem) and St John's Mead in Hackney and Bishop's Egney and South Egney in Stepney, all originally granted to John Cokk by King Henry VIII.[57] Five years earlier, Henry's father Thomas Offley, then Mayor of London had been pressed by the Queen to lease the *'house'* called The Stocks in The Poultry, London, to William Haynes, Nicholas's brother. This might have been a similar situation, with pressure on Henry to demise coming from on high.

Nicholas's last house was in Mare Street,[58] the main road from the south where most Hackney residents then lived, and consisted of an acre of pasture, main house, stables, gardens, orchards and additional land. Nicholas purchased the property in 1586 from Sir George Phillipott [Philpot] who had inherited it[59] on the death of his father, Thomas Phillipott, that year. A hundred years earlier the lands, part of the manor of Hoggeston [Hoxton] in Hackney, had been held by John Philpot who died in possession of it in 1485, his son, John being his heir.[60] It was held of the then Bishop of London in free and common soccage by a quit-rent of 12 shillings. By Nicholas's time the superior landlord was Thomas, Earl of Cleveland, who owned the manor of Stepney. Their home was destined to house at least two more generations of Nicholas's family.[61] Mare Street was a distinct settlement in 1586, probably with houses near the Well Street junction, possibly including the inn known later as the Flying Horse. The residents included moneyers and merchants who worked in London. Fellow Royal Household yeoman Hippolitus [Ippolite] Lynnet [the son of George Lynnet, constable of Hackney], who, with Nicholas, was granted the 31 years' lease on part of the Kentish manor of Raynehurst, lived nearby in Homerton. In 1589 Lynnet was involved in a dispute with Edward Elmer over land supposedly in the manor of Grumbolds, Hackney, whose boundaries were queried by Lynnet. This led to a lengthy suit in which Lynnet on one occasion was accused of perjury. It appears that the argument centred on a brook which seems to have served as a boundary that had been stopped up and diverted. Of those called to answer interrogatories, one was *'Nicholas Heynes of Hackney, Middlesex, gentleman'*, aged

59, the only reference still extant enabling an accurate determination of Nicholas's year of birth.[62]

Hippolite Lynnet and Nicholas Haynes had more in common than work and place of domicile, for each of them had a daughter who married a son of John Kay and Dorothy Mauleverer of Woodsom, Yorkshire. Lynnet's widowed daughter Ellin Welshe, married Arthur Kay who lived in Hackney, and Nicholas's daughter, Joan, married Arthur's younger brother, John,[63] who also resided in Hackney parish. It is likely that this John Kay was the person of that name listed as clerk comptroller of the Green Cloth from 27 August 1585[64] and Clerk of the Green Cloth from 25 April 1588. John Kay of Hackney esquire, died in 1589 and his will dated 7 May was proved on 22 May 1589.[65] Another important Household resident of Hackney was Thomas Wood, sergeant of the pantry, who lived in Clapton House, near Clapton Pond.

Ann and Nicholas are mentioned in connection with property in Essex granted on Tuesday 18 April 1581, by her father Richard Enowes and stepmother, his third wife Edith, to Ann's brother, Edmund for life, with the reversion to Ann.[66] The property comprised an aldercarr called Gossfenn and three crofts of land in Colne Engaine and White Colne called Colecroft. On Edmund's death, the property reverted to Ann whose holding was held by her husband, Nicholas. After Nicholas's death, it was inherited by their son, Richard:[67]

> 'Rich Haines gentleman son and heir of Nich Haines deceased to the moiety of one aldercarr called Gosses Fenn and three crofts of land lying in the parishes of Colne Engaine and White Colne called Cotecroft lying contiguous between the queen's highway leading from Colford Hill to Colne Engaine and land belonging to the Manor of Sheriffs and also one croft of land formerly Jn Bridge and lying upon Colneford Hill to hold to the said Rich and his heirs in fee'.

As was not uncommon for married daughters who had received gifts in their fathers' lifetime, Ann was not remembered in her father's will in January 1589.

She seems to have taken an active interest in the Essex property as, according to the Colne Priory Manorial Court Roll of 24 September 1599, a *'licence* [was] *granted by the lord to Agnes Haines widow to fell and stub up two old decayed oaken bollingers or pollingers'*.[68]

Unlike fellow yeoman, Hippolite Lynnet and kinsman by marriage, Arthur Kay, Nicholas does not appear listed as a member of the Vestry of St John at Hackney although it was probably he, mentioned only as Mr Haynes, who was the subject of a memorandum recorded in the Long Memorandum Book in April 1593[69] as having retrieved a legacy left 17 years earlier by a Dr Huett to St John's church for £40. He was paid 3s 4d for his services in this regard. The legacy was put to various uses including the care of children, gravelling Kingsland Road and repairs to the church in which Nicholas was actively involved, especially in supervising the leadwork to the roof of the south chancel.

Nicholas died in January 1594 in Hackney aged 63, and he was buried at St John at Hackney, his parish church. From September 1593 to January 1594, the plague had passed through Hackney and according to the list in the parish records, had claimed 269 victims[70] and it is acknowledged that only the more notable adults were registered. Some 15,000 people had been killed by the disease as it swept through London during 1592/3. Nicholas is not mentioned as dying from the plague, but he appears just after the end of the list of plague victims, so is likely to have. His family was certainly not immune as at least two of his grandchildren died from the disease in 1607. His entry in the Parish register reads: *'Nicholas Haynes purveyour for her mate [majesties] grayne was burried the xxijth Daye of January Ao 1593'* [22 January 1594].[71] Probate dated 25 January 1594, was granted to his son Richard,[72] 'natural and lawful son of Nicholas Haines', for the *'administration of the goods, rights or credits of the said deceased'*. Richard was not mentioned as his father's executor, so Nicholas may have died intestate, an event consistent with him succumbing rapidly to the plague. The four years following Nicholas's death were characterised by excessively heavy rains, floods and crop failures. Bridges were washed away in many places and the price of grain and other food rose in consequence.[73] Throughout this time Ann lived in the Mare Street house while Richard and his family lived in nearby Cambridge Heath.

When Ann Haynes died in 1605, against the entry for her burial in the register at Hackney on 22 June was written *'Greate Bell'*.[74] The passing-bell, soul-bell or saunce-bell was believed to be intolerable to, and to ward off, evil spirits. To the superstitious, the bells were rung to keep these malicious spirits away from the souls of the deceased as they made their way from this life to the next. More appropriately, it also called for those who heard it to pray for the soul of the departed.[75] The church charged a much higher fee than the usual 'bell-penny' for ringing its great bell, as being louder, the evil spirits must go further away to be clear of its sound; besides, being heard further off, it would procure the dying person a greater number of prayers.[76]

Notes and references

1. He was 59 in 1588; TNA: STAC5/E10/7 (30 Eliz).
2. 'The Church-Wardens' Account Book for the Parish of St Giles, Reading', Part I (1518-1546), transcribed from the Manuscript by W. L. Nash. The omission of the 's' from the Haynes surname was quite usual and even Thomas Haynes referred to his brother as Nicholas Hayne.
3. Hackney [St John] parish register: *'Nicholas Haynes purveyour for her mate [majesties] grayne was burried the xxijth Daye of January Ao 1593'* [i.e. 22 January 1594]. It appears that the abbreviation 'mate' had been wrongly transcribed as malt; Cracknell [note 5].
4. TNA: E115/205/4.
5. E.M.W. Cracknell, ed. (1934), *The Barbadian Diary of Gen. Robert Haynes 1787-1836*, The Azania Press, Medstead, Hampshire, UK [The word 'herbage' was from the French *herberge* (hostelry); the old English word for harbinger being 'herbergere'. As with message and messenger, passage and passenger, the 'n' intruded much later].
6. See e.g. C. Manners Rutland, J.J.R. Manners Rutland, R.Ward and J.H. Round 1905, 'The Manuscripts of His Grace the Duke of Rutland etc.', preserved at Belvoir. Records of the Borough of Leicester: HMC, HMSO, Vol. 4, p. 431; *'to the harbingers, xs'*.
7. Allegra Woodworth, 1946, 'Purveyance for the Royal Household in the Reign of Queen Elizabeth', *Trans Amer Phil Soc* 35, part 4, p.71.
8. Ibid. p. 13.
9. Cal Pat Rolls Eliz Vol. III, 1563/6, 2097 (p. 371); Arch Cantiana 21, 1895, p.166.
10. W.H. Ireland, 1829, *A New and Complete History of the County of Kent; etc,* London, C. Virtue, pp. 212-13. The fees-simple of these manors and estates, which were granted for several terms to different people, remained vested in the Crown until 1630, when Charles I passed them to the City of London.
11. Cal Pat Rolls Eliz Vol. IX, 1580-1582, 2125 (p. 290): Amyas Paulet (1536-1588), the son of Sir Hugh Paulet and Phillipa Pollard, was Mary Queen of Scots' last jailer. Paulet, a Puritan, disliked Mary intensely. He prevented her outings to Buxton, stopped most of her correspondence, and broke into her apartments while she was lying ill in bed and confiscated her money on Queen Elizabeth's instruction. He was entrusted with several

letters from Mary to Elizabeth, but he delayed their dispatch in case Elizabeth was minded to revoke the death warrant. Paulet was present at Mary's execution and was knighted after it.
12. Cal Pat Rolls Eliz Vol. IX, 1897, mm. 36-7 (p. 258).
13. 'Historic Manuscripts Commission: Calendar of the Manuscripts of the Most Hon. The Marquis of Salisbury, KG, etc.', 1889, London HMSO, 1883-1976, p. 304.
14. TNA: C66/1237 m.37-38; subject to reservations of 13s 4d and 6s 8d from the church of Owston to be paid to the Archbishop and Dean of York respectively. The cost was £26 13s 4d payable to the exchequer, the yearly rent being £17 16s 8d.
15. Nicholas's taxation certificates: TNA: E115/205/4 (1567 & 1571); E115/197/75 (1586); E115/187/95 (1587); E115/200/10 (1589).
16. 'Calendar of the Manuscripts of the Most Hon. the Marquis of Salisbury', Preserved at Hatfield House, Hertfordshire.Vol. 4, entry 262: 1590-1594.
17. R. Folkestone Williams, 1660, *Domestic memoirs of the Royal Family and of the Court of England*, Vol. 2, London: Hurst & Blackett, p. 131.
18. TNA: E115/197/75 (1586); E115/187/95 (1587).
19. Allegra Woodworth, 1946, 'Purveyance for the Royal Household in the Reign of Queen Elizabeth'. *Trans Amer Phil Soc* 35, part 4, 1-89.
20. British Library: Lansdowne 56, No.27 f.79.
21. A. Woodworth, 1946, 'Purveyance for the Royal Household in the Reign of Queen Elizabeth'. *Trans Amer Phil Soc* 35, part 4, 1-89.
22. British Library: Lansdowne 34, No.31 f.86.
23. British Library: Lansdowne 46, No.91 f.91.
24. TNA: SP12/120, 29.
25. Three hundred quarters were kept there in winter and 250 quarters in summer; TNA: SP12/120, 29.
26. A. Hassell Smith and G. M. Baker, eds, 1990, *The Papers of Nathaniel Bacon of Stiffkey*, Vol. III, Centre of East Anglian Studies and the Norfolk Record Society, Vol. LIII, 1586-1595, p. xix. In fact, the wheat lost each year in faulty weighing, cleaning and spoil by vermin was in 1576 eighty-nine quarters worth £29 13s 4d; The notional 1s 9½d daily allowance set out in the Queen's book in 1576 was in reality almost doubled (3s 3d) that year; J. Nichols, 1823, *The Progresses and Public Processions of Queen Elizabeth*, London: John Nichols & Son, p. 45.
27. British Library: Lansdowne 46, No.92. ff.193-193v.
28. Chapter 3 note 80: At the sessions at Newgate held in January 1889, Nicholls one of the Queen's purveyors was sentenced to be hanged for infringing the law prohibiting the taking, buying, or commandeering carriage for the same, not contained in his commission; or for taking more than was delivered to the Household, or for taking goods for which payment was not made; E. Coke, 1797, *The Second Part of the Laws of England: Containing the Exposition of many Ancient and other Statutes*, Vol. II, London: E & R Brooke, p. 545.
29. A. Hassell Smith and G. M. Baker, eds, 1990, *The Papers of Nathaniel*

Bacon of Stiffkey, Vol. III, Centre of East Anglian Studies and the Norfolk Record Society Vol. LIII, 1586-1595, pp. 142-3.
30. Ibid. p. 325.
31. Ibid. pp. 144-152.
32. Ibid.
33. Ibid.
34. Ibid.
35. Ibid.
36. Allegra Woodworth, 1946, 'Purveyance for the Royal Household in the Reign of Queen Elizabeth'. *Trans Amer Phil Soc* 35, part 4, 1-89.
37. Cal Dom Eliz, vol. ccxl, p. 136.
38. 'The Visitation of Berkshire 1665-6', Harleian Society Vol LVI, p.144, 1907.
39. W.N. Clarke, 1824, *Parochial Topography of the Hundred of Wanting with other Miscellaneous Records relating to the Co. of Berks*, Oxford, p. 34.
40. Quoted by P. Erlanger, 1967, *The Age of Courts and Kings, Manners and Morals 1558-1715*, Weidenfeld & Nicolson, London, p. 70.
41. R. Greene, 'A Quip for the Upstart Courtier', Sig. A[iii] v, quoted by G. Greer, 2007, *Shakespeare's Wife*, London, Bloomsbury, p. 204.
42. P. Erlanger, 1967, *The Age of Courts and Kings, Manners and Morals 1558-1715*, Weidenfeld & Nicolson, London, p. 163.
43. William Harrison, 'Description of Elizabethan England, 1577' (from *Holinshed's Chronicles*): Chapter 1, 'Of Degrees of People in the Commonwealth of Elizabethan England'. http://www.fordham.edu/halsall/mod/1577harrison-england.html#Chapter%20I. The term 'Master' was abbreviated to the prefix 'Mr' to denote a gentleman – see e.g. the death entries for: William, son of *Mr* John Haynes, *Mr* John Haynes, William son of *Mr* William Haynes, and John Haynes servant to *Mr* William Haynes, in the parish records of St Nicholas Cole Abbey (Chapters 4 & 7).
44. Hackney Archives: D/F/TYS/47; microfilm XP399.
45. Allegra Woodworth, 1946, 'Purveyance for the Royal Household in the Reign of Queen Elizabeth'. *Trans Amer Phil Soc* 35, part 4, 1-89.
46. Copied from Sloan 1971. fo.61; Hackney Archives D/F/TYS/70/6/7, microfilm XP 408.
47. Will of William Copyn, of Earls Colne, Agnes's uncle, dated 6 November 1541; Essex Record Office: D/ACR4/182.
48. Earls Colne Manor Court Rolls, 26 May 1534 (Essex Record Office: D/DPr91). This probably means she sold bread and beer, as in an inn; *'tipeler'* meant a barman in medieval English.
49. He gave evidence about a sex scandal involving the bigamous marriage of the Earl of Oxford and its horrific consequences in responses to interrogatories taken in 1585 (Huntington Library MS EL5870). Approximately two years before her death in 1548, Dorothy née Nevill, who married John de Vere in 1536, had separated herself from the 16th Earl in the grounds of *'the unkynde dealing of the Earle'*. Richard Enowes testified in January 1585 when aged about 92 years. He had been prevailed upon

by the Duke of Norfolk to attempt to persuade Countess Dorothy to return to her husband, but she wanted none of it. He testified that the Earl had bigamously married one Joan Jockey, at White Colne Churche around the end of May 1546. When the Countess received confirmation of the Earl's bigamy, she left him. The Earl was separated from Joan Jockey after a number of men (including Sir Thomas Darcy, Lord Sheffield [the Earl's brothers-in-law], John Smith, Richard Enowes [the Earl's servants] and another servant), broke into her house at Earls Colne. The gang *'spoyled'* her (probably meaning they tore off her clothes and raped her): then, in the words of Enowes, *' Iohn Smyth cutt her nose'* (cutting off the nose was a traditional punishment for a prostitute). Joan survived the attack and outlived Countess Dorothy. It seems likely that the Earl had a hand in her disfigurement as a way of forcing her into seclusion; he was never charged with bigamy. The Earl retained Smith in service until he (the Earl) died in 1562; Enowes testified that he had been a servant to the 16th Earl, though he had left his service before the Earl died. Smith was mentioned in the Earl's will of 1562 [TNA: PROB11/46; 28 July 1562]; See also the account of the attack on Joan Jockey in: A. H. Nelson, 2003, *Monstrous Adversary: the Life of Edward de Vere, 17th Earl of Oxford*, Liverpool UP, pp 14-19, for a fuller account. Enowes' responses to the interrogatories can be accessed on: http://www.oxford-shakespeare.com/HuntingtonLibrary/EL_5870.pdf.
50. Essex Record Office: D/DPr269.
51. Essex Record Office: D/ACW2/33, 'Will of Richard Enowes of Earls Colne', 25 January 1589.
52. 'Visitation of London 1568', ed. Sophia W. Rawlins. Harleian Society, London, 1963.
53. D. Richardson, K.G. Everingham, D. Faris, 2004, *Plantagenet Ancestry: A Study in Colonial and Medieval Families*, Genealogical Publishing Company Inc., p. 244.
54. Cal Pat Rolls EI Vol. VII 1575-8, 2880, (p. 435).
55. 'Middlesex Sessions Rolls: 1574', Middlesex county records: Vol. 1, 1550-1603 (1886), pp. 85-90.
56. *Miscellanea Genealogica et Heraldica*, ed. Joseph Jackson Howard, 2nd Series, Vol. 3 p. 38; see also *The Genealogist,* 1904, p.198.
57. *The Gentleman's Magazine,* Vol. 65, June 1795, p. 453; John Cokk may have been alias Hayne of Wotton, Surrey (Chancery Inq. P.M. Jac 1, part 2, No. 175).
58. R. Simpson, 1882, *Memorials of St John at Hackney*, Joseph Billing. After his parents' deaths, Richard Haynes lived in Mayre Streete [Mare Street], Hackney, where in 1605 he is recorded as donating 2s 8d towards the repairs of St John's Church.
59. *Curia Wardorum et Libaconum* [24 February Chas 13]; TNA: WARD 7/89/51: WSRO: Add Mss 52,205.
60. Daniel Lysons, *The Environs of London: being an account of the towns, villages and hamlets within twelve miles of that capital, interspersed*

with biographical anecdotes, vol. II, part 1, 'County of Middlesex, Acton-Heston, 2nd edn (1811), London, T. Cadell & W. Davies, pp. 299–300.
61. *Curia Wardorum et Libaconum* [24 February Chas 13]; TNA: WARD 7/89/51: WSRO: Add Mss 52,205.
62. TNA: STAC5/E10/7 (30 Eliz).
63. 'Visitation of Rutland 1618-1619', pp. 23–24; C. R. Haines, 1901, *Notes and Queries,* 9th S. VII, p.172.
64. Draft Calendar of Patent Rolls 1584-5, pp. 84, 184.
65. Will of John Kay of Hackney, esquire, proved 27 May 1589: PROB 11/73 f.48.
66. Essex Record Office: D/DPr41 (Earls Colne Project: Abstract Document 53902488); http://linux02.lib.cam.ac.uk/earlscolne//cprolls2/53902488.htm
67. Ibid: (Abstract Document 54100234); http://linux02.lib.cam.ac.uk/earlscolne//cprolls2/54100234.htm.
68. Ibid: (Abstract Document 54100841); http://linux02.lib.cam.ac.uk/earlscolne//cprolls2/54100841.htm
69. R. Simpson, 1882, *Memorials of St John at Hackney*, Joseph Billing, vol. ii, pp. 4–5.
70. St John at Hackney parish register for 1593/4; Stow (note 73) gives this number of plague deaths for 1603 and 42 for 1593.
71. London Metropolitan Archives: St John-At-Hackney parish register: 22 January 1594.
72. PCC Administrations 1581-1595, p. 77: Nicholas Haines, Hackney, Mdx 1594, F. 83.
73. Stow, quoted in J. B. Black 1994, *The Reign of Elizabeth 1558-1603*, OUP, pp. 109–410.
74. According to Cracknell [note 5], but this was not confirmed in the copy of the record seen by this author. It might have been a confusion with the word 'gent' (gentlewoman) written after the name of the deceased.
75. J. Brand, 1810, *Observations on Popular Antiquities: Including the Whole of Mr Bourne's Antiquitates Vulgares*, Chap. 1, p. 1; D. Cressy, 1997, *Birth, Marriage and Death, Ritual, Religion, and the Life-Cycle in Tudor and Stuart England*, OUP p. 423.
76. The Sexton's and Churchwardens' charges for tolling of the bells at funerals at this time at St John's [then St Augustine's] Parish Church, Hackney, are given in Simpson (1882; note 29). [Sexton's followed by Churchwardens']: *'Knells for ringing the great bell: 12d; 6s 8d. For the 4th bell: 12d; 6s 8d. For the 3rd bell: 6d; 5s. For the 2nd bell 4d; 2s 6d. For the 1st bell: 2d; 2s'*.

7

William Haynes of London, Purveyor of Sea-fish [c. 1532-1598/9]

Fishmonger - Perks - The Stocks Market - Marriage and Family - Rye - Dissent and Disorder - Replacement - Other Williams - Last Years

William was the sixth son of Richard and Thomasine Haynes of Whitley in the Parish of St Giles, Reading, although the order of his mention in three of his brothers' wills is consistent with him being older than Nicholas. His elder brothers, Richard and Thomas, remained in the Reading area, Thomas eventually inheriting the family estate. Born about 1532, William, like his brothers, John, Christopher and Nicholas, had therefore to seek his livelihood elsewhere and he eventually settled in London.

Fishmonger

William entered the employ of the Royal Household. A William Haynes appears first on 28 January 1553 as a yeoman of the guard, but clearly had been in service some time before then.[1] If this was the same person, he subsequently became a fishmonger as had his brother, John, before him and he established a shop in Queenhithe ward, the most important area of the London fish trade, which in due course was patronised by the servants of the noblest of families. Thus the accounts of the household of Robert Dudley, Earl of Leicester prepared by Richard Ellis his financial officer for 1559-1561, record a payment of 38 shillings to '*William Haynes fyshemonger*'. Also, among the expenses recorded for the burial of John Dudley,

Leicester's father, at St Mary's Stoke Newington, Hackney, in 1580 is found: '*To Mr Haynes for freshe fishe, £2 5s*', which is possibly another reference to William.[2] Meanwhile John Haynes had worked up the various levels of the Queen's Household in the acatry, which was responsible for the purchase and preparation of fish, meat and salt. In his need for large supplies of fish, John would almost certainly have sought William's help and William in due course joined his brother in the acatry, where he struck up a firm friendship with John Dunning, then the sergeant, or head man of that department, who was from Sussex.[3] In addition to his job as sergeant, Dunning was also the customer of the Port of Chichester. When Dunning resigned from the acatry having become a Freeman of Rye in 1565, then its mayor, John Haynes became sergeant in his place,[4] leaving a vacancy for yeoman purveyor which needed to be filled by an experienced fishmonger. Who better than his brother William already employed in the acatry? So William became purveyor of sea-fish to the Queen, reporting at first to John Dunning then to his brother, John, and it is because of this position, and the difficulties, dangers, odium and aggravation associated with it, that a few glimpses of his life have been preserved for posterity.

Perks

William's pay as yeoman would have been £10 a year, not a considerable sum at the time, but it would have been enhanced in a number of ways. Certainly most gentlemen could live well on their wider court income. The yeoman purveyor of sea-fish also received £22 11s 8d for '*losses*'[5] and food, clothing and lodging when at Court, usually at Richmond or Greenwich, and there were many other perks.

Not the least of these were rewards for services from Queen Elizabeth and her predecessors in the form of intermediate-term leases on manors and other properties at low annual rentals, which could then be let by the recipients at higher rates to provide additional income. Such profit could be substantial and William, John and Nicholas all benefited in this way. The first such 'reward' recorded for William was by the Privy Council on 27 January 1553 requiring the Chancellor of the Augmentations to make out a lease to William Haynes, one of the '*Yeomen of the Garde*', of the manor and parsonage of Stanton Lacy in Shropshire for 21 years at the same rent that he was already paying on the same property under an earlier lease, the unexpired part of which he was required to surrender.[6] Then, on the 29 May 1568 William Heynes, yeoman purveyor of sea-fish alias *'Yoman Purveyor of the Freshstore'* was granted the lease for a mansion in Porchester, Hampshire, known as Moralles and of the rectory of Porchester based on 31 quarters of wheat worth £10 6s 8d and 40 quarters of barley valued at £6 13s 4d. At the same time he was also granted the lease on the Rectory of Shipton, Hampshire formerly held by John, the late Bishop of Winchester, for a yearly rental of £12. For all these rents William paid a single payment [fine] of £50 to the Exchequer.[7] The following year (17 May 1569) he obtained leases on four, potentially more useful, properties, previously owned by religious orders, which he himself appears to have discovered in Winchelsea. These, which were located not far from his place of work in Rye, included the house and land of the Black Friars, St Anne's Chapel and associated land, the house and land of the Grey Friars and the Holy Cross Chapel.[8] In total the yearly rents were assessed at £60 6s 8d – a considerable sum.[9] The leases were awarded to him because *'the premises were concealed, but were discovered at Haynes own expense as appears from an inquisition taken on Jan. 19 last at Old Winchelsey, Co. Sussex, before Edward Myddleton, Mayor there, and others by virtue of a Commission out of the Exchequer'*.[10] Winchelsea, once the chief port of Sussex, had originally more than 700 houses. By the middle of the sixteenth century its harbour became so heavily silted up that most of the sea-trade was lost to the town. By the early fifteenth century its southern and western suburbs were virtually abandoned and at the time of William Haynes's discovery, there were only about 60 houses left in the borough, none of which was near to the old monasteries. It looks as though William had paid for the clearance of the undergrowth which had appeared during the 30 years since the Dissolution, with the result that he acquired some valuable properties. It happened that the Queen paid Winchelsea a visit within a year of this acquisition and, as a member of her Household, it is tempting to speculate that William might have participated in her progress there, or made one of his monastic properties available to her entourage. Smuggling was rife in Winchelsea during Elizabeth's reign, and the remoteness of William's monastic buildings with their cellars, like that of the houses of his brother Christopher at Climping and Ford, would lend themselves ideally to such enterprise.

On 25 May 1580 another parcel of land acquired by the crown came William's way jointly with William Lingard, *'ordinary yeoman officers of the Queen's household'*, this time in Blisworth, Northamptonshire, for their *'good service'*, the yearly shared rental being £20 2s 8d.[11,12] On 20 August the same year he was further granted for his service three lots of tithe rents from the rectory of Baschurch, previously held by the Monastery of St Peter and St Paul, Shrewsbury, totalling £19 7s 4d.[13] This was a year when difficulties were mounting with the fisherman in Rye and his efforts on behalf of the Queen had clearly not gone unnoticed or unrewarded. Indeed, William Lyngarde and William Haynes *['two of your ordinarie Servauntes']* petitioned the Queen for the 21-year lease in reversion at £34 per annum of the parsonage of Bisley, Gloucestershire, to the use of the tenant, and, according to a note written by Valentine Dale (then standing in as lord high admiral and as a special commissioner of oyer and terminer for Middlesex), the Queen granted the petition in July 1585[14] and the grant was duly recorded by writ under the Privy Seal on 31 October 1586.[15]

The Stocks Market

In June 1576 Sir Thomas Offeley and other City Aldermen had been instructed to discuss the terms of a lease with William (*'William Heynes Fishmonger'*) following the Queen's request that he be granted a lease of the City's house known as *'the Stockes in the Powltry'*, then in the tenancy of Joan, the widow of Raffe White.[16] This market occupied a large tract of ground at the junction of Cornhill, Threadneedle Street, Lombard Street and the Poultry adjacent to the church of St Mary Woolchurch. William was granted the lease, the following month, but its term could only start after the death of Widow White which didn't occur for at least another six years, as in the London Subsidy of 1582 (Cornehill Ward), *'Wyddowe Wayght'* was still being assessed in lands at £15 on which she paid 20s tax. However, it appears that Joan White relinquished the lease in 1580 (she was last recorded as a lessee in 1579), but remained in part of the accommodation after William took on the lease in midsummer 1580 for 21 years at an annual rent of £56 4s 8d.[17] The Stocks was a market enclosure, some 230 feet by 108 feet, at one end of The Poultry, where the Mansion House now stands. It was originally built for the sale of fish and meat by Henry le Waleys, mayor in the tenth year of the reign of King Edward I, and apparently named from the site of the stocks that earlier had stood there. The annual rent paid by William to the City Aldermen was practically the same as that for the same market in 1312/3, when the lessee of the 'House called the Stocks could sublease places' [stalls] to fishmongers and butchers.[18] This was William's position as well, but under the terms of the lease he had to covenant with the City not to put out any of the then tenants or occupiers, or to raise their rents. The washing of fish in the market was also banned even though there was a stone water conduit, essential for such a market.[19] William also had to bear the full cost of repairs to the fabric of the building, and the money for this would come from his profit on the rentals. By 1543 there were 25 fishmongers in The Stocks Market paying yearly rents totalling £34 13s 4d and 18 butchers paying yearly rents of £41 16s 4d and 15 let chambers that brought in £5 13s 4d a year, and presumably similar numbers of sub-tenants in William's time.[20] Even assuming no increase in rents paid by the Stocks' tenants over the intervening 40 years, this represented a 46 per cent annual profit to William of £25 18s 4d. It also meant a considerable administrative undertaking – the rents were paid weekly, at a time when his problems with the Rye fishermen were reaching their climax. It would appear that William held the lease of the Stocks until 1598, when presumably he died.[21] The market was destroyed in the Great Fire of 1666.

Marriage and family

William had married Ellen Bonne, or Bonner, by 1568. According to the Visitation of London 1568, she was the daughter of Harmon Bonne of Wolverhampton.[22] In the 1576 and 1582 lay subsidies, William's main home was in St Nicholas Cole Abbey and St Mary Magdalene parishes in Bread Street Ward, London. In 1576 he paid 35s; in 1582 he was assessed at £50 on which he would have paid 50s but was partly exonerated of this because of his separate assessment of 20s in the Queen's Household.[23] In 1589, the year of William's 'retirement' from the Household, they were still in the Parish of St Nicholas Cole Abbey, and in that year he was assessed at £20 in goods, on which he paid 20s (see note 62; TNA: E115/198/

125). Although the couple lived in London, they may well have spent some of their life together in Rye.

William and his wife were said to have been childless, because there is no mention of his children in any of his brothers' wills and his own will has not been traced. However, in the parish register of St Nicholas Cole Abbey appears the following entry: '*William ye sonne of Mr William Haynes was buryed ye 31th of December 1575*'.[24] There is an important usage of 'Mr' (Master) here to signify that this William Haynes was a 'gentleman' as there was another William Haynes in the parish, a '*comfot maker*' (sweetmeats manufacturer) who had died two months earlier, who was never so prefixed. Thus, it would appear that William had had at least one child who predeceased him. William's son had probably succumbed to the plague which was still present in London that year. It wasn't the first such death to affect William, for on 23 October 1575 was buried '*John Haynes servant to Mr William Haynes*'. Who this '*servant*' was is unknown, but sergeant John Haynes in his will described his nephew William Gunnell as his servant, suggesting that John Haynes, William's servant, might also have been a kinsman in his employ or apprenticeship, possibly an unrecorded son of Nicholas or even a grandson of William. The date of his death precludes him being John Haynes merchant tailor or his son, John, who also lived in the same parish. In the absence of any information on the date of birth of William's son, William, it is entirely possible that by the time of his premature death, the latter had married and had issue. Indeed, this might explain one or other of the young William Hayneses in royal employ from 1585 onwards (see below; p.145). Furthermore, a child or grandchild would be an understandable reason for William in 1578 applying for arms bearing elements of both his paternal and maternal forebears.

Ellen and William, when in Sussex, would have seen something of William's brother Christopher Haynes, in Arundel. Perhaps they stayed in Christopher's tavern there. The two brothers were very close and Christopher appointed William his executor and his main beneficiary. Ellen, if it was she – she wasn't mentioned by name – was left a valuable bequest of 10 silver-gilt cups in Christopher's will.[25] This was in November 1586, by which time she and William had been married for about 30 years.

Rye

Much of William's time, or that of his deputy, would have been spent away from home, for his main job was to purchase sufficient fish for the Court, which meant that he had to be at the markets in London, Rye and elsewhere to negotiate the enormous quantities needed and to arrange for its transport to London or wherever the Court happened to be. The large fishing fleet in Rye provided much of the deep-sea fish, such as cod and ling, for the Royal Household. Wherever he was buying, safeguards were in place to protect the fish supply to the royal purveyor, who by law had first pick of the best fish from the daily catches as caught. Furthermore the purveyors, through the royal prerogative, had the right to requisition supplies below market prices, often on credit, and they, or rather the clerks of the Green Cloth who were in charge of the finances, were notoriously slow and poor payers, which was the subject of considerable disquiet on the part of the fishermen. In carrying out his duties, William inevitably attracted the blame for the poor rates and inevitable delays in settling accounts, and the invective and not a little obstruction from the fishermen. This last could take many forms: secret deals with local buyers, or with the ostes, the fishmongers who were mainly supplying the London Markets. Sometimes the fishermen would ride off-shore and wouldn't land their catches until after William had left for London.

In view of the perishable nature of his fish, and the need for the purveyor or his deputy to be in the fish market to label his purchases before the market got under way,[26] William often needed to stay overnight in order to get to the market early. London posed no problem in this regard because he had a home and a shop there. Rye on the other hand meant that overnight accommodation was necessary and because of the frequency of his visits, he clearly felt that this warranted a house there. In January 1576 his name appears in the Rye records as a non-resident ratepayer for Baddinges Ward[27] when he paid 10 shillings. His house, possibly situated in Watchbell Street, was but a few minutes' walk from the Strand and the fish market, where there was later reported to be the *'King's Shop'*, from which the royal purveyor operated.[28]

On 12 January 1563 Parliament introduced a law banning the importation of fish in barrels or casks because of the '*deceitfull packing used in codd and linges brought in barroles*'. From the 1

April 1564 cod and ling could only be imported loose. However, the fishing ports in the south-east couldn't provide all the fish required by the household and in 1564, William Haynes, *'yeoman, purveyour of our seafishe'* and John Dunning, gentleman, were granted a licence to import into the country cod, ling and other fresh fish, caught by foreign or English vessels and packed in barrels or casks,[29] an activity still specifically prohibited to anyone else. They and only they or their assigns had the right to import and sell such fish *'to their moost profitt, gayne and advantage'* for the duration of the licence, up to a cumulative maximum not to exceed 5000 lasts of fish in barrels, a last being 12 barrels, which quantities had to be recorded at their time of importation on the reverse of the licence. This licence was extended, for on 23 August 1574 William Haynes and John Dunning who was by then living in Rye but still working with William, were again granted a licence for a further ten years this time to import 6000 lasts of cod and ling and all other fish called *'grene fyshe or grene codd'*[30] in barrels or casks.

These licences gave the pair the power to seize as forfeit all barrels of fish illegally imported by others and to enjoy part (half; one of two moieties) of the proceeds of the sale of such forfeitures (the Queen received the other half), after the customs and other duties on the imports had been paid. Up until 1572 the Rye fishermen used to fish for cod and ling in the North Sea off Scarborough, but foreign competition all but put an end to much of this trade,[31] which is probably why the purveyor had to obtain increasing quantities of his fish from foreign fishermen. By letters patent on 13 June 1569, a similar licence had been granted to one, John Brook, for eight years to import into England and sell 3000 lasts of cod and ling in barrels, paying the customs due on the import. A similar licence was granted to William Hunt on the next day, but Brook bought the rights of Hunt thus enabling him to import 6000 lasts. On 22 January 1578 Brook's grant was specifically made conditional upon not interfering with the aforementioned grant to John Dunning and William Haynes.[32] John Dunning was a jurat (town councillor or burgess) of Rye and was the Collector of Customs there until his death in 1579. In fact he was one of the customers for the port of Chichester under which Rye came[33] and he worked from the customs office on the Strand Quay, along with Robert Wells the searcher. Thus, not only was he a colleague of William Haynes and the former boss of both John and William, but he would have known their brother,

Christopher, who was the customer of Arundel, which appointment also came under the jurisdiction of the Chichester customs later headed by the dishonest John Yonge. Dunning served as Mayor of Rye for the years 1567, 1570, 1571 and 1573, an office for which he was particularly suited as he played no part in any of the factional infighting common among the other jurats at that time. William Haynes and John Dunning were granted their second import licence the year after John's last year as mayor. They were kindred spirits, both being outsiders and as well as their work and homes in Rye, had doubtless much else in common.

Queen Elizabeth determined to encourage people to eat more fish and had introduced compulsory fish days. This was intended to lead to the development of the fisheries and thereby increase the supply of trained seamen for the navy,[34] as a naval war with Spain had become a distinct possibility. Much of the populus of Rye was involved in fishing one way or other and by 1580 there were 31 fishing boats between 10 and 22 tons working out of the town employing 200 men as well as boys;[35] yet the demand outstripped the supply. For every catch that was landed there were three groups of buyers: the royal purveyor and his deputy, the ostes and the market men. The oste [oast or oost] was a middleman, the equivalent of a wholesale fishmonger,[36] who bought from the fishermen to supply others. Most of the fish bought by the ostes was destined for the three London fish markets and the ostes in Rye made fish purchases on behalf of members of the London Company of Fishmongers. The 'market men' were people from outside Rye who bought fish for local consumption in the neighbouring towns and villages.[37] The ostes and marketmen had ready money unlike William Haynes as royal purveyor, yet through the royal prerogative, he had first pick of the catches.

Once purchased, the fish was gutted, sometimes salted and packed into large baskets [dossers] and transported by packhorses to London by the rippiers [rippers; ripiers], some 300 horses being used on the Rye to London journey alone. The journey itself took about four hours. The gutting and filleting left considerable rotting waste, which was piled up outside the shops of the ostes and there is no reason to believe that the King's shop would have been very different from the others in this respect. The ostes eventually had to make financial provision for the removal of the fish residue, such provision depending on the amount of fish sold. The fish waste was supposed to be thrown into the sea from a platform at the Strand provided for this purpose.[38] The dung from the rippiers' packhorses caused a serious nuisance too, so that from 1578, the occupants had to clear away their own dung twice yearly.

From 1576 onwards, disquiet among the hard core of troublesome fisherman began to surface again. The mayor and jurats wrote to both William and his brother, John, soliciting their help in having a trouble maker called Richard Blacke arrested and imprisoned in the Bridewell as an example to others.[39] The Rye fishermen preferred to sell their fish locally for immediate and more profitable return, rather than wait for payment from the Crown, and to do so was a serious offence. The fishermen's grievance was simple to understand. Not only had William bought their fish at less than its true market value, but they were hardly ever paid on time or in full. William had lost his one true friend, provider of intelligence and local helper with the death of John Dunning in 1579. By now the Rye Council were informing him about the fishermen clamouring for their payment from the Crown and of their moves to deal with the local disorder arising from it. On Friday, 9 February 1581,[40] William dispatched a letter to the mayor[41] and jurats of Rye from his home in London (Figure 7.1), from which it is clear that the news from Rye had been of significant disorder and that the fish for sale had been illegally diverted by even the well-known, better-off fishermen from the royal purveyor to other buyers including some of the more notable ostes. He appreciated that the mayor's actions to stop the illegal selling of fish in the market were likely to inflame the situation further and offered what help he could. William knew the identities of the wrongdoers and made it quite clear he expected the mayor to take the responsibility of policing the market to ensure that the purveyor's rights were maintained. He requested a note of what was owing, estimating that it was unlikely to be more than £100 and urged the mayor to speed up the payment to the fishermen, the money for which he was sending with Henry Gaymer, the Rye jurat, who deputised as purveyor while Haynes was in London, but he was emphatic that they couldn't expect to be paid before the money had been received by him from the Board of Green Cloth.

Furthermore, in writing to the jurats, he wasn't addressing an objective disinterested group, as more than 40 per cent of them were involved in fishing and associated occupations.[42] But William himself wasn't an objective, disinterested party either. Eight months

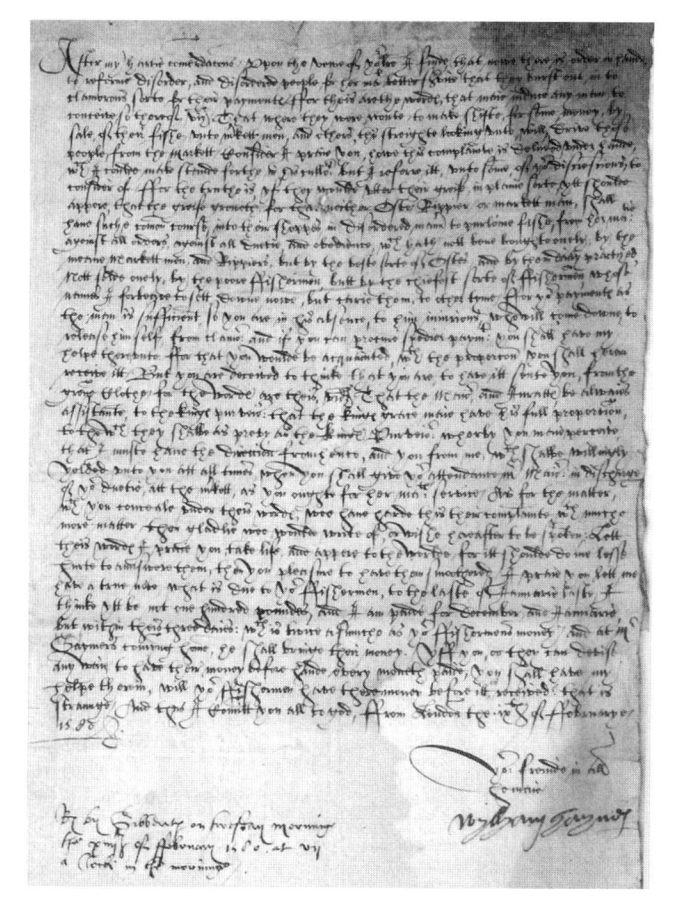

Figure 7.1 Letter signed by William Haynes to the Mayor and Jurats of Rye
(RYE 47/25 (184); reproduced by kind permission of East Sussex Record Office)

earlier, he had taken up the lease of The Stocks, one of the premier London fish markets, where he would have had his own shop, just as his brother John had in Old Fish Street, and there can be no doubt that the Rye fishermen suspected William of lining his own pockets. It has to be said that the huge savings made on fish purchases by the acatry, following William's eventual departure from the purveyor's post, lend support to the suspicion that some of his purchases were surplus to requirements and possibly intended for his own business. The Household required that the purveyors of fish be skilled fishmongers and that they could keep *'a shoppe onlie for the sale of sea fishe'*. This was reiterated in 1588.[43]

As the row with the Rye fishermen escalated, it was brought to the notice of Lord Burghley himself, as effective head of the Royal Household, and early in May 1582, he agreed to receive a deputation from Rye at his home, Cecil House, led by William Appleton, the Town Clerk, and including William Coxson, an elderly fishing master, and jurat, and another fishing master, Nicholas Fowler.[44] They wrote out their grievances against the Haynes brothers, signing it *'Nemo, for fear of offending them'*.[45] Burghley received them diplomatically and led them to believe that the large amount due to them would be paid on Haynes's next visit to Rye scheduled for 24–25 of that month. There were in fact only enough funds available for William to be able to pay the amount owed for the previous month's fish, a fact that Burghley must have known when he fobbed off the Rye deputation. Suspicious that they were not going to be paid in full, some of the fishermen refused to sell the purveyor the amount of fish he required on that visit and thereby challenged the royal prerogative of pre-emption, an offence tantamount to rebellion. This was summarily dealt with by Burghley, who organised the preparation of interrogatories by the Privy Council, which were to be answered on oath by those summoned to give evidence of alleged riotous behaviour, verbal abuse and threats towards William Haynes, before Richard Barrey Esq., Lieutenant of Dover Castle. The examination took place on 26 July 1582,[46] by which time William Appleton, Thomas Chesswell [one of the deponents examined] and Robert Bett [fishing master and jurat] were being held in prison.[47]

From the interrogatories one can gather something of the process of payment of the fisherman by the purveyor when in Rye. William first contacted the mayor of Rye and several leading fisherman to make them aware of what money he had brought for them. The next day the mayor issued a proclamation that those who still had money due to them from the purveyor for fish bought the previous month should go to the house of Henry Gaymer where William Haynes or his deputy would pay them what they were owed. Henry Gaymer, a controversial jurat and magistrate, was one of the wealthiest fish merchants in Rye. His house in Middlestreet Ward, where the claimants were asked to convene, was the most highly rated of the whole town.[48] Gaymer, who was Searcher of Rye[49] also served in the capacity of royal purveyor of fish[50] in William's absence and held the leases on a number of other enterprises.

Dissent and disorder

It appears that the rumour had been put about Rye that the Lord Treasurer [Burghley] had told Appleton, Coxson and the other petitioners, that Haynes had been given £1000 to settle their bills. However, Nicholas Fowler the 33-year-old fisherman who had been one of the petitioners at Cecil House, was more specific, saying that the Lord [High] Treasurer [Burghley] had told them that the day before their meeting he had delivered a thousand pounds *'unto Master Treasurer [Knollys] and Master Comptroller'*.[51] This has the ring of truth about it. The petitioners, not knowing the financial mechanisms within the Household had clearly wrongly assumed that all this was money for their payments. In reality, only the amount due for fish purchases made in Rye during the month of April 1582 was to be disbursed from this sum, a misunderstanding that doubtless Burghley had deliberately permitted to arise. A number of those who expected payment for the current month's fish and backpayment for sales prior to April as well were angered at what they considered duplicity on the part of the purveyor, and several said they would accept no payment except the full amount due them. As the excitement mounted, some said that unless they received their due arrears, they would not sell Haynes any more fish or even put to sea after that week. They even threatened that the fishermen and land-based men would combine forces to this end.

Later on in that visit, while Haynes and his servant, William Clarke of St Michael's Parish, Queenhithe, London, were getting the fish they had managed to buy ready for the journey to London, they learned that the other osts had 20 or 30 seams[52] more than they had, ready for transporting to London. Seeing this, William Haynes commandeered two baskets [dossers] of fish from one of the buyers. There followed an ugly scene in which William Haynes was surrounded by angry fishermen and subjected to invective and threats of every description. This could not go unchallenged and the Privy Council acted.

This notwithstanding, two months later Henry Godsmarke, a 34-year-old feter, a local wholesale fishmonger, had taken on the mantle of resident troublemaker and was still attempting to stir the osts into ganging up against the purveyor by withholding their fish, and quite prepared to face Haynes down, had apparently told him personally during his next visit to Rye, that if he had his way, Haynes would not get so much as a fish-bone while he was in the town.

WILLIAM HAYNES OF LONDON

Interrogatories based on Haynes's testimony attempted to get at the truth of what had gone on in the May stand-off. Richard Barrey heard the responses of Thomas Chesswell [or Chiswell] a 46-year-old fisherman [fishing master]; William Clarke, 50, William Haynes's servant; Thomas Harman, 42, William Beard, 40, and Nicholas Fowler, 33, fishermen; Stephen Welche, 40, and Henry Godsmarke, feters. Clarke, who had helped Haynes pay the fishermen, agreed with the accuracy of the points of which he was aware in the interrogatories. Chesswell was one of those who admitted saying that he would not receive any payment except the full amount due, although he wasn't prepared to withhold his fish from the purveyor or not put to sea. Although he himself at the time had not heard the speeches urging the uniting of the landsmen and fishermen in opposition to the purveyor, he was aware that Butcher and Ballard had since been committed to prison for such words. Harman's testimony was in general agreement with that of Chesswell, but in addition he denied that anyone had clamoured around William Haynes and abused him verbally, which was clearly untrue. Beard swore testimony against Harman, saying that when Henry Gaymer had approached Harman to buy fish from him for the Queen, Harman said that he should first know how he and others were going to be paid what they were already owed, and if they didn't receive payment, then they would supply the Queen directly themselves. Fowler admitted that he knew of the mayor's proclamation about payment at Gaymer's house, but didn't go himself as he wasn't owed anything for April (he had no tally sticks). He then gave what appears to be an accurate account of his meeting with Lord Burghley [above]. He denied the accusation that he had said the Queen should have no more fish until the money in arrears had been paid, but he agreed he had heard Chesswell say that unless he was paid he wouldn't put to sea after that week. He, like the other fishermen, said that he had brought his fish to market on the occasion in question as usual. Welche said that he had heard Fowler and Coxson talking at the Strand about a thousand pounds provided by the Lord Treasurer to pay them, or something like it. He himself had gone to the market as usual, and after the Queen's fish was taken up, he had ten baskets left. It was from him that Haynes had commandeered two more dossers. He had then heard Henry Godsmarke remark at The Strand that if our fish be culled like that and all men be ruled by him [Haynes], no-one should be able to buy a fish-tail as long as Haynes was in town. Godsmarke in his own testimony admitted that on Friday 13 July while in his own shop, he had said to William Haynes that *'yf he should have his fysshe culled in this sorte that yt were best for him to buye noe fyshe so longe as M[aster] Haynes were in Towne'*. This plausible explanation of what Godsmarke had said was a clever if slippery reversal of what Haynes had reported, and gives a measure of the intellect and cunning of those with whom the purveyor had to deal in Rye.

The failure (alluded to in William's letter of 1581) of the Exchequer to pay for the fish supplied to the Royal Household was a matter of longstanding and acrimonious complaint in Rye,[53] yet matters continued to get out of hand. William Haynes and his brother, John, were suspected of underhand dealing and were accused of taking more fish than was necessary for the Household and of withholding payment without good reason.[54] The fishermen had no means of knowing the parlous state of the Green Cloth's finances, and probably wouldn't have believed it had they been told. In fact for two centuries it appears that the fishermen, whether in Norfolk, Lincoln, or Kent, had been paid the amount received by the purveyor from the crown; but the purveyor and not the Crown was the one blamed for the shortfall.

It has to be said that the expenditure required of the Board of Green Cloth, usually exceeded its income and this inevitably led to delays in payment. The trouble in Rye was nothing new. In October 1539 one of William's predecessors, King's Purveyor John Fletcher, had complained to the comptroller of the Household that Thomas Birchet, the newly re-elected mayor of Rye, had so set the Rye fishermen against him that he would be unable to procure fish for the Household. Birchet was imprisoned in Dover castle for a few weeks, but was released on Cromwell's order following his discovery that there was more personal animosity behind Fletcher's complaint than fish.[55]

In 1586 William Haynes complained to the Lord Warden of the Cinque Ports of his *'mislikes'* of what he had found in Rye. This was again to do with the fishermen bypassing the purveyor, due, they claimed, to his repeated lateness of arrival. He also claimed that the fishermen were holding fish back in their boats. This, claimed the fishermen, was permitted of the inshore fishermen and was to *'kepe [the fish] the longer freshe and swete'*.[56] There was every incentive for the fishermen to sell their fish on the open market as demand had escalated following the introduction of compulsory fish days.[57]

Replacement

The allegation of abuses made against the purveyor of sea-fish coincided with those against purveyors in other Household departments and culminated in the Privy Council establishing a *'Commission for household causes and reformation of abuses'* fully to investigate the matter. It would appear that as a result of the Commission's findings, William Haynes was eventually one of those, along with his brother, John, who were replaced in the acatry.[58] He was succeeded by William Angell[59] as *'her Majesty's cheife purveyor of sea fyshe'*, who was having exactly the same difficulties and aggravation with the Rye fishermen, by which time, the *'queen's price'* was about a quarter of the then market value.[60] Sir Francis Knowles had to take up the matter again with the mayor of Rye.[61] William Haynes was still purveyor of sea-fish from June 1586 to May 1589[62] and the timing of his removal from the acatry is far from clear, although Woodworth[63] dates his departure as 1589 the year William Angell was appointed. This must have been after 12 May 1589, as of that date for taxation purposes he was still certified by Sir Francis Knowles and others as *'being most resyaunt and abiding here at the Courte'*, and valued at the rate of £20 in goods on which he paid 20s, rather than his rate in St Nicholas Cole Abbey, Breadstreet Ward, of £60 and 60s respectively.[64] His brother John was still sergeant of the acatry in December 1591,[65] and it is possible that William was also still employed within the Household but in a different role. Like his brother, John, he appears in the opinion of at least one observer to have been one of those who, as a result of the Privy Council's inquiry, were *'displaced wth to muche favor'*.

Other Williams

There appear to have been at least two William Hayneses in royal employ at the same time. The second was, in November 1582, a *'Grome of her Ma^{te} Stable'*, for in that month he was certified by Robert, Earl of Essex, Master of the Queen's Horses, as living in Court at the time 1582 subsidy.[66] This William was also named in a letter of 6 March 1586, as one of the *'Yeomen of the Toyles'* [Toils] who was to be provided from the Queen's store of *'soch horses, cartes, carriages, deale borde'* and other supplies necessary for conveying fallow deer to the King of Scotland (James VI) for which this William had been given responsibility.[67] Whether or not this William and William the purveyor were related remains to be established,[68] but they were definitely not the same individual; his assessment as groom in goods at £3 in 1582, on which he paid 8s,[69] was at the lowest taxable level, suggesting a younger person at the start of his career. According to the Lord Chamberlain's Accounts of 15 March 1604, a William Haines was still employed as one of the *Gromes Haylemen*.

Yet another William Haynes, then a pensioner (presumably of the Royal Household, for he certainly knew to whom he should write), wrote to Secretary Francis Walsingham in August 1589 petitioning for a pardon for his part in the actions of Sir William Stanley's English Legion, which despite its title was made up mainly of Irishmen. He had apparently accompanied Lt Roger Billings, who reported to Stanley's Lt-Colonel, to Calais in order to serve *'her magestie faythfullie as a triue subiect ought to doe'*,[70] but matters had got out of hand in the Netherlands, when Stanley, a recusant and supporter of Lady Arbella Stuart, went over to the Spanish in January 1587. Many of his officers were appalled and, on discovering their commander's treason, left him immediately.[71] Hence Haynes's petition to Walsingham seeking a pardon for past faults, stating that he had always been faithful to his Queen and country, as he would be until the day he died. He and Billings, by that time a captain, were pardoned by the Queen and had already returned home by 28 July 1590, when their names were recorded with those of others pardoned,[72] Haynes and T. Reynolds being listed as pensioners. As he mentioned in his petition that *'I hav bine subject to the errors of youth, and hav offended her magesties laws in som thinges'*, this does not seem to be what an elderly former purveyor might say, but more consistent with what might have been said by one of the young *'Gentlemen Pensioners'* with whom the Queen surrounded herself. He may in fact have been the abovementioned William Haynes, the groom.

There is one more reference to a William Haynes. This one was accused along with Pattrick Turner of piracy in Sussex, for which his arrest by the High Sheriff of that county and trial at the Marshalsea were ordered by the Privy Council on 29 September 1596.[73] The Marshalsea was at that time the court and prison for members of the Royal Household. Although the 1596 date certainly fits with our

William's period out of office and the ending of his lease on 'The Stocks' after his 1598 payment,[74,75] the likelihood of this being the elderly former purveyor must be considered remote. It is more likely that this was the '*Captain Hains*' of whom in 1582 the French had complained of plundering their shipping.[76]

Last years

William Haynes was second in line to John in the will of his brother Thomas of which he was also a witness. Had he not predeceased John, he would have inherited Thomas's Berkshire properties. In 1586 he was also principal life beneficiary and sole executor of the considerable Sussex estate of his brother Christopher,[77] which was willed in tail, and subsequently passed, to Nicholas's son Richard.

No record of William's will, if any, has so far been found. However, it is certain that he inherited Christopher's Arundel estate in tail for their nephew, Richard, as Richard Haynes did indeed come to own these properties. It is of interest therefore that, in the 1597 London Subsidy assessed on 1 October 1599, Richard Haynes appears for the first time in Allhallowes parish, Bread Street Ward, in partnership with William Price, as though he might have taken over William's business there, as the Gunnells appeared to have done with his brother John's. The cessation of William's lease of the Stocks after 1598 can be assumed to have followed his death. Thus, William must have died sometime between July 1598 and June 1599, aged about 66, which is consistent with the fact of his death being mentioned in April 1602 in relation to the 21-year lease he (William Haynes, citizen and fishmonger of London, deceased) formerly held on the rectorial tithes and parcel of land previously held by the monastery of St Peter and St Paul, at Baschurch, Shropshire.[78] He predeceased his brother, John, to whom he may have passed on his seal with his arms upon it. The arms in question were those of Haynes quartered with the modified arms of Foxley which along with the stork crest (argent beaked or) were confirmed on William on 10 June 1578[79] as well as on his brother Nicholas[80] by the same herald (R. Cooke, Clarenceux) and William thereafter sealed his official letters with these arms.[81] Judging from the seal impression on William's letter mentioned above, the engraving was of such exquisite detail that it can only have been the work of one of the finest London goldsmiths. As in the case of Nicholas, William's parents were also given as *'Richard Haynes of Reddinge in barkeshire Gentleman, and of Thomazine his wife, daughter and coheire of John ffoxley of Barkeshire, gentleman'.*

Notes and references

1. Acts of the Privy Council, Vol. 4, 1552-3, p. 209; TNA: PC 2/4 f.675.
2. Simon Adams, ed., 'Household Accounts and Disbursement Books of Robert Dudley, Earl of Leicester, 1558-1561, 1584-1586'; fo. 12, Royal Historical Society, Camden Fifth Series, Cambridge Univ. Press, p. 132. See: Ancient Funeral Charges: New England Historical and Genealogical Register: Vol. XI, July, 1857, p. 279.
3. East Sussex Record Office: RYE 1/4/228.
4. Allegra Woodworth, 1946, 'Purveyance for the Royal Household in the Reign of Queen Elizabeth', *American Philosophical Society*, New Ser. 35, pp. 1-89.
5. R. Folkestone Williams, 1660, *Domestic memoirs of the Royal Family and of the Court of England*, Vol. 2, London: Hurst and Blackett, pp.133-4.
6. Acts of the Privy Council, 1552-4, pp 209-10.
7. Cal Pat Rolls Eliz Vol IV 1566-1569, 1597, m. 24 (p. 278).
8. The Grey Friars and Black Friars monasteries and other religious houses in Winchelsea had been dissolved in 1538.
9. Cal Pat Rolls Eliz Vol. IV 1566-1569 2360, (p. 399).
10. Cal Pat Rolls Eliz Vol. IV 1566-1569 2360, (p. 399). Edward Myddleton (Midelton) was Mayor of Winchelsea for the years 1568, 1571, 1580 and 1582.
11. Cal Pat Rolls Eliz Vol. IV 1566-1569 2360, (p. 399).
12. Cal Pat Rolls Eliz Vol. VIII, 1578-1580, 1443, mm. 1-2 (p. 177).
13. Cal Pat Rolls Eliz Vol. VIII 1578-1580, 1480, mm. 35-6 (p. 184). William Haynes subsequently legally conveyed the tithes to one Bellamy. This is clear from interrogatories in 1601 after William's death, when the original lease had expired and clarification of ownership was sought; TNA: E134/44Eliz/East24.
14. 'HMC Calendar of the Manuscripts of the Most Hon The Marquis of Salisbury, KG etc.', 1889, Vol. 3, London: HMSO 1883-1976, No. 179, p. 104.
15. Berkshire Record Office: D/EE/T35.
16. Sydney Perks, 1922, *The History of the Mansion House*, Cambridge University Press, pp. 51, 84.
17. Ibid.
18. *Calendar of Letter Books of the City of London, A-L*, 11 vols, ed. R. R. Sharpe (London, 1899-1912), D.281-2.
19. W. Thornbury, *Old and New London: A Narrative of its History, its*

People and its Places. Illustrated with Numerous Engravings from the Most Authentic Sources, Vol. 1, p. 436.
20. Sydney Perks, 1922, *The History of the Mansion House*, Cambridge University Press, p. 46.
21. A William Haynes also had accommodation and a shop, near Old Fish Street Market, as in 1582 he paid 50 shillings in the London Subsidy for Bread Street Ward, St Nicholas' Parish [R.G. Lang, ed., 1993, 'Two Tudor subsidy rolls of the city of London', pp. 159-164, London Record Soc.] This might, however, have been William's namesake, as a William Haynes and his wife, Joane, leased a tenement in St Nicholas's Shambles for £6 a year from the Masters and Governors of the Bridge House. By December 1582 the couple, described as aged and in poor health, had already leased the property for 28 years.
22. 'Visitation of London 1568', ed., Sophia W. Rawlins, Harleian Society, London, 1963.
23. TNA: E179/145/252o; R. G. Lang, ed., 1993: '1582 London Subsidy Roll: Bread Street Ward', in *Two Tudor subsidy rolls for the City of London: 1541 and 1582* (1993), pp. 159-164, London Record Society.
24. London Metropolitan Archives: Parish Records of St Nicholas Cole Abbey (St Nic CA).
25. Will of Christopher Haynes gentleman of Arundel, 20 November 1586 PCC TNA: PROB 11/69.
26. Graham Mayhew (1987), *Tudor Rye*, Falmer, University of Sussex, pp. 39-41. [Two other Royal Purveyors of Fish, living in Rye while William Haynes was Royal Purveyor, Mathew Millis and Henry Gaymer, were both jurats and ostes; they served as William's deputies in Rye].
27. East Sussex Record Office: RYE 1/4/228; see also Mayhew, *Tudor Rye*.
28. Graham Mayhew (1987) *Tudor Rye*, Falmer, University of Sussex, p. 40.
29. TNA: SP12/35/36; Cal State Papers, Domestic 1547-1580 Vol. XXXV, No. 36, p.247.
30. Cal Pat Rolls Eliz Vol. VI, 1103, m. 37, 1572-5. Green usually referred to fresh fish that had been heavily salted while wet; stockfish was air-dried fish. Both processes served to preserve the fish from decay during its slow journey from the coast.
31. Historic Manuscripts Commission Report xiii, App. pt iv, 18: County History of Sussex vol. II, p. 267.
32. W. Notestein, F. H. Relf and H. Simpson, 1935, *Commons Debates*, Appendix B, Vol. VII, p. 342, Yale Historical Publications.
33. Mayhew, 1987, *Tudor Rye*, p. 103.
34. J. B. Black, 1994, *The Reign of Elizabeth 1558-1603*, 2nd edn, OUP, p. 263.
35. Historic Manuscripts Commission Report xiii, App. pt. iv, 71.
36. County History of Sussex vol. II, p. 266.
37. Mayhew, 1987, *Tudor Rye*, p. 25.
38. Ibid. pp. 42-3.
39. East Sussex Record Office: RYE 47/15/4. The Bridewell was a London prison where small-time crooks from outside the city, apprehended by the

authorities, could be imprisoned and put to work for seven days, and beaten to help correct their ways. They were then returned to their own parishes.

40. Ibid. RYE 47/25 (184); the letter, reproduced here by kind permission of the East Sussex Record Office, reads: *'To the Worshipfull Mr. Maior and Jurattes of the Towne of Rye, this be delivered [on the outside]. After my hartie commendacions: Upon the Vewe of your lettre I finde (that now there is order in hande, to reform disorder, and disordered people, for her Majesties better service) that they burst out in to clamorous sorte for their payments. For theis are the wordes that maie induce any man to conceive so therof, viz: That when they were wonte to make shifte for some money, by sale, of their fishe, unto markett men and others, this streighte looking unto will drive those people from the Markett. Consider I praie you howe this complainte is delivered under hande, which I coulde make stande forthe in his cullours: but I reserve itt unto some of your discrescions to consider of. For the truthe is, yf they woulde utter their greife in plaine sorte, ytt shoulde appere, that the greife groweth for that neither Oste, Rippier or markett man shall have suche common course into their Shoppes in disordered manner to purloine fishe from her Majestie: against all orders, against all duetie and obedience, which hath nott bene boughte onely by the meane Markett men, and Rippiers, but by the best sorte of Ostes and by them daily practized, Nott solde onely by the poore Fishermen but by the chiefest sorte of Fishermen whose names I forbeare to sett downe nowe, but carie them to other tyme. For your payments as the man is sufficient, so you are in his absence, to him iniurious, who will come downe to release himself from clamour: and if you can procure spedier payment you shall have my help thereunto: for that you woulde be acquainted with the proporcion you shall herein receive itt. But you are deceived to thinke that you are to have itt sente you frome the Grene Clothe for the Wordes are theis: viz. That the Maior and Jurattes be alwaies assistante to the kinges perveiour; that the Kinges grace maie have his full proporcion to the which they shalbe as prevy as the Kinges Purveiour, whereby you maie perceive that I muste have the direccion from hence, and you from me, which shalbe willingly yelded unto you att all times, when you shall give your attendance, Mr Maior, in discharge of your duetie att the markett as you oughte for her Majesties service. As for the matter which you conceale under theis wordes, wee have herde this their complainte with muche more matter then gladlie wee woulde write of, or wishe hereafter to be spoken: Lett theis wordes, I praie you, take life, and appere to the worlde, for itt shoulde do me lesse hurt to aunswere them, then you pleasure to have them smoothered. I praie you lett me have a true note what is due to your Fishermen to the laste of Januarie laste. I thinke ytt be not one hundred poundes, and I am paide for December and Januarie but within theis three daies, which is twice asmuche as your Fishermens money, and at Mr Gaymers coming home, he shall bringe their money. Yff you or they can devise any waie to have their*

money before hande every moneth paide, you shall have my helpe therein. Will your Fishermen have there money before itt is received? That is straunge. And thus I comitt you all to god, From London the 9th of Februarye 1580. Youre freinde in all he maie, Wyllyam Haynes'. The letter was delivered on Tuesday 14 February 1581 at 8 o'clock in the morning. The letter was sealed with the seal of the arms of Haynes quartered with Foxley together with the heron crest as granted to William and Nicholas in 1578; about 2cm diameter.

41. Robert Jackson, 1580-81.
42. Mayhew, 1987, *Tudor Rye*, p. 124.
43. British Library: Lansdowne 56 No. 31 ff.88-89v.
44. Ibid. 34 No. 27 ff.56-61.
45. Ibid. 155 No. 4; Woodworth, 1946.
46. Ibid. 34 No. 28 ff.62-74.
47. Ibid. 34 No. 27 ff.56-61.
48. Mayhew, 1987, *Tudor Rye*, p. 285.
49. British Library: Lansdowne 48 No 70; 1586; In his capacity as Searcher in 1586, while he was serving as mayor, Gaymer detained one called Julio Marino who was trying to enter the country and who claimed to have poisoned the Queen of Navarre.
50. Mayhew, 1987, *Tudor Rye,* p. 126.
51. Anthony Crane or Sir James Crofte; it was about this time that Crane died.
52. A seam was a packhorse load of fish, usually two large panniers; a pack horse in Elizabethan times could carry up to 120kg [264 lbs] or one third of its own weight in two wicker pannier baskets, for up to 25 miles [M. G. Lay, 1992, *Ways of the World*, Rutgers Univ. Press, p. 23]. Long strings, or 'drifts', of horses were sometimes needed. Regular packhorse services operated along the fish routes to and from London, the one-way journey taking four hours.
53. See East Sussex Record Office: Rye 47/27/57 (5 Apr 1582) and Rye 47/27/18 (29 May 1582); Rye 47/51/67 (10 Jan 1594); 13 H.M.C. pp. 40, 64, 77. All refs quoted in: Suss. Rec. Soc. 64, 'Rye Shipping Records 1566-1590', Lewes, 1964, pp. 53-54.
54. Corruption on the part of royal purveyors was nothing new. In 1362 the Commons complained of *'outrages and grievances committed by the purveyors of victuals'* even suggesting that the name 'purveyor' be dropped in favour of 'buyer'; Rot Parl vol. ii, p. 269, quoted by M. Prestwich, 1980, *The Three Edwards; War and State in England 1272-1377*, Weidenfeld & Nicholson, London. In the early 1400s the fishermen of Norfolk complained that year after year they were being paid £2 a hundred less than the market rate for saltfish [Given-Wilson, 1986, *The Royal Household and the King's Affinity; Service, Politics and Finance in England 1360-1413*, New Haven and London: Yale University Press, pp. 45-46].
55. Mayhew, 1987, *Tudor Rye*, pp. 64-66.
56. See: East Sussex Record Office: Rye 33/10/1v-2; Rye 1/4/7; Rye 1/4/347v; Rye 1/5/88; quoted by Mayhew, 1987.

57. J. B. Black, 1994, *The Reign of Elizabeth 1558-1603,* 2nd edn, OUP, p. 263.
58. East Sussex Record Office: Rye 47/27/57 (5 Apr 1582) and 18 (29 May 1582); Rye 47/51/67 (10 Jan 1594); 13 H.M.C. pp. 40, 64, 77. All refs quoted in: Suss. Rec. Soc. 64, 'Rye Shipping Records 1566-1590', Lewes, 1964, pp. 53-54.
59. British Library: Lansdowne 83 No. 53 f.142v; in 1601 Angell lived near John Haynes in Old Fish Street, London; East Sussex Record Office: RYE/47/60/7.
60. J.J.N. McGurk, 'Royal Purveyance in the shire of Kent, 1590-1614', *Bulletin of the Institute of Historical Research,* 1977, 1, p. 58.
61. East Sussex Record Office: Rye 47/50/10; Rye/47/57/1; VCH Sussex II, 266.
62. British Library: Lansdowne 46 No 91 f.91 (June 1586); see also TNA: E115/197/22 (20 August 1586), TNA: E115/204/112 (24 October 1587) and TNA: E115/198/125 (12 May 1589).
63. Allegra Woodworth, 1946, 'Purveyance for the Royal Household in the Reign of Queen Elizabeth', *American Philosophical Society*, New Ser. 35, pp. 1-89.
64. British Library: Lansdowne 46 No 91 f.91 (June 1586); see also TNA: E115/197/22 (20 August 1586), TNA: E115/204/112 (24 October 1587) and TNA: E115/198/125 (12 May 1589).
65. TNA: SP 12/240 f.148.
66. TNA: E115/221/16.
67. Acts of the Privy Council, 1586-7, p. 22; the toils were nets or pens placed along high hedgerows, into which deer or other game were driven and trapped. The Yeoman of the Toils was responsible for producing deer for hunting, placing them in royal parks, or providing venison for the Queen's table.
68. A different William Haynes was in the employ of Robert Dudley, 1st Earl of Leicester, who on 12 July 1587 was appointed Lord Steward responsible for the *Domus Providencie.* This William Haynes had been Dudley's personal servant since before the death of Dudley's first wife Amye, under suspicious circumstances in September, 1560. Indeed the two appeared to be close – close enough for Dudley to borrow from Haynes to settle his gaming debts [Alan Haynes, *The White Bear; the Elizabethan Earl of Leicester*, London: Peter Owen 1987, p. 29]. As part of Amye's funeral expenses was a note that it was William Haynes who collected the shroud from London, for which he was paid £3 0. 6d [ibid. p. 33]. Dudley was well regarded by his servants, especially William Haynes, who was promoted to gentleman of the Bedchamber [ibid. p. 123]. In less than fourteen months as Lord Steward Dudley took to his bed and, after being looked after by William Haynes and his wife, Lettice, he died on 4 September 1588, possibly of a fever, but rumoured to be poisoned. One contemporary story quotes William Haynes as the source of the implausible rumour that a 'fatal cup' had been handed to Dudley [*The Eclectic*

Magazine, 1850, pp. 367-8; L. Stephen, *Dictionary of National Biography*, 1888, Macmillan & Co., on Leicester, p. 120].
69. TNA: E115/221/16.
70. TNA: SP 15/31 ff.61-2.
71. TNA: SP 84/14 f.63.
72. TNA: SP63/153 f.170.
73. Acts of the Privy Council, 1596-7, pp. 207-8.
74. From midsummer 1599 a new lessee, John Catcher, is recorded.
75. Sydney Perks, 1922, *The History of the Mansion House*, Cambridge University Press, p. 85.
76. TNA: SP 78/8 f.31: *'great and notable piracies and depredations constantly committed by a number of Englishmen'*.
77. Will of Christopher Haynes gentleman of Arundel, 20 November 1586 PCC TNA: PROB 11/69.
78. TNA: E 134/44Eliz/East24.
79. British Library: Add MS 14. 495; J. E. Rayne, E. L. Chapman and T. H. Hubbard, *New Engl Hist & Genealogical Register*, 1878, p. 312, quoting Harl MSS.1438 fo.10B; John Wesley Haines, 1966, *Richard Haines and his descendants*, Boyce, Va. USA, p. 15.
80. Bodleian Library, Ash MSS. 840, f.399.
81. See East Sussex Record Office: Rye 47/25 (184) for a perfect example of an impression made by William's new seal. The image made by William's seal matches in every detail the description cited in the 'Exemplification of Arms' and Crest to Nicholas Haynes, 'Visitation of London 1568', ed., Sophia W. Rawlins. Harleian Society, London, 1963.

8

Birth, Marriage and Death in the Haynes Family

Birth – Marriage – Death – Causes of Death

There are only two records I have come across that indicate when any of the Haynes siblings were born. The first is from an interrogatory answered by Nicholas Haynes dated *'30 Elizabeth'*. As Elizabeth's accession was on 17 November 1558, it is likely that the interrogatory in question was dated 1588, in which year Nicholas declared his age to be 59,[1] making his birth year 1529. The second is the record of a deposition dated 1601 in which John Haynes was said to be 89 years old, making his birth year 1512,[2] which leaves the just plausible gap of 17 years between the births of John and Nicholas. Christian, John, Nicholas and William had children and, although reasonably complete baptismal records have been obtained for Christian's children in Wargrave, the state of the records in Hackney was such that the number of children born to Nicholas and Ann (Agnes) remains uncertain, and only one child each is known to have been fathered by John and William. Nicholas's first child of whom we have record, Joan, was baptised on 9 November 1562, when Nicholas was about 33 years old. We know that Richard and Christian were married in February 1549 and November 1551 respectively, from which it may be inferred that they were probably born around 1524 at the earliest, allowing 25 years between birth and marriage, as most Tudor marriages took place during the couple's mid-late twenties. This does not accord with the date of birth of 1512 for John, and assuming it is correct, suggests that the order of birth of the seven siblings was not as hitherto supposed. If, on the other hand, the birth order in the Visitation of London was correct then, allowing two years between births, this would give Thomas's birth year as about 1510. Richard isn't mentioned in this Visitation as he had died two years earlier, and we have positioned him as first-born for no better reason than he was given his father's name. Thus, Richard would have been an implausible 39–40 when he was married and about 58 when he died 17 years later, although he could have been up to 16 years younger.

Christian was married for 20 years before her death in February 1572. She had had a baby on average every 21.9 months, the norm for the age[3] and died within a few weeks of her last delivery. She was therefore still of childbearing potential and likely to have been in her mid to late 40s, making her birth year around 1525. It is highly likely that other children would have been born to Thomasine. Taking into consideration the stated order of the male births and the likely birth interval, Thomas [b. c. 1510] and Christopher [b. between 1514 and 1527] were respectively about 72 and 58–70 when they died, Nicholas was 64 [b. 1529], William about 66 [b. c. 1532] and John 95 [b. 1512]. However, William's birth poses a real puzzle, as the wills of both Thomas and Christopher put William before Nicholas in order of legatees, and Harl 1438 fo.10.b, places 'William Haynes of London gentleman' as *the first sonne of Richard Haynes of Reddinge in Barkesheire Gentleman ... etc*', which would make him about 90 years old when he died, yet in the Visitation of London of 1568, he is clearly stated to be the last son.

The Queen had turned 57 in 1590, and she lived for another 13 years. Although this lifespan is not unusual today, the Elizabethan age lacked adequate medical, obstetric and surgical care, modern pharmaceuticals, adequate methods for sewage disposal, attention to personal hygiene, or a balanced diet. It is of interest, therefore, that it was not only the three brothers recorded with certainty to be in royal service, and enjoying the food and other facilities of Court, who apparently lived to see old age. John in particular was remarkable for his longevity, living as he did through the reigns of six monarchs and keeping clear of the plague which accounted for many of his younger kinsmen.

The birth date estimates for Thomas and John indicate that their mother, Thomasine, would have been born in the late fifteenth century. In support of this and assuming Thomasine and her husband were of similar age, we learn that Richard Haynes was a churchwarden in 1520, which fits in well with a birthdate in the late fifteenth century. This also accords with the year 1500 mentioned by Lieutenant-General Robert Haynes[4] as when Richard and Thomasine Haynes were living.

Christian's marriage to Thomas Gunnell was likely to have been her first and only marriage, as her maiden name was given in the parish records. On the basis of her being of childbearing potential at the time of her death, we have estimated that Christian was positioned between John and Christopher in birth order, but there is no evidence that this was so – she might have been born later for all we know. We have hitherto assumed that Richard's marriage to Anne Ford was his first marriage mainly because it took place only two years before that of Christian. Given the less likely earlier date of birth suggested for him, it could have been his second marriage.

Apart from the dates of Richard and Christian's marriages, next to nothing is known of the marriages of the six brothers and their sister other than some of their spouses' full names. It was usual for marriages within families of the minor gentry to follow the medieval practice of the spousal, or promise to marry, followed by the marriage proper. Spousals could be, and often were, made by parents while their children were still younger than the legal age of 14 for males and 12 for females. The spousal tended to be a public affair and took place either at the bride's house or at the church door, where hand to hand, vows were exchanged (handfasting and betrothal by words) along with gifts. However, teenage marriages were extremely rare and most couples waited until their mid to late twenties.[5] This was because in Tudor times married couples seldom lived in a parental home after marriage and had therefore to be in a position to set up home on their own, and it took time to accumulate the necessary resources. A marriage was usually augured by the calling of banns on three separate occasions during a church service, so signifying that an agreement, or pre-nuptial contract, had been reached, which was binding on both parties to go ahead with the marriage;[6] although banns could be avoided by the purchase and granting of a licence, a favoured route taken by some of those wishing to avoid parental consent. The marriage subsequently took place at the church door followed by the blessing and nuptial mass inside the church itself. For a family whose involvement in the church was active, their marriages, even after the Reformation, would not have taken place during Lent or Advent. If their spouses had been from a farming background, as was the case of Christian, Richard and Nicholas, the marriages would not have occurred during harvest time; so these three would most probably have married in the autumn or early months of the year before Lent. We know that the former was the case for Christian (8 November 1551) and the latter for Richard (5 February 1549) at Wargrave Church. The other members of the family could have married at any time outside the major religious festivals, although September to November was common for them too. Whether this had to do with the possibility of early pregnancy with a preference for the birth to take place in the warmer months is unclear. This would not be a factor when the wife was already pregnant at the time of marriage, which for the first marriage ran at about 20 per cent, possibly because sexual relations commenced after the spousal in most such cases of older couples.[7]

It is known that Richard, and Christian, were married to spouses from their immediate locality, as Richard and Christian both married and lived in Wargrave. Thomas probably married Agnes Wheatley from the local Reading environs, as he lived in the family home at Whitley, then a small village. John, Christopher, Nicholas and William seem to have married away from home. John's wife, Agnes Thomson, was from London; Christopher's wife Elizabeth Colbroke was from Sussex; Nicholas's wife Agnes Enowes was from Essex and William's wife, Ellen Bonne was from Wolverhampton. John and Agnes presumably met in London where they both attended churches in Old Fish Street. How the others came to meet is not known, and research into the backgrounds of these wives would undoubtedly contribute greatly to the knowledge of this Haynes family. Remarriage of widows and widowers was very common at this time, and it appears that at least John married twice. We assume that this was not a factor for Richard, Christian, Thomas and Nicholas, all of whom predeceased their only known spouses; also the wife of William, who may or may not have been Ellen, was still alive in 1586 some 12 years before her husband's death. The wording of Christopher's will suggests that he might have remarried, although his widow bore the same forename as his wife mentioned 18 years earlier in the Visitation of London.

Causes of death

Causes of death, other than by the plague, are seldom given, but in the sixteenth and seventeenth centuries it was often the case that wills were prepared immediately prior to the testators' demise, suggesting that they knew that their end was not far away. Most wills were written by 'secretaries' in the presence of witnesses and, if the testator's health was poor, this was mentioned.

There were few cures other than inherent immunity. Infection was a major cause of death. In 1510, 1557 and 1580, there were pan-European influenza epidemics. It was also known as the sweating disease, English disease, or grip. In 1557 the sittings of the Paris law courts had to be suspended because of it. Smallpox was rife: Henry VIII's fourth wife, Anne of Cleves had survived the disease before she married the King; Queen Elizabeth I was dangerously ill from smallpox in 1562, which is said to have left her face pock-marked, although not as much as Lady Mary Sidney, who never appeared in public after the facial disfigurement suffered when she contracted smallpox while caring for the Queen throughout her illness. It was a major cause of death at the time. Bubonic plague was the most common recurrent epidemic disease, killing 20,000 people in London in 1563 and 15,000 in 1592. By 1593 it had spread to Hackney where between September and the following January it accounted for 269 deaths. There were other serious outbreaks of plague in England in 1528, 1563, 1569, 1574, 1578, 1581, 1603, 1625 and in 1636.[8] In fact, plague and smallpox epidemics occurred in England every four to eight years.

Before the Queen embarked on her summer progresses, checks were repeatedly made to ensure that all places along her intended route were free from outbreaks of the plague. The Queen drew up preventive ordinances against the plague, but the causes of the plague were invisible and therefore, to the populace, didn't exist. The people bribed those in authority to take down the 'bills' fixed to infected houses; they refused to carry the required signs of infection and mixed with others while infectious. They refused to sluice the streets as ordered. No funeral gatherings were permitted and all those without lawful business in London were to be expelled, but there was no-one left to supervise the ordinances, so the epidemic persisted. Treatments prescribed included blisters, clysters [enemas], poultices, bleeding, purgatives and emetics, all of which were likely to have weakened the subjects even further. Other remedies included treacle, angelica root, vervain, scabious and other herbal products. The objective appears to have been to induce sweating. The remedy favoured by Queen Elizabeth was that of Sir Francis Bacon, who had a fund of herbal remedies: Malmsey wine reduced by boiling to which crushed cardamom seeds and treacle had been added. Two or three spoonsful as often as possible, washed down with copious Malmsey induced the appropriate sweating.[9] Pain responded well to the opiate, laudanum, which was also advocated for Lord Burghley *'to procure sleep'*.[10]

Cholera, tuberculosis, erysipelas, typhoid were all rife. In 1569 in Lisbon 40,000 inhabitants died from an epidemic of *'carbuncular fever'* (said to have been anthrax, although this could have been the bubonic plague that struck England that year). Traumatic injury often proved fatal. The causes of most conditions were ill-understood, and the cures, medical as well as surgical, were often more dangerous than the ailment.

The causes of death of most of the Haynes family will never be known. It is unlikely that Richard Haynes of Wargrave [d.1566] died from the plague as some nine months passed between the date of the will and the granting of probate, 12 days after his burial. It is possible that a bout of ill-health triggered the preparation of his will from which Richard made a short-term recovery. Nicholas Haynes of Hackney [d. 1594], however, is a different matter. His death occurred at the time of a serious plague epidemic in Hackney and, although he wasn't listed as a plague victim in the parish records,[11] his burial was recorded immediately after the end of this list and he left no will, suggesting a terminal illness of very short duration. Certainly two of Nicholas's grandchildren, John and Ann Haynes, died in childhood of plague in the 1607 epidemic and their baby sister, Elizabeth, died in the same month.[12] Nicholas's son, Richard, may have died of smallpox, the notable epidemic of 1634.[13] William's son, William, possibly died from plague in 1575. Christian Gunnell's eldest son, Moses, died of the plague in 1603 at the age of 49, along with two of his children. Christopher would appear to have died of an acute illness, as probate was granted to William in London two days after the preparation of his will. He may have had another stroke, for he had suffered a right-sided weakness ('right hand lame') three years earlier from which he subsequently recovered.

The average size of a family towards the end of the sixteenth century was four and a half.[14] Many more were born only to die in infancy and this is highly likely to have been the case with the Haynes family of Reading. We know of only two of seven Haynes siblings, Christian and Nicholas, having children who outlived them. Even when Richard died in 1566, only the children of Christian and Nicholas were mentioned. Mayhew,[15] commenting on the lack of male heirs amongst the Rye magistracy in Tudor times, as indicated by the wills of 60 jurats, noted that 15 [25 per cent] left no surviving children. It may have been exceptional, therefore, even in the sixteenth century, for five out of six [83 per cent] married male siblings to die childless, and it may be assumed that any issue they might have had must have predeceased them, as in the cases of William the son of John Haynes and William the son of William Haynes, been omitted from baptismal registers, or been overlooked in their wills. (The possibility that William had grandchildren surviving at the time of his death has still not been excluded.) Complications of pregnancy and the puerperium were also common causes of maternal and child death and, as occurred with Christian and will be seen in later Haynes generations, this sometimes occurred after several children had been born and thrived. The death of Thomasine Dartnall, Nicholas's daughter, seems likely to have been due to puerperal fever as she died within three weeks of the birth of her fourth child.

Notes and references

1. TNA: STAC5/E10/7 (30 Eliz).
2. Survey of London, 1938 v.19, p.17; LCC; TNA: E134/43 Eliz/Hil4.
3. P. Glennie, 2004, 'Life and Death in Elizabethan Cheshunt', Ch. 5, pp. 65-91, in *Hertfordshire in History*, ed., Doris Jones-Baker, University of Hertfordshire Press.
4. E.M.W. Cracknell, ed. (1934), *The Barbadian Diary of Gen. Robert Haynes 1787-1836*, The Azania Press, Medstead, Hampshire, UK.
5. D. Cressy, 1997, *Birth, marriage and death; ritual, religion, and the life-cycle in Tudor and Stuart England*, OUP, p. 312.
6. Ibid. p. 307.
7. P. Glennie, 2004, 'Life and Death in Elizabethan Cheshunt', Ch. 5, pp. 65-91, in *Hertfordshire in History*, ed., Doris Jones-Baker, University of Hertfordshire Press.
8. W. Besant, 1903, *London in the time of the Stuarts,* London, Adam & Charles Black, Ch. I, pp. 215-39.
9. Ibid.
10. British Library: Lansdowne 57, No. 39 f.96.
11. London Metropolitan Archives: St John-At-Hackney, Parish Register: 22 January 1594.
12. Ibid. 22 March, 1607; 23 March 1607; 9 April 1607.
13. P. Erlanger, 1967, *The Age of Courts and Kings, Manners and Morals 1558-1715*, Weidenfeld & Nicolson, London, pp. 183-185.
14. Ibid.
15. Graham Mayhew (1987), *Tudor Rye*, Falmer, University of Sussex, p. 116.

9

In the Queen's Service

Domus Providencie – Past Hayneses in royal service – Hurst – The diet: allowances to household staff – Entertainment at Court and elsewhere – Dispersal of the family

Robert Haynes, the diarist, writing in 1787, two centuries after the event, had briefly mentioned that four of the Haynes brothers were *'known to have held office in the household of Queen Elizabeth, one being the Groom of her Bedchamber, and another, Nicholas, Purveyor of Her Majesty's Grain and Malt at Hackney'*,[1] a statement inaccurate in one detail in that malt was purchased by the buttery[2] and grain by the bakehouse, therefore it is most unlikely that Nicholas was responsible for both and suggests a mistranscription by someone undertaking genealogical research of the word '*ma^{te}*' (majestie's) in Nicholas' burial entry in the Hackney parish records. However, it is beyond doubt that John, William and Nicholas were employees of the household, and it has yet to be discovered what the occupations were of Thomas, who in his will described himself as gentleman, and Richard, who at the time of his early death was styled *'yeoman'*; if there was a fourth brother in the Royal Household, it is likely to have been one of them. Richard is the more likely candidate, as yeoman was a common status within the Royal Household. Perhaps he was the elusive groom of the bedchamber. There is also a record of a Richard Haynes being appointed Queen's gunner in the Tower of London in 1565. Thomas may have worked in the Reading cloth industry – his mother's family had included fullers – as Christopher was allegedly involved in the trading of Reading cloth in Arundel and Thomas was his only brother still in Reading. Furthermore, Thomas's wife, Agnes, made bequests of lengths of uncoloured [Reading] broadcloth in her will. Christopher was undoubtedly in the Queen's *service* as customer of Arundel, but he was clearly not a member of the Queen's Household, although there is one reference to him being a purveyor of sea-fish.[3]

Earlier Hayneses in royal employ

The *Domus Providencie* consisted of the staff that ran the court, and was composed of lower gentry,[4] most of whom were both educated and literate and came from families that had served the monarch for generations, and maintained a strong presence in the Household. It is of interest, therefore, to learn of a grant to four king's servants on 23 May 1414, one of whom was called Richard Haynes,[5] and in 1447/8 another Richard Haynes was one of the King's [Henry VI] Commissioners of Customs.[6] This Richard, later styled *'armiger'*, with John Popham, knight,[7] and Thomas Maderst, or Madehurst, a leading citizen of Winchester, was on 12 August 1462 required by King Edward IV to summon all those people in Hampshire who could assist with a dispute over the ownership of Winnall manor, Winchester.[8] They were to call witnesses under oath and return their findings to Chancery. This Richard Haynes, who possibly lived in Bassingshaw Ward, London,[9] could conceivably have been the father of Richard Haynes of Reading. A third Richard Haynes, esquire, who held the lifelong appointment of the office of Constable of the Castle of Salisbury, under King Edward IV from 1466,[10] appears to have resigned from this office in order that it could be granted to John Baret, clerk of the king's kitchen, on 3 December 1471. There was also a Richard Haynes, possibly one of these three, who was active in local affairs in Wokingham, Berkshire,[11] and this Richard might well have been of the same family as that of Richard of Reading, bearing in mind the close proximity of Reading and Wokingham. Lastly, a Nicholas Heynes was also employed in the Royal Household as a messenger (one of 13 listed) of both King Henry VII, whose funeral he attended on 11 May 1509, and King Henry VIII,[12] when in October 1512 he was recorded as *'riding with a letter to my lord Steward from my lord Chamberlain'*. If this Nicholas was related to Richard Haynes of Reading, he might have provided the entrée into the Household for Richard's sons.

It was evident that Nicholas Haynes, son of Richard and Thomasine, was not the only member of the Royal Household to live in Hackney. His lifelong friend and fellow royal harbinger, Hippolite Lynnet, lived there as did his son-in-law John Kay, the clerk comptroller of the Green Cloth and Thomas Wood, sergeant of the pantry (see Chapter 5). Likewise it is unlikely to have been a coincidence that a similar cluster individuals employed in the Royal Household lived in the Hurst area near Reading during the same period. Thus Richard Ward (1480-1577), who owned Hurst House, had been granted the manor of Whistley by the King in 1540. He was cofferer,[13] and, according to his memorial brass in St Nicholas' Church, Hurst, served four monarchs and reared 17 children. The three Haynes brothers, one sergeant of the acatry, the second purveyor of sea-fish, the third purveyor of grain, were born in the Reading area, and John eventually inherited the house in Hinton, Hurst, once owned by his brother, Richard. The Haynes brothers must have been well known to Richard Ward. John Barker (b.1540; d. 23 January 1620), who built Hurst Lodge in about 1580, was for 34 years a *'Servant and Gentleman Usher'* to Queen Elizabeth.[14] Lawrence Austen of Hurst, who died in 1593, was yeoman of Her Majesty's larder.[15] He reported to Thomas Durham of Hurst, who was sergeant of the larder and married to Jane, daughter of Richard Ward. He appears in the list of taxpayers of Brodehinton for the subsidy of 1576 as '*Thomas Deron gentleman one of the queenes majesties ordinarie servauntes of household*'.[16] Finally, the Windebank Family, who became the first recorded occupants of the estate known as Haines Hill, as correspondence from Thomas Windebank (Clerk of the Signet to Queen Elizabeth), addressed to Sir Robert Cecil, at or mentioning '*Haynes Hill*' in 1596 and 1597, has survived.[17] It seems possible that some of the manors or lands in the Hurst area were granted for loyal service to the sovereign from the royal forest of Windsor, which may explain why so many servants of the Household had come to live, or held property, there.

The cofferer, the position held by Richard Ward, was the highest position that a man who began his career within the Household could hope to reach by promotion. It carried a salary of £100 per annum in 1540.[18] He handled much of the money that passed through the Board of Green Cloth, and was the working head of the countinghouse. Richard Ward, in order to become cofferer, must have been apprenticed among the clerks of the *Domus Providencie*. The clerks advanced between departments instead of from the bottom to the top of one department. According the regulations, a clerk started in the pastry and worked his way up successively through the larder, scullery, woodyard, bakehouse, poultry and acatry. Having served his apprenticeship, he would have been appointed to a junior clerkship in either the ewery, spicery or kitchen, depending on which had a vacancy. It was from these three departments that the

Board of Green Cloth in turn appointed its junior clerks, and the cofferer was in turn selected from the ranks of the clerks of the Board of Green Cloth. The cofferer ranked just below the white-staves[19] in authority, so if a man had political ambitions for his career in the Royal Household he would attempt to gain that position. Because the cofferer handled money, and since there was no accountability from year to year, he could benefit quite handsomely by lending the Queen's money until he was called to account for it. The cofferer was accountable to the Lord Treasurer, who, for part of the time of Richard Ward, was William Paulet, first Marquis of Winchester, previously, Lord St John of Basing. The positions of sergeant and purveyor were those of career servants. As occupiers of these roles, the Haynes brothers were not therefore courtiers. The *Domus Providencie* was separated into departments, the actual day-to-day command of each of which fell to its sergeant, who was promoted from the lower ranks (yeomen). The sergeants were also the recipients of the fees of a particular department. In addition, the sergeant had the authority to manage his men as he saw the workload required. The drawback to these benefits was that, once a man was named sergeant of a department, he usually remained there until retirement or death. Apparently he had no chance of being promoted to the Board of Green Cloth and headed the department for the remainder of his term of service. There were, however, a number of recognised perks and bonuses for the sergeant of the acatry and his men. In addition to his annual sergeant's pay of £11.8s, John Haynes would have received cash bonuses at Christmas or Easter of a sum greatly in excess of his annual income. He was also the recipient of special privileges awarded in much the same manner as cash. In May 1560 the Queen gave *'Licence for the officers of the household to sell the raw hides of oxen slaughtered before 1ˢᵗ April next, for the provision of the household, to any persons at the queen's slaughterhouses or wherever else the hides shall be, without carrying them to fairs or markets'.* Thus, John and the other officers of the acatry may have got to keep the money they received by selling the hides; possibly their reward for good service as with the leases they were also granted. However, the hides of calves and stirks eventually became the recognised perquisite of the clerk of the acatry.[20] The fish purveyor was permitted to set up shop in London to dispose of fish left over from the Household.[21] John and William did so, their shops reputedly carrying the finest fish available.[22]

The work of the acatry was heavy. When the Queen was in residence at the larger palaces, there could be up to 1500 people also in residence, all of whom had to be fed. Elizabeth's Court, which was not extravagant compared with that of Henry VIII, ate 1240 oxen, 8200 sheep, 2330 deer, 760 calves, 1870 pigs and 53 wild boar, 560 flitches of bacon in one year while John was sergeant [probably 1591].[23,24] In addition, venison was provided by the royal parks. All this food had to be obtained in the first place, stored and cooked. With so many to feed, there were always those ready to take the risk of pretending to be members of the Household in order to obtain free lodgings and food. This was evidently such a problem, that the officers of the Board of the Green Cloth were instructed to make every effort, at least once a week, to root out strangers.

The royal purveyors were well known, not only for misrepresenting the prices being offered, but also for deliberately buying too much and then re-selling the excess at full market price. They were generally despised and the subject of much invective; *'perverse and crooked; hurtful to many, odious to all'*. The main bone of contention appears to have been the royal right of pre-emption, although underpaying and late payment despite having the funds, were common causes of complaint.[25] Purveyors were accused of taking more than was needed for the Household, the implication being that it was purchased for their own use. Over the years, the purveyors came under stricter control, and in Queen Mary's time changes were made to try to ensure that suppliers were paid faster. However, William Haynes's letter to the mayor of Rye shows that this wasn't entirely successful. The legendary complaints of suppliers against the royal purveyors were dealt with by the members of the countinghouse, including the cofferer. In the case of William Haynes and the Rye fishermen, this had got so far out of hand, that Lord Burghley himself became involved. It appears that several members of the acatry staff were removed from office in the 1590s, John Haynes and possibly William being among them. The ethical standards of the Haynes brothers were undoubtedly those accepted at the time throughout the Household; their occupations providing irresistible opportunities for personal gain. The *Domus Providencie* was under the control of the Lord Steward. His power within the Household, when he chose to exercise it, was absolute. The fourth Lord Steward in Elizabeth's reign was Robert Dudley, Earl of Leicester, who became more than a nominal head of the *Domus*

Providencie and took an active part in the deliberations of the Board of Green Cloth and seems to have been interested in reorganising certain Household offices right up until his untimely end.

The diet: allowances to Household staff

For those working at the Court, food was part of their remuneration. To make matters more complicated, food was also an important status symbol. It was no wonder, then, that the Eltham Ordinances,[26] the regulations for the Royal Household, went as far as giving sample menus for the different grades. The *'Wardrober of the Bedds, Groome Porter, Yeomen and Officers of the House'* could expect roast beef, veal *'or other rost'*, and *'pig, goose or cony'* (rabbit) plus their bread, cheese and ale allowance at dinner.[27] This was certainly good food by Tudor standards but it paled compared to the sumptuous meals served to the more senior grades, such as the gentlemen of the Privy Chamber. Their dinner had two courses rather than one, and included luxuries such as veal, capons, pigeon, game birds and tarts. They were also served wine as well as ale. Although some vegetables were plentiful, legumes and roots especially, they were not popular, and some degree of malnutrition, such as scurvy, was experienced by most people to some extent in the winter months. A great deal of salt was used in the preservation of meat and fish, which must have taken its toll in terms of the health of its consumers. Paradoxically, the poor enjoyed a more balanced diet by today's standards. They ate vegetables (leaks, onions, peas, beans, lentils, roots, spinach and cabbage) in soups, along with some meat, chicken, cheese, eggs and wholemeal bread. Tudor etiquette was complex to say the least, and table manners were considered extremely important, children with aspirations to the gentry having to learn and practise the rituals of service associated with each meat, poultry, fish and vegetable, together with their sauces. In general, this was part of their training when sent into the service of an aristocrat. Those in royal service had to be particularly knowledgable of every ceremony from daybreak to bedtime and what they could and could not handle, and with which hand.

Provision and preparation of this food was particularly difficult during the Queen's annual summer progresses which took place up until about 1580. The organisation of a summer progress started early in the year. Court officials would start the process by inspecting houses the Queen intended to visit, checking at the same time that there were no local outbreaks of plague. The first group of officials would find places suitable for the Queen herself; later, the harbingers would seek suitable lodgings for the rest of the Court. Nicholas Haynes spent much of his working life as a harbinger. Finally, the proposed itinerary would be published in the form of a list of places where the Queen would stay, although without specific dates. These were then sent to the mayors of towns and the Lords Lieutenant who then had to confirm that their areas were free from plague. The next stage must have been dreaded by many, as it was now that the royal purveyors arrived. Towns and villages were ordered to provide stocks of food, fuel and fodder.[28] A yeoman purveyor and his assistant then moved in to buy up stocks for royal use. Queen Elizabeth was modest in her own dietary requirements and ate only a limited range of meats; the wine she drank was diluted with three parts water. She never overate. Elizabeth had always enjoyed good health, thanks to her *'exact temperance both as to wine and diet, which, she used to say, was the noblest part of physic'*, and her active habits.[29]

Entertainment at Court and elsewhere

It is not easy to separate Elizabethan entertainments from Elizabethan sports. Both were very popular. Entertainment generally accompanied celebrations of espousal, marriage or church feast. The Haynes brothers at Court would have been familiar with the readings, music and dancing that accompanied dining. Away from Court they would in addition have enjoyed bowls, theatre, singing, dancing, juggling, drinking, fairs, and above all, gambling. Gambling on the outcome of games of chance e.g. dice, cards, board games such as chess and shove ha'penny, skittles, etc. was very popular, whether at home, in taverns or dedicated gaming houses. Christopher Haynes ran a gaming house in Arundel in his later years. Blood sports featured highly among Elizabethan entertainments. Every town and village had its cock-pit. Bating in its various forms was common, especially the bating of tethered bulls with dogs.

The Court held mock battles and the tilt. Tournaments or jousts were enjoyed by all classes, although the participants tended to be knights and the upper classes. Jousts took the form of knock-out competitions over a few days, each heat consisting of three turns at the lists with each opponent. Participants used blunted weapons; even so, injuries were not uncommon. There can be no doubt that the Haynes brothers would have been frequent spectators of the joust.

Archery was a popular competitor activity with frequent contests for prizes. This gradually gave way to practise with various firearms, especially work with hand guns and the repeat loading of the caliver with powder, shot and touch. The broad-bladed sword had been replaced by the rapier, and skill at fencing became a self-defence prerequisite, especially for the traveller who would wear a rapier as a life-preserving accessory. The theatre was popular when time permitted as were the annual fairs and religious feasts. Games and sports would have contributed greatly to the urban hubbub as the lower orders threw horseshoes, batted shuttlecocks, competed at quarterstaffs, hammer-throwing and other feats of strength such as wrestling, on the outcome of which much money would be won and lost.

Dispersal of the family

Was it work, family connection, marriage, or chance that took the Haynes brothers to their various places of domicile? Richard owned the house at Hinton, Hurst, subsequently inherited by Thomas, although he and his wife lived in Wargrave, near to his sister and her family. This suggests that Richard might have inherited the house in Hinton as it was not his main home. After the death of his parents and of his brother Richard, Thomas lived in the family house at Whitley, Reading. The other brothers moved away from Berkshire, possibly because they were all in the Queen's service, but also because they may have moved to be near kinsmen. There was a large Haynes family in the Chichester area of Sussex and one of the Constables of Arundel Castle was a John Haynes. Christopher lived in Arundel and his wife Elizabeth Colbroke came from that area. Hackney, the fashionable village' attracted Nicholas, as it did many of those connected with the Royal Household. It was near the main grain market, but was also within a day's ride of his wife's family in Essex. It is even possible that Nicholas had met his wife, Agnes Enowes, during one of his visits to the Essex Haynes family. One of the largest Haynes contingents at that time was in Hackney (Hoxton) and environs, where they had been for about 150 years, by the time Nicholas settled there. It is interesting to speculate that they were also kinsmen, as their coats of arms suggest this to be plausible. John eventually settled in Marylebone almost certainly because of the house and land he held from the Crown and because of his involvement with the Queen's herds at Tottenham Court. His home was within easy reach of his business interests in London and his brother William and those London kinsmen who settled in both Finchley and the City, from which Marylebone was an easy ride. Lastly William lived in London with his wife Ellen, originally from Wolverhampton. He also had a house in Rye, Sussex.

One striking feature of the brothers was how closely they kept in touch. It was possibly the case that, with the upheavals of the state religion throughout the century and the Draconian punishments meted out to the unwary, the brothers knew that nothing but their own blood could be absolutely trustworthy. This closeness was reflected in their readiness to intervene or assist when necessary, stand surety for debts, witness each other's wills – all of which indicated their adjustment to the 'new' religion – and all their main bequests ensured that their possessions remained in the nuclear family right up until the last one died.

Notes and references

1. E.M.W. Cracknell, ed. (1934), *The Barbadian Diary of Gen. Robert Haynes 1787-1836*, The Azania Press, Medstead, Hampshire, UK.
2. Allegra Woodworth, 1946, 'Purveyance for the Royal Household in the Reign of Queen Elizabeth', *American Philosophical Society*, New Ser. 35, p. 58. In fact, Geoffrey Duppa, a London brewer, was in charge of malt purveyance to the end of Elizabeth's reign.
3. Acts of the Privy Council 1578-80, HMSO, 1895, p. 250.
4. The style 'gentleman' referred to those who generally did not work with their hands – more than 5% of the population in the 16th century.
5. The others were William Malbon, Robert Couper and Thomas Porter; Cal Pat Rolls Hen V, vol. 1, pp. 198, 288.
6. Sir Harry Nicolas, ed., 1837, *Proceedings and Ordinances of the Privy Council of England*, p.xxxiv.
7. Popham died in 1463; see entry in *Dictionary of National Biography*.
8. Joan Greatrex, 1978, 'The Register of the Common Seal of the Priory of St

Swithun, Winchester, 1345-1497, Items 395-397', Hampshire County Council.
9. Near St Paul's. R.R. Sharpe, ed., 'Calendar of letter-books of the City of London: L: Edward IV-Henry VII (1912)', Folios 161-172, June 148, pp. 178-190.
10. TNA: C66/531 m. 15; Cal Pat Rolls 1467-1477; 13 Edw 14: p. 389, m. 15. Mention was made on 4 February 1474 of the surrender of a grant to Richard Haynes, esquire, of the office of Constable of the Castle of Salisbury, co. Wilts and the custody of the castle, made by letters patent dated 29 March 6 Edw IV [1466]. Richard appears to have resigned the constableship in favour of Baret and sent the letters patent back to the Chancery for cancellation.
11. This Richard Haynes [Heynes] was a trustee of the Westende Charity for almshouses for the poor in Wokingham, Berkshire, founded in 1451 by John Westende, clerk of Wokingham [Berks Record Office: D/QWO 35/1/1-7 & D/QWO 35/2/1-2]. On Christmas Day 1452 he was also enfeoffed with others by Roger Landen of Wokingham of all the latter's goods, debts and chattels within the realm and elsewhere [Cal Close Rolls 1453, p. 419, m. 21d]. This Richard, together with a John Haynes, was a witness of two deeds in Wokingham in 1454 and 1455 respectively [Berks Record Office: D/EE T1/1/17-18]. Again, he witnessed another grant in 1458 [British Library Add. Ch 76531] of a different Wokingham property.
12. TNA: SP 1/229 f.1 (f.133); Exchequer. Accts., 55 (28).
13. Ward was first recorded in the post in the funeral list of Queen Mary I dated 14 December 1558 (TNA: LC 2/4/2 f. 22r), but appears to have relinquished the position in favour of Weldon nine months later after Elizabeth's accession. Weldon died in 1567 and Ward then resumed as Cofferer until his death in office on 11 February, 1578 (History of Parliament 1558-1603, iii, 580).
14. Lorna Fox, *Journal,* Twyford and Ruscombe Local History Soc., 52, pp. 20-21: St Nicholas Hurst, booklet; Farrar, 1984.
15. Amanda Goode, *Berks Family Historian*, 28, pp. 14-17.
16. He paid for lands and fees £20 and 53s 4d: against his name were the words: *'exoneratur per billam residencie'* – see 'Two Taxation Lists 1545 & 1576', manuscripts of the Marquis of Salisbury, 1895 vol., p. 392 and 1883 vol., p. 184.
17. In HMC Cal MSS of Marquis of Salisbury, 1895 volume p. 392, there is a letter addressed at *'Haynes Hill'* 19 Sept 1596', probably from Thomas Windebank, Clerk of the Signet, father of Francis. Also the 1899 vol, p. 184, refers to *'Haynes Hill'*, 2 May 1597 in a letter to Sir Robert Cecil.
18. 'Letters and Papers of Henry VIII', xvi, 394(56).
19. The Lord Treasurer, Lord Steward, and Comptroller were collectively known as the 'whitestaves' because of the white rods they carried as a badge of their office.
20. R. Folkestone Williams, 1660, *Domestic memoirs of the Royal Family and of the Court of England,* Vol. 2, London: Hurst and Blackett, pp.133-4.

21. British Library, Lansdowne 155, no. 4; Allegra Woodworth, 1946, 'Purveyance for the Royal Household in the Reign of Queen Elizabeth', *American Philosophical Society*, New Ser. 35, p. 63.
22. British Library, Lansdowne 56, No. 31 ff.88-89v. See also, Woodworth, 1946, p. 65.
23. Alison Sim, 'The royal court and progresses', *History Today*, May, 2003.
24. Woodworth, 1946.
25. See complaints against the purveyor of seafish: British Library: Lansdowne 34, No. 27 ff.56-61.
26. The Eltham Ordinances were drawn up by Cardinal Wolsey in 1526.
27. Dinner was the largest meal taken between 11 a.m. and 1 p.m; ale was drunk by everyone, as water, often polluted, was considered unhealthy.
28. For example, Richard, son of Nicholas Haynes, was with others in 1621, ordered to source sufficient hay (8 loads taken from over 600 acres) *'for the King's Household'*.
29. E. S. Beesly, 1892, *Queen Elizabeth*, London, Macmillan, Ch. 12, 'Last years and death; 1601-3'; p 235. [http://englishhistory.net/tudor/beesly chaptertwelve.html].

10

Comparisons of Coats of Arms of Possibly Related Families

Arms with three Crescents – Bezants and Annulets

Haynes was a common name in the sixteenth century. It is a locative name and it is therefore likely that there were (and are) very many unrelated families of this name. Tudor people called Haynes and its variants lived mainly in Berkshire, Essex, Middlesex, Gloucestershire, Shropshire, Sussex, Warwickshire and Wiltshire and the early Reading family had links with most of these counties.

There are three main groups of coats of arms associated with individuals and families called Haynes to be found in the Heralds' Visitations or in the College of Arms. First, and most important for our consideration, are the arms with three crescents and for a crest, a heron volant, that is, rising with wings extended ready to fly. The second and third groups comprise those with annulets or bezants instead of crescents in the escutcheon and different crests. Most of the second and third groups appear to be unrelated to the Hayneses of Reading and will not be considered further.

Arms with three crescents

The origin of the crescent in the Haynes arms is unknown but may be derived from the earlier bezant or annulet. There are many variants several of which have been reproduced and can be found in the Thyssen papers in the Hackney Archives Department.

Haynes of Reading

In Berkshire, a possible ancestor of Richard Haynes of Reading is the Richard Haynes [Heynes] who was a trustee of the Westende Charity for almshouses for the poor in Wokingham founded in 1451 by John Westende, clerk of Wokingham. On 1 September 1451, Haynes and the other trustees received the feoffment of the properties donated by Westende including the eight almshouses in Peach Street and two tenements near the chapel's cemetery, the rents of which were to pay for the upkeep of the almshouses.[1] This Richard Haynes with John Haynes were witnesses of two deeds relating to a quitclaim and confirmation of grant of a property in Becke Street, Wokingham in 1454 and 1455 respectively.[2] Again with two of the same witnesses, he witnessed another grant in 1458[3] of a different Wokingham property. It is possible that this Richard was of the same family of that of Richard of Hinton, bearing in mind the close proximity of Hinton and Wokingham. The oldest records that we have found of people called Haynes in Berkshire are those of Elias Haynes and his wife Agnes dated 1340.[4] It is also possible that Richard Haynes of Reading was of the Essex, Gloucestershire or Hertfordshire family [below].

According to Clarke [1824], the plain coat of arms of Haynes of Reading was gules (red), three crescents paly wavy argent (silver) and azure (blue),[5] i.e. the background colour of the shield was red and the tones of the crescents alternated silver and blue separated by vertically directed wavy lines. Papworth, Robson and Berry[6] agreed and the description in Burke (1884)[7] was also identical, but in the latter, the crest was added as in Deuchar (1817): *'on a crescent an arrow in pale all proper'*. In the Burrell manuscripts on Sussex there is ascribed to Haynes of that county the original *'Reading'* coat.[8] It is possible that these were the arms of Christopher Haynes of Arundel, but there is no evidence to this effect. Three of the children of Richard and Thomasine Haynes, John, Nicholas and William, are known to have held and used coats of Arms on seals, although all six brothers were entitled to do so.

a. John Haynes of Marylebone and Reading

A seal in the possession of a contributor to Notes and Queries for 17 November 1860[9] bore the Haynes of Reading Arms, except with different tinctures (colours; argent, three crescents barry wavy of six azure and argent), with a mullet as a differencing mark. It was surmounted by an esquire's helm and the crest was *'a stork, heron, or crane rising'*. The motto was *velis et remis*,[10] which was used by the Barbadian descendants of the Reading family on their arms quartered with Foxley, so it appears that the motto was originally that of the Haynes family, rather than of Foxley. There are two reasons for believing that this was the coat of arms of John, sergeant of the acatry. First, the mullet (star) indicates that these were the arms of the third son. A respondent identified the arms as those of Nicholas Haynes that were quartered with those of Foxley in 1578,[11] but ignored the fact that the arms of Nicholas bore a martlet as his differencing mark as fourth son. Second, John was the only brother of the six who was styled *'esquire'* and therefore entitled to the esquire's helm. Finally, there is no record of John ever seeking permission to quarter the Haynes Arms with those of Foxley. It is tempting to speculate that this might have been the seal passed on by John Haynes in his will to his nephew, Richard Haynes of Hackney, perhaps to remind him of his genuine coat of arms. In passing it should be noted that the silver colour of the shield was a common feature of the three Haynes brothers of Reading known to have borne arms.

b. *Nicholas Haynes of Hackney and Reading*

Nicholas Haynes had two coats of arms. His original coat was argent three crescents barry wavy azure and gules[12]. This was identical with the arms supposedly of his elder brother, John, also with the original red of the plain shield of Reading changed to silver, and the silver tincture of the crescents replaced by red although alternating the opposite way round (argent and azure being replaced by azure and gules). Changes of tincture and/or addition of brisures (small differencing marks such as a mullet (star) or bird (martlet)) were common devices for distinguishing different members of the same family. Apart from the colour and additions, the direction of the wavy lines on the crescents on the arms of both Nicholas and John differed from that of their father, being barry (horizontal in direction) as opposed to paly (vertical in direction).

Well into his middle age, Nicholas in 1578 set out with his brother, William, to distinguish himself more clearly from other Hayneses in Hackney and elsewhere, by applying to the Clarenceux King of Arms for the quartering of the arms of Haynes of Reading with those of Foxley which he claimed on his mother's side. For this to have been permitted, his mother must have been an heraldic heiress, entitled to transmit her father's (or other male forebear's) arms to her own children. This tells us that she either had no brothers or that her brothers had died without issue. Nicholas would have been required to provide convincing information in support of this claim. For Nicholas, the description of the Haynes quarters is: argent three crescents unde[e] (wavy; direction not mentioned, but *paly wavy*) azure and gules. However in the illustration of the exemplified arms in the Hackney Archives,[13] the Haynes quarter is presented as: argent three crescents azure, each charged with two barrulets wavy gules (two small horizontal wavy lines [little bars]), much closer to the original arms depicted for Nicholas, which difference he presumably wished to keep. The Foxley quarters also differed in bearing two horizontal silver bars humite, or hummetty, that is to say cut short, instead of the two complete bars of the original Foxley arms. Such modifications were generally made when the arms of an extinct family were resumed by a distant heir. The crest on a torse (wreath made of a twisted roll of fabric) argent and gules (no helm, although one was mentioned in the original exemplification in 1578) was a heron, volant argent, body in proper colour, wings silver, beaked and legged or (gold), holding up one of its feet mantled gules.

c. *William Haynes of London*

William's early crest must have been similar to that of his brothers. The quartered shield for William in the exemplification of arms, dated 10 June 1578, was similar to that of Nicholas: for Haynes: argent three crescents paly wavy azure and gules; for Foxley: gules two bars hummetty argent; crest a stork [heron] rising argent beaked or[14]. This crest differs from that of Nicholas in that only the beak and not the bird's legs were coloured gold and one foot is not mantled gules.

Comment

There was no obvious reason why Nicholas or William would wish to acquire these new arms, so late in their lives. There is no evidence that their elder brothers did so. Nicholas and William were clearly entitled to bear the arms of Haynes, and it is unlikely that the new arms quartered with those of a family long since defunct, would be considered more advantageous or prestigious in the eyes of their peers. What motivated the brothers to differentiate their arms in this way remains a mystery, but in the absence of significant advantage to the applicants, it is assumed that the process must have been legitimate – and probably on behalf of their descendants in order to have armigerous elements of both paternal and maternal forebears represented – children in the case of Nicholas and children or possibly grandchildren in the case of William. In which case either a male forebear or spouse of their mother was a quasi-Foxley, i.e. a male having married a female Foxley who decided to use the Foxley name as an alias, or that their mother was a female descendant of a Foxley. The fact that in the 'Exemplification of Arms', Thomasine, Nicholas's mother was deliberately described as the daughter and coheir of John Foxley of Berkshire, rather than the daughter of John Folkes which she was, suggests that either John Folkes had used the alias, for which there is absolutely no evidence, or that the herald, in reassigning the Foxley arms, had used heraldic short-hand coupled with modifications to the arms in order to save the effort of elaborating on a more complex female descent from the remoter Foxleys – '*daughter of*' meaning simply 'descendant of' – the more probable explanation.

Haynes of London, Hackney and Chessington

People bearing the name Haynes had lived in Hackney since 1405. The wider Haynes 'families' seemed to have their fingers in every aspect of Hackney life, from churchwardens to inn-keepers, grain merchants and wool men. Michell[15] provides the intriguing possibility that the Haynes of Hackney may have been related to the family of that name at Frampton on Severn. If so, this may have been through Thomas Haynes, who held rectorial tithes there in the fifteenth century.

The arms of William Haynes, citizen and merchant tailor of All Hallows, Barking, the father of Benedict, William and Elizabeth, were: argent three crescents paly wavy [of six] azure and gules as found in St Dunstan in the East.[16] These were identical with those of William Haynes that were subsequently quartered with those of

Foxley. Similar arms appear under Haynes of Surrey.[17] It would be William's eldest son, Benedict, who would become the founder of the main branch of this Haynes family in Hackney in the early 1600s, and the similarity of their arms suggests that the family of William of All Hallows was also closely related to the family of Haynes of Reading.

This family is recorded in Hackney some 40 years after Nicholas Haynes of Reading. Benedict had probably acquired the manor of Wick (Wyke) by 1602, when he was assessed in Hackney.[18] He is described as Benedicke Haynes of Well Streete and Grove Streete,[19] from which address in 1605 he donated 15s towards the repairs of the parish church. As a manor with 90 acres in Hackney and Stepney, formerly held of the Hospitallers (Templars from 1185) and of the Crown, Wick passed on Benedict's death in 1611, to Henry, his eldest son who grew up and was a lifelong friend of Richard, son of Nicholas of Reading. Wick was eventually sold by the executors of Henry Haynes[20] (d. 1627). In 1633 Thomas Haynes, eldest son and heir of Henry, surrendered it to John Bayliffe, a Hackney lawyer.[21]

This family also had possession of the manor of Chessington-at-Hoke, Surrey, which Nicholas Saunders of Ewell, kinsman of Thomas Saunders, the husband of Benedict's sister, Elizabeth, conveyed to Benedict in 1601.[22] In 1610-11 the manor was held by Benedict's brother, William.[23] William was succeeded by two sons who died before marriage and four daughters, one of whom, Thomasine [b. 1604], married Sir John Evelyn of the famous gunpowder family, by whom she had seven children. His arms impaled those of Thomasine: Argent, a fesse wavy azure between three annulets gules. The crest was a heron. It is possible that William, her father, had differenced himself from his elder brother's very large family by changing the crescents to annulets [rings] for, as C. R. Haines[24] noted, William Haynes (son of William the merchant tailor) was taxed £50 for leaving London for Croydon in 1590, the year of his father's death, and there in Croydon Church in a window concealed since 1598, were the arms: Argent, a fess nebuly azure (a blue horizontal band with bulbous cloud-like protrusions across the centre of the shield) between three annulets gules.

William Haynes of Chessington-at-Hoke, Surrey, was *'one of his Majesty's Farmers of the Custome of money'*, i.e. a collector of customs duty or taxes.[25] On his death in 1611, the Chessington estate briefly passed to his son Matthew who died in 1617; the estate was

177

then divided among William's four daughters.[26] William had also acquired the manor of Friern from William Harvey in 1594. In Surrey he held manors, lands and tenements not only in the parish of Chessington but also in the parishes of Long Ditton and Kingston. The Long Ditton and Kingston properties passed via his daughter Alice Maria to the Hattons.

Haynes of London and Hoxton (Hackney)

The Haynes Arms [Haynes of London] given in the Visitation of London 1568[27] were: *'Argent, on a fesse nebuly azure between three crescents gules three fleurs-de-lis or'*, the crest: *'on a torse a bird (stork, crane or heron) rising with wings displayed russet, the inside of the wings argent, beak and crest or'*. The latter crest more closely resembles that given to Nicholas Haynes in his 'Exemplification of Arms' ten years later, with slight differences in tinctures.

In the Visitation of London 1634, appears an illustration[28] of the Haynes arms similar to those described in the Visitation of 1568 with the legend *'The Armes in Allderman Smith's house at London Stone'* which may be summarised: argent on a fesse azure three fleur-de-lis argent, between three crescents paly wavy six azure and gules. They accompany a pedigree for Haynes of Walbrook Ward (City of London) showing descent from *'Richard Haynes of Hogsdon (Hoxton, then part of Hackney) in com. Middx'* (wife omitted) to *'Thomas Haynes of ffinchley in com. Middx = Elizabeth da. of William Lennard of London mercer'* who had a son *'Lyonell Haynes of London liveing ao 1634. 2 sonne'*. Lyonell, who was a citizen and grocer of London of the parish of Saint John Walbrook, married *'Mary da of Robt Briggs of London'*. He died in 1644, leaving Mary his wife and an only daughter, Elizabeth. Information was clearly provided by Lionell Haynes, the name appended.[29] This Richard was therefore of the same period as Richard of Reading, the father of Nicholas Haynes of Hackney, or his eldest son. There was also a Robert Haynes of Finchley, yeoman, who was the son of John Haynes, probably Thomas's brother, who lived first in the Hoxton area, then in Finchley.[30] There are many notices relating to this Robert, including one when he appeared at the Sessions held at '*Le Castle*' St John Street, London, on 10 July 1609, accused of *'not punishing rogues'*.

These arms appear on the shields of Cage and Hart, through the marriage of Anne Haynes of Hoxton, to Anthony Cage and subsequently to Sir John Hart, and through the marriage to Cage they occur in the multiple quarterings of Hastings of Durham, Buckinghamshire, given as argent on a fesse azure between three crescents of the last as many fleurs-de-lys of the first; the crest was three moors' heads joined on one neck.[31] It is plausible that the families: Haynes of Reading; Haynes of London, Hackney and Chessington and Haynes of London and Hoxton, were all branches of the same family.

Haynes [Haines] of Daglingworth, Gloucestershire

This family had its seat at Daglingworth, Duntisborne and Bagendon. There is the record of the will of William Heynes being proved in 1528 and Maud Haines in 1590. There are several later Haines' monuments in Daglingworth Church down to the twentieth century. The family was one of the major landowners, in the nineteenth century, farming 3000 acres. The arms of this Haines family are still to be seen in Daglingworth Church, and are the same as those of William, son of Richard Haynes of Reading (argent three crescents paly wavy of six azure and gules), suggesting common ancestry. An account of the family is given by Rees.[32,33]

Haynes of Frampton on Severn, Gloucestershire

The Haynes of Reading may conceivably have had ties with the Haynes family of Frampton on Severn as mentioned above. There, one Thomas Haynes held rectorial tithes in the late fifteenth century to his death in 1543. In 1616 Giles Haynes settled a house and two plough-lands on his marriage with Anne, daughter of Richard Haynes, and she and their son and heir, Richard, survived at Giles's death in 1629. The family house, later called the Grange, contains a central chimney stack behind the front door, a room with carved panelling and a decorative plaster ceiling of the early 17th century, and a staircase of similar date with turned balusters. According to George B. Michell, *'the arms over the fireplace of the panelled room appear to be those of Haynes of Hackney'*.

Haynes of Essex and Hertfordshire

There was an important Haynes family in Essex that originated in Much Hadham, Hertfordshire. John died in 1606 having had 11 children [2 boys and 9 girls] by his wife Mary Mitchell [d. 1619]. His son and heir, John,[34] married twice, the first time in England to Mary Thornton, by whom he had three sons and three daughters. He bought Copford Hall and manor, Essex in 1624. This John was a devout Puritan who, after the death of his first wife, emigrated to New England in 1633 allegedly to escape the Laudian persecution, where he met and married Mabel Harlakenden, originally from Earls Colne, Essex (baptised 27 December 1614), in Newtown (Cambridge), Massachusetts Bay, in 1635. They had five children. John Haynes was the first Governor of Massachusetts Bay and subsequently of Connecticut. He remained in America until 1651 and during his absence his brother Emanuel acted as Lord of Copford. John's sons, Robert Haynes (b. 1618) and his younger brother Hezekiah (b. 1619), were on opposing sides during the Commonwealth.[35]

Some genealogies record John Haynes the elder as the son of Nicholas Haynes of Hackney, but this is certainly not the case,[36,37] the evidence indicating that he was the son of George Haynes and his wife Blithe Eliott of Much Hadham, Hertfordshire. However, the arms used by the Governor's great grandson, Hezekiah Haynes (d.1763) of Copford Hall (argent three crescents barry undée azure and gules, with a stork rising proper for a crest) were similar to those used by William Haynes, father of Benedict Haynes of Hackney and even closer to those of Nicholas Haynes of Hackney in his quartered arms following the exemplification, which, if true, is the only evidence to link the two families.[38] However, it was noted that the evidence presented for his coat of arms to the herald, Edward Byshe by Hezekiah Haynes, was a seal on the top of a silver inkhorn which he alleged to be his father's. In the Visitation of London 1634, there is a pedigree of Haynes which refers to this entry, but without arms.[39] The arms are those of Haynes of Hackney [paly wavy] with a crescent for a difference indicating the second son, which Hezekiah Haynes of Copford Hall indeed was.[40] Too much rests on the reliability of the provenance of the inkhorn to permit much credence in its veracity, and therefore the validity of the claim, at this distance. It can however be noted that arms matching these were impaled with those of Barlee, by John Barlee, the husband of Mary Haynes, the sister of the aforementioned Governor. According to Haines,[41] these can be seen on Barlee's tomb, dated 1633. Furthermore the arms appear from Hezekiah's seal on a letter in the Bodleian Library[42] to have been: argent three crescents barry of six azure and gules, with a crescent as a cadence. C. R. Haines pointed out that the arms of Hezekiah Haynes differed from those of Nicholas Haynes only in having the crescents barry instead of paly. However as noted earlier, before Nicholas had his arms quartered, the original arms were argent three crescents barry wavy azure and gules, making those of Hezekiah indeed identical with those used by Nicholas Haynes of Hackney 100 years earlier. Hezekiah's motto was also *Velis et Remis*.

If the Haynes families of Reading and Much Hadham were related, then it is likely to have been at or before the level of the fathers of Nicholas of Hackney (i.e. Richard) and George of Much Hadham.[43] C. R. Haines suggested that Richard Haynes of Reading very possibly came from Essex or Hertfordshire.[44]

As indicated, with the onset of the Civil War, this Haynes family was divided in its loyalties. 'Governor' John's second surviving son by Mary Thornton, Robert, was a staunch and devoted royalist, whereas, Hezekiah, his third son, fought for Cromwell as a Major-General and Military Governor of the Eastern Counties. He was the captor of the royalist poet, John Cleveland.[45] He later became a member of Cromwell's Council and was described as a *'presbyterian Elder'*. By 1656 he warned Secretary John Thurloe that Norwich was controlled by people disaffected with the Protector's regime. With the Restoration he was reduced and imprisoned in the Tower, 1657-1662, from which he obtained his release on giving a bond for £5000 and two sureties.[46]

In 1654 John Haynes was succeeded by his son, Robert,[47] and after Robert's only son died in May 1659,[48] by Hezekiah who lived until 1693. Hezekiah's son John, on whom Copford had been settled, predeceased his father in 1692 and the manor passed to Hezekiah's grandson John, who died childless in 1713, and then to John's younger brother, another Hezekiah.[49] In the Glossary to the *Diary of General Robert Haynes*, attention was drawn to the statement in letters of General Hezekiah Haynes in Thurloe State Papers, to the effect that in Barbados in 1647, *'a man was buried from the Haynes Estate'*.[50] Other than the arms, this appears to be the only potential link we have come across between the Copford Hall Hayneses and those who settled in Barbados. Both Robert and Hezekiah Haynes crossed the Atlantic to visit their father in Connecticut. In the seventeenth century many ships made landfall first in Barbados, before completing their voyage to America. However, it is not known if the Copford Hayneses ever stayed over on the island. There is no reason to believe that Robert Haynes of this family was the one of that name found on the island in the 1640s who left descendants there.

Bezants and annulets

Haynes of Westbury on Trym, Wick, Abston and elsewhere in Gloucestershire and Dorset

The history of the family of Haynes of Westbury on Trym would not appear to have links with the Haynes of Reading. Its genealogy from Thomas Haynes in the time of Henry VII and his wife Agnes Wall of Westbury was extensively documented by Revd F. J. Poynton,[51] who concluded that this family was most likely descended from the Heynes of Shropshire, because of the close similarity of their coats of arms: those of Haynes of Westbury and Wick being: *argent on a fesse gules three bezants, between as many demi-hinds azure*; those of Haynes of Shropshire having *three demi greyhounds* but being otherwise identical. Although the forenames of the Haynes of Westbury are similar to those of the contemporaneous family at Reading, nowhere is there any suggestion of a connection between the two and it can therefore be concluded that such a connection did not exist.

For the arms and other Haynes families having coats of arms with either bezants or annulets, the reader is referred to the review by C.R. Haines.[52]

Notes and references

1. Berkshire Record Office: D/QWO 35/1/1-7 & D/QWO 35/2/1-2.
2. Berkshire Record Office: D/EE T1/1/17-18.
3. British Library Add.Ch 76531.
4. Berkshire Record Office: H/RTA 34.
5. W.N. Clarke, 1824, *Parochial Topography of the Hundred of Wanting with Other Miscellaneous* ... p. 34.
6. Mentioned in C.R. Haines, 1899, *A complete memoir of Richard Haines (1633-1685): a forgotten Sussex worthy, with a full account of his ancestry and posterity*, London, Harrison, p. 138.
7. Sir Bernard Burke (1884), *The General Armory of England, Scotland Ireland and Wales*. See also: Alexander Deuchar (1817), *British Crests:*

POSSIBLE RELATIONSHIP OF THE HAYNES FAMILY OF READING WITH OTHERS

Containing the Crests and Mottos of the Families of Great Britain and Ireland, London: Kirkwood & Sons, p. 136 and pl. 81, cr. 7. The crest is a vertical arrow pointing downwards in a crescent in their natural colours.

8. Gules, three crescents per pale argent and azure - C.R.Haines, 1899, p. 142.
9. *Notes & Queries,* 2nd Series, Vol. X, p. 387.
10. *Velis et remis,* literally 'with sails and oars', the motto means full speed ahead or making an all-out effort.
11. J.G.N., *Notes & Queries,* 2nd Series, Vol. XI, 12 January 1861, p. 38.
12. Hackney Archives Department: D/F/TYS/47, microfilm XP399.
13. Copied from Sloan 1971. fo. 61, Hackney Archives Department D/F/TYS/70/6/7, microfilm XP 408.
14. British Library: Harl. MS 1438 fo. 10 b; British Library: Add. MS 14295 f. 10b; Camden's Grants: Coll Arms I, 15; Hackney Archives Department: D/F/TYS/70/6/7, microfilm XP 408.
15. G. B. Michell, 1929, *Pages from the Past at Frampton on Severn,* Glouc. & Lond., p.35.
16. Hackney Archives Department: D/F/TYS/47 XP399.
17. British Library: Harl. MS 1147. f. 170.
18. C.G. Paget, 1937, *Croydon Homes of the Past,* 63-64; VCH Surr. iii. 264; Hackney Archives Department: D/F/TYS/1.
19. R.Simpson, 1882, *Memorials of St John at Hackney,* Joseph Billing.
20. TNA: PROB 11/155, f.145; R. Simpson, 1882, *Memorials of St John at Hackney,* Joseph Billing; TNA: WARD 5/26, no.5.
21. W. Robinson, 1842, *Hackney,* i. 320.
22. Feet of Fines Surrey; East. 43 Eliz.
23. Chan. Inq. p.m. (Ser.2) cccxxiv, 168.
24. C.R. Haines, 1899.
25. In a document dated 28 January 1610, [TNA: E 214/802], Thomas Estcourte, citizen and leatherseller of London, assigned a bond, dated 29 September 1608, whereby William Gunnell, citizen and skinner of London and William Gunnell, of Marylebone co. Middx., gent., grandson of Richard Haynes of Reading, were indebted to Estcourte for 100L (£100), unless they had repaid the £51 5s they owed him by 31 December that year, on which they appear to have defaulted. Estcourte owed the Exchequer, via Haynes, £60, so in order to clear it he assigned the bond in £100 made to him by the Gunnells, over to William Haynes. It was Haynes's job then to collect payment from the Gunnells.
26. Chan. Inq. p.m. cccxxxiv, 55 & ccclxi, 101.
27. 'Visitation of London 1568', ed., Sophia W. Rawlins. Harleian Society, London, 1963.
28. 'Visitation of London 1634', Harl 1476 fo. 280, p. 371.
29. Hackney Archives Department: D/F/TYS/47, microfilm XP 398.
30. London Metropolitan Archives: Middlesex Sessions Register. 5 & 10 October 1609 (7 James I), p. 174. In Finchley a John Haynes of this family left the church 6s 8d, charged on land, for an obit in 1536 (VCH Middx; Guildhall MSS. 9171/6, f. 184v., 8, f. 17, Prob. 11/7 (P.C.C. 9 Logge).

According to Boyd's Marriage Index 1538-1840, John Haynes married Margaret Man at Finchley in 1573.
31. Robson, quoted by C.R. Haines (reference 24), p. 141.
32. G. E. Rees, 1932: *The History of Bagendon*, Cheltenham: Thomas Hailing Ltd., Oxford Press, pp. 100–102, 170.
33. Interest in this family was initially heightened by the existence of a family called Foxley in Bagendon, including one John Foxley at the end of the fifteenth century. If Richard Haynes of Reading really was from Daglingworth, the possibility at least had to be considered that Thomasine, whom he married, was a daughter of John Foxley of Bagendon, and sister of Thomas [d.1525]. However, this possibility is unlikely to be true as no Thomasine Foxley has been discovered in Bagendon and Thomasine, the wife of Richard Haynes of Reading, is no longer believed to be the daughter of an individual called Foxley.
34. Born 1 May 1594; the inquisition post-mortem on his father's death gave the son's age as 11 years, 11 months and 21 days on 22 April 1606 (NEHGR 49, 310).
35. C. Durston, 2001, *Cromwell's Major-Generals. Godly Government during the English Revolution*, Manchester University Press, p. 105.
36. See C. R. Haines, 1901, 'Governor Haynes's Grandfather'. *Notes & Queries*, 9th S. VII, Mar 2, p. 172.
37. A. G. Davis, 2004, 'The Hertfordshire Background to Settlements in the Connecticut Valley', Chapter 8 in Doris Jones-Baker, *Hertfordshire in History*, University of Hertfordshire Press, pp. 127–148.
38. *The Heraldic Journal; Recording the Armorial Bearings and Genealogies of American Families*, Vol. 1, (IV), 1865, p. 51; J.B. Felt, 1870, *New England Historical and Genealogical Register and Antiqu J.*, Vol. xxiv, p. 423; see the 'Visitation of London begun in 1687', 2004, Part 1, p. 197, Harleian Soc for discussion of these arms.
39. Hackney Archives Department: D/F/TYS/47, microfilm XP399.
40. Although Hezekiah Haynes may have been the third son of John, his elder brother, John having died young, the position in the family as far as cadency marks was concerned appears to reflect the number of sons still living or mentioned at the time of grant. Nicholas Haynes was actually the fifth-born son, but used the martlet for a difference [i.e. for fourth son] after his eldest brother's death. The arms of John Haynes, Governor of Massachusetts was given as *'Argent three crescents paly wavy azure and gules'* on pp. xxv and 165 (footnote 3) of S. E. Morison 1998, *The Founding of Harvard College*, ISBN 0674314514.
41. C.R. Haines, 1899, p.140.
42. Bodleian Library: Rawlinson A 24, 388.
43. Probably John; TNA: SP 1/153 f.130.
44. There is reference to a George Haines or Heynes who worked as a groom in the buttery of the Royal Household. He was granted a corrody [allowance of food, clothing and lodgings] in the Priory of Worcester in 1532; 'Houses of Benedictine monks: Priory of St Mary Worcester' (Hist. County Worcs, Vol. 2 (1971) pp. 94–112). Could he have been a maltster

as purveyance of malt was within the province of the Buttery? Was this the George Haynes who had the mill at Much Hadham, the one who married secondly Agnes Allis in London in November 1571? Or could he have been another child of Richard and Thomasine of Reading, who died before the Visitation of London? George seems to have been a Christian name popular with the Much Hadham Haynes family, for William Haynes, yeoman of Much Hadham, in his will dated 10 May 1598, left sons William and George. His will is of interest for its connection with the Royal Household, for James Quarles, Chief Clerk of the Kitchen, Sergeant of the Poultry and Victualler of the Queen's ships, who was appointed keeper of Creslow Pastures after John Haynes, was appointed supervisor of this will (Essex Record Office: D/ALSW 11/241).

45. See C. H. and Thompson Cooper, *Notes & Queries,* Vol. 4, 3rd S. (84) 8 Aug, 1863, p.115.
46. C. Durston, 2001.
47. Died 1657, buried 21August, aged 39; see 48.
48. Alan Macfarlane, ed., 1991, *The Diary of Ralph Josselin, 1616-1683*, 19 May 1659.
49. *History of the County of Essex* (2001), Vol. 10, 'Lexden Hundred (Part) including Dedham, Earls Colne and Wivenhoe', pp. 143-6.
50. Cracknell, ed. (1934). In 1647 what is likely to have been yellow fever killed more than 5000 people in Barbados. A further outbreak in 1691 killed many of the British settlers, who had arrived, or had been born, since the earlier epidemic, whereas older inhabitants had acquired immunity. Yellow fever became known as the 'Barbados distemper'.
51. Revd F. J. Poynton 1884-5, 'The Family of Haynes of Westbury on Trym, Wick & Abston and other Places', *Transactions of the Bristol and Gloucestershire Archaeological Society* 9, 277-97.
52. C.R. Haines, 1899, Chapter XVII.

11

The Next Generation

Haynes of Reading and Hackney: Children of Nicholas Haynes Thomasine Haynes; children by William Dartnall - Joan Haynes - Richard Haynes - Alice Haynes - Robert Haynes, son or grandson?

Gunnells of Wargrave: Children of Christian Haynes Sara Gunnell - Moses [Moyses] Gunnell - Daniel Gunnell - Christopher Gunnell - William Gunnell - Note on William Gunnell junior and Ralph Gunnell - Other children of the Gunnells of Wargrave

Haynes of Reading and Hackney

Nicholas Haynes died in January 1594, his wife Ann (Agnes) in June 1605. They were both buried at St John at Hackney, at that time called St Augustine's. They had at least five children, but the poor state of the Hackney parish records prevents us from learning how many they actually had. A Robart Haynes, the son of Nicholas Haynes, was baptised in St Clement Danes Church, Westminster, on 12 April 1579 and he has been assumed to be another son of Nicholas of Hackney, although if indeed related, he might have been Nicholas's grandson.

Thomasine

Thomasine was born before 1568, as she was mentioned in the Visitation of London of that year. She was most probably the first-born child as she is listed as the first daughter. This would give her birth date somewhere between 1556 amd 1561. Being the first

THE NEXT GENERATION

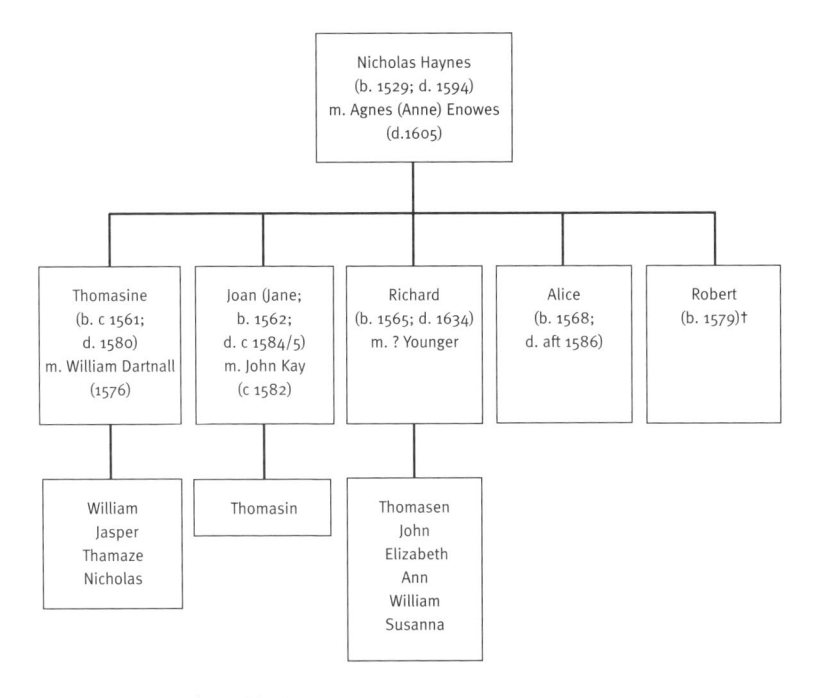

Figure 11.1 Family of Nicholas Haynes and Agnes Enowes of Hackney

Thamaze = Thomasine; †Possible Son

daughter would be consistent with her being named after her paternal grandmother, Thomasine Folkes. She may be the one referred to as Nicholas's daughter in her uncle Richard's will in March 1666, although her sister Joan was born in 1562. She is not mentioned in family documents after 1568 because although she married and had children, Thomasine had died by the time her uncle Thomas wrote his will in 1582.

Thomasine married William Dartnall [Dartnoll or Darknall], son of wealthy landowners Edward and Elizabeth Dartnall of Slinfold, Drungewick and Rudgwick, Sussex, on 14 May 1576;[1] she could have been no more than 20 years old and possibly as young as 16. The Dartnalls may have originally come from Kent, but William, his three brothers and three sisters[2] continued to live in Sussex. The marriage took place at the parish church of St Peter ad Vincula, Wisborough Green, near Billinghurst in Sussex, where Thomasine was to live and rear her family in Drungewick[3], until her death just four years and five months later. William had obtained the house at Drungewick from his mother, Elizabeth, under a covenant whereby she could live with William, Thomasine, and their family there for as long as she wished.[4] Whether or not this arrangement was satisfactory is not known; however in the detailed bequests of her mother-in-law, Thomasine was left two sacks of toe,[5] or tow, the short fibres of flax left over after combing out the longer flax fibres used to make linen. Tow was used either to make cheap cloth (hence 'tow-rag'), or as fire kindling, an odd bequest for a daughter-in-law to receive, although her husband was handsomely provided for. Exactly nine and a half months after her marriage Thomasine gave birth to her first son, whom she named William after his father. William was baptised in the Parish Church of Wisborough Green on 29 March 1577. One of his godfathers was Christopher Haynes of Arundel, his great uncle, who in his will dated 20 November 1586, left *'William Dartnall my godsonne five pounds in money'*.[6] William was also remembered in the will of his paternal grandmother.[7] Their second son, Jasper was born the following year and baptised on 23 August 1578. A daughter Thomasine [Thamaze] followed in 1579 and was baptised on 13 September. Her last child, Nicholas, named after her own father, was baptised on 9 October 1580 and this birth appears to have led to Thomasine's death as her burial was registered on 26 October 1580, seventeen days after Nicholas's baptism. However this was the year of the first epidemic recognised as influenza to which she might also have succumbed.[8] William Dartnall [Darknoll] remarried by licence on 4 September 1581 Agnes Chaper of Lodesworth[9] at Wisborough Green Church. The baptism of a child of this marriage, a daughter Elizabeth, was registered on 24 February 1583. William was still alive in March 1600 when he was left lands in Slinfold and Billingshurst in the will of his brother, John.[10]

THE NEXT GENERATION

Children of William Dartnall and Thomasine Haynes

```
Thomasine Haynes
(b. c 1561; d. 1580)
m. (1576) William Dartnall
or Darknall
(m. (1581) Agnes Chaper)
```

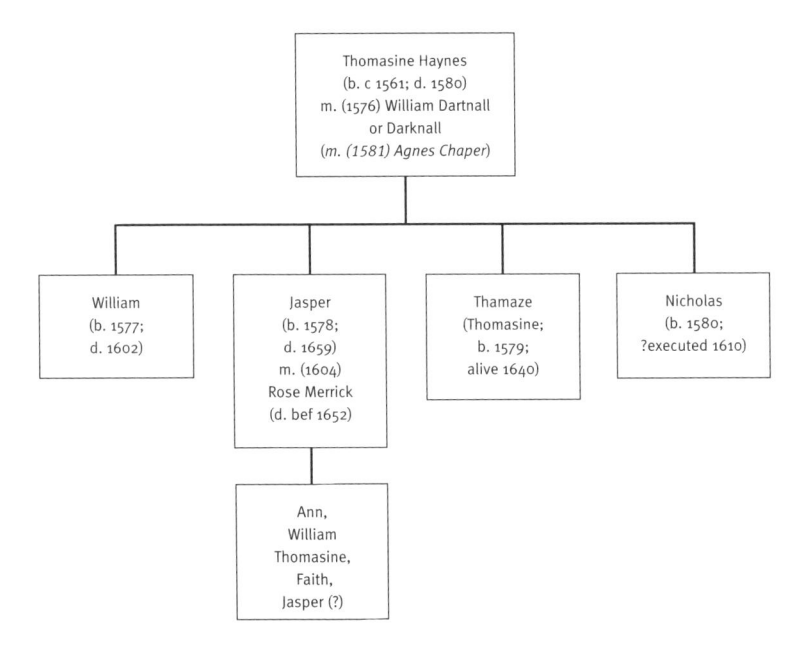

- William (b. 1577; d. 1602)
- Jasper (b. 1578; d. 1659) m. (1604) Rose Merrick (d. bef 1652)
  - Ann, William Thomasine, Faith, Jasper (?)
- Thamaze (Thomasine; b. 1579; alive 1640)
- Nicholas (b. 1580; ?executed 1610)

Figure 11.2 Family of Thomasine Haynes and William Dartnall

William Dartnall

William was born in 1577 and was baptised on 29 March in the Parish Church of Wisborough Green, Sussex. At the age of nine, he received a £5 bequest from his godfather, Christopher Haynes of Arundel. He inherited ten pounds in the will of his uncle John Dartnall (Darknall) of Kirdford, Sussex, in 1600.[11] He was also remembered in the will of his paternal grandmother Elizabeth Darkenall, who died when William was ten months old.[12] He appears to have moved to or been visiting Hackney, for a William Dartnell was buried in the graveyard of St John at Hackney on 8 February 1602, less than two months after the death of his aunt Alice Haynes. There were no other inhabitants of that name then in Hackney, so he may have been visiting his grandmother Haynes when he fell sick. He was described in the parish records as a master of arts (*in artibus magister*) and so had presumably received a higher education, thereby distinguishing him from his father.

Jasper Dartnall

Jasper was born in 1578 and was baptised in the Parish Church of Wisborough Green on 23 August. He was only 26 months old when his mother died. Nothing is known of his upbringing, whether by his stepmother in Sussex or his mother's family in Hackney. He is next heard of when he inherited ten pounds in the will of his uncle John Dartnoll (Darknall) of Kirdford, Sussex in 1600.[13] He, like some of his Haynes great uncles, was destined to live a long, and at times, public life. He moved to London where, like so many of his wider family, he became a citizen and fishmonger.[14] He possibly stayed with his great uncle John Haynes who still held property in Old Fish Street, for it was there, at the Church of St Mary Mounthaw, that Jasper was married to Rose Merrick on 6 January 1604.[15] It was unlikely to have been an occasion of great celebration, for several of his fishmonger Gunnell cousins had died of the plague there barely ten weeks earlier. The marriage was presumably attended by his brother Nicholas, said to have been living in London, and his sister Thomasine; his great uncle, John Haynes, his grandmother Agnes Haynes and Uncle Richard Haynes might also have been there.

Jasper and Rose had three daughters: Ann, born in 1606 (bap.29 November), who died and was buried at St Michael Crooked Lane, on 15 August 1610, Thomasine and Faith,[16] who were reportedly alive in September 1625 and a son, William, baptised at St Michael Crooked Lane on 10 November 1608. Jasper Dartnall, a child, was buried at St John at Hackney on 3 July 1622. This location, the home of Jasper's Haynes forebears and cousins, suggests that the deceased might have been another child of Jasper and Rose.[17] Thomasine, was appointed executrix of the will of Sebastian Dartnall in September 1625, *'if shee bee alive otherwise my Cosin ffaith her Sister'*.[18] But Thomasine was a minor when Sebastian died and so her father, Jasper, was on 13 October 1625 commissioned by the probate court to administer the will on her behalf during her minority, Jane, Sebastian's widow, at the same time renouncing any part in the administration of her husband's will.[19] No further information as to what happened to Thomasine and her sister has been pursued.

Jasper's son, William, was the godson of Henry Timberlake, the wealthy London merchant, sea-captain, ship-owner, famous traveller and good friend of the Hayneses of Hackney (Timberlake acted as executor of Henry Haynes of Hackney and his uncle William and

THE NEXT GENERATION

aunt Anne Haynes of Chessington); Timberlake remembered William (his kinsman) in his will in 1625. William was in Hamburg in January 1631, from where he penned an informative letter on the military successes of Gustavus Adolfus, King of Sweden,[20,21] against the German Imperial army and of the meetings soon to be held at Leipzig with the Duke of Saxony, The Marquis of Brandenburg and other leaders of the Hansa towns. He wrote also of the Catholic League based in Frankfurt and his concern at what it might do, of which he promised to inform his father. William's letter indicates that his mother, Rose, was also still alive and well. In 1638 a William *Darknall* was listed as an inhabitant of St Mary Somerset, London, with a rent valuation of £9.[22] It is more likely that it was his second cousin, rather than he, who married Rebecca Rawleigh on 15 May 1649.[23] Both William and his mother were dead by 1652.

Jasper and Rose eventually moved to the Parish of All Hallows [Saints], Barking[24] where they leased property including the New and Old Wool Quays and Hartshorne Quay, the tenancy of which he acquired from Sir William Cope of Hauwell, Oxfordshire, in 1625.[25] These quays exchanged owners frequently but remained in private hands until 1721. The New Wool Quay had been renamed Custom House Quay in 1558, but the old name persisted in Jasper's time.

Jasper became a merchant with a lighterage business and was wharfinger at Custom House Quay; notices of several of his business deals have survived. Thus, on 9 June 1611 Jasper Dartnall, citizen and fishmonger of London, with Henry Gibb, esquire, a Groom of the King's bedchamber, and two others, entered a 12-year agreement with Sir John Bruce of Airth, Stirlingshire, Scotland, to receive Scottish coal at various points on the Thames. The contract was for 4000 tons in the first year and 8000 tons thereafter at a price of £13 per 22 tons (i.e. 11s 10d a ton).[26] The amount of coal to be received by each of the four recipients was not specified, but it serves to indicate that Jasper was by that time already dealing in large sums of money.

As wharfinger at the Custom House, Jasper was ever seeking increases in his lighterage allowances for various goods transported on behalf of the East India Company, whose House was situated in Leadenhall Street, next to the flesh market which he himself held at farm. In the Company's minute book for 29 November 1626, appears his petition to the Court of the East India Company for enlargement of his allowance for lighterage from 3d to 4½d a bag (of pepper), alleging that the freight of a lighter laden with pepper at the rate now given came to 30s and the cost to 40s. The Court of the East India Company saw no cause to alter the rate. Dartnall desired the Court to provide other lighters.[27] Much of his work for the Company was in the transport by barge of ordnance and timber between ships.[28]

His ownership of lighters led to a ground-breaking case relating to the responsibility of those transporting goods by water. A merchant called Simonds sued Jasper [Darknall; defendant] on the grounds that his goods had been damaged during transit in one of Dartnall's lighters, and that in common law every lighterman transporting goods for money should ensure that the conditions of his lighter were such as to render damage to goods impossible. Jasper was said to have '*so negligently governed his lighter, that the goods were damnified*' as a result (the lighter had presumably sunk). The case was found for the plaintiff despite an appeal by Jasper that Simonds had not averred that he [Jasper] was a common lighterman and had not mentioned how the goods had been spoiled.[29,30]

On September 15 1624 Jasper Dartnall and others relinquished to the Crown certain leases of the Tower Wharf in their possession. They received £2483 6s 8d in recompence for the leases and the cost of the repairs which they had undertaken.[31] Jasper's cousin, Sebastian, son of Edward Dartnall, also originally from Sussex, was a grocer of St Peter Paul's Wharf, London in 1623, living in the Parish of St Clement Danes. He and Jasper were clearly close as Sebastian left Jasper an English silver girdle with buckles and black silk stockings in his will, in which he also named Jasper an overseer, and the guardian of his under-age son-in-law Francis Phillips, the husband of Sebastian's daughter, Hester.[32]

On 16 October 1634 Jasper appears in the administration of the will of Anthony Hurst of Eltham, Kent, as a creditor and was granted the deceased's estate after Mary his widow, renounced it in Jasper's favour.[33] He was involved in a large number of legal actions, some of which were by virtue of his role of farmer of the Flesh Market in Leadenhall.[34] He was, for example, sued by Robert Peere and other butchers in 1638.[35] On Thursday, 16 November 1643 the Court of Great Ward was petitioned by the Master and Wardens and Commonalty of Butchers, to look into abuses in the running of all markets within the City. A Committee of Aldermen and others was set up to consider '*the matters of difference*' between the petitioners and

'Jasper Dartnall Farmer of the Flesh Markett in Leaden hall' and abuses in all other markets.

One of Jasper's first forays into litigation was to contest the will of his great-uncle, John Haynes,[36] which he petitioned through his proctor, the notary public, Richard Stubbs. The grounds for his action are unknown – Jasper may have been working for John after the death of his cousin, Moses Gunnell, and believed he had come to some arrangement over John's business; he could also claim to be an heir of Nicholas, lawfully begotten of his body, a condition of Thomas Haynes's will[37] when he passed on his lands to John; but the sentence pronounced on Wednesday, 1 June 1608 was in favour of William Gunnell, John Haynes's executor. One of his last legal challenges was against Lady Elizabeth and Sir Anthony Cope,[38] the former of whom he was again to sue over the All Saints Barking properties in his seventy-sixth year.

Jasper was still fighting injustices as a lonely old man after the deaths of both Rose, and their son, William. This was in October 1652 when he, citizen and fishmonger of London, was in his seventy-fifth year. His action was brought in the form of a petition to the Lords Commissioners for the Custody of the Great Seale of England against 57-year-old Sarah Baleere [Balier or Blyer] the daughter and eldest surviving child of Henry Timberlake (his son Thomas Timberlake having died in India in 1643), who had died some 27 years earlier,[39] who had bequeathed 40s each to both Jasper and Rose and £10 to their son, William. None of these bequests had been settled and he determined to put the matter right in his role as administrator of the goods and chattels of his son, William, as well as of his late wife. In his petition as complainant on 26 October 1652 he details all the circumstances as to how the non-settlement of Henry Timberlake's bequest came about, and claimed not only payment of the legacies from Sarah Baleere, then widow of Timothy Baleere, Clerk, former vicar of Tichfield, but also a sum of £20 for his work undertaken in helping her regain possession of certain tenements near the Tower of London, part of her late father's estate, which had been misappropriated. Apparently at Sarah's request six years earlier, and after agreeing to recompense him, Jasper Dartnall had assisted in the preparation for her successful suit against Sir William Balfoure, then Lieutenant of the Tower of London, and others, in order to recover these tenements.[40] He pointed out that, despite Sarah's substantial wealth, she had urged Jasper to postpone this action against her on the promise of full settlement, but as a significant time had passed, witnesses were becoming fewer, and no payment had been forthcoming, Jasper had no option but to sue. In her answer and demurrer sworn on 5 November 1652 Sarah denied that she had gained significantly from the court action against Sir William Balfoure because of the high costs incurred, denied that she had promised anything to Jasper for his assistance ['*I didn't employ him as a solicitor*'], and claimed that the settlement of the legacies had been the business of Jasper Dartnall and the late executors and her late brothers, and was none of her business. In any case, Jasper had been so '*very intimate*' with her late mother, Margaret Timberlake, after her father's death, that Sarah found it hard to believe that the legacies had not been paid. Indeed she had heard or surmised to the contrary. She believed this was an unnecessary and vexatious suit, which should not have been heard in this way, but that the complainant should have sought remedy in common law, as the action was '*not examinable in this court, neither has this court any power or jurisdiccion to assesse damages for the pretended care and paynes of the said complainant*'.[41] It is quite clear that the relationship between the lonely, embittered, elderly Jasper and the daughter of his former friends, Henry and Margaret Timberlake, had broken down irretrievably; that Sarah Baleere never had any intention of paying Jasper anything is clear from her demurrer. One forms the impression that she was probably trying to delay settlement until Jasper had died (he was buried on 7 March 1659 in Southwark).

Thomasine Dartnall

Thomasine was born in 1579 and was baptised in the parish church of Wisborough Green on 13 September. Thomasine appears not to have married. Jasper Dartnall and '*his sister*' were mentioned in an indenture dated 1640 relating to the surrender of a lease in Leadenhall, indicating that Thomasine was still living at that time.[42] She had a niece named after her.

Nicholas Dartnall

Apart from his birth in 1580 and his baptism on 9 October that year, and the fact that he also inherited ten pounds in the will of his uncle John Dartnall (Darknall) of Kirdford, Sussex in 1600, we know nothing for certain of Nicholas the third son of Thomasine and

William. There was however a Nicholas Dartnall (Dartnoll), said to be of London, who was tried on 26 July 1610 with 29 other *'English Pirates'* in Edinburgh before the Vice-Admiral of Scotland, his Deputes and certain Lords of the Privy Council, accused of piracy off the coast of Ireland the previous year.[43] All thirty were found guilty as charged and twenty-seven were executed, three being pardoned for providing King's evidence. Nicholas Dartnall was not one of them. He was sentenced to be hanged on the gibbet on the shore by Leith pier *'within the sea-flood and mark thereof'*,[44] the traditional place of execution for pirates.

Joan

Joan Haynes, also known as Jane, was born in 1562 and was baptised on 9 November that year at St John at Hackney. She received a bequest in her Uncle Thomas's will in 1582, but not in that of her Uncle Christopher in 1586, in which her sister Alice and brother were mentioned. Joan had probably died in the interim.

Thomas left her ('Jayne') a joined bedstead from his parlour with all the painted drapery belonging to it, along with a table from the hall together with stools, a cupboard, linen press and a wooden case for glassware. He also left her a settle with a painted covering, but she would only receive this bequest after the death of her aunt, Agnes. However there is good reason to suppose that Joan predeceased her aunt.

It is recorded that one of the daughters of *'Haynes of Hackney'* married into the Kay family of Woodsom, Yorkshire,[45] and if one of Nicholas's daughters, by exclusion it must have been Joan who married John, the eighth son of John Kay and Dorothy Mauleverer, as Thomasine had married William Dartnall and Alice was an unmarried minor. The marriage must have taken place soon after the 2 February 1582 when she was mentioned as Joan *Haynes* in her uncle Thomas's will,[46] as John Kay had five children by two wives between then and his death eight years later. Joan would have been 20 or 21 when she married John Kay. They had a daughter, Thomasine,[47] no doubt named after her aunt and adding more certainty to the likelihood that she was Nicholas' granddaughter. The couple had presumably met in Hackney where John Kay's elder brother Arthur (second son) had settled *('Arthur Kay of Hackney')* and married Ellen Welshe, the widowed daughter of Hippolite Lynnet, a fellow

Hackney resident and yeoman of the Royal Household with Nicholas Haynes. John Kay of Hackney was also a member of the Royal Household being a clerk of the averye in April 1579[48] and becoming an esquire and clerk controller of the Green Cloth from August 1585. There is an illustration of the arms of Kay impaling *'Heynes of Hackney'* in Hackney Archives Department.[49]

In the last years of his life, John Kay (*'scrivenor of Hackney'*) spent much of his time in the Bell Tavern in King Street, Westminster, in accommodation leased from the Dean and Chapter of Westminster. This lease was renewed after his death by his then widow Bridget on 3 December 1590.[50] Sadly, his detailed will of May 1589[51] mentions only his widow, Bridget, and their four children (John, Bridget, Dorothy and Lucy). His daughter, Thomasine, is not mentioned and therefore, like her mother, may be presumed to have predeceased her father.[52] John Kay's widow, Bridget, and her son, John, continued to live in Hackney, where on 1 October 1600 they were separately assessed in the third Subsidy for the Middlesex Hundred of Osulston, Hackeney,[53] she at £8 in lands and fees, he at £6 in goods. Bridget Kay died in Hackney and was buried at St John at Hackney on 2 May 1601. Her son appears to have become Sir John whose son, Arthur, was *'christened the 8 daye of August 1613'* at which ceremony, Richard Haynes would have officiated as churchwarden [below] by making the appropriate entry in the parish register. Sir John Kay [Keys] was buried on 4 May 1624 and his wife on 13 April 1632. His son, John, was created baronet in 1641, and served as a colonel on the Royalist side in the Civil War. Although his son Arthur died without issue, Sir John Kay had a number of distinguished descendants.[54]

Richard

Nicholas and Agnes had their first son, Richard in 1565. He was baptised on 18 June 1565, at St John at Hackney. Richard was destined to assume the mantle of heir apparent not only to his father but also to his uncles. He was brought up in Hackney and probably attended the grammar school there.

Richard is first mentioned in the will of his Uncle Richard, dated 11 March 1566, when he was barely nine months old. His uncles, observing the provisions of Richard and Thomas Hayneses' wills, bequeathed to him their lands in Berkshire. Uncle Richard Haynes left him the house at Hinton, Hurst, presumably one that was destined to become part of the estate known as Haines Hill:

> 'Item I give and bequeth unto my Brother Thomas Haynes fouretie shillinges and vi silver spoones and after his decesse the same spones to remeyne to my Brother Nicholas children Also I give to the saide Thomas my howse with the appurtenances sett lyinge and being in Henton (Hinton) and after the decease of the saide Thomas the same to remeyne unto Richard Hayn the sonne of Nicholas Hayne my brother and if he dye withoute yssue the same to remeyne to his Daughter'.[55]

However, his uncle Thomas did not bequeath his properties directly to Richard, but indirectly through uncles John and William, and although under the terms of the will of his Uncle Richard [d. 1566], Richard should have acquired the house at Hinton on Thomas's death in 1582, he may not have done so, as a property at Hinton subsequently appears in the will of his uncle John. Furthermore, John, who seems to have survived his youngest brother William, appears to have gone back on this entailment, instead leaving Richard only his '*greate Seale Rynge of gould with my Armes uppon yt*', and all his lands to Richard's cousin, William Gunnell.[56] The ring seal might have been the '*best*' gold ring originally bequeathed by Thomas to John, but it is much less likely to have been one on which the arms of Haynes were quartered with those of Foxley, as there is no record of John ever applying for such arms. John appears to have made an earlier will which he subsequently revoked, and it is not known whether or not in this earlier will he left all his Berkshire properties to Richard. This change of heart on John's part may have had something to do with a lawsuit in which Richard was a defendant, for which only a fragment of what appear to be his responses to interrogatories remains.[57] According to The National Archives, this document predated 31 December 1600 and therefore the subject events were in John's lifetime. It is just possible to make out that Richard had either acquired some of the property left him by his uncle Thomas, whose will he mentioned, or that someone else had taken possession in contravention of this will and effectively disseised him. Richard clearly regarded this litigation and the complainants as vexatious and said as much ('... *the compl[ainan]ts doe sue this deff[endant] in this courte meerely for vexac[ion]* ... '); in other words, he felt the suit was without merit and brought solely for the purpose of giving him trouble. Under the terms of Thomas Haynes's will dated 2 February 1582,[58] brought up in Richard's responses, all his lands and tenements held in England were to go to his brother, John Haynes for life, then likewise to William, then to Nicholas and his heirs, but *'provided alwayes and it is my will that my sayd brother John Haynes nor William Haynes shall not sell lett nor laye to mortgage ... , do or putt a waye the sayd lands tenements possessions and hereditaments or anie parte or p[ar]cell thereof without the speciall licence consent and agrement of the sayd Nicholas and his heyres'*. This was the nub of the issue, the property consisting of a messuage, tenement and garden, had been let to, or was in the possession of, the complainants without Richard's approval, his father Nicholas having meanwhile died. Furthermore, these tenants had not only refused to pay him rents due, but they themselves had sub-leased the property to one Michaell Wade,[59] effectively disseising Richard. It can only have been his elderly uncle John Haynes who had allowed this to happen, and this is consistent with Richard's statement that he: '... *doth not claime the said Rent p[re]served by the said John Haines uppon the said lease of the said messuage [tenement and] garden ...*', so the outcome of this case may be what led John to draw up a will in favour of his sister's children rather than Richard. The document emphasises that the messuage was "*in taile to the said Nycholas Haines his father by force of the last will and testament and of the said Thomas Haines and not in anywise as heyre to the said John*", which seems to have been at the heart of the legal decision whereby the property could eventually be willed to Richard's cousin.[60] But why didn't Richard contest John's will? Why was it left to his sister's son [Jasper Dartnall] to do so? Could it have been that John Haynes was one of the complainants in the suit, and that Richard's defence was unsuccessful? It is likely that the answers to these questions have been lost in the damaged and missing parts of the document.[61]

However, his uncle Christopher had also made Richard his heir in tail after the death of William Haynes and left him all his messuages, houses, lands tenements and hereditaments in Arundel and elsewhere. This was a considerable bequest, and as a condition of receiving it, Richard was to pay £20 to his youngest and last surviving sister, Alice. Richard would also inherit his parents' property in Hackney and Essex.

Richard's occupation is unknown, although in November 1594 he appears to have acted as solicitor on behalf of his uncle John at a case brought by John before the Court of Requests in which John was countering a common court action by William Page for payment of an outstanding debt. The £5 left owing of this debt was awarded by the court which was paid to Page through Richard Haynes acting on behalf of John.[62] This shows that Richard and his uncle were on good terms, five years before the abovementioned court case in which the two were on opposing sides.

Whether or not he was the same person, in 1597 a Richard Haynes is listed as a clerk attending in the office of Her Majesty's ordnance for which he received quarterly remuneration on 31 March of £5.[63] He was listed along with a John Haynes as a Gunner in the Lord Chamberlain's Accounts of 15 March 1604: *'Houshoulde'*. His identity remains to be established.

The Court of Requests was the usual court for bringing suits against servants of the Royal Household. On 12 June 1624, Thomas Anyon, President of Christ Church College, Oxford filed a Bill in Chancery against Robert Davison, a groom of the chamber. Evidence was given in Chancery that year by Richard Haynes, another groom of the chamber.[64] So, perhaps Richard, an attorney by profession, had followed in the family's service to the Household. Could this be the missing *'son'* to whom the 'Diarist' referred: *'... four are known to have held office in the household of Queen Elizabeth, one being the Groom of her Bedchamber ...'*?

When his daughter Ann was baptised in November 1601 he is described in the parish records as Richard Haines of Cambridge Heath, gent. It was here where, in the Middlesex Subsidy of 1597, he was assessed for tax on £3 in goods, the threshold for a householder in Hackney parish.[65] Ten years earlier [29 Eliz.] he had been assessed in the lay subsidy on £3 8s. After both parents had died, Richard moved to the family home in Mayre Streete [Meerstreete; now Mare Street], Hackney, where in the lay subsidies of 1620/21 and 1623/24 he was assessed respectively for goods at £3 10s 8d and £4.[66] However, he may have held on to his former dwelling for it was in Cambridge Heath that Richard's daughter, Susanna, subsequently held property.

In 1605 he is recorded as donating 2s 8d towards the repairs of St John's Church, Hackney.[67] There is also mention in August 1601 of a number of Hackney townsmen being out on the marshes in Mr Heynes' wheat field *'deciding where the boundary between*

Hackney and Stepney parishes lay'.[68] This could be a reference to Richard's field, or, less likely, to a field belonging to Benedict Haynes, who began to hold property in Hackney around this time.

He was an important member of the Vestry of St John at Hackney, the local governing body of the parish, whose principal responsibilities were the maintenance of the church fabric, management of property owned by the parish and of bequests to the church, the repair of highways and collecting and meeting the requirements for the poor of the parish.

Richard appeared in the vestry minutes probably before he was a member, his first mention being at the meeting held on 29 June 1602 in the *'A Seassment ffor Stocke for the House of Correction'* when £7 13s 4d was raised; Richard Heines, gent, was assessed at 16d.[69] In fact, the stocks were not built until 1614 when Richard was a churchwarden. In 1605 he appears twice in the account of moneys received for repairs to the church. In the first he is entered as Mr Richard Haynes, Mayre Streete, 2s 8d; in the second on 22 December, the churchwarden, Francis Maylkyn, listed him as giving a further 2s.

On 12 December 1605 he signed a memorandum of payments and disbursements amounting to £35 7s 7d and other sums in the hands of the churchwardens towards payment of the carpenter's and plumber's bills. At the vestry meeting on the 22 December Richard Haynes was involved in the assessment of the unpaid amount outstanding for the church repairs, including a sum of £14 still owing to the plumber and 33s 4d to George Willson, former churchwarden. Haynes and four others had to chase up those behind with their subscriptions towards the repairs.

On 1 June 1606, as one of the appointed overseers of the poor in the parish, he received 40 shillings from John Shellye the elder, to be disbursed among the poor of Hackney Parish. A further £20 was forthcoming on 20 October that year and handed into the care of Mr (Henry) Banister until the decision could be made as to its distribution.

On 28 July 1613 he was assessed at 4d for provisions for His Majesty's household.[70]

Members of the Vestry included the two churchwardens, two sidesmen, two surveyors of the highways, four collectors for the poor, overseers of the poor, the three clergy and a number of elected parishioners. Richard Haynes became a churchwarden at St John's for the year 1613[71] and again on 26 April 1614, for the second year being joined by Henry Haynes. Both men's signatures are clearly written in the minutes of the meeting. Also appointed at the meetings were the four Collectors for the Poor which in 1614 included William Lynnet, son of Hippolite. They were appointed *'according to the statute by Sir John Kaye Knight and Henry Thurston Esquire, Justices of the Peace'*. The two surveyors were also appointed for the year.[72]

Prior to 1613 in Hackney the meetings of the Vestry were open to all, but following a petition to the Bishop of London by many persons including Richard Haynes as churchwarden, seeking the right to have a *'new Vestrie'*, a closed or select vestry *'of the better sort'* (i.e. to keep out the riff-raff), it was minuted at the meeting on the 9 December 1613 that the Bishop had acquiesced.[73] After this date, the vicar and wealthier parishioners, Richard Haynes being one, established a monopoly on the *'new vestrie'* meetings and the decision-making which occurred. In effect, the new vestry membership was limited to the clergy, churchwardens and 32 named parishioners, a quorum being any 10 of them. Replacement members were co-opted. In the Vestry Minutes of 9 December 1613 it was agreed that the churchwardens (Richard Haynes and Walter Halliley) should identify all the wills of recently deceased Hackney residents in which legacies had been left to the parish or for the poor and then to organise payment to the church coffer.[74] This was important, because most wills included small charitable bequests for the care of the poor and such bequests reduced the burden of charges on those inhabitants who were regularly called upon for donations. The money was distributed in small sums to poor widows, orphans and other deserving parishioners at the discretion of Hugh Johnson, the vicar. Such contributions came under the Act of 1597/8, making payment of a poor relief rate thereafter compulsory. Several of these wills were copied into the Vestry Minute book. It must have been around this time that Richard Haynes and Walter Halliley collected £20 from the Countess of Oxford. They retained £10 apiece, out of which they both paid 20 shilling every year to the poor of the parish, a bond for which was handed to the churchwardens. According to a vestry minute dated 12 August, they were still paying this in 1622.[75] Another decision arising from the vestry meeting of 9 December 1613 was the appointment of James Chrychton BA as schoolmaster in the grammar school, the Vestry setting out his conditions of appointment including the weekly fees he could charge each child and the amount of notice required on termination that had to be given to the churchwardens of the time, who for his first two years included Richard Haynes. It was in the schoolhouse that some of the vestry meetings were held.

Richard Haynes as churchwarden oversaw the assessment made on the 17 July 1614 of contributors to the eight loads of hay for the King's Household in response to a warrant issued on 14 June that year. The other assessors were Walter Halliley, Miles Preswicke and Roger Rose. Richard Haynes wrote down the names of the 25 who were assessed, their acreage and contribution [loads or trusses] and appended his signature.[76]

In the vestry meeting of 22 August in 1614, his second year as churchwarden, the assessment was made for the great composition for his Majesty's household and for repairing the armour belonging to Hackney. Richard's contribution was 6d. In December of that year he was one of the assessors for the House of Correction.[77]

One of the responsibilities of Richard Haynes as churchwarden would have been to present the cases of couples living together out of wedlock to the Ecclesiastical and Manorial courts and, subsequently, to supervise their sentence of penance in Hackney Church which usually took the form of a marriage at which time the couple would upon their knees before the vicar and congregation confess their sins (*'that they have lived incontinentlie togethere before that time'*).

As an ordinary vestryman, he appeared active until July 1621. Thus, he voted for the parish surveyors (11 April 1615); voted a pension to Francis Collumbell, the clerk (31 August 1617) and was an assessor for the eight loads of hay to be provided for His Majesty's Household (8 July 1621). Thereafter he disappears from the vestry records until 1 May 1632, when he again appears amongst the vestrymen. His last attendance was recorded on 8 April, 1634, two weeks before his death.[78,79]

On 30 September 1606, following his mother's death, Richard acquired her half of the property around Earls Colne, Essex, including several messuages and almost 17 acres of land, of which in 1595 he had inherited a half share on his father's death. The property included land in Perse Myll Lane adjoining the manor of King's Garden and three crofts with land in the parish of Colne Engaine and White Colne called Cote Crofte, near Gosse Fenn and the manors of Sheriffes and Barwick Hall. He also held the aldercarr called Gosse

THE NEXT GENERATION

Fenn between Cote Crofte and Little Sixpence adjoining Inglestorpe manor. There is a map of Richard's Gosse Fenn holding, dated 1598.[80] It is noted in the Colne Priory Manorial Court Rolls[81] that, on Saturday, 7 January 1609 Richard surrendered (i.e. sold) his Essex landholdings, the new tenants (i.e. purchasers) being William Prentis and his two daughters, Helen and Elizabeth.

Richard is listed with Henry Haynes and Francis Lombley, farrier, all of Hackney, in the Sessions Records for Middlesex, 1616-18[82] where they stood surety for Christopher Thacker of Hackney for assaulting and beating Edward Willoughby, surgeon, in the King's highway, causing a serious, life-threatening head wound. Thacker was fined 5s.

There is another entry in the Middlesex Sessions Records for 1615-16 in which John Taylor was accused of *'breaking into the shop of Richard Haynes and stealing a cloak of his'*.[83] Taylor was found not guilty of house-breaking, but guilty of felony to the value of 11d, and sentenced *'to be whipped'*. This is the only record of Richard Haynes owning a shop in Hackney, the nature of which is unknown.

Much of Hackney was laid to pasture, hay being an important cash crop grown to supply London's horses. In July 1621 Richard, with others (*Walter Hallilie, Miles Preswicke, John Nicoll, Thomas Allen and Willm Linet* [son of Hippolite Lynnet], *with the 3 constables*) had again to source eight loads of hay for the King's Household under a composition agreement[84] [provision of an agreed quantity every summer at an agreed fixed price]. This huge amount of hay was the amount mown that year from 619 acres.

Richard was married, but the identity of his wife has yet to be discovered. There is evidence in the form of a coat of arms in which the arms of a Haynes male impaled those of Yonger,[85] but the identification of this lady with the wife of Richard Haynes, though plausible, lacks proof at this stage. They had a large family of whom the following are recorded: John (b. 1600), Ann (b. 1601), William (b. 1603), Elizabeth (b. 1606) and Susanna (b. 1612). Some of the parish records for the intervening dates are in such a poor state of preservation as to be illegible, and in view of the gaps between births it is possible that Richard had more children than those mentioned.

Lt-Gen Robert Haynes claimed to be *'a descendant of Richard Haynes, Royalist, forced to emigrate to Barbadoes in the time of Cromwell'*. The editor of the *Diary* suggested that it was Richard

Haynes, son of Nicholas, or his son, Richard, that first went to Barbados.[86] That it was not Nicholas's son, Richard, is now certain, as he died aged 68/9 years in Hackney where he, '*Rich. Haynes, gt.*', was buried on 21 April 1634. He had attended his last meeting of the Vestry only two weeks before on 8 April, indicating that his death was sudden. He probably succumbed to smallpox as 1634 was the year of a major epidemic of that disease,[87] which accounted for about 12 per cent all deaths that year. His only surviving daughter Susanna was to be his sole beneficiary.

Alice

Nicholas's youngest daughter, Alice, was born in 1568 and was baptised on 9 November at St John at Hackney. She was still unmarried in 1586.

She was remembered in the wills of two of her uncles. Thus, she received a useful bequest in her uncle Thomas's will of his best featherbed, bolster, pillows and white coverlet after the death of her aunt Agnes: she was also left twenty shillings.[88] Alice was also left £5 in the will of her uncle Christopher in Arundel (26 November 1586)[89] with the rest of his estate eventually going to her brother Richard, with the proviso that Richard should pay Alice an additional £20 from the Arundel estate proceeds. She certainly would have received both bequests, as her uncle William was executor of both Thomas and Christopher's wills, and would have made sure of it. Alice is not heard of again until she appears in the burial records of St John at Hackney for 20 December 1601, from which it may be inferred that she remained unmarried.

Robert – son or grandson?

A Nicholas Haynes is recorded as father of Robert Haynes at his baptism at St Clement Danes in April 1579. However, this child, if a son of Nicholas Haynes of Hackney, appears to have been unknown to (or ignored by) his uncles Thomas, Christopher and John as he wasn't mentioned in any of their wills. It is possible that he was the son of Nicholas Haynes, the painter and stainer, known to have lived at that time and presumably the one assessed at £3 in the London Subsidy Roll of 1597 for Aldgate Ward, St Katherine Crechurch parish.[90] No other Robert can be accounted for in the other Haynes family then in Hackney. However, in the nearby Parish of St Leonard,

Shoreditch, the baptism was recorded on 27 July 1586 of '*Robt Haynes ye sonne of John Haynes*', presumably a member of the Hoxton branch, John probably being a brother of Thomas Haynes of Finchley (Chapter 9). We have speculated that Nicholas and Agnes Haynes had other children, the proof being lost within the damaged Hackney Parish records; one of these might have been a son, Nicholas, the father of Robert born in 1579.[91] Certainly the birth date is consistent with Robert being either a son or grandson of Nicholas and Agnes Haynes, and it may not be without relevance that Agnes had a brother called Robert.

Land transactions involving a Robert Haynes took place in the early to mid 1640s on Barbados.

Gunnells of Wargrave; children of Christian Haynes

Christian Haynes married Thomas Gunnell, yeoman, at Wargrave Church on 8 November 1551, by whom, according to the Wargrave parish records, she had ten children: Dorothy, Moses [Moyses], Richard, Daniel, Christofer, Susanna, Christian, Sara, William and Thomas. Thomas Gunnell, their father, mentions by name only Sara, Moses, Daniel, Christopher and William in his will of 10 October 1580, although several of those not mentioned were indeed alive and flourishing.[92] Their uncle Thomas Haynes left 50s to Christian's children,[93] which would be consistent with only five being still alive in 1582, 10s being a respectable bequest for nephews and nieces of the time.

Thomas Gunnell was clearly a man of substance who, in due course, was to leave £45 10s in monetary bequests alone, in addition to a £25 8s annuity to a servant. Six children were born to Thomas and Christian before 11 March 1566, and their uncle Richard left £4 in his will of that date to be divided equally between them. The money was retained by their father, Thomas who, in his will dated 1580, passed on £1 to each of the four survivors, born in their uncle Richard's lifetime.[94]

The Gunnells were part of an extraordinarily large local family[95] that had been present in the village for several generations and Thomas's brothers, William, and John and their numerous children, lived nearby. There are several wills of the Gunnells in the seventeenth century, one of whom, John lived at Castlemans, Hare Hatch, named after Thomas Castleman, a contemporary of Thomas and Christian, who had earlier lived there. A house bearing this name still survives. Gunnells were still appearing in the Wargrave parish records in the eighteenth century.

Moses [Moyses]

Moses Gunnell [Moyses Gonnell] was his father's executor and residuary legatee,[96] who took over the family home and appurtenances. He was the eldest male child having been baptised on 7 January 1553/4 and therefore was one of those who received the 20 shillings left him by his uncle, Richard Haynes in 1566. On 26 November 1586 he was left £5 by his uncle Christopher Haynes.[97] Moses followed in the footsteps of his uncles William and John Haynes and became a fishmonger with them in the parish of St Nicholas Cole Abbey in the 1590s, possibly around the time of their departure from the Royal Household. He and his wife Katherine had six children: John, Margaret (bap 2 Jun 1594), Hellen (bap 27 Mar 1597), Sara (bap 1 Oct 1598), Moses (bap 31 May 1601) and Edward (bap 24 Oct 1602).[98] Hellen lived for only five months; Margaret and John died of plague in 1603 as did their father Moses who was interred in the Church of St Nicholas Cole Abbey on 21 October that year aged 49. Sara died in July 1608 and Moses's widow Katherine was buried in the south aisle of the church on 16 May 1611 by the pews *'marked 1.2'*.[99] Only Edward survived to adulthood; he married Joan Pettiman of Kent in 1630.

Daniel

Daniel Gunnell was baptised on 15 March 1559 in Wargrave church. He received 20s on his father's death in settlement of the bequest of his uncle, Richard Haynes, in 1566. He also received £6 (*'whyche ys in the handes of my brother wyll[ia]m Gonnell to be payde unto hym at the expiration of syxe yeares as apperythe by an obligac[i]on whearein the sayde wyll[ia]m standith bounde'*) under the terms of his father' will, which is noteworthy only in that the next three siblings were each left £6 13s 4d.[100] A Daniel Gonnell was living in Surrey in 1593/4 when he was assessed at £3 in the Surrey Subsidy Roll[101] of that year.

Christopher

Christopher Gunnell's baptism was not recorded in Wargrave. He was doubtless named after his uncle Christopher Haynes. He received the 20s from his uncle Richard Haynes and from his father: *'syx poundes thyrtene shyllinges fowre pence nowe in the handes of John Gonnell to be payde at the ende of tenne yeares nexte ensuinge as apperythe by an obligac[i]on wherein the sayde John standythe bounde'*. The John mentioned as holding the money was possibly Christopher's uncle, the testator's brother, who inherited the income from the estate of grandfather Thomas,[102] but he might also have been Thomas's cousin. No more is heard of Christopher.

Sara

Sara was born in February 1566, being baptised on seventeenth of that month. Sara Gunnell was Thomas and Christian's only daughter known to have survived them. Christian, her sister baptised on 24 August 1564, possibly died in infancy. As well as the £1 from her uncle Richard, Sara was left by her father: *'xiijli vjs viijd (£13 6s 8d) nowe in the handes of Wyll[ia]m Forde,*[103] *and to be payde unto her at the ende of viij yeares nexte ensuynge as apperythe lykewise by an obligac[i]on wheerein the sayde Wyll[ia]m standithe bounde'*. She would however, do rather better from the will of her uncle John Haynes.

On 10 May 1590, at the age of 25 years, Sara married Thomas Bigge, a London fishmonger, whom she doubtless met through her uncles William and John, for the marriage took place in the Church of St Nicholas Cole Abbey.[104] On the 20 July the following year was baptised her first and only child registered in that parish, a son, Alexander. Her husband had died by the time her uncle John Haynes penned his will in January 1506, in which she is mentioned as Sara Bigge, widow, by then aged 40.

Sara was bequeathed an engraved gold ring and five pounds in money in her uncle's will,[105] along with three rooms in Old Fish Street, London, for which she had to pay twenty shillings a year to her brother in rent. These rooms were near the shop and warehouse where her brother William Gunnell and uncle John Haynes had carried out their business as skinner and fishmonger. Another condition set out in John Haynes's will was that William Gunnell had to keep an eye on Sara's maintenance. She was obviously a much-loved niece.

William

William Gunnell was baptised in Wargrave Church on 5 September 1568. He received the £1 from his uncle Richard Haynes and from his father he received a bequest of *'vili xiijs iiijd'* [£6 13s 4d], which was left in the safe keeping of the vicar of Wargrave, Robert Blotheman *'yt to be payde unto hym at the ende of nyne yeares nexte ...'*[106]

William Gunnell appears next as his uncle John Haynes's principal beneficiary in 1606, inheriting the Haynes lands, etc. in Reading, Hinton and Twyford and elsewhere in Berkshire and Wiltshire. William was also the sole executor and residuary legatee of his uncle's estate and thereby inherited his property in London and Marylebone. He was 38 years old.

William married Elizabeth and by 1606 had two children, Thomasine and John, both of whom were mentioned in their great-uncle's will. He appears to have been in business with his uncle John Haynes, who, in his will, provided for the shop and warehouse in Old Fish Street, London, to pass in tail to William's son John in the event that William predeceased his wife. In view of its location in the heart of one of the most important fishmarkets, the nature of the business was most probably to do with fish, yet animal hides, furs, or leather, are also possibilities as two of John Haynes's witnesses were skinners.

After his uncle's death, William styled himself *'William Gunnell of Marylebone, gentleman'*.

It seems he was joined in business by another family member of the same name, styled *'William Gunnell, junior'*, who was a witness of John Haynes's will, as on 28 January 1610,[107] there was an assignment of a bond originally dated 29 Sept 1608 whereby William Gunnell, citizen and skinner of London and William Gunnell, of Marylebone co. Middx., gent., were indebted to Thomas Estcourte, citizen and leatherseller of London for £100. Estcourte assigned the bond to *'William Haynes of Chessington, one of his Majesty's Farmers of the custome of money'*, in settlement of £60 that Estcourte himself owed to the Exchequer. It was Haynes's job then to collect payment from the Gunnells. By coincidence this William

Haynes was the son of William Haynes of Hackney and brother of Benedict.

Note on William Gunnell Junior and Ralph Gunnell

William Gunnell junior appears to have been a cousin of William Gunnell, gent., who was baptised at Wargrave on 24 January 1574. He was the son of another Thomas Gunnell of Hare Hatch, Wargrave. As William Gunnell, citizen and skinner of London, he made his will on 17 September 1622, while in good health, prior to departing overseas. He was in debt to his *'brother Raphe and divers others'*. His brother Raphe [Rafe] was his executor. He mentioned but did not name his wife and children, although from the probate dated 20 May 1626, we learn that her name was Margerie. There is little doubt that this was the William Gunnell, citizen and skinner of London, who was party to the Estcourte debt with his cousin.

Raphe Gunnell, the penultimate child of Thomas of Hare Hatch, was baptised in Wargrave church on 9 October 1580. He married Lydia Halley on 14 November 1614 at St Stephen Walbrook and St Benet Sherehog, London and had a daughter Lydia baptised on 17 September 1615 St Mary Aldermary, London, and other children. From Raphe's will dated 20 April 1636,[108] we discover that he too was a *'Citizen and Skinner of London'*. He left a third of his considerable estate to his children equally, and the residue after debts etc. to his wife, Lydia, who survived him by 17 years, and did not remarry. From her will dated 20 March 1652,[109] we learn the names of their eight children: two sons, Mathew and Thomas (Mathew was married to Elizabeth; Thomas had gone overseas) and six daughters: Lydia, Mary, Anne, Elizabeth, Hannah and Parnell. Parnell Gunnell was her mother's executrix. She mentioned seven grandchildren.

Other Children of Gunnells of Wargrave

The baptisms of Dorothy (28 October 1552), Richard (27 March 1556), Susanna (20 June 1562), Christian (24 August 1564) and Thomas (31 August 1569) are all recorded in the Wargrave parish register.[110] They were the children of Christian (Haynes) and Thomas Gunnell. Some may not have reached adult life, as they were not mentioned in their father's will.

Notes and references

1. Wisborough Green, Sussex, parish register. The name was spelt variously Dartnall, Darknall, Darkenall, Dartnoll. Dartnall emerged as the preferred spelling of William and Thomasine's children.
2. William's brothers were Christopher, John and Edward; his sister Mary married John Naldect, or Naldret, on 21 January 1579; Dorothy married John Waller on 12 Jun 1571 and the third sister was married to Richard Gayneford. See Will of Elizabeth Darkenall, widow, late wife of Edward Darkenall of Slinfold, dated 15 December 1577, probate dated 28 January 1578: TNA, PROB 11/60.
3. A thirteenth-century fortified manor house founded by the Bishops of Chichester and occupied since 1256. In December 1583, three years after her death, William *Darkenoll* is described as *'of Drungewek'*, Sussex. Drungewick is also mentioned in the will of William's mother, Elizabeth – note 2 – from which it appears that she had transferred Drungewick to William for £100 less the cost of looking after her there.
4. TNA: Will of Elizabeth Darkenall, widow, late wife of Edward Darkenall of Slinfold, dated 15 December 1577, probate dated 28 January 1578: PROB 11/60.
5. Ibid.
6. TNA: Will of Christopher Haynes gentleman of Arundel, 20 November 1586 PCC PROB 11/69.
7. TNA: Will of Elizabeth Darkenall, widow, late wife of Edward Darkenall of Slinfold, dated 15 December 1577, probate dated 28 January 1578: PROB 11/60.
8. F. H. Garrison, 1929, *An introduction to the history of medicine*, p. 243.
9. Sussex Record Society, vol. 9 p. 3; recorded as *'Chap'* in the parish register.
10. Will of John Darknall yeoman of Kirdford, Sussex, dated 19 March 1600; probate 19 September 1600: PROB 11/96. He remembered the three sons of his brother William and left William the interest in his lands called Nether Tote at Slinfold and lands known as Symons at Billingshurst.
11. Ibid.
12. TNA: Will of Elizabeth Darkenall, widow, late wife of Edward Darkenall of Slinfold, dated 15 December 1577, probate dated 28 January 1578: PROB 11/60.
13. TNA: Will of John Darknall yeoman of Kirdford, Sussex, dated 19 March 1600; probate 19 September 1600: PROB 11/96.
14. William Bruce Armstrong, 1892, *The Bruces of Airth and their Cadets*, Privately Printed.
15. Harleian Soc., 1977–9, 'Marriage Records', IGI.
16. Parish records of St Michael Crooked Lane, 29 November 1606, 15 August 1610. Will of Sebastian Darknall of St Clement Danes, Middlesex, 13 October 1625. This also mentions another Ann Dartnall.
17. Parish records of St Michael Crooked Lane, 10 November 1608. TNA: SP 82/7/f. 5.

THE NEXT GENERATION

18. Will of Sebastian Darknall of St Clement Danes, Middlesex, 13 October 1625. Sebastian and Jasper were first cousins.
19. She became Mrs William Awberry in 1642.
20. TNA: SP 82/7/f. 5.
21. Gustavus Adolphus, later to become known as Gustav Adolf the Great, had landed with his Protestant forces, many from the British Isles, in Germany on 6 July 1630 and soon overran Hamburg and the surrounding area.
22. 'The Inhabitants of London 1638 (1931)': http://www.british-history.ac.uk/report.aspx?compid=32036.
23. Parish records of St Peter Paul's Wharf, London 15 May 1649.
24. T.C. Dale, 'The inhabitants of London in 1638 – All Hallowes Barking', from Lambeth Palace MS 272 [London, 1931] pp. 3–8 (1931). His rent was valued at £8 in 1638.
25. J.M. French, 1939, *Milton in Chancery*, New York, The Modern Language Association of America, pp. 191–2: Sir William Cope was indebted to Henry Timberlake, a kinsman of Jasper Dartnall, who also lived in the Parish of All Hallows, Barking, to the tune of £3947 secured on lands in Little Wakering, Essex, and a mortgage of £3000 on Withycombe ground in Drayton, Oxfordshire, and because of a defect in Cope's title on those lands, it took until November 1627 for a the decree in Chancery to be obtained for receiving repayment of the debt out of rents due to Sir William out of Custom House Quays and wharves with tenements, Barking, London. The decree ordered quarterly payments of £181 10s from receipts of the Quays for satisfaction of the debt. This would presumably have been paid by Jasper Dartnall as Cope's tenant. However, the debt does not appear to have been paid (Hampshire Record Office: 43M48/317; Hampshire Record Office: 43M48/374, Will of Henry Timberlake, 10 July 1625; Commissary of London, Vol. 29, f.211; Henry died in September, 1625 [TNA: C10/15/37]). The Custom House Quay was acquired by William Coltman, whose rent was valued at £15 in 1638. The Copes eventually acquired the Bramshill estate, Hampshire, the ancestral home of the Foxleys, the alleged forebears of Jasper Dartnall.
26. William Bruce Armstrong, 1892, *The Bruces of Airth and their Cadets*, Privately Printed.
27. 'East Indies: November 1626', *Calendar of State Papers Colonial, East Indies, China and Persia*, Vol. 6: 1625–1629 (1884), pp. 258–75.
28. Ibid. pp. 486–94.
29. *B.R. Simonds v Darknall*, in C. Viner, 1791, *A General Abridgement of Law and Equity*, 2nd edn, vol. 1, p. 222: Palm 523, Pasch. 4Car.
30. Addison (1847) described the situation as follows. In the case of the carriage of goods by water, it is often difficult to determine whether the loss has heen occasioned by the act of God or the act of man. If the carrier by water overloads his boat or vessel, and by so doing causes it to founder in a gale of wind, and the goods on board are lost, the loss is occasioned by the negligence of man, and not by the act of God; but the opposite is true if the boat has not been overloaded, but sinks solely through the

violence of the winds and waves. If a lighterman or barge owner shoots a bridge in stormy weather or at a dangerous period of the tide, and the barge is sunk and the goods lost, the loss is occasioned by the negligence and misconduct of the lighterman; but if the latter has shot the bridge at a proper time and in proper weather, but the barge is then taken aback by a sudden gust of wind, and driven against the abutments of the bridge and sunk, and the goods on board lost, the loss is deemed to have been occasioned by the act of God, and the owner, consequently, is exempt from responsibility in respect thereof. C. G. Addison, 1847, *A treatise on the law of contracts and rights and liabilities* ex contractu, Ch. 24, Section II, London: W. Benning & Co., p. 797.

31. 'Calendar of State Papers', Domestic, Jas I, 1623-1625, p. 340.
32. Will of Sebastian Darknall of St Clement Danes, Middlesex, 13 October 1625. Francis Phillips had married Hester Dartnall on 18 December 1623, at Holy Trinity Minories by licence.
33. L.L.Duncan, 1893, *Archaeologia Cantiana 20*, pp. 1-48.
34. S. Perks, 1922, *History of the Mansion House*, Cambridge University Press, p. 63.
35. P. E. Jones, 1976, *The Butchers of London: A History of the Worshipful Company of Butchers of the City of London*, Secker and Warburg, p. 91.
36. Sentence of the Will of John Haynes, 1 June 1608, Windebanck Quire Numbers: 56-114; TNA: PROB 11/112.
37. Berkshire Record Office: D/A1/76/205; Will of Thomas Haynes of Whitley, Parish of St Giles, Reading, 1582. Text reproduced by permission of the Berkshire Record Office.
38. TNA: C8/120/24; J.M. French, 1939, *Milton in Chancery*, New York: The Modern Language Association of America, pp. 192.
39. TNA: C10/15/37.
40. This may have been what was written up in the House of Lords Journal on 30 April 1647 as 'Dartnall's Cause', a petition which the Lords referred to the Court of Chancery, to have such *'Relief as shall be according to Justice and Equity'*. From: 'House of Lords Journal Volume 9: 30 April 1647', *Journal of the House of Lords*, vol. 9, 1646 (1802), pp. 160-165. Sarah won the case in Jan/Feb 1648.
41. TNA: C10/15/37.
42. J. C. Davies, 1972, *Catalogue of Manuscripts in the Library of the Most Honorable Society of the Inner Temple*, f.113. OUP, p. 540.
43. Robert Pitcairn, 1833, *Ancient criminal trials in Scotland*, Vol. III, Part 1, 1609-1615, Edinburgh, Bannatyne Club, pp. 123-6.
44. *Extracts from the Records of the Burgh of Edinburgh, 1931*, Vol. 7, 1604-1626, p. 349, Scottish Burgh Record Soc. A full record of this remarkable trial has been preserved.
45. 'Visitation of Rutland 1618-1619', pp. 23-24.
46. Berkshire Record Office: D/A1/76/205; Will of Thomas Haynes of Whitley, Parish of St Giles, Reading, 1582. Text reproduced by permission of the Berkshire Record Office.
47. C. R. Haines, 1901, *Notes & Queries*, 9th S., VII, p.172.

48. British Library: Lansdowne 34, 31 ff.78-88v.
49. Hackney Archives Department: D/F/TYS/47; XP 399: The legend reads: *'John Kaye mar ... d of ... Heynes of Hackney in com. Midd.* (see ped Ag Coll)'.
50. C. S. Knighton, 1999, *Acts of the Dean and Chapter of Westminster 1543-1609*, Boydell and Brewer.
51. Will of John Kay: PROB 11/73.
52. John Kay's brother Arthur had also died by 1589, as his wife Ellen was by that time Ellen Addye. There was another Kay [Kaye or Key], Thomas, working in the Royal Household as Avenor and *'accomptant'* in March 1577 (British Library: Lansdowne 34, 31 ff.78-88v); the Avenor was an officer of the stables charged with the provision of oats for the horses.
53. Middlesex Subsidy Roll 234; TNA: E179/142/234.
54. Daniel Lysons, *The Environs of London: being an account of the towns, villages and hamlets within twelve miles of that capital, interspersed with biographical anecdotes*, vol. II, Part 1, 'County of Middlesex, Acton-Heston', 2nd edn, (1811), London, T. Cadell & W. Davies, p. 319.
55. Will of Richard Haynnes 14 December, 1566, TNA: PCC 33 Crymes & Morrison, PROB 11/48.
56. Will of John Haynes of Marylebone, 1 January 1605/6, Windebanck Quire Nos 1-55: PROB 11/111. Probate granted 12 January 1608.
57. TNA: C4/27 (2)/40.
58. Berkshire Record Office: D/A1/76/205; Will of Thomas Haynes of Whitley, Parish of St Giles, Reading, 1582. Text reproduced by permission of the Berkshire Record Office.
59. A person of this name is listed as a Haberdasher of Aldersgate Street; in the 1582 London Subsidy he appears as Maurice Wade, St Botulph's Parish, Aldersgate Ward.
60. TNA: C4/27 (2)/40.
61. The names of those mentioned in this manuscript are Richard Haynes, John Haynes, Michaell Wade, Nycholas Haines, Thomas Haines and John Thorne. The events mentioned date it to between 1594 and 1600. As William Haines can also be presumed to have died, it most probably postdates 1598.
62. TNA: REQ1/18/73; REQ1/18/133; REQ1/18/159.
63. *The Monthly Review or Literary Journal 1779*, p. 206.
64. Frank Marcham, 1931, *William Shakespeare and his daughter, Susannah*; Grafton & Co.
65. Middlesex Subsidy Roll 234: Hundred of Osulston, Hackeney; TNA: E179/142/234.
66. Hackney Archives Department: D/F/TYS/19, microfilm XP387.
67. R. Simpson, 1882, *Memorials of St John at Hackney*, Joseph Billing, vol. ii, p. 72. London Metropolitan Archives (LMA), P97/JN1/137: Vestry Minutes, Saint John at Hackney:, 9 December 1613; for disbursement of the donation of the Countess of Oxford, see 12 August 1622.
68. M. Taylor, 1999, 'Naboth's Vineyard: Hackney Rectory in the 17th Century', *Hackney History,* Vol. 5, Friends of Hackney Archives.

69. Hackney Archives Department: D/F/TYS/1 (Transcript of Vestry Minute Book 1581-1613, pp. 45-7).
70. Ibid. D/F/TYS/10-14.
71. Appointed by the 15 attendees at the *'new Vestrie'* on 24 February 1613.
72. Attendees of the vestry meeting on 26 April 1614 were: Hugh Johnson, vicar; Walter Halliley, outgoing churchwarden; Richard Haynes, churchwarden; Pawle Crooke; Henrye Row; Henry Banister; Bartholomew Smith; William Harvie (this is likely to have been William Harvey, who first described the complete circulation of the blood, and who was physician at St Bartholomew's Hospital, London, between 1609 and 1643. His father, Thomas Harvey had moved to Hackney after the death of his wife in 1605, where he died and was buried in 1623. William and his wife, Elizabeth Browne, presumably settled in Hackney which was but a short ride away from St Bartholomew's); John Shellie; Edward Cull; Nicholas Millard; Miles Preswick; Roger Rose; John Steward; William Linnett.
73. Hackney Archives Department: D/F/TYS/10-14.
74. R. Simpson, 1882, *Memorials of St John at Hackney*, Joseph Billing, vol. ii, p. 72. London Metropolitan Archives (LMA), P97/JN1/137: Vestry Minutes, Saint John at Hackney:, 9 December 1613; for disbursement of the donation of the Countess of Oxford, see 12 August 1622.
75. Ibid.
76. Hackney Archives Department, microfilm XP 386.
77. R. Simpson, 1882, *Memorials of St John at Hackney*, Joseph Billing, vol. ii, p. 72. LMA: P97/JN1/137: Vestry Minutes, Saint John at Hackney:, 9 December 1613; for disbursement of the donation of the Countess of Oxford, see 12 August 1622.
78. Ibid.
79. Hackney Archives Department, microfilm XP 386.
80. Terrier to accompany Map of Earls Colne and Colne Priory Manors [Essex Record Office: Temp Acc 898; Earls Colne Project document 44701995; http://linux02.lib.cam.ac.uk/earlscolne/survey2/44701995.htm].
81. Essex Record Office: D/DPr21[Earls Colne Project document 33300033: http://linux02.lib.cam.ac.uk/earlscolne/cprolls1/33300033.htm].
82. Middlesex Sess. Roll 560/134, 165; Sess. Reg. 2/442, 448.
83. Calendar to the [Middlesex] Sessions Records, vol. 3, 1615, p. 248.
84. Hackney Archives Department: microfilm XP 386.
85. Younger; argent on a bend between two dolphins nayant (embossed) sable, three eaglets displayed of the first; Hackney Archives Department: D/F/TYS/70/6/7; microfilm XP408; Lincolnshire Church Notes made by Gervais Holles, A.D 1634 to A.D. 1642, p. 54. A family of this name lived in nearby Stoke Newington.
86. E.M.W. Cracknell, ed. (1934), *The Barbadian Diary of Gen. Robert Haynes 1787-1836*, The Azania Press, Medstead, Hampshire, UK.
87. P. Erlanger, 1967, *The Age of Courts and Kings, Manners and Morals 1558-1715*, London: Weidenfeld & Nicolson; p. 185.
88. Berkshire Record Office: D/A1/76/205; Will of Thomas Haynes of Whitley,

THE NEXT GENERATION

Parish of St Giles, Reading, 1582. Text reproduced by permission of the Berkshire Record Office.
89. TNA: Will of Christopher Haynes gentleman of Arundel, 20 November 1586 PCC PROB 11/69.
90. TNA: E179/146/372b. On 27 April 1580, Nicholas Haynes of St Andrew, Eastcheap, painter-stainer, married Elizabeth Dawes, spinster of St Andrew in the Wardrobe, at that church; Hackney Archives Department: D/F/TYS/71/13.
91. There is also a record of a Nicholas Haynes being baptised at St Clement Danes Church, Westminster on 23 April 1576; was this Robert's brother?
92. TNA: Will of Thomas Gunnell or Gonnell, Yeoman of Wargrave, 1582, PCC Windsor, PROB 11/69. Sara, Moses, Daniel and Christopher; Richard is believed to have died and there was no record of daughter Christian in 1580.
93. Berkshire Record Office: D/A1/76/205; Will of Thomas Haynes of Whitley, Parish of St Giles, Reading, 1582. Text reproduced by permission of the Berkshire Record Office.
94. TNA: Will of Thomas Gunnell or Gonnell, Yeoman of Wargrave, 1582, PCC Windsor, PROB 11/69. Sara, Moses, Daniel and Christopher; Richard is believed to have died and there was no record of daughter Christian in 1580.
95. See: Parish Records of St Mary's Church, Wargrave.
96. TNA: Will of Thomas Gunnell or Gonnell, Yeoman of Wargrave, 1582, PCC Windsor, PROB 11/69.
97. TNA: Will of Christopher Haynes gentleman of Arundel, 20 November 1586 PCC PROB 11/69.
98. LMA: Parish Records of St Nicholas Cole Abbey (St Nic CA) for dates stated.
99. Ibid. 16 May 1611.
100. TNA: Will of Thomas Gunnell or Gonnell, Yeoman of Wargrave, 1582, PCC Windsor, PROB 11/69.
101. S362 7v; TNA: E179/186/362 Surrey.
102. Will of Thomas Gunnell the elder of Hare Hatch, Wargrave, 1579, Fo. 170 Archdeacon of Berkshire's Reg 1574–84.
103. The Fordes were a well-known family with land in Wargrave and Hare Hatch; William Forde witnessed a number of deeds around the time of Thomas Gunnell's death. He was probably the brother of Richard Haynes's wife, Ann.
104. LMA: Parish Records of St Nicholas Cole Abbey (St Nic CA): 10 May 1590.
105. Will of John Haynes of Marylebone, 1 January 1605/6, Windebanck Quire Nos 1–55: PROB 11/111. Probate granted 12 January 1608.
106. TNA: Will of Thomas Gunnell or Gonnell, Yeoman of Wargrave, 1582, PCC Windsor, PROB 11/69.
107. TNA: E 214/802.
108. Witnessed by Andrew Aldworth, Robert Hanson and Sir William Salisbury.
109. Will of Lydia Gunnell: TNA: PROB/11/229.
110. Berkshire Record Office Parish Records of St Mary's Church, Wargrave for dates stated.

12

Children of Richard Haynes of Hackney [c.1600-1686]

Plague victims: John, Ann and Elizabeth – Richard, putative émigré or figment – William – Susanna, father's heiress – Edward Hopkinson – Nicholas Meade – Thomas Ballard – Susanna's children

The parish registers of St John at Hackney record the baptisms of five children of Richard Haynes. Sadly, Richard lost three of his children to the plague in March and April 1607.[1] His eldest son John, born in 1600, was buried on the twenty-second of that month. His burial entry in the parish records reads: *'John Haynes the son of Richard Haynes was buried the xvijth daye of March 1606'* [i.e. 1607]. By the entry is written *'plag: 9'* indicating the ninth death to have resulted from the plague during that particular visitation. John was not yet seven years old and may have been Richard's eldest son. Three-month-old Elizabeth also perished the day after John was buried. She was listed as the tenth plague victim. John and Elizabeth's deaths were to be followed within three weeks by that of their 5-year-old sister, Ann, in April 1607 who was buried on 9 April. Her entry in the parish register of St John at Hackney reads: *'Ann Haynes the daughter of Richard Haynes was buryed the ixth day of Aprill 1607'*, a note in the margin, *'pla'*, confirming that she was also a plague victim. Thus, in a single visitation, Richard's known family had been all but wiped out. Of the children whose names appear in the St John at Hackney parish records only William remained in 1607, or at least no indication of his death survived.

It is not possible to be certain how many children Richard fathered between Elizabeth in 1606 and Susanna, born in 1612, although none is evident from the parish register. Had his wife survived the 1607 plague that took her children, it is likely that they would have had more. It is possible therefore that she had also died and that Susanna, whose life was documented well enough to show that she survived to lead a full life and become Richard's heir 22 years later, was Richard's daughter by a second wife.

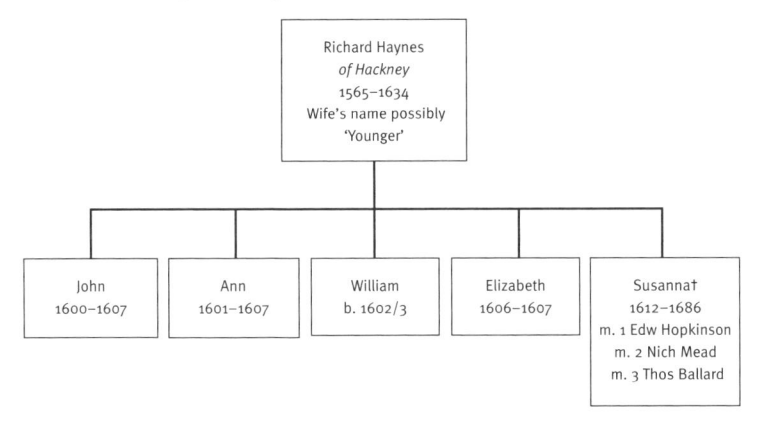

```
                    Richard Haynes
                      of Hackney
                       1565-1634
                  Wife's name possibly
                       'Younger'

    ┌──────────┬──────────┬──────────┬──────────┐
   John        Ann      William   Elizabeth   Susanna†
 1600-1607  1601-1607   b. 1602/3  1606-1607  1612-1686
                                            m. 1 Edw Hopkinson
                                            m. 2 Nich Mead
                                            m. 3 Thos Ballard
```

† Possibly the daughter by a second marriage

Figure 12.1 Family of Richard Haynes of Hackney

Richard, putative émigré or figment?

According to family legend, Richard was the putative son of Richard Haynes of Hackney who migrated to Barbados and founded the Haynes dynasty there.[2] However, evidence for a son of this name simply does not exist. In 1638 Susanna was accepted as being Richard's sole heir, which would have been most improbable had a son also survived.

A Richard Hannis who, given the orthography of the time, must be considered a possibility for our Richard, arrived in Barbados soon after settlement and became a planter. He was aged 21 and sailed on the ship *The Expedition* (Master, Peter Blackler), which left London for Barbados[3] on 20 November 1635, making his year of birth 1614, but there is no record of his baptism in Hackney where the records for this year did survive. Richard Hannis established himself as a planter in St Peter, for it was under that name that he is recorded as taking on indentured servants from England: one was John Feason, a labourer from Cawson in Cornwall, whose three-year-long indenture was dated 28 November 1655, and the other the following year was

Thomas Ilicke of Dunkerton, Wiltshire, indentured for seven years on 18 July 1656.[4] Richard Hannis retained the surname Hannis and never became a Haynes: he and his family are mentioned in the Will of John Hannis, gent of 31 May 1675.[5] Another possible candidate is one Richard Hames, who at the age of 18 years [b.1617], was put on board the *Falcon* for Barbados on 25 December 1635,[6] on whom, however, there is no further information.

Thus there is no evidence for the existence of a Richard Haynes, son of Richard Haynes of Hackney who left England for Barbados, and no record has emerged for the inheritance of the latter's extensive property holdings being contested by any surviving descendant. There is also no evidence to link either of the above-mentioned émigrés Richards with the Haynes family that subsequently emerged in St John's Parish, Barbados. This means that there is no evidence for the *Diarist's* assertion that his forebear was '*Richard Haynes, Royalist, forced to emigrate to Barbadoes in the time of Cromwell*'. However, there can be no doubt that a number of wealthy Royalists did move to Barbados during the political upheavals in the 1640s, which coincides with the years of the earliest land transactions of Robert and William Haynes there.

William

William Haynes son of Richard was baptised at St John at Hackney on 9 January 1603.[7] He may also have died young as there is no record of him in Hackney after his infancy. At the start of the *'plague'* in 1603, the following was written in the parish burial records: "*Here began the great plague but I have set down none but men or women of note. I have left out all children and vagrants*". If William died during this, his first year, it would explain his absence from the parish records. The William Haynes who was found in Barbados in 1641 and who died in 1679 with property in St John's Parish and was clearly a close member of the St John family was almost certainly the son of Robert Haynes by his first wife and not the son of a Richard. His notices will be dealt with in the next chapter.

Susanna

Susanna was born in 1612 and was baptised on 26 July that year at St John at Hackney, the daughter of Richard Haynes. Susanna was possibly named after Lady Susanna, wife of Sir Henry Rowe, who had died two years earlier leaving a number of descendants of that name in Hackney. The Rowes were noted for their extravagantly lavish funerals in Hackney,[8] which would have undoubtedly made a deep impression on Richard Haynes, a member of the Vestry of increasing importance.

Susanna Haynes is so far apart from her nearest sibling as to suggest that she may have had a different mother. Although this is purely conjectural, it is entirely possible that Richard's first wife succumbed to the plague along with her three infants in 1607, although there is no record of this and the plague records in Hackney were reasonably comprehensive that year.

On 23 June 1630, by his indenture enrolled in the Chancery,[9] Richard Haynes transferred his property in Hackney and all of the properties inherited from Christopher Haynes in Arundel to Edward Hopkinson of London, Chapman, and Susanna Haynes, his only daughter and heir, for and in consideration of the marriage to be solemnised between Edward and Susanna. After the marriage the properties were passed to be held by them, and the heirs of the body of Susanna begotten by Edward Hopkinson, and in default of such issue, to the heirs of the body of Susanna, in default of which to the right heirs of Susanna, on the death of her father.

Richard Haynes died on 21 April 1634, and Susanna, who had married Edward Hopkinson, son of Edward Hopkinson of Halifax, Yorkshire at Lambeth on 7 June 1630, was Richard Haynes's only daughter and next heir.[11] Thus, she acquired on her marriage all her father's property in Hackney (worth annually in revenues after deductions 40s), and Arundel, Sussex (worth annually in revenues after deductions £3), previously belonging to her grandfather Nicholas Haynes and great uncle Christopher Haynes respectively, thence by descent to Richard Haynes, her father. This is evident from the record of the inquisition after Richard's death at the Questhouse in High Holborn on 7 May 1636, held before the escheator for the County of Middlesex, Robert Pigeon, esquire. As the jurors were not able to ascertain of whom and by what services Richard had held the various properties, a second inquisition was held on 20 October

1638, at which it was declared of whom the various properties were held. The final settlement came about by order of the Court of Wards and Liveries dated 24 February 1639.[10] It is clear that her father, Richard, had intended Susanna to have all the indentured possessions on her marriage to Edward Hopkinson, who appeared before the court on her behalf. The document proves beyond doubt the relationship between Susanna, her father, Richard Haynes, and grandfather, Nicholas Haynes. It also shows that Susanna was the only daughter and the description '*next heire*', implies that Richard had no surviving male heirs, at the time of his death in April 1634. In other words, William and the putative Richard, if not imaginary, had predeceased him. Richard's property holdings were enumerated at the start of the Order:

> Whereas by an inquisic[i]on taken at the Quest house in high Holborne in the countie of Midds the vijth daie of May in the xith yeare of his Ma[jes]tie[s] raigne [1636] that nowe is by vertue of his Ma[jes]t[i]e[s] writt of Mandamus to the Escheator of the said county directed to enquire after the death of Richard Haynes gent deceased It was found that the said Richard Haynes was longe before his death seized in his demesne as of a fee of and in one close of pasture with thappurtenances conteyninge by affirmation one acre w[i]th all the messuage, barnes, stables gardens and orchards to the same belonginge situat lyinge and beinge in the p[ar]ishe of Hackney in the said Countie of Midds ... And of and in five messuages and ten[emen]te[s]] and one shopp called a storehouse with thappurtenances situat lyeinge and beinge in Arundell in the County of Sussex ... And that he beinge thereof soe seized did by Indenture bearinge date the xxiiith daie of June in the vith yeare of his Ma[jes]t[i]e[s] raigne [1630] inrolled within vi monethes in his Ma[jes]t[i]e[s] high Co[u]rt of Chancery in considerac[i]on of a marriage had and solemnized between Edward Hopkinson and Susanna his wife to have and to hold the same to them and to the heires of the body of the said Suzanna by the said Edward to be gotten Remainder to the heires of the body of the said Susanna Remainder to her right heires forever And the said Edward and Susanna by vertue of the Statute for thansserringe v Rx into possession were seized therof accordingly ... And that the said Richard Haynes ... being seized aforesaid the xxith day of Aprill Anno Domini 1634 of such his estate died thereof soe seized And that the said Susanna in the said Indenture menc[i]oned was his daughter and next heire but of what age the Jurors were ignorant ...' [She was 21].

The Order then goes into the acquisition of the Hackney properties by Nicholas Haynes and their previous ownership back to the time of

Richard III [1485]. The Arundel properties were held of Thomas, Earl of Arundel and Surrey in free burgage as of his borough of Arundel.

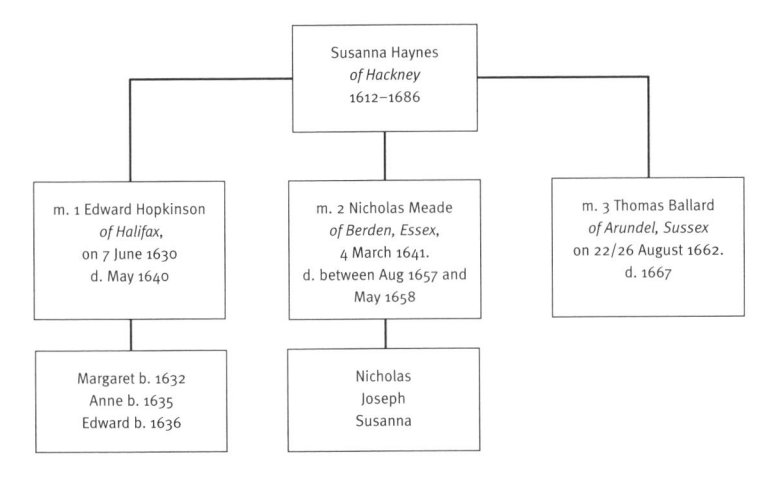

```
                    Susanna Haynes
                      of Hackney
                       1612–1686
┌───────────────────────┼───────────────────────┐
m. 1 Edward Hopkinson   m. 2 Nicholas Meade     m. 3 Thomas Ballard
    of Halifax,            of Berden, Essex,       of Arundel, Sussex
  on 7 June 1630            4 March 1641.       on 22/26 August 1662.
   d. May 1640          d. between Aug 1657 and        d. 1667
                               May 1658
│                              │
Margaret b. 1632            Nicholas
Anne b. 1635                Joseph
Edward b. 1636              Susanna
```

Figure 12.2 Family of Susanna Haynes

Susanna and Edward had their first child, a daughter Margaret in 1632 (baptised 6 May), followed by another daughter, Anne, in 1635 (baptised 9 July) and a son Edward in 1636 (baptised 14 April), all of whom were baptised at St John at Hackney. Edward Hopkinson senior died in May 1640, being buried on the twelfth of that month at St John, leaving behind Susanna, (to whom he referred as Susan, his loving wife), a 28-year-old widow and three infants. He left the Haynes properties in Hackney and Sussex to his wife for her lifetime and then to his son Edward and his heirs along with his own possessions in Halifax, Yorkshire.[11] He left his daughters £100 each on deposit. His will was witnessed by his cousin, Nicholas Enos, who also received a small bequest. In fact the Enos (or Enowes) were his wife's kin (second cousin; Susanna's, grandmother Agnes Enowes was the sister of Nicholas Enos's grandfather) originally from Earls Colne, Essex. Probate was granted on the 2 May 1640.

Susanna married second, Nicholas Meade of Berden Hall, Berden, Essex on 4 March 1641 [1642] at St John at Hackney. Nicholas was the elder son of Nicholas and Mary Meade and grandson of Edward [d. 1577] and Agnes.[12] The Meade family had lived in Berden since the fourteenth century. The family included Joseph Mead, Fellow of Christ's College, Cambridge, a man of varied learning, author of *Clavis Apocalyptica*, for a long time a standard theological work.[13]

Nicholas Meade had three children: Nicholas, named after his father; Joseph after the theologian, and Susanna after her mother.[14] The marriage lasted 16 years, when Nicholas Meade died some time between August 1657 and May 1658.

He left a considerable estate including copyhold lands and tenements held of the manor of Berden Hall in the County of Essex and mansions in Essex, to his *'now wife Susanna'*, thence to his eldest son Nicholas and his heirs. All his freehold lands and leased lands with appurtenances in Berden and Manuden and elsewhere in Essex, he bequeathed (after the death of Susanna) to his two sons Nicholas and Joseph and to his daughter Susanna in three equal parts and to their heirs respectively.[15] All these lands were by descent from his grandfather, Edward Meade.[16] He then appointed Master Thomas Walley[17] a clergyman, to convey to Susanna and her heirs absolutely the two newly built brick messuages with appurtenances situated at Cambridge Heath in the parish of Hackney,[18] in accordance with an agreement drawn up between Walley and the testator, leaving Susanna to pay the outstanding amount of £300 out of an agreed total of £540. His *'loving wife Susanna'* was his residuary legatee and executrix. So for the second time, Susanna and her children were very well provided for in their bereavement.

More than four years passed before Susanna married her third husband, Thomas Ballard of Arundel. He was a Sussex man, born in West Firle. He almost certainly met Susanna in Arundel, where in about 1646 he had been mayor as evidenced by his bequest of a silver bowl still in the possession of the corporation. As mayor and a Presbyterian, he was nominated as one of Cromwell's commissioners for securing the peace of the Commonwealth in Sussex.[19] He was again mayor of Arundel in 1655. He married Susanna Meade on 26 August 1662 in St Dionysius Backchurch, London, and the couple kept their links with Sussex; two years after their marriage, Thomas Ballard purchased a house, garden and stable from the corporation at Bulverhide.[20] Ballard had substantial property holdings in Broadwater, Laughton, Millcroft, Tarring Neville, West Firle and Worthing, all in Sussex. He died in London between 3 August and 2 November 1677, the date probate was granted.

Thomas Ballard left Susanna *'my loving wife'*, all the house and lands and household things that were hers before their marriage.[21] He also left her the old hospital lease that *'came by her during the residue of years yet to come and unexpired'*, and he gave her his 'new hospital lease commencing the twenty ninth September 1679 for nine years next following and all the rents and profits thereof during the term therein to come if she so long live she paying doing and performing whatsoever I and my executors stand [i.e. are committed] to do and save my executors harmless from the same whilst she enjoy it and my will is that whilst she holds the said lease that she pay unto her daughter Margaret Watson [née Hopkinson] the moiety or half part of the yearly rents and profit that she makes thereof the same every half year by equal portions'.

The only hospital in Hackney at that time and to which this might therefore refer, was the former Kingsland Leper Hospital, a mediaeval foundation run as a satellite of St Bartholomew's, Smithfield, and fronting onto Kingsland Road. At the time of Thomas Ballard's will it had become a general hospital for women of six wards. Patients were said to have been provided with the best food available. It also had its own chapel. The hospital was closed in 1760.

Susanna died at Hackney in October 1686 and was buried at St John at Hackney on the thirteenth of that month. She was 74. Like Pepys, she had lived through the Civil War, regicide, the Commonwealth, the Restoration, settlement of the North American colonies, the plague, Great Fire of London, wars with Holland, the exile of James II and finally saw the accession of William and Mary. She had also outlived those of her generation of the Haynes family who first settled in Barbados.

Notes and references

1. Parish Registers, St John at Hackney, 1607.
2. E.M.W. Cracknell, ed. (1934), *The Barbadian Diary of Gen. Robert Haynes 1787-1836*, The Azania Press, Medstead, Hampshire, UK.
3. Hotten's Lists, p. 139; the men had taken the oaths of Allegiance and Supremacy, and had been examined by the minister of the town of Gravesend *'touching on their conformitie to the orders and discipline of the Church of England'*.
4. Bristol Database of Indentured Servants,16541686: http://www.virtualjamestown.org/indentures/search_indentures.cgi?search_type=individ&id=430&db=bristol_ind; http://64.233.183.104/search?q=cache:XNgvOkjOAAJ: www.virtualjamestown.org/indentures/search_indentures.cgi%3Fsearch_ type%3Dindivid%26id%3D737%26db%3Dbristol_ind+%22Bristol+database %22+%2B+Indentured+%2B+Dunkerton&hl=en&ct=clnk&cd=2&gl=uk& lr=lang_en.

5. 'Will of John Hannis, Gent; 1675', Barbados Department of Archives, RB6/9, p. 341.
6. http://209.85.229.132/search?q=cache:aPkghU5RBcMJ:www.packrat-pro.com/ships/falconDec1635.htm+Falcon+%2B+%22Master:+Thomas+Irish%22+%2B+Hames&cd=1&hl=en&ct=clnk&gl=uk.
7. London Metropolitan Archives: Parish Records of St John at Hackney
8. Daniel Lysons, *The Environs of London: being an account of the towns, villages and hamlets within twelve miles of that capital, interspersed with biographical anecdotes*, vol. II, Part 1, 'County of Middlesex, Acton–Heston', 2nd edn (1811), London, T. Cadell & W. Davies, p. 318.
9. 23 June 1630 [6 Char I] indenture enrolled in the Chancery, mentioned in TNA: WARD 7/89/51.
10. TNA: Curia Wardorum et Libaconum [24 Feb Chas 13], WARD 7/89/51, WSRO: Add Mss 52,205.
11. Will of Edward Hopkinson, Gentleman of Hackney, 22 May 1640, TNA: Coventry Quire Numbers: 54-116; PROB 11/183.
12. Agnes Meade d.1601; see the Will of Agnes Meade of Berden, 31 March 1601, Commissary of the Bishop of London, Essex Record Office: D/ABW26/118.
13. Joseph Mead dedicated his whole life to the study and the theory of theology, philosophy, classical philology and many other sciences. His versatile interests included anatomy, logic and mathematics, Egyptian astrology, mythology and the religion of the Semitic people. He was admired during lifetime because of his encyclopaedic knowledge as the greatest scholar of history in his college. Meade was a close, good-natured and charismatic mentor of John Milton and Henry More.
14. The identity of the mother(s) of the three children has yet to be confirmed, although Susanna was almost certainly the mother of Susanna. The manner in which Nicholas subsequently bequeathed his property is consistent with all three being Susanna's children.
15. Will of Nicholas Mead, Gentleman of Hackney, Middlesex, 19 August 1657, Wootton Quire Numbers: 210-263, PROB 11/276.
16. Will of Edward Meade, Yeoman of Berden 10 Sept 1577, proved 19 November 1577; PCC44 Daughtry; PROB 11/59.
17. Thomas Walley was a famous divine at that time.
18. Cambridge Heath is at the top of the present Mare Street.
19. C. Durston, *Cromwell's Major-Generals: Godly Government during the English Revolution*, Manchester University Press, p. 61.
20. Sussex Archaeological Collections Vol. IX, 1862, p. 118.
21. Will of Thomas Ballard, Gentleman of London, 22 November 1677, PROB 11/355.

Part II
Barbados

13

Settlement on Barbados [c. 1635-1680]

Settlement - Origins - Known facts - Acquisition of land - The Earliest Hayneses - William and Robert Haynes - Elizabeth Haynes [- c.1674]- Edmund Haynes [- bef Oct 1685] - John Haynes [?1639 - 1678] - William Haynes [- 1679] - Other early Barbadian Hayneses - The Dutch Connection

On May 14 1625 a ship under the command of the British captain John Powell stopped to explore the island. After verifying that it was uninhabited, Powell returned to England to formalise a plan to establish a permanent settlement on Barbados. Two years later, on February 17 1627, a British ship under Captain Henry Powell (brother of John) carrying ten African slaves and more than 80 British colonists landed on the western side of the island, at a site later named Holetown Village. By 1639 the population of the island had grown, and the political turmoil in England was such that the colonisers felt able to create their own 'independent' legislature free of any control from London. This grew from 16 landowners chosen by the Governor in 1639 to a General Assembly of elected representatives the following decade.[1] At first, there were few colonists who could afford to purchase slaves, so most had to work the land themselves. European indentured servants were the primary source of labour during the seventeenth century, and as has been noted already Richard Hannis acquired two of them and Robert Haynes, mariner, a third (see Table 13.1). Poor, uneducated labourers were recruited in England, Scotland, and throughout Europe to work on tobacco and cotton plantations. These crops, however, were unable to compete with those produced elsewhere, and a drop in world tobacco prices in the early 1640s weakened the economy of Barbados. This led eventually to the adoption of sugar as the main source of revenue. By 1645 Barbados, with a total arable acreage of less than 100,000 acres, had more than 18,000 white colonists, of whom

11,200 were proprietors with an average holding, therefore, of less than 10 acres: there were only 1680 black slaves.[2]

Origins

At least three generations of Hayneses in Barbados 'knew' of and believed in their descent from Nicholas Haynes of Hackney and of the coat of arms confirmed on him, which they continued to adopt. This is clear from the Diarist's account in 1787[3] and the subsequent use of the coat of arms in the Hayneses' monumental inscriptions in St John's Church, Barbados. Despite this, there remains considerable doubt as to which of Nicholas's descendants, if any, was or were the first to migrate to Barbados. This is due to the lack of certainty as to the identities of all of Nicholas's children, the poor state of preservation and completeness of the parish records of St John at Hackney, and the loss of many of the early records in Barbados. What can be said with reasonable certainty is that the one from whom they claimed descent, Richard Haynes of Hackney, never left England. He died in Hackney in 1634 leaving as his sole heiress his daughter, Susanna. Thus no *proof* has so far been found as to from whom the early Barbadian Haynes were descended. It remains possible that they were not of the Hackney family thus far described and it may be that their origins were from a different family with names coinciding with those of the Hackney Hayneses with the latter being 'positioned' on the family tree at a later date.

Known facts

- A Robert Haynes, son of a Nicholas Haynes was baptised in April 1579 but not in Hackney.[4]
- The first Hayneses actually to be recorded in Barbados were Robert and William who must have arrived around 1640; Robert was styled gentleman.
- The names of Robert and William Haynes do not appear on any of the surviving passenger lists, which means that no ages are available that would have helped to link them back to a family in England.
- Robert and William Haynes purchased property in St John's Parish.
- This property formed the nucleus of what would become Newcastle estate.

SETTLEMENT ON BARBADOS

- The matriarch of the 'Newcastle' family was Elizabeth Haynes who died around 1674.

As will become clear, the evidence points to Robert and William, rather than others listed in the following table being the forebears of the family of Lieutenant General Robert Haynes in St John's Parish. As the analysis will reveal, however, the story had been complicated in the past by the application of twentieth-century definitions of family relationships to those appearing in seventeenth-century wills without consideration of the alternative meanings applicable at that time. For example, brother (sister; son; daughter) in 1670 was used for either a blood relative or an in-law: in-law could mean a relative by marriage as today, but was more commonly used for a step-relative; cousin could mean cousin, second cousin, nephew or niece, grandchild, indeed any non-sibling relative.[5]

Table 13.1 Persons called Haynes† recorded travelling to or on Barbados by 1660[6]

Name	Ship	Date of Departure	Barbados
Richard Hannis* (21)	*The Expedition*	20 November 1635	Settled in St James. He was a planter. His family was not obviously related to the Haynes family of St John.
Richard Hames (18)	*The Falcon*	25 December 1635	–
William Haynes			1641; land purchase 1643; land sale 1644; witness of land purchase by Robt Haynes 1645; land purchase
Robert Haynes, gent‡			1643; land purchase. 1644; land transfer to William Haynes
Elizabeth Haynes			1652; witness to a will
Richard Haines*			1655/6; planter; took on indentured servant
John Haynes[7]**	*John*	17 March 1656	Royalist prisoner sentenced to servitude; High Sheriff of Wiltshire's Trumpeter.
Anne Haynes			Marriage to William Byack 24 December 1656

Robert Haynes‡			1657; mariner; took on an indentured servant
Anne Haynes daughter of Elizabeth			Marriage to Robert Stewart 22 January 1659

†Including orthographic variants.
*Possibly the same individual. He retained the name Hannis and was a planter in St Peter: see Ch. 12 note 5 (Will of John Hannis, Gent, 1675), in which Richard Hannis's family are mentioned.
‡ No unequivocal evidence that they were the same individual: a '*Robt Haynes ye sonne of John Haynes*' was baptised in the Parish of St Leonard, Shoreditch, on 27 July 1586. John, son of Robert Haynes mariner of Shadwell (the Thameside maritime hamlet) and his wife Elizabeth, was baptised on 6 October 1639 at St Dunstan and All Saints, Stepney.
**Not obviously linked to Robert and William.

Acquisition of land – the earliest Hayneses

If the Hayneses had already arrived in Barbados by 1638, they don't appear to have acquired land amounting to more than 10 acres, as all inhabitants owning this acreage or more were listed in that year;[8] Robert Benson and John Cheeswright with whom the Hayneses were subsequently recorded as doing business were recorded as two of them. Really substantial land holdings, those above a hundred acres, were obtained in the 1640s by Captain James Drax, a distant kinsman of the Haynes family of Hackney, who purchased 200 acres in St George, in 1640. In St John's Parish the largest land purchases were those of Richard Estwicke (100 acres; 1640), Captain Thomas Hothersall (1000 acres; 1641), Richard Peers (400 acres; 1641), Thomas Batt (100 acres; 1644) and Captain William Hawley (820 acres; 1644).

The Hayneses were typical of the vast majority of settler landowners, who acquired a few acres at a time. Their first land acquisitions were recorded in the early 1640s, well before the execution of King Charles I, indeed the earliest being before the outbreak of the Civil War.

William and Robert Haynes

In 1641 William Haynes purchased from James Dillon 6.25 acres *'being part of a Plantation where John Mills now liveth'* in the Parish of Christ Church.[9] The deed of sale was signed by James Dillon in the presence of John Weston and Charles Morish. This is probably the first surviving record of land purchased in Barbados by a Haynes. William had also obtained other land in Christ Church as, two years later, he sold two parcels of land, each of about 12½ acres. The first was to Robert Morgan, merchant, to whom he sold '*my plantation situate in the parish of Christ Church*' for 2600lb of cotton and tobacco. The land adjoined that of [Peter] Edney to windward, Lieutenant William Pead to leeward and between the corn plantation and John Weston. The bill of sale, which was witnessed by Peter Edney and Robert Sidall, included all houses and other buildings and chattels including pigs, goats, poultry and farm implements.[10] The second parcel was sold to John Weston, one of the witnesses of the earlier transaction in 1641 and owner of nearby property, for 3000lb of neat clear cotton wool: '*twelve and one half acres of land situated in Christ Church Parish adjoining Capt [Peter] Edney's land, bounded on both sides by land of John Mills*'.[11] The sale deed was signed on 22 October 1643 by William Haynes in the presence of Thomas Cooper and Thomas Barnes and the transfer was recorded two months later on 28 January 1644.

In 1643, on 14 June, Francis Dickenson[12] and Robert Haynes, styled gentleman, jointly bought, from Alexander Lindsey, merchant,[13] 60 acres of land, '*fallen and unfallen*', lying in the Parish of St John, together with buildings, four negro slaves and two English servants,[14] livestock including goats, pigs, poultry and four assinegoes (young asses), a variety of tools and hardware, and standing crops of tobacco, cotton, corn and indigo, for 20,000lb of cotton wool, well cleared of seed and plant waste.[15] The land in question abutted the lands of Richard Ashton and other land of Francis Dickenson to the south-east, and the lands of St John's Church[16] to the south-west, other land of Francis Dickenson to the north and land of Thomas [?] to the north-east. The bill of sale was signed, sealed and delivered in the presence of John Johnson, and Christopher Starton. This was the probably the nucleus of what would become the Newcastle Plantation, which, after the acquisition of Dickenson's share, remained the property of the Haynes family until 1960, increasing in size to more than 500 acres. The purchase price in pounds of cotton wool was typical of the barter currency of the time. Although the first settlers used the English monetary system they were used to, there was very little actual money in the island. For a long time, therefore, cotton, tobacco and sugar were legal tender.

Robert Haynes and Francis Dickenson were next recorded as transacting business with Ensign Robert Benson[17] and John Cheeswright of St John. Benson was among the first to be recorded as owning land in Barbados. By March 1641 his plantation in St John's Parish, by this time grown to 50 acres or thereabouts adjoining the lands of Thomas Hothersall and Clement Roberts, was mentioned as the subject of a mortgage from Adam Thompson, gent. Ensign Benson and Katherine, his wife, had four sons and one daughter. Their eldest son, Captain Robert Benson married Mary Cheeswright, the only daughter of Richard Crossing of St John[18] and widow of John Cheeswright by whom she had two children, Elizabeth and John Cheeswright. After Captain Benson's death, his widow married Captain Thomas Downes.[19] The John Cheeswright who signed the following deed was probably Mary Crossing's first husband. The descendants of Thomas and Mary Downes were destined to become inextricably linked with the Haynes family of St John. Indeed several would become the direct ancestors of Lt-Gen. Robert Haynes.

Robert Benson and John Cheeswright signed a deed on 13 January 1644[20] binding them

> 'in the sum of 24,000lbs good [indigo] dye[21] and well cleared cotton wool[22] together with one indigo workhouse and 3 assinegoes lately purchased with all ye whole estate of either plantation or servants or cattle if we have or enjoy in the said island of Barbados, to be paid unto Robert Haynes and Francis Dickenson the full and just sum of 12,000lbs of ye like cotton wool or ye quantity of so much indigo as it will yield in cotton from man to man at ye sea side due to be paid at or before ye last day of June 1645 at some convenient storehouse at Indian Bridge'.

The witnesses to the signing were Thomas Kitchin and William Haynes.[23,24]

On 25 May 1644 Robert Haynes transferred to William Haynes:

> 'All that my Plantation which I formerly bought of Frances Dickinson situate in the said Island between ye Plantations of Richard Hackett to windward and John Reeves to leeward, Richard Adamson north, Captain Cotrington[25] south, 60 acres or thereabouts now in tenure and occupation of me the said Robt Haynes', together with all buildings, for 64,000lbs tobacco.[26]

The sale document was signed on 10 May 1644 by Robert Haynes and witnessed by Thomas Simpson and Cyprian Goare. This is clearly a different parcel of land from the 66 acres bought by Robert Haynes and Francis Dickenson the previous year, as the boundaries were different. It was possibly from the 'other lands of Frances Dickenson' mentioned in the 1643 transfer. The mention of Richard Adamson and Captain Codrington positions Robert Haynes's land in St John Parish.

On 3 June 1645 William Haynes with William Senets (William Sennott who left London aged 20 on the *Falcon* on 3 April 1635) purchased from John Ashurst[27] lands *'Adjoining ye Plantation of (Samuel) Richardson, Coroner, together with houses and appertenances'*. The sale agreement was signed by John Ashurst and was witnessed by Samuel Richardson and Nathaniel Garatt. The parish location of this land was not given.

So, by 1645, it is clear that William Haynes and Robert, had already acquired significant land holdings in the Parish of St John and William held land elsewhere. The precise relationship between Robert and William appears to be one of close kinship and the evidence suggests that they were father and son. Robert might have been the youngest son of Nicholas Haynes of Hackney or Nicholas's grandson [son of Nicholas; or perhaps of another son, John]. For reasons already discussed, he is unlikely to have been a son of Richard Haynes of Hackney. This uncertainty requires us to also consider that they may have had no kinship at all with Nicholas's family, but the family legend has proved correct in so many other details as to suggest that this line of enquiry should not readily be rejected without further evidence, given the circumstantial information linking a Nicholas, a Robert and Hackney. Furthermore, it is now known that Nicholas's brother, William, had at least one son, after whose death William also applied for the heraldic shield combining both his paternal and maternal arms. It has been argued that both he and Nicholas would have done so only for some perceived benefit to their descendants. In which case, the possibility must be considered that the Barbadian descent might have been through William. Although this narrative therefore continues by assuming the ancestral link between Robert and William Haynes of Barbados and a member of the original family of Haynes of Reading to be true, it is acknowledged that the link might have been more remote, such as descent from the Hoxton (Hackney) branch of the family.

Nothing else is heard of Robert Haynes on the island after the transfer to William recorded in May 1644, until a person of that name took on an indentured servant in Barbados, Edward Thomas of Brecknock, Wales, on 24 January 1657.[28] This Robert Haynes's occupation as the *'Agent'* is given as *'mariner',* and being a mariner might explain long absences from the island. Presumably he would only have taken a servant in Barbados if he had a home there. There is a little circumstantial evidence, summarised in Table 13.2, consistent with the two Roberts being one and the same, for a Robert Haynes, mariner, and his wife Elizabeth, lived in Shadwell in 1639, the year in which their son, John, was baptised at St Dunstan and All Saints Stepney on 6 October that year. This was the year that John Haynes of St John, Barbados, the son of Elizabeth, was said to have been born.[29]

Table 13.2 Circumstantial evidence linking Robert Haynes of Barbados and Robert Haynes of Shadwell

Fact	*Robert Haynes of Barbados*	*Robert Haynes, mariner of Shadwell*
Connection with Barbados	Yes: Land purchase/disposal, 1641-4	Yes: Took on indentured servant to work in Barbados, 1657
Wife	? Elizabeth: d. St John, Barbados c.1674	Elizabeth: mentioned with husband at son's baptism
Son	John: d. St John, Barbados, 1678	John: bap. 6 October 1639, St Dunstan and All Saints, Stepney
Son, John's birthdate	?1639 [see Oliver, note 44]	1639, year of baptism
Origin in England	Unknown: believed to be Hackney	Living in Shadwell in 1639, Parish of St Dunstan, Stepney
Styled	Gent	Mariner

The family of Lt-Gen Robert Haynes, the diarist, understood that the Newcastle Estate had been founded in the 1640s, 1647 being the date provided by Arthur Percy Haynes (1864-1928).[30] However, 1647 was the year that the islanders were visited by an epidemic, probably yellow fever, from which thousands of settlers died, ten males for every female, among whom one or more of the early Hayneses might well have numbered. Richard Ligon, on his arrival found the living were unable to bury the dead, so huge were the numbers.[31]

The next time that the name of Francis Dickenson, mentioned with Robert Haynes in the above deeds, occurs again alongside that of a member of the Haynes family is as a witness to the will of Richard Wallis of Barbados on 18 January 1652.[32] One of his co-witnesses was Elizabeth Haynes. Francis Dickenson was one of the gentlemen chosen as Trustee for the Parish of St John on 10 January 1652 and continued to play an active part in parish affairs for the rest of his life.

Elizabeth Haynes (–c.1674)

It so happens that the Haynes property in St John is next found in the hands of this Elizabeth Haynes, described as an elderly widow in 1674.[33] She and her family must have endured extreme difficulties over the previous twenty years, as did most of the early planters, when a series of natural disasters struck, including a plague of locusts in 1663, the Bridgetown fire and a major hurricane on 19 August 1667, drought in 1668 and torrential rain in 1669. Furthermore, from 1663 a series of taxes took their toll including 4½ per cent tax on all exports, a tax of 3lb sugar per acre of land and 10lb sugar per head on slaves;[34] import duties were imposed on alcoholic liquors. The fact that Elizabeth served as a witness of Wallis's will in 1652 suggests that she was already a widow and a landowner.

The identity of Elizabeth's former husband is uncertain although Robert Haynes must be the likeliest candidate. It cannot have been William for in 1674 we learn from Elizabeth's will that William was her *son-in-law*.[35] Significantly, we learn that he was to receive the same one-third share in her St John plantation as her own sons John and Edmund Haynes. Elizabeth also mentioned her daughter Anne, the record of whose marriage to Robert Stewart on 22 January 1658 [i.e. 1659] at St John's Church was one of the earliest on the island. Indeed, this is the only marriage of the Haynes family that must have taken place in the original timber parish church, for in 1660 the building of the first stone church began. Assuming Anne's age at marriage was about 18 years old, her birth date would probably have been around 1640. If her brother John had also been born within a year or so of this date, he would have been married by 1660 or thereabouts. There is no indication that Edmund ever married.

The description of William Haynes as Elizabeth's son-in-law should not be taken as it has been in the past to indicate the existence of another daughter. As it does today, the term *'in-law'* meant related by marriage: however from the fifteenth to nineteenth centuries it was also applicable to children, i.e. step-children.[36] So, the son of Elizabeth's husband by a former wife, her step-son, would properly be termed a *'son-in-law'*, and this is almost certainly what she meant the relationship to be when she mentioned William in this way. Had she meant William was the husband of her daughter, she would have phrased the relationship exactly the same as she did for her daughter Anne Stewart: '*(Christian Name) Haynes my daughter and wife of William Haynes*'.

Elizabeth Haynes also mentioned her son John's five children in her will:[37] Anne, Robert, John, Richard and Elizabeth, to each of whom she bequeathed 2000lb of muscovado sugar at 21 years of age or on marriage, whichever came first. The 10,000lb of sugar was owed her by her *'son-in-law'*, William Haynes, whom she called upon to pay this debt at the appropriate times.

Elizabeth also remembered three other *'granddaughters'* in her will: Elizabeth, Susanna and Jane Haynes, to each of whom she left 8000 lbs Musco[vado] sugar at age 21 or on marriage. These girls were said to be the daughters of John and Elizabeth Haynes. The relationship of this John to Elizabeth to the testatrix becomes clear only if the children were her late husband's grandchildren, i.e. her step-grandchildren, and not those of Elizabeth herself, as we would take grandchildren to mean today. They were not the children of Elizabeth's own son John, otherwise she would have said so, as she did for the children of John and Anne. Also, the fact that these three girls were not mentioned in the will of her natural son John,[38] which was also drawn up in 1674, suggests that they were not his children by a former wife. It may or may not be of significance that John and Elizabeth's three daughters were to receive a larger bequest of sugar from their grandmother than John and Anne's five children; this could have been because her own son John was to inherit a third of the St John plantation as well and of the likelihood of Edmund's share being left to the sons of John and Anne in due course. It was clearly Elizabeth's intention to do right by everyone in her will. It would therefore have been truly odd for one son-in-law to be treated differently from another. The fact that William Haynes received a third share of the plantation and Elizabeth's true son-in-law Robert Stewart did not is additional evidence that the two men differed in their relationship with the testatrix and her late husband – William must have had a greater right to a third share; a son's share. For this reason, the added fact that neither John nor his wife Elizabeth, the parents of the three *'granddaughters'* were themselves down to receive any bequest from Elizabeth Haynes can be taken to mean that they were already dead. Add to this that Elizabeth is unlikely to have given two of three sons the same name, and the conclusion must be that John, the deceased father of Elizabeth, Susanna and Jane Haynes, was also the son of an earlier marriage of the testatrix's husband, and full brother of William. The possibility has not been excluded that he was the royalist supporter *'barbadoed'* for his part in the Penruddock rising.

Given her children's likely birth dates, it is entirely possible that Elizabeth married around the time of, or not long before, their arrival on Barbados. From an estimate of her childbearing age range and assuming her three children were all born in wedlock, Elizabeth would probably have been born around 1610 (best guess range 1600 to1620). Elizabeth might of course have married a much older man.[39] She was described as an *'aged'* widow in her will, so was probably at least 60 years old in 1674. It remains only to name her husband.

The hypothesis that seems most plausible is that Elizabeth, the elderly widow and matriarch of the Haynes family of St John's Parish Barbados, was the second wife of a Haynes who by a previous marriage had at least two children, John and William; these had (or in the case of John, his family had) accompanied their father to Barbados. As Robert was involved with William in the land transactions in St John in the 1640s, and William was Elizabeth's step-son, then Robert Haynes must have been Elizabeth's late husband.

It is not known when Elizabeth died but it might have been in the devastating hurricane of 31 August 1675, which *'almost laid waste the whole island'* – the worst natural event since its settlement, according to Schomburgk;[40] St John's Church was so badly damaged that it had to be rebuilt the following year. This catastrophe had a major impact on the confidence of planters in seeing a return on their capital and little sugar was produced for the next two years.

READING TO BARBADOS AND BACK

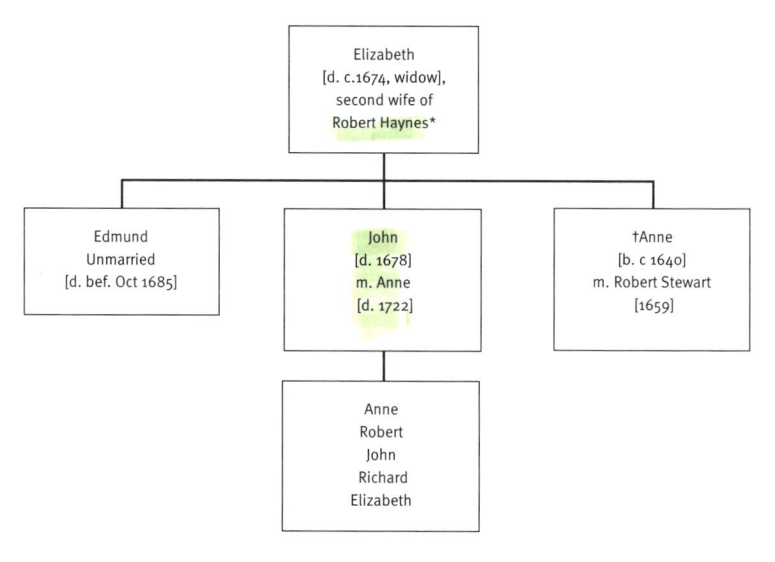

```
                    Elizabeth
                [d. c.1674, widow],
                 second wife of
                  Robert Haynes*
     ┌──────────────────┼──────────────────┐
   Edmund             John              †Anne
  Unmarried          [d. 1678]         [b. c 1640]
 [d. bef. Oct 1685]  m. Anne        m. Robert Stewart
                    [d. 1722]            [1659]
                        │
                      Anne
                     Robert
                      John
                     Richard
                    Elizabeth
```

* Possibly dead before January 1652.
† It seems likely from William Haynes's will that Anne married Richard Perriman Snr after her first husband's death some time between 1675 and 1679.

Figure 13.1 The Earliest Family of Newcastle Plantation – Family of Elizabeth Haynes, second wife of Robert Haynes

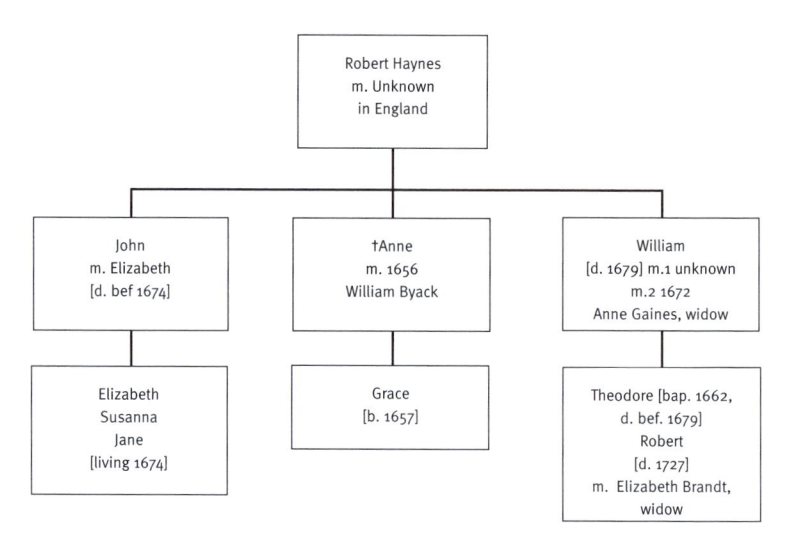

```
                 Robert Haynes
                  m. Unknown
                   in England
     ┌──────────────────┼──────────────────┐
    John              †Anne              William
 m. Elizabeth        m. 1656        [d. 1679] m.1 unknown
 [d. bef 1674]    William Byack         m.2 1672
                                    Anne Gaines, widow
      │                 │                   │
  Elizabeth           Grace          Theodore [bap. 1662,
  Susanna            [b. 1657]           d. bef. 1679]
   Jane                                   Robert
 [living 1674]                           [d. 1727]
                                     m. Elizabeth Brandt,
                                           widow
```

†This is the more likely position for this Anne Haynes, but her relationship to Robert is entirely speculative.

Figure 13.2 Putative step-family of Elizabeth Haynes of St John's Parish, Barbados

238

Edmund Haynes (– bef. Oct 1685)

Elizabeth's son, Edmund Haynes, had appeared earlier in the will of Allen Cammelland dated 28 June 1666[41] in which Edmund's brother-in-law, Robert Stewart (and presumably his wife Anne), is mentioned as living at Major Bell's. This was a reference to Philip Bell, a former Governor of Providence, who, then a Captain, was appointed Lieutenant Governor of Barbados in June 1641. Because of his judicious administration, Lord Carlisle appointed Bell Governor-in-Chief of the island in 1645, in which roll he was succeeded by Lord Willoughby in 1650.[42] After Willoughby's death in the hurricane of 4 August 1666, the island was governed by a commission of three, to which Philip Bell was appointed one of two extra Commissioners in March the following year with the brief to revise and set down all 58 Barbadian laws then in force, which were eventually sent to England for royal approval. Stewart was left sugar (by Cammelland) which was *'in his brothers hands, Lt Edmund Haynes'*. 'Edmond' Haynes was a witness of the will of George Foster (husband of Hester), planter of St John's Parish dated 23 November 1670.[43] Another witness of Foster's will was Samuel Forte who was said subsequently to have owned land known as Haynes Hill Plantation in St John's Parish and where his tomb and memorial dated 1711 still remains.

Edmund received a third share of his parents' estate on his mother's death and he effectively ran the plantation with his nephew Robert after the deaths of his brother and step-brother. He was dead by October 1685. There is no record of him being married.

John Haynes[6] (?1639–1678)

The birth year of John Haynes, Elizabeth Haynes's son, is not known, but he is probably the John Haynes said to have been born in 1639, mistaken by Oliver (1919)[44] as the birth year of his son, John of New York. In support of this, the fact that two of his children married in the early 1680s, is consistent with them being born in the early 1660s and John himself therefore would have been born in the 1630s–1640s. This must have been about the time his parents settled on the island. A John Haynes, presumably he, was a Constable of St John's Parish in 1661 and Surveyor of Roads in 1671.[45] He died in 1678, his will being signed and sealed on 23 January 1674[46] in the presence of Thomas Phillicott, his brother, Edmund Haynes, and Thomas Brett. It was proved on 18 July 1678 and entered on 20 July 1678 as '*John Haynes of the Island aforesaid (Barbados)*'. He was clearly a sugar plantation owner, but had not yet achieved the eminence of his neighbours.[47] Half of the plantation was left to his '*dear wife Anne Haynes*' and his son, Robert was appointed executor and residuary legatee. His two daughters Anne and Elizabeth each received a named '*negro woman*'(Anne's slave was called Jane and the bequest included her two children: Crab and Will), a cow and 6000lb of muscovado sugar when they reached 21 years of age or on marriage. They were '*to be maintained on the Estate*' until marriage. His two sons, John and Richard, were to be '*apprenticed to some trade, as judged by his wife, Capt John Leslie and Mr Thomas Phillicott*'. John and Richard were also each left 10,000lb of muscovado sugar, half at the age of 21 and the remainder a year later. We now know that the business of his son, John, took him to New York. Nothing else is heard of Richard, who is assumed to have died young.

Thus, Robert, his mother Anne and his uncle Edmund Haynes would appear to have been in effective possession of two-thirds of the St John [Newcastle] Plantation by 1678. William Haynes of St John's Parish, who held the remaining third, died in 1679, and probably left his share either to '*Edmund, my brother*' [brother-in-law] and to '*Robert, my cousin*' [nephew],[48] or more likely, to his own son from whom it was subsequently purchased. That one such interpretation is correct can be surmised from the Barbados Census of 1680, which lists Edmund and Robert Haynes of St John's as possessing a plantation of 77.5 acres and 29 negroes.[49]

Anne Haynes, John's widow, may have been younger than her husband. She married secondly into the Beachamp family, and her will dated 1722,[50] some 44 years after the death of her first husband, is invaluable in enabling some reconstruction of John Haynes's family, as she lists a large number of her grandchildren, and in one case, her great grandchildren, reflecting her advanced years. Thus, we learn that her grandson, John [the son of John Haynes], was already dead, but that John's three children, Robert, John, and Elizabeth Haynes, were under the guardianship of her granddaughter, Ann Critchlow, née Perryman, the daughter of Anne Haynes and Richard Perryman. Some of the children of her son Robert Haynes, mentioned in his own will, can be identified in that of Anne (Haynes) Beachamp. Widow Anne Beachamp's daughter, Elizabeth, appears in

SETTLEMENT ON BARBADOS

the will as *'daughter, Elizabeth Steward'*, an error for Seaward, the name of her second husband. This Elizabeth Haynes had married first, Samuel Branch who died in 1692, and their son, Samuel Branch, also appears in Anne Beachamp's will as her grandson. The identification of some of the children of Anne Beachamp's son Robert Haynes with those in his own will of 1696,[51] enables all the generations occupying Newcastle from Elizabeth Haynes onwards to be confirmed with reasonable certainty.

Anne [Haynes] Beachamp's final act by her will, signed with a cross, was to free her woman slave *'Asshee'*. Her executors in trust were her granddaughter, Ann Critchlow, and John Mower. The will was proved on 23 February 1722.[52]

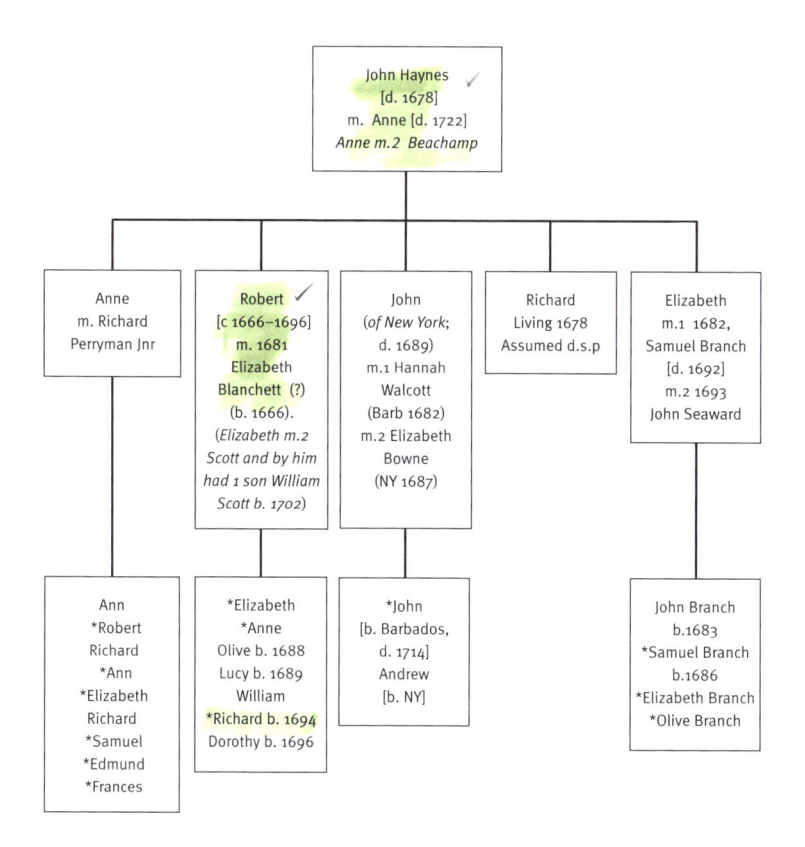

```
                    John Haynes
                     [d. 1678]
                  m. Anne [d. 1722]
                  Anne m.2 Beachamp
```

Anne	Robert	John	Richard	Elizabeth
m. Richard Perryman Jnr	[c 1666–1696] m. 1681 Elizabeth Blanchett (?) (b. 1666). (Elizabeth m.2 Scott and by him had 1 son William Scott b. 1702)	(of New York; d. 1689) m.1 Hannah Walcott (Barb 1682) m.2 Elizabeth Bowne (NY 1687)	Living 1678 Assumed d.s.p	m.1 1682, Samuel Branch [d. 1692] m.2 1693 John Seaward
Ann *Robert Richard *Ann *Elizabeth Richard *Samuel *Edmund *Frances	*Elizabeth *Anne Olive b. 1688 Lucy b. 1689 William *Richard b. 1694 Dorothy b. 1696	*John [b. Barbados, d. 1714] Andrew [b. NY]		John Branch b.1683 *Samuel Branch b.1686 *Elizabeth Branch *Olive Branch

*Ones mentioned in grandmother, Anne (Haynes) Beachamp's will in 1722.

Figure 13.3 Descendants of John Haynes and his wife Anne

Confusion had arisen in the past by the identification of Anne Beachamp with the Ann Perryman who married Thomas Beachamp in 1693. This Ann Perryman was most probably another granddaughter of Anne (Haynes) Beachamp, and possibly the first child of Richard Perryman and Anne Haynes. Alternatively, as the will of William Haynes (9 March 1678)[53] mentions his *'brother'* Richard Berriman [Perryman] Snr, who must have been married to a Haynes, it is possible that this Perryman's wife was Anne, the widow of Robert Stewart. On Perryman Snr's death, she would have been free to marry Thomas Beachamp.

William Haynes (– 1679)

The evidence for William being Elizabeth's step-son[54] would explain his active involvement in the St John's Plantation and why he and not Robert Stewart, Elizabeth's true son-in-law, was left one-third of it in her will. As William's property transactions have been examined earlier, only his later notices and an interpretation of his last will will be considered here.

William married late in life a widow, Mrs Anne Gaines at St Michael's Church on 2 June 1672. This is likely to have been his second marriage for William Haynes had a son, Theodore, who was baptised on 8 February 1662 at St Michael's Church. As neither Theodore nor his second wife were mentioned in William's will they had presumably predeceased him.

William witnessed a deed of sale of slaves and farm animals by Robert Bowman to Captain Robert Benson, on 1 May 1670, and on 2 October 1672 he witnessed an indenture between Captain Benson and Mary Pyott of Tring, Hertfordshire relating to a lease of a plantation of 39.5 acres in St John.[55] The penultimate notice of William Haynes was dated 18 September 1674, on which day he transferred a slave called Mary to Robert Gibbson.[56] William was also a witness of the will of Robert Benson, Gent, of St John's Parish dated 27 November 1675[57] and the will was *'Proved by Oath of Henry Walrond Jr and William Haynes'* on 7 February 1676.[58] It was possibly one of his last recorded acts, for his own will was dated 9 March 1679[59] and proved on 27 January 1680. In it William left bequests to Richard Berriman [Perryman] senior, brother (this must refer to Richard Perryman the father of the Richard Perryman who married

Anne the daughter of John and Anne Haynes); Richard Cockman, brother (brother-in-law, aged 63, husband of an as yet unidentified sister or half-sister) who left London for Barbados on the *Falcon* on 3 April 1635, aged 20; Elizabeth Wyatt (relationship unknown, possibly another half-sister or niece by marriage); John Haynes, godson (most probably nephew, son of his half-brother, John Haynes); Edmond Haynes, brother (half-brother), his son, Robert Haynes and Robert Haynes his cousin (nephew; son of John and Ann).

The likeliest contender to have been William's son, Robert, was Robert Haynes of St Thomas's Parish, although this attribution remains speculative. He would have been living and of the right age to have witnessed the will of Thomas Chander on 17 October 1718.[60]

In a deed executed on 15 March 1693, Roger Thomas of St Andrew's Parish sold to Robert and Elizabeth Haynes of the Parish of St Thomas '*all that Plantation or Parcel of land situated in St Thomas*', estimated to be 6 acres'.[61] The deed was signed and sealed by Roger Thomas on 16 March 1693 in the presence of Hanna Rogers and Mary Kidney. Robert appears from his will (below) to have become a wealthy planter in St Thomas. On the map of the island of about 1730, by the geographer Herman Moll (who died in 1732) subsequently printed by T and J Bowles, London, the plantation (marked Hayns) is situated at the eastern end of the parish adjoining the boundary with St Joseph's.

Some other records of property transactions have survived. Thus, on the 10 August 1695 Robert Haines of St Thomas and Elizabeth his wife, formerly Elizabeth Brandt(?) widow, sold to James Burrowes of St James about 10 acres of land, and everything on the land in the Parish of St Thomas.[62] Burrowes paid five shillings, presumably the ensealment charge. The transaction was signed by Robert and Elizabeth Haynes in the presence of Judith Jenkins, Martin Griland and Robert Hyampsill on 16 September 1695, and '*Elizabeth was examined personally in accordance with law*'.

In 1726 Archibald Reid discharged a debt to Robert Haynes by handing over 33 acres, 34½ perches of land as set out in the survey of Thomas Stevenson, Surveyor, to the value of £498.4.2½d.[63] Three tenements and appurtenaces were mentioned as was Henry Gibbs, who was also a creditor of Archibald Reid to the tune of several thousand pounds. This deed was signed on 22 December 1726 by Archibald Reid and witnessed by Lawrence Trend and William Timbrell (Chief Justice). It was proved on the following 4 February

(1727). The Henry Gibbs may have been the father-in-law of Robert and Elizabeth's only daughter, Elizabeth.

Robert died on 9 October 1727 and was buried in St Philip's; his monumental inscription was recorded in the *Gentleman's Magazine* of November 1863.[64] Elizabeth his wife was not mentioned in his will of 24 October, 1722[65] and presumably had predeceased him. His daughter Elizabeth, the wife of Joseph Gibbs, and Robert's executrix, received a legacy of one hundred pounds per annum, ten negroes of her choice and all profits of his land. His grandson, Haynes Gibbs, received the rest of the plantation's slaves and one thousand pounds. His six remaining grandchildren Jane, Mary, Alice, Katharine, Joseph and Thomas Gibbs each received sums varying between two and four hundred pounds. He left twenty pounds to Elizabeth Burrowes his sister. Plantations under the names of both Haynes (Hayns) and Gibbs (Gibbes) were to be found in both St Thomas's and St Peter's Parishes by 1730.

A note of possible significance in the abovementioned article in the *Gentleman's Magazine*[66] referring to this Robert, was that *'a Colonel Haynes of Cromwell's army, who was killed April 13 1655, at Hispaniola, was possibly his ancestor'*. This was a reference to the Major General James Haynes who was killed on Santo Domingo during an heroic, single-handed combat against insuperable odds, partly because of the cowardice of his soldiers recruited in Barbados.[67] In recognition of his heroism, his wife Elizabeth was awarded a pension of £150 by the Commonwealth Council of State. It has not proved possible to confirm the antecedents of Robert Haynes of St Thomas including the link with the Major General. Clearly much more work is needed to verify and extend these still hypothetical relationships arising from the consideration of the beneficiaries mentioned in William's will.

Other early Barbadian Hayneses

There were other Hayneses who appear in Barbados in the mid-seventeenth century. The first was Anne Haynes who married William Byack[68] on Christmas Eve 1656 at St Michael, and subsequently had a daughter, Grace, baptised on 13 September the following year. It has always been assumed that Anne Haynes, who would have been born about 1630–40, was of the St John family, even though she was married in St Michael. She has been provisionally placed as the daughter of Robert Haynes by his first wife. There was also Bartholomew Haynes who was a witness of the will of Nathaniell Leddra on 20 September 1664 and presumably the same person who was married to Sarah Cavare on 2 May 1671, also in St Michael. It is possible that Bartholomew and Sarah had a son, also Bartholomew, for a Bartholomew Haynes married Martha Bartlett on 1 December, 1698. The third was James Heayenes, who is mentioned in the will of Henry Arrion dated 13 February, 1670[69] as being in possession of a boat. Another Haynes, Francis, was listed as a passenger on an unnamed ship bound for Barbados from Bristol between 1654 and 1663: it is not known whether or not he reached his destination.

The Dutch connection

The Enowes [Enos or Ennew] were of Dutch ancestry; they appear to be among those who emigrated from Holland to Essex, many of whom worked in the cloth industry. It was Agnes, the daughter of Richard Enowes of Earls Colne, Essex, who married Nicholas Haynes of Hackney. By the seventeenth century they were spelling their name Enos. Nicholas Enos and Susanna Haynes [Nicholas Haynes's granddaughter] who were probably second cousins, kept in touch, and Susanna's first husband, Edward Hopkinson, left in his will dated 29 April 1640, his *'cosens Nicholas Enos and his wife twentie shillings apeece to buy each of them a ringe to weare in memoriall of me'*.[70] Nicholas Enos in his own will, dated 20 April 1678, left five pounds to his *'Dutch cousin Edward Enos'*. Enos's wife, Anne, was sister of James Drax, one of the first sugar planters on Barbados. The Draxes had a son, Henry, who had clearly received financial help from Nicholas Enos in order to establish a sugar plantation in Barbados. Nicholas Enos had rights and title to this plantation according to the terms of the indenture made with Henry. The loan interest was *'fiftie pounds Sterling and one hundredweight of Sugar yearly issuing and payable out of the Plantation and Estate of my nephew Henry Drax esqr in the Island of Barbados'*, which Nicholas bequeathed to his daughter, Mary Archer.[71]

James Drax and his brother, William, were recorded as owning more than ten acres of land in Barbados in 1638. They set up the Drax Hall Plantation at St George from 200 acres of land acquired in

1640 from the London Merchant House. The House at Drax Hall was built in the 1650s; of classical Jacobean architecture, it survived the hurricanes and remains one of the oldest properties on Barbados and, indeed, in the Western hemisphere. The brothers were truly innovative and introduced new milling machinery based on a model they had obtained in Holland. By 1654 James Drax had shares in two slave ships. James is generally regarded as the one who made *'the first-ever sugar fortune'*. It was in 1654 that Drax returned to England for good leaving his brother, William, and his son Henry in charge of the plantation. William Drax left Barbados in 1669 to set up Drax Hall Estate in Jamaica.

James Drax was involved in some irregular activity in which he removed sugar from the already loaded vessel, the *Samuel* of London, docked in Barbados, and transferred it to a Dutch ship. This action appears to have come to light at the instigation of Nicholas Enos, for in depositions made by the ship's crew in July and October 1651, reprinted in *'The Complete Book of Emigrants 1600–1660'* by Peter Coldham: John Ladd of London, mariner, deposed that he was master of the *Samuel* at Carlisle Bay (Bridgetown, Barbados) in April 1651 *'when sugar loaded by Robert Arundell was taken off by Lieutenant-Colonel James Drax and put on board a Dutch ship'*. Bartholomew Howard of Ratcliff, Middlesex, mariner, Master's mate of the *Samuel* of London, and William Tucker of Shadwell, Middlesex, boatswain, deposed *'at the request of Nicholas Enos of London'*, merchant, that *'sugar loaded on their ship in Barbados was taken off by Lieutenant-Colonel Drax'*[72]. These depositions show that Enos's interest had been from the very early years of the Draxes' time in Barbados, probably through his marriage to Anne.

This action of James Drax contravened the first Navigation Act of 1651 which required all colonies to ship their exports in English ships to English ports. The object was to cut off trade with the Dutch in the West Indies. In return, the prices of sugar and tobacco were protected in England. However, Drax's contravention seems to have done him no harm, as in the 1650s he was knighted at the instigation of the influential merchant Martin Noell. James Drax was said to have an Anglo-Dutch background, which would fit with his uncle being an Enos. James's wife, Margaret, died in 1681/2.[73]

Henry Drax of Barbados was married to Dorothy. Their son James was christened on 5 May 1673. Another son, also James was baptised on 17 October 1677. Their daughter Frances was baptised on 15

October 1675, all at St Philip. By 1673 the Drax plantation amounted to 800 acres. Henry Drax returned to London in the *Honor*, which set sail from Barbados for London in April 1679[74] along with his servant Christopher Lowther. As Henry Drax of St Giles in the Fields esq., he died in 1682.[75]

Notes and references

1. R. H. Schomburgk, 1848, *The History of Barbados*, Longman, Brown, Green & Longmans, p.267.
2. Ibid.
3. E.M.W. Cracknell, ed. (1934), *The Barbadian Diary of Gen. Robert Haynes 1787-1836*, The Azania Press, Medstead, Hampshire, UK.
4. Unfortunately the existence of another Nicholas Haynes, a painter and stainer, of St Andrew, Eastcheap, at the same time as Nicholas of Hackney doesn't simplify matters, as he could also have been Robert's father. Nicholas Haynes of Eastcheap was married to Elizabeth Dawes: Allegations for Marriage Licences Issued by the Bishop of London, 1520-1528, Joseph Lemuel Chester, 1887, Harleian Soc., p. 96.
5. W. Little *et al.*, 1969, *The Shorter Oxford English Dictionary on Historical Principles*, 3rd edn, Oxford. See proof of such use of son-in-law for step-son in the Will of Captain Robert Benson of St John, Barbados, witnessed by William Haynes in 1675, in James C. Brandow, 1983, *Genealogies of Barbados Families from Caribbeana and the Journal of the Barbados Museum and Historical Society*, Genealogical Publishing Co Inc, Baltimore, pp.168-169, fns 7 and 9 (ISBN 0-8063-1004-9). It was used in this way by William Cecil on 9 January 1564 in a letter to the Countess of Rutland about *'Lord Rutland your sone in law'* who was in fact her step-son (A. H. Nelson, 2003, *Monstrous Adversary: the Life of Edward de Vere, 17th Earl of Oxford*, Liverpool UP, pp. 41-42).
6. J. C. Hotten, ed., 1874, *The Original Lists of Persons of Quality: Emigrants; Religious Exiles; Political Rebels; Serving Men Sold for a Term of Years; Children Stolen; Maidens Pressed; and Others who went from Great Britain to the American Plantations 1600-1700*, p.139.
7. A John Haynes, whose relationship to the St John Hayneses, if any, is unknown, was *'delivered to Barbados'* in 1656 aboard the ship called *John* of London, commanded by John Cole, with eighty other named (indentured) *'servants'*. The bill of lading signed by Cole was dated at Plymouth 17 March 1655 [i.e.1656]. These were some of the Royalist prisoners who were exiled to the island for their participation in the Salisbury Plot, led by Col. John Penruddock, an insurrectionary struggle against Cromwell's government in 1655, by residents principally of Wiltshire, especially Marlborough, and Dorset. John Haynes, whose occupation was given as the *'Sheriffe of Wilts' trumpeter'* (a quasi herald

who transmitted proclamations, publications, denunciations, etc. from the county sheriff), was one of nine tried together before the Grand Jury in Exeter for High Treason, and convicted ['The Clarke Papers. Selections from the Papers of William Clarke, Secretary to the Council of the Army, 1647-1649, and to General Monck and the Commanders of the Army in Scotland, 1651-1660', ed. C.H. Firth (Camden Society, 1901, vol. XLIX), vol. 3, Newsletters 1655.1, para. 206 (f. 81)]. Haynes and six others were sentenced to be hanged (Penruddock and Hugh Grove to be beheaded), their death warrants being signed by the Lord Protector on 3 May 1555 (Wiltshire and Swindon Record Office: 332/265/21), but the sentence was commuted to transportation. Five of them were shipped together on the same vessel.

The High Sheriff of Wiltshire at the time was John Dove, a staunch parliamentarian, who had been captured in 1655 by the royalist rebels at Salisbury during the uprising and was saved from hanging only by the intervention of the rebels. He showed no such mercy when the latter were rounded up, John Haynes among them. Seventy were sent to the West Indies. Dove purchased property at Winterbourne, where coincidentally a large family of Hayneses lived. Of interest to us is the existence of both a Susanna and a Richard Haynes son of Nicholas Haynes and Elizabeth Attwood [The records are available on: http://www.frenchay museumarchives.co.uk/Archives/Parish_Records/Winterbourne/Winter bourne_1600-1643_BapMarBur.rtf.]. If this family was somehow linked to Robert of Barbados, it would at least be consistent with the family's lore invoking royalist credentials.

In December 1661, after the Restoration, Cole deposed in the Mayor's Court of London at the request of Colonel Marsellus Rivers [Rivers, one of the transportees, had petitioned parliament in March 1659 when full details of events leading to the deportation, the crossing to Barbados had subsequently, emerged: 'Diary of Thomas Burton, Esq. Member in the Parliaments of Oliver and Richard Cromwell, from 1656 to 1659: Now First Published from the Original Autograph Manuscript. With an Introduction, Containing an Account of the Parliament of 1654: from the Journal of Guibon Goddard, Esq., M.P. Also Now First Printed', London: H. Colburn, Vol. IV pp. 254-60] and other transportees, that in March 1655/6, Captain Henry Hatsill, merchant, had transported prisoners to Barbados for his account and that of Captain Thomas Alderne and Mr. Martin Noel. Enclosed with the deposition was a copy of the original bill of lading listing the prisoners ('Lord Mayor's Court of London, Depositions Relating to Americans, 1641-1736' by Peter Wilson Coldham, 1980). They arrived in Barbados on 7 May 1656, having been transported, locked up below deck amongst horses. The prisoners, who included clergymen, officers and gentlemen, were sold to Barbadian planters for 1550lb of sugar each, to do menial physical work, during which they were harshly ill-treated. With Cromwell's death two years later and the subsequent Restoration in 1660, the survivors were freed.

8. R. H. Schomburgk, 1848, *The History of Barbados*, Longman, Brown,

Green & Longmans, p. 141. For landowners of St John see: http:www.rootsweb.com/~brbwgw/Deeds1.htm.
9. Barbados Department of Archives RB3/1/65-66.
10. Ibid. RB3/1/55-6.
11. John Weston arrived at Barbados from London in the Peter Bonaventure in the spring of 1635 aged 26: John Mills was Speaker of the House of Assembly for 1691-1692.
12. In 1649 Capt Francis Dickenson was one of the Gentlemen of the Vestry and Trustee for the Parish of St John (*J Barb Mus Hist Soc*, Vol. XXIII, 1934, pp. 32-33); he was buried in St John on 9 June, 1678.
13. Alexander Lindsay, Lord Balcarres, in 1643 renounced his obedience to the Commonwealth; see Samuel Rawson Gardiner, 2000, *History of the Commonwealth and Protectorate 1649-1660*, p. 391. As early as 1638 in Barbados, Lindsay served several London agents by handling their affairs on the island, acting as purchasing agent and debt collector. He entered a long series of land purchases and sales, through which he eventually acquired a plantation with 22 slaves in St Philip (L. D. Gragg, 2003, *Englishmen transplanted: The English Colonization of Barbados 1627-1660*, OUP, p. 135.).
14. Indentured servants could not own the land they worked and were unable to leave the plantation without permission in the form of a pass from their employer. The harsh conditions of indentured servitude made it increasingly difficult for Barbadian tobacco and cotton planters to recruit such servants. The unexpired times of such servants could, as in this case, be 'sold' with the land on which they were contracted to serve.
15. Barbados Department of Archives RB3/1/56.
16. St John's Church must therefore have already been built by June 1643.
17. Ensign was rank of the lowest commissioned officer of foot. In 1649 Ensign Robert Benson was one of the Gentlemen of the Vestry and Trustee for the Parish of St John (*Journal of the Barbados Museum and Historical Society*, Vol. XXIII, 1934, pp. 32-33).
18. See Barbados Department of Archives: RB6/12/2.
19. This is clear from her father's will dated 23 May 1679 (Barbados Department of Archives: RB6/12/2); Mary Downes was not the only daughter of Captain Robert Benson and Mary Cheesewright as sometimes asserted (*Journal of the Barbados Museum and Historical Society*, Vol. XXIII, February 1934, p. 141).
20. See James C. Brandow, 1983, *Genealogies of Barbados Families from Caribbeana and the Journal of the Barbados Museum and Historical Society*, Genealogical Publishing Co. Inc., Baltimore, p. 170 (ISBN 0-8063-1004-9).
21. Indigo, a vegetable dye from *Indigofera tinctoria*: a shrubby perennial with silvery branches, and thin pinnate foliage; it came to Europe first in 1516.
22. Cotton wool with the cotton seeds and plant waste removed.
23. It may have been used to settle the debt for the land purchased from

Alexander Lindsay, as many such transactions were conducted on a credit basis.

24. Indian Bridge was the original name for Bridgetown, so-called in 1628 after the old bridge built over the river allegedly by the Carib Indians, who had disappeared from the island by the time of British settlement. The bridge was replaced about ten years after this deed was signed when the town was renamed.
25. Christopher Codrington, who as Colonel Codrington, was appointed Deputy Governor of Barbados in 1668 and 1670; Schomburgk, 1848, p. 293.
26. Barbados Department of Archives. See Brandow, 1983, p. 170 (ISBN 0-8063-1004-9).
27. See J. C. Hotten, ed., 1874, *The Original Lists of Persons of Quality: Emigrants; Religious Exiles; Political Rebels; Serving Men Sold for a Term of Years; Children Stolen; Maidens Pressed; and Others who went from Great Britain to the American Plantations 1600-1700*, p. 74.
28. Edward Thomas remained in Barbados after his indenture period had been completed and on 6 November 1669 he took on his own indentured servant for four years, John Lewis from Bristol. He may have married Elizabeth Kent of St Philip on 6 August 1694 and had a son, Edward, baptised on 7 July 1700. Elvira Thomas who married Richard Haynes might have been of this family.
29. Vere Langford Oliver, ed., 1919, *Caribbeana: being miscellaneous papers relating to the history, genealogy, topography, and antiquities of the British West Indies*, London: Mitchell Hughes & Clarke, 1919: Reprinted: Candoo Creative Publishing Inc., 2000, p. 96.
30. Correspondence dated 27 June 1923, now in the author's possession.
31. Schomburgk, 1848, p. 80.
32. Barbados Department of Archives, RB6/13/177.
33. Will of Elizabeth Haynes, St John's Parish, 3 December 1672; Barbados Department of Archives: RB6/9, p. 178.
34. Schomburgk, 1848, p.171.
35. Will of Elizabeth Haynes, St John's Parish, 3 December 1672; Barbados Department of Archives: RB6/9, p. 178.
36. W. Little *et al.*, 1969, *The Shorter Oxford English Dictionary on Historical Principles*, 3rd edn, Oxford. See proof of such use of son-in-law for step-son in the Will of Captain Robert Benson of St John, Barbados, witnessed by William Haynes in 1675, in James C. Brandow, 1983, *Genealogies of Barbados Families from Caribbeana and the Journal of the Barbados Museum and Historical Society*, Genealogical Publishing Co Inc, Baltimore, pp.168-169, fns 7 and 9 (ISBN 0-8063-1004-9). See note 5.
37. Will of Elizabeth Haynes, St John's Parish, 3 December 1672; Barbados Department of Archives: RB6/9, p. 178.
38. Will of John Haynes, St John, Barbados dated 23 January 1674, proved 18/20 July 1678: RB6/13 p. 497. The name Susanna, in the will of John's mother Elizabeth, is of interest in view of this being the name of the daughter of Richard Haynes of Hackney, still living at that time.

39. In England there was a marriage recorded between Robert Haynes and Elizabeth Morgin on 19 August 1636 at the Church of St Gregory by St Paul, London. Coincidentally it was noted that in 1643, William Haynes sold some of his Christ Church land to a merchant called Robert Morgan. A William the son of Robert Haynes was baptised on 10 April 1614 at St Mary The Virgin, Aldermanbury, London.
40. Schomburgk, 1848, p. 45.
41. Barbados Department of Archives, RB6/15/108.
42. Schomburgk, 1848, pp. 266-71.
43. Barbados Department of Archives, RB6/8/330.
44. Vere Langford Oliver, ed., 1919, *Caribbeana: being miscellaneous papers relating to the history, genealogy, topography, and antiquities of the British West Indies*, London: Mitchell Hughes & Clarke, 1919: Reprinted: Candoo Creative Publishing Inc., 2000, p. 96.
45. Ibid.
46. Schomburgk, 1848, p.171.
47. W. Noel Sainsbury, ed., *Calender of State Papers, Colonial Series* (Vol. 7), 'America and West Indies, 1669-1674', preserved in Her Majesty's Public Record Office (Vaduz: Kraus Reprint Ltd., 1964), First Published London: HMSO, 1889. pp. 496-7.
48. Will of William Haynes 1678; Barbados Department of Archives: RB 6/14 p.13.
49. John Camden Hotten, ed., 1874, *The Original Lists of Persons of Quality; Emigrants; Religious Exiles, Political Rebels; Serving Men Sold for a Term of Years; Apprentices; Children Stolen; Maidens Pressed; and Others Who Went from Great Britain to the American Plantations, 1600-1700*, Public Records Office. State Papers, Colonial Office General, CO1/44/47/f141-379, State Papers, CO28/16/2/f100-375.
50. Will of Ann Beachamp, Christ Church, Barbados, Widow; Barbados Department of Archives: Wills: RB6/6, p. 520. See also Joanne M. Sanders 1981, *Barbados Records: 1701-1725*, Vol. 3, p. 24.
51. Will of Robert Haynes of St John's Parish, Barbados, 12 November, 1696. Barbados Department of Archives: RB6/11/ 404.
52. Will of Ann Beachamp, Christ Church, Barbados, Widow; Barbados Department of Archives: Wills: RB6/6, p. 520. See also Joanne M. Sanders 1981, *Barbados Records: 1701-1725*, Vol. 3, p. 24.
53. Will of William Haynes 1678; Barbados Department of Archives: RB 6/14 p.13.
54. W. Little *et al.*, 1969, *The Shorter Oxford English Dictionary on Historical Principles*, 3rd edn, Oxford. See proof of such use of son-in-law for step-son in the Will of Captain Robert Benson of St John, Barbados, witnessed by William Haynes in 1675, in James C. Brandow, 1983, *Genealogies of Barbados Families from Caribbeana and the Journal of the Barbados Museum and Historical Society*, Genealogical Publishing Co Inc, Baltimore, pp.168-169, fns 7 and 9 (ISBN 0-8063-1004-9). See note 5.
55. Indenture between Captain Benson and Mary Pyott of Tring, in James C. Brandow, 1983, *Genealogies of Barbados Families from Caribbeana*

and the Journal of the Barbados Museum and Historical Society, Genealogical Publishing Co., Inc., Baltimore, p. 171 (ISBN 0-8063-1004-9).
56. Transfer of slave Mary to Robert Gibbson, ibid, p. 551.
57. Barbados Department of Archives, RB6/9 p. 365.
58. *Journal of the Barbados Museum and Historical Society,* vol. 27-30, 1959, p. 132.
59. Will of William Haynes 1678; Barbados Department of Archives: RB 6/14 p.13.
60. Barbados Department of Archives, RB6/4 p. 386. However, a Robert Haynes married Katherin Manley at St Michael's Church on 30 October 1684; Barbados Department of Archives, RL1/1, p. 313.
61. Barbados Department of Archives: Land Registry, Vol. 17 f. 498; Grantee Robert and Elizabeth Haynes, Grantor Roger Thomas.
62. Barbados Department of Archives: Land Registry, Vol. 21 f. 230; Grantee Robert and Elizabeth Haines, Grantor James Burrowes. The transfer was recorded on 20 September 1695. James Burrowes was probably Robert's brother-in-law by marriage; Elizabeth Burrowes was named as Robert's *'sister'* in his will, although when she married Burrowes on 15 January 1685, she was styled Mrs Elizabeth Haynes.
63. Barbados Department of Archives: Land Registry, Vol. 35 f. 4: Grantee Robert Haynes, Grantor Archibald Reid. Land in settlement of debt. Signed 22 December 1726, proved 4 February 1727.
64. *Gentleman's Magazine,* November 1863, 'The Monumental Inscriptions of Barbadoes', p. 570.
65. Barbados Department of Archives: Wills: Vol. 2, p. 449; Will of Robert Haynes Parish of St Thomas, Planter, dated 18 September 1834, proved 2 January 1728. He died 9 October 1727.
66. *Gentleman's Magazine,* November 1863, 'The Monumental Inscriptions of Barbadoes', p. 570.
67. Edward Long, 1774, *The History of Jamaica: General Survey of the Antient and Modern State of that Island,* Vol.1, London, T. Lowndes, pp. 230-231; the General's body was retrieved by his son, Captain Haynes.
68. Or Byrch; a family of this name held a plantation in St Michael in 1750.
69. Barbados Department of Archives, RB6/8 p. 270.
70. TNA: Will of Edward Hopkinson, gentleman of Hackney, dated 29 April 1640, probate 22 May 1640; Coventry Quire Numbers: 54-116 PROB 11/183.
71. Will of Nicholas Enos dated 20 April 1678, probate 30 May 1678; PROB 11/356.
72. P. W. Coldham, 1988, *The Complete Book of Emigrants 1607-1660,* Genealogical Publishing Co., p. 255.
73. TNA: The will of Margaret Drax, Widow of Kensington, Middlesex, was dated 11 February 1682, probate 17 February 1682: Drax Quire Numbers: 1-51; PROB 11/372. As well as a son, Henry, James Drax had two daughters, Meliora and Phalatia. Meliora married Robert Pye and Phalatia married Thomas Gomeldon (or Gumbleton). Martin Noell was a London

SETTLEMENT ON BARBADOS

   merchant whose main business during the Commonwealth was in shipping arms, horses, prisoners and other goods from England to Barbados.
74. His ticket was granted 22 April 1679; Hotten's Lists [reference 47].
75. TNA: Will of Henry Drax of St Giles in the Fields esq., dated 30 June 1682, probate 20 September 1682; PROB 11/370.

14

Children of John and Anne Haynes of St John, Barbados [c. 1660-c.1730]

Robert Haynes [c. 1666-1696] - John Haynes of New York [-1689] - A Victim of Leisler's Rebellion? - Elizabeth Haynes - Anne Haynes

Robert Haynes (c.1666 –1696)

Edmund appears to have died without issue before October 1685, for on the eighth of that month an indenture was agreed between Robert Haynes of the Parish of St John, Gent, and Francis Dickenson, Gent, on the other part.[1] It referred to the halving of the plantation (*'parcel of land'*) with Francis Dickenson, where the *'said Haynes and Dickenson now live'*, consisting of 150 acres in St John's Parish bounded north and south by land of Sir Peter Colleton, westerly by Major John Leslie, easterly by William Crossing, Mr Edward Beale and Mr Thomas Pooler and the sea. It allowed for one half of all servants, negroes, cattle, coppers, stills, boilers, ladles, basins, utensils and plantation implements. Dickenson was to pay rent of £200 p.a. to Robert Haynes, the rent to begin and to be made the last day of May 1686. At expiry of the rental, all had to be handed back in proper condition, *'hurricanes only excepted'*. A condition of the rent was that no trees were to be cut down. Another condition was that Dickenson should build a stone sugar boiling house for which dimensions were defined. The indenture, signed by Robert Haynes on 6 April 1686, was witnessed by Archibald Carmichael, Andrew Leslie and John Moore. The following inventory was annexed: men slaves £220; women negroes £295 10 0; boys £34 10 0; girls £35 5 0; bulls (named) £36; cows (named) £20. Total £641 5s 0d plus utensils and stills, etc. The properties abutting the Haynes-Dickenson plantation are also described in the following extract of an indenture dated 17 May 1688[2] in which William Crossing and Ann his wife assigned a plantation (against debts owed) to John Burston in

> 'the Parish of St John containing by estimation one hundred and thirty five acres of land ... butting and bounding upon the lands of Col. Thomas Colleton upon the lands of the said William Crossing upon the lands of Robert Haynes and Francis Dickenson and upon the sea as also one other parcel of land containing thirty acres be the same more or less to the said plantation adjoining together with the mansion house windmill boyling house styll house and all other buildings thereon ...'.

On 4 August 1681 Robert Haynes married Elizabeth Blanchett of St Philip by licence,[3] by whom he had seven children, two boys and five girls, all of whom, except son, John, were subsequently mentioned in their father's will. His wife was not Elizabeth Bourne [Bowne] as erroneously given in the IGI, who had in fact married Robert's brother, John, in New York on 9 November 1687.[4]

Robert's family, with the possible exception of John, was fortunate to have survived the ravages of the virulent illness that occurred in 1693, described as a *'violent Distemper'* in *'Barbadoes'* which killed many inhabitants and visitors there over a three-year period, *'... but [now] it has pleased God to stop his Afflicting Hand ... this Island being at present in a perfect State of Health and in a very flourishing Condition.'*[5]

Robert was recorded as being a sidesman of St John's Church in 1683-4, Surveyor of Roads 1684-5 and *'Surveyor'* of the Vestry in 1695, in which year, on 19 November, he was mentioned in the will of his neighbour, the widow Accuna Beale.[6] It was in October, that year that the island was visited by another hurricane which drove ashore most of the ships then in Carlisle Bay.[7]

Robert Haynes signed his own will in November 1696 in the presence of Francis Dickenson,[8] Thomas Chickell (or Clinkett) and John Hoppin (Robert's future son-in-law). He couldn't have been much more than 30. He probably died of the same virulent illness as mentioned as easing three years earlier in The Times, for Oldmixon[9] reported that it carried off more than a third of the island's inhabitants. It appears to have been a highly contagious fever, called by some the *'spotted fever'*, said to have been brought to Barbados by troops who embarked at Cadiz in 1691[10] and were destined to act against Martinique[11] and it rapidly spread across the whole island and its presence lasted for at least twelve years. By 1695 it had killed off most of the crews of the *Tiger* and *Mermaid*, two naval vessels in Carlisle Bay, and Colonel Francis Russel, appointed Governor only the previous year, died of the fever in 1696. The situation was not helped by injuries suffered from the hurricane of 1694 and the reduction in food production from the resulting labour shortage.

Robert's will was proved after his death on 7 December 1696[12] by Messrs Dickinson and Chickell. His executors were Col. John Leslie, Major Christopher Estwicke, Lt-Col. Richard Downes (his *'trusty and well beloved friends'*) proprietors of neighbouring properties, who were also named guardians of his infant children. He describes himself as Robert Haynes in the Parish of St John. He was *'sick and weak in body but of sound and perfect memory'*. He made the following bequests. To his wife, Elizabeth, he left a male slave called Arran, his black riding horses, two named cows and her bedroom furniture. Elizabeth also received a third of his real estate in dower (during her natural life). His daughter, Elizabeth, was left twelve acres of land, with the house on it then occupied by Margaret Prendergrass, adjoining the boundaries of Col. John Leslie, Col. James Colleton and Francis Dickenson. This land was also to pass on to her children, in the absence of which it would revert to her two brothers. Elizabeth was also left *'one Mallatto girl (Cinaine Hagar), one negroe girl (Minbah), one milch cowe, one feather bed, one boulster and pillows'*. His daughters, Anne, Olive, Lucy and Dorothy were each to receive £200, a cow, bed furnishings and a slave girl (called Flora, Ambrabah, Abba and Ashee respectively). Elizabeth and her sisters were to receive their bequests when they reached the age of 18, or on marriage. The residual estate including personal effects was to pass to his sons William and Richard Haynes at the age of 21. Robert stipulated that all his children were to be maintained and educated out of his estate until they obtained their legacies.

Robert's wife, Elizabeth, aged 30 when her husband died, married secondly a man called Scott, probably James Scott, who was a planter in St John's Parish, but she remained at the St John's Plantation house looking after her unmarried children, who were recorded in the 1715 census for St John: Elizabeth Scott aged 49; Richard Haynes aged 21; William Scott aged 13; Olive Haynes aged 27; Lucie Haynes aged 26; Dorothy Haynes aged 19; Margarett King aged 17. She had

had a son, William, by her second husband in 1702, so had presumably married him around 1700.

John Haynes of New York [–1689]

John Haynes, the second son of John and Anne Haynes, was mentioned in the Barbados Census of 1680. He was a witness of the wills of Nicholas Walcott, a Barbadian merchant, on 16 April 1680[13] and of the will of Marvin Hales of St Michael's Parish on 3 January 1685.[14] John's occupation is unknown, but under the terms of his father's will, he should have been apprenticed to a trade under the watchful guidance of Captain John Leslie and Thomas Phillicott, neighbouring planters and family friends. He was also left ten thousand pounds of muscovado sugar to invest and possibly a similar amount left by his brother Richard, who appears to have died young.

It is likely that it was through John Leslie's family connections that John Haynes came to have interests in North America. John Leslie's sister, Elizabeth was the wife of Sir Peter Colleton, owner of another plantation bordering the Hayneses; Leslie's wife was also a Colleton. Sir Peter Colleton was a member of the Hudson Bay Company and the Company of Royal Adventurers Trading to Africa, which became the monopoly slave-trading Royal African Company in which he inherited stock worth £1800. He had become a planter in 1650 and his plantation was to become one of the largest in the island. He was one of the Lords Proprietors of Carolina, for which he recruited settlers from Barbados.[15] It is therefore entirely plausible that Colleton assisted his brother-in-law in ensuring that John Haynes acquired his mercantile credentials – perhaps as an agent of a shipping company, or even with his own vessel; and John set about trading with the American mainland.

Examination of the overriding needs of New York at that time reveals an acute labour shortage – *'the greatest want'*, according to Governor Andros. In 1678 few black slaves had been imported and their value had rocketed to around £35.[16] Barbados, on the other hand, was the first port of call and reception centre for many of the slaves after their long transatlantic voyage on the ships of the Royal African Company. Although this is entirely conjectural – it might have been in their onward transportation where John Haynes saw his trading opportunity. As an agent acting for the company, he could earn 7 per cent commission on all sales. Barbados to New York was an important trade route. The outward bound cargoes included slaves, sugar and molasses; cargoes for Barbados on the return trip included flour, fish, livestock and timber. Frederick Philipse, with whom John was well acquainted in New York, was one of the city's largest traders, with his own vessels travelling between both the East and West Indies; he certainly imported slaves from Africa.[17]

John married Hannah Walcott (Woolcot; née Evans), widow of Nicholas, on 30 November 1682 at St Michael, but she appears to have died not long after giving birth to their son, whom they named John after his father. John's business, which necessitated travelling to and from New York, was probably initially related to the family's sugar business, but the scope widened, for a bill from New York relating to his general trading with the island dates from as early as January 1684. Trading between the Hayneses in Barbados and the Bowne family in New Jersey, is likely to have taken place at an early stage. The death of his wife was probably the reason why he felt he could leave the island for New York where he soon became a prominent businessman and lieutenant of militia. But the price of this was to leave infant John to be brought up by his family on Barbados.

There is no record as to when he finally left Barbados, as his ticket (permission from the Secretary's Office) to leave has not survived. However, it must have been around 1686, for on 9 November 1687 John Haynes married Elizabeth Bowne of New York in the Dutch church there.[18] Elizabeth was the only daughter of Andrew Bowne by his wife Elizabeth. Andrew Bowne had earlier been a sea captain on the North Atlantic run; he made his home in, and became an Alderman of, New York City and an officer of the Council in 1685[19]. He eventually settled with his family in Chingaroras in Monmouth County, New Jersey and was appointed Governor of East New Jersey based at Perth Amboy in 1699. A considerable amount of his correspondence has survived. John and Elizabeth had only one son, whom they named Andrew after his maternal grandfather.

While in New York, John Haynes continued to trade with merchants in Barbados and a bill of exchange has survived dated 9 January 1684, for £25 sterling in favour of Haynes from his Barbadian *'friend and servant'* Joseph Cox. Passenger and freight traffic between, Barbados and New York, Boston and elsewhere in New England, was high in the late seventeenth century and the surviving passenger lists for the 1670s and 1680s – 583 travellers in 1689 alone

CHILDREN OF JOHN AND ANNE HAYNES OF ST JOHN

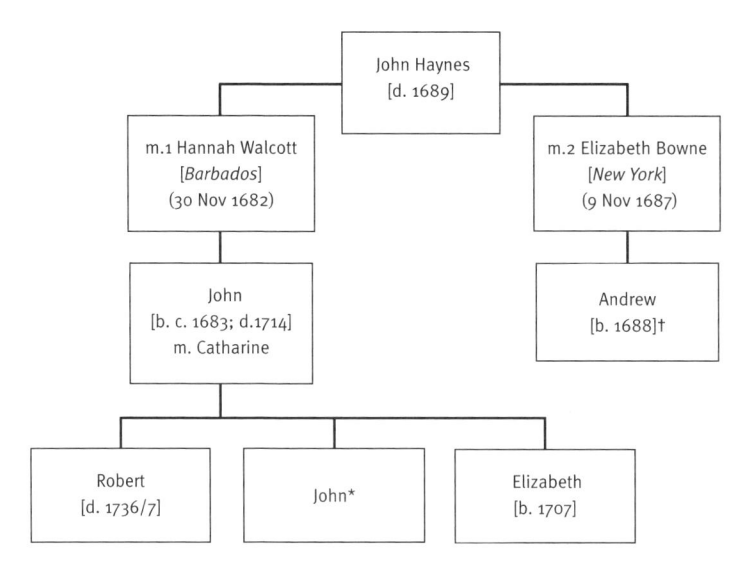

```
                        John Haynes
                         [d. 1689]
         ┌──────────────────┴──────────────────┐
   m.1 Hannah Walcott                     m.2 Elizabeth Bowne
      [Barbados]                              [New York]
     (30 Nov 1682)                           (9 Nov 1687)
         │                                        │
        John                                   Andrew
  [b. c. 1683; d.1714]                       [b. 1688]†
     m. Catharine
         │
   ┌─────┼─────┐
  Robert  John*  Elizabeth
 [d. 1736/7]    [b. 1707]
```

† Although born Andrew Haynes, he may have been known as Andrew Bowne jr; * less than 21 years of age in September 1734.

Figure 14.1 Suggested Family of John Haynes of New York

– include several that can be identified from St John's Parish, notably Francis Dickenson. Haynes's trading with John Bowne of Monmouth County developed and this connection was possibly what led John Haynes to meet Andrew Bowne and his daughter Elizabeth, John Bowne's cousin, whom he subsequently married – although it might have been the other way round. This family connection, strengthened by baby Andrew's half-brother, paternal grandmother, uncles, aunts and cousins being in Barbados, continued long after John's death: for instance, it is recorded that John Bowne was indebted for varying sums in the ordinary way of business to the Hayneses as late as 1697,[20] by which time John's son, John, would have been old enough to have taken over his father's business interests in Barbados. Captain John Bowne, merchant of Amboy, East New Jersey was also recorded as importing flour, sugar, lamp oil and other consumables from John Smith of Barbados as late as May and June of 1707[21].

Victim of Leisler's Rebellion?

With the accession of King James II in 1685 and the possibility of England returning to Catholicism, New York too became a centre of religious unrest in which John Haynes was to become ensnared. In November 1688 William and Mary of Orange deposed King James II and prevented a Catholic succession. This news reached New York (a former Dutch colony but then a royal province) in April 1689, and on 31 May Jacob Leisler, a prosperous protestant German immigrant merchant, a long-standing resident of the city and one of five captains of militia, decided upon a course to secure the colony against those he perceived as papists. He succeeded in capturing Fort James on Manhattan Island and New York harbour and in supplanting the Royal Lieutenant Governor, Francis Nicholson, who was packed off on a ship for London. Leisler appointed himself in Nicholson's place, intending to hold New York for William and Mary, disregarding the fact that Nicholson had deputed Frederick Philipse, Stephanus van Cortlandt (New York's mayor) and Nicholas Bayard in his place.

On 24 June 1689 Stephanus van Cortlandt, obtained a copy of William and Mary's proclamation that had been delayed in Boston, confirming all Protestant officers in their places in the colonies. In it the authority of Philipse, van Cortlandt and Bayard, who held their commissions from the Crown, was confirmed. Van Cortlandt immediately convened the aldermen and citizens to proclaim it. Leisler was incensed. The next day, van Cortlandt convened a meeting of the councillors and Common Council at his house, who thought it best to remove Matthew Plowman, a Catholic collector of revenues, for the peace of the restless community, or as Philipse and van Courtlandt in their letter of 5 August, sent to England with the ousted Nicholson, described Plowman: '*a papist and obnoxious to the people*'.[22] They appointed four commissioners in his place to become receivers of the King's customs revenues, which also involved the provision of harbour permits and clearance of vessels to land and depart, until a successor could be appointed. Colonel Nicholas Bayard, Thomas Wenham, John Haynes (Haines) and Paulus Richards, all prominent merchants, were chosen, sworn in and received the keys to the Custom House. They daily set about their work when suddenly Leisler burst into the Custom House with a company of armed men, questioned the four's authority, and ordered them out of the building. He scoffed at the resolutions of the mayor and council that had by that time been pasted over the door of the Custom House,[23] and put in Peter de la Noy in place of the commissioners.[24] All of them were violently turned out into the street, Wenham being physically assaulted in the process and Bayard only being spared worse by the intervention of the crowd.[25] Bayard was relentlessly pursued and, his life being constantly in danger, with the help of friends he made his escape to Albany and the protection and hospitality of Mayor Peter Schuyler. He was eventually captured and with others, held in a dungeon for 13 months in appalling conditions. Van Cortlandt was seriously threatened and escaped to Connecticut.

What happened to John Haynes after his ejection from the Custom House on 25 June 1689 has not yet come to light, but the fact that he and Thomas Wenham were together in New Haven, Connecticut, two months later (August 1689),[26] where he became seriously ill, suggests that the two men had also had to make their escape from New York. How they arrived at New Haven, whether by land or sea is unknown, but there is no doubt that the wives and children of the principal families left behind in New York were subjected to searches and all manner of indignities by Leisler's soldiers. It is possible that a message had reached John Haynes in New Haven that Elizabeth and baby Andrew had left the city, for some time during the next three months John doubled back and made his way to his father-in-law's estate in Monmouth County, East Jersey, where he was to spend his last days with his family.

Largely oblivious of the turmoil afflicting New York, King William had set up a government framework for the colony consisting of a governor and council appointed by the Crown, and an assembly appointed by a majority of the freeholders in the several counties of the province. The King appointed Colonel Henry Sloughter as Governor in September 1689, and reappointed many of the former officers of the colonial government (some of whom were in gaol or in hiding) including Frederick Philipse, Stephanus van Cortlandt (then Mayor), Colonel Nicholas Bayard, William Smith (a former Governor of Tangier, he was commonly known as 'Tangier Smith'), Gabriel Minvielle, Chidley Brooke, William Nicolls, Nicholas de Meyer (aged and ill), Francis Rombouts, Thomas Willett (of Long Island), William Pinhorne and John Haynes, as councillors to the new governor, to await his arrival from England.[27] This appointment of John Haynes to the Council of New York was made in London on 11

January 1690,[28] at the height of the unrest, but the ratification of this appointment arrived from England just too late, as John had died the previous month. To what extent the anxiety, strain, and suffering and possibly bodily injury under Leisler's regime had contributed to his death is not known. He was certainly *'sick and weak'* following an illness or injury during his escape.[29,30] Nicholas de Meyer and Francis Rombouts, who were aged and frail, died around the same time as John, none of them living to assist the new Governor.

The rapid rise to prominence of the relatively young John Haynes in New York alongside the likes of Philipse, Van Courtlandt and Bayard – the richest men in New York – the resident members of the previous Governor's Council, well-known and respected in London, and the part played by him in these historic events, may well reward further study.[31]

The identification of John Haynes of New York as John the son of John and Anne Haynes of Barbados is evident from a revised version of the Pelletreau transcript of his will[32] in which the phrase *'his brother-in-law Mr Andrew Bowne and Mr Thomas Winslow of New York executors'* is replaced by 'his *father-in-law* Mr Andrew Bowne and Thomas Wenham of New York, merchant, also his brother Robert Haynes and *brother-in-law Samuel Branch both of the Island of Barbadoes* to be his executors' [my italics]. This firmly places him as the brother of Robert Haynes and his sister Elizabeth who was married to Samuel Branch. John died on 9 December 1689 and on the same day an inventory of his estate was taken by Elizabeth's cousin, John Bowne and William Richardson of Monmouth County, East Jersey. Logistically, this could only have been done if John had died in that vicinity, presumably with or near to his wife's family. The appointment of Thomas Wenham as an executor identifies the testator as the John Haynes ejected from the Custom House and subsequently nominated as councillor. John Bowne, William Richardson and John Haynes were, with others, named as overseers of the will of Colonel Lewis Morris of New York who actually died after John Haynes.[33]

In his will, dated 24 August 1689, written in New Haven, in which he describes himself as John Haynes, formerly of Barbados, then of New York, merchant, he left his wife, Elizabeth, all his household goods and plate (worth £26), including a tankard (£10) and 16 silver spoons (£9 12 0d). His pewter weighing 77 lbs, included four porringers and two dozen plates. He also left a large quantity of brassware and a sword with a silver handle. The total value of household effects came to £100. He left his son John Haynes in Barbados, one negro woman, and all his '*estate in New York and in Barbadoes*' he left to his wife Elizabeth, and his sons, John (in Barbados) and Andrew (in New York). The witnesses were Richard Rosewell, John Smiles and Joseph Alsop.[34] The will was proved in New Haven, where it had been drawn up, before James Bishop Esq., Deputy Governor of Connecticut, on 16 May 1690. Letters of administration on John's estate were granted to his father-in-law, Andrew Bowne, on 17 July 1691; a quietus (discharge of his duties) being granted after the appraisal of his presentation on 27 November 1695.[35]

Elizabeth Haynes married secondly her first cousin Obadiah Bowne (1666–1726) of Monmouth County New Jersey, by whom she had three more children: John, Ann and Lydia. She predeceased her second husband who remarried and had further issue.[36]

John, the son of John Haynes, left behind in Barbados may have inherited his father's business interests on the island and to have had continued mercantile links with the Bownes in New Jersey (above). According to naval records, a John Haynes imported 301 Africans into Barbados in June 1700[37] and, even though he could have been little more than 18 years old, no other John Haynes has been identified as being on the island at that time. John married and had three children: Robert, John and Elizabeth. Elizabeth, who was baptised in 1707, is said to have been the daughter of John and Catharine Haynes. Catharine's parentage is unknown. He is believed to have been the John Haynes whose death was recorded in 1714; in 1722 he was said to be dead and his three children were still under the guardianship of Ann Critchlow, the daughter of Anne Haynes and Richard Perryman, so presumably his wife, Catharine, had also died. Robert the eldest of the three children eventually married a daughter of John Hein, but she wasn't mentioned in his will signed 18 September 1734, so can be presumed to have already died.[38] In his will Robert mentions his father-in-law but no children. He made his brother, John Haynes, who was not yet 21 years old, his residuary legatee, but the estate (possibly Pilgrim where the will was signed) was to pass to his sister Elizabeth Haynes, if John died before reaching his majority. Elizabeth inherited in her own right five acres of land in Christ Church with a negro woman called Sue and her children, and some negro men. Robert remembered his wife's siblings, one of whom, Charles Hein, was his godson. Robert died before 4 February 1737 when the will was proved. Although the cause of Robert's death was not recorded, this was a year when an epidemic of yellow fever was at its height in Barbados.[39] An important link with his family's past in St John was Francis Dickenson, who witnessed Robert's signature and was described as his *'kinsman'*.

Andrew Haynes, who was no more than one year old when his father died, continued to live with the Bowne family in New Jersey. It seems that he assumed the Bowne family name, possibly after his mother's remarriage, for on 17 December 1705, John Bray and Susanna his wife conveyed to Obadiah Bowne and Jarret Wall land in Middletown, the conveyance being witnessed with others by Andrew Bowne Jr. Governor Andrew Bowne in his will, dated May 6 1707, bequeathed his estate at Matawan, New Jersey, to his wife Elizabeth, and after her death to his grandchildren, John, Anne and Lydia, in equal shares.[40] These were the children of Obadiah Bowne and Elizabeth his daughter; Andrew (Bowne Jr), Elizabeth's son by her marriage to John Haynes was not mentioned and may possibly have predeceased his grandfather. The further descent of their family has not been researched although it is of interest that the name Andrew Bowne continued in Matawan, New Jersey, to the twentieth century.

Elizabeth Haynes

Elizabeth Haynes, daughter of John and Anne Haynes married Samuel Branch on 16 March 1682 at Christ Church, Barbados. They had four children: John, baptised in 1683, Samuel baptised in 1686, Elizabeth and Olive, both born before 1692. The family settled in St Michael where Samuel died in 1692, probably a victim of the virulent illness that led to the death of more than a third of the island's population.

The year after Samuel's death, Elizabeth married John Seaward. As Mrs Seaward, she subsequently served as (her grandchildren's) godmother at a group baptism in 1727 of three children of her son Samuel who had recently died. Their maternal grandmother, Elizabeth Price (mother of Barbara, widow of Samuel Jr., whom she had married in 1719) also served as godmother on this occasion. Elizabeth's son, Samuel, was mentioned in the will of his grandmother, Anne Beachamp (former wife of John Haynes) in 1722, but by that

CHILDREN OF JOHN AND ANNE HAYNES OF ST JOHN

time her elder son, John Branch, who had married his step-sister, Elizabeth Seaward, had already died. John and Elizabeth Branch had five children: John [b. 1710], Samuel [b. 1713] and triplets James, Olive and Anne [b. 1715]. Samuel and Barbara Branch also had five children: Richard [b. 1719], Samuel [b. 1721], Stephen [b. 1722], Elizabeth [b. 1723] and Richard [b. 1725].

Elizabeth had many great grandchildren and her descendants continue to this day.

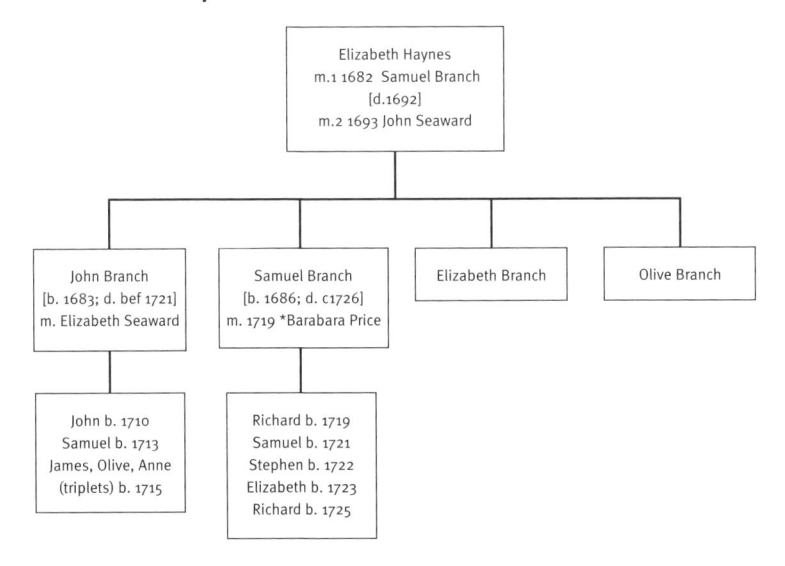

```
                    Elizabeth Haynes
                 m.1 1682  Samuel Branch
                          [d.1692]
                  m.2 1693 John Seaward
   ┌──────────────────┬─────────────┬──────────────┐
John Branch      Samuel Branch   Elizabeth Branch  Olive Branch
[b. 1683; d. bef 1721]  [b. 1686; d. c1726]
m. Elizabeth Seaward    m. 1719 *Barabara Price
   │                       │
John b. 1710           Richard b. 1719
Samuel b. 1713         Samuel b. 1721
James, Olive, Anne     Stephen b. 1722
(triplets) b. 1715     Elizabeth b. 1723
                       Richard b. 1725
```

*Barbara Branch married secondly John Miskett in 1731.

Figure 14.2 Family of Elizabeth, daughter of John and Anne Haynes

In the will of Elizabeth's son John Branch, dated at St Michael's on 23 January 1720,[41] he left his mother the house in which she then lived with five acres of land. His son Samuel was left five acres of land adjoining those of his grandmother. His son, James, was to receive ten acres bordering on the estate of William Codrington and Hannah Wilson on his twenty-first birthday. His daughters Olive and Anne Branch were also beneficiaries. His son, John was the residuary legatee. His wife Elizabeth and brother Samuel were executors along with Col. James Thorne,[42] and Robert Fitt. John's signature was witnessed by Sarah Codrington, Elizabeth Carrington, Jacob Maynard and John Browne. It was proved on 16 May 1721.

Elizabeth's daughters Elizabeth and Olive Branch may have

married by 1722, for in the will of their grandmother appear the names of three grandchildren: Elizabeth Luke, Elizabeth Fitt and Olive Critchlow. One of the Elizabeths would have been Robert's daughter, but, Olive Branch was the only 'Olive' of marriagable age in the wider family whose husband is not accounted for, so it is reasonable to assume that it was she, who like her cousin Ann Perryman, also married into the Critchlow family.

Anne Haynes

Anne Haynes married Richard Perryman junior, by whom she had ten children: Ann, Robert, Richard, Ann, Elizabeth, Elizabeth, Richard, Samuel, Edmund and Frances.[43] Only Ann, the fourth-born, and Samuel, Edmund and Frances were mentioned in the will of their grandmother, Anne [Haynes] Beachamp. This fourth-born Ann married James Critchlow on 4 May 1704, and they had three sons, James, Henry and James. Only Henry lived to marry and have children which eventually led to an important Critchlow dynasty. Her brother Edmund married Mary Howard by whom he had four children. Edmund's son Robert had issue from whom a major Perryman dynasty descended. From the will of her grandmother in 1722, we learn that Ann Critchlow had taken on the responsibility of bringing up Robert, John and Elizabeth Haynes, the three children of her cousin, John Haynes, son of John Haynes of New York, who had died in 1714.

CHILDREN OF JOHN AND ANNE HAYNES OF ST JOHN

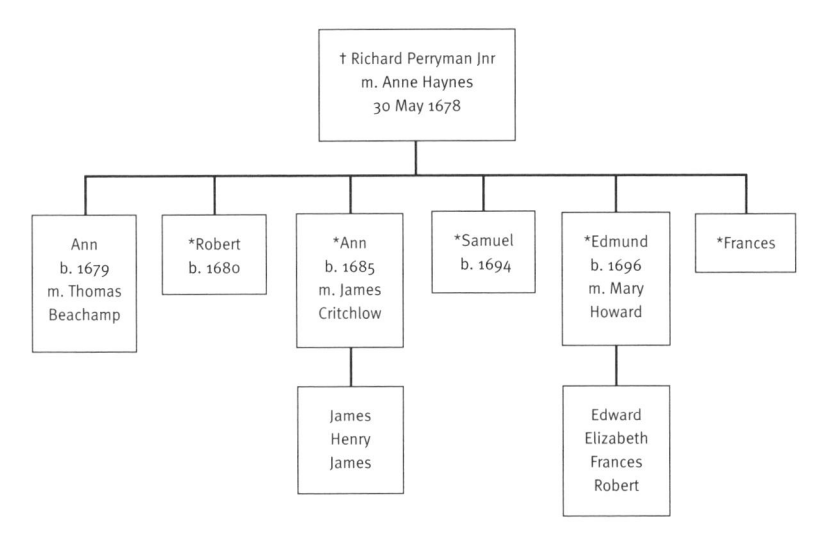

```
                    † Richard Perryman Jnr
                       m. Anne Haynes
                        30 May 1678
```

Ann	*Robert	*Ann	*Samuel	*Edmund	*Frances
b. 1679	b. 1680	b. 1685	b. 1694	b. 1696	
m. Thomas		m. James		m. Mary	
Beachamp		Critchlow		Howard	

Under *Ann (m. James Critchlow):
James
Henry
James

Under *Edmund (m. Mary Howard):
Edward
Elizabeth
Frances
Robert

*Grandchildren of Anne [Haynes] Beachamp mentioned in her will 1722. †Richard Perryman and Ann Haynes had ten children.

Figure 14.3 The Family of Richard Perryman Jnr and Ann Haynes

Notes and references

1. Indenture made 8 October 1685 between Robert Haynes and Francis Dickenson, gent., Barbados Department of Archives, Land Registry References, Vol. 14, f. 34; William Crossing was the brother of Mary Crossing, wife of Captain Thomas Downes. Thomas Pooler died in 1687 (RB6/40, p. 566); Sarah Pooler, his daughter, was John Haynes's god-daughter. She married Arthur Arscott.
2. Entered 1688; Barbados Department of Archives: RB3/17/ 35.
3. IGI. Barbados Department of Archives: RL1/24, p. 4.
4. William Montgomery Clemens, 1926, *American Marriage Records before 1699*, reprinted by the Genealogical Publishing Company in 1864, p. 106. The Dutch church was probably that built within the fort in 1642.
5. *London Gazette* of 17 April 1693.
6. Barbados Department of Archives: RB6/11/307.
7. R. H. Schomburgk, 1848, *The History of Barbados*, Longman, Brown, Green & Longmans, p. 45.
8. Barbados Department of Archives: Wills: RB6/11 p. 404; RB6/ 13 p. 249. Francis Dickenson was the son of the Captain Francis Dickenson, who had acquired land with Robert Haynes, in 1643. In 1679 Capt John Leslie owned 4 white servants, 55 negroes and 60 acres in St John. His sister, Elizabeth (widow of William Johnston) was the wife of Sir Peter Colleton (1635-1694). Richard Downes was the island's Treasurer. Robert couldn't

267

have chosen three more interesting characters as his executors than Leslie, Estwicke and Downes, for all three subsequently had the 'distinction' of being expelled from the House of Assembly for opposing the actions of the Governor Sir Bevill Granville [John Poyer, 1808, *The History of Barbados, from the First Discovery of the Island in the Year 1605, till the Accession of Lord Seaforth, 1801*, London: J. Mawman, pp. 185-189].

9. John Oldmixon, 1708, *British Empire in America*, Vol. ii, p.14.
10. Spotted fevers are frequently caused by micro-organisms called Rickettsia which infect insects, particularly ticks, which then transmit the disease to man. Rickettsial disease is endemic to the Cadiz area of Spain where even recently, new groups of Rickettsia have been isolated from the tick, *Ixodes ricinus* (Marques F J *et al*. 1998: Am. J. Trop. Med. Hyg., 58, 570-577). The obvious source of the spotted fever was *Rickettsia africae* which causes African tick-bite fever; this could have been transported to Barbados on the slave ships. There was also said to be an outbreak of yellow fever in 1691 in Barbados; another insect-borne disease.
11. Schomburgk, 1848, p. 84.
12. Barbados Department of Archives: Wills: RB6/11 p. 404; RB6/ 13 p. 249.
13. Will of Nicholas Walcott of Barbados, merchant; Barbados Department of Archives: RB6/14/252; proved July 1681.
14. Will of Marvin Hales of St Michael's Parish; Barbados Department of Archives: RB6/40/180.
15. B. D. Henning, 1983, *The House of Commons, 1660-1690*, Vol. 1, London: Secker & Warburg, p. 106.
16. Martha J. Lamb, 1896, *History of the City of New York: its origin, rise and progress*, Vol. 1, Ch. XV: '1673-1678; Admiral Evertsen', New York, A. S. Barnes & Co., p. 279.
17. Ibid., p. 271.
18. William Montgomery Clemens, 1926, *American Marriage Records before 1699*, reprinted by the Genealogical Publishing Company in 1864, p. 106.
19. Edwin Salter and George Crawford Beekman 1887, *Old Times in Old Monmouth: Historical Reminiscences of Old Monmouth County*, p. 275, Baltimore : Reprinted for Clearfield Co. by Genealogical Publishing Co., 2006; Orra E Monnette, 1930, *First settlers of ye plantations of Piscataway and Woodbridge, olde East New Jersey, 1664-1714: A period of fifty years*, p. 698, Leroy Carman Press. For passenger list between Barbados and North America, see: H.G. Somerby, 1854, State Paper Office. Papers of Barbadoes, Board of Trade No 1, *New England Historical and Genealogical Register*, vol. 8, pp. 206-7; also 'A list of persons who left Barbados in the year 1689', *Journal of the Barbados Museum and Historical Society*, 1934; 1: 155-180. The Bowne family is said to have borne the following arms: ermine, a lion rampant or, on a canton of the second a mullet sable: crest: a demilion rampant sable, holding in its dexter paw a sprig of laurel leaves proper. Motto: *Fama nominis bona* (M. D. Ogden, ed. 1917, *Memorial Cyclopedia of New Jersey*, Vol. III. New Jersey: Memorial History Co.). Andrew Bowne's appointment as Alderman and officer of the Council of New York City is to be found in: H L Osgood, AB

Keep and CA Nelson, 1905; Minutes of the Common Council of the City of New York, 1675-1776, Volume 1 pp. 170, 297, New York Historical Society.
20. John Edwin Stillwell, 1970, *Historical and genealogical miscellany: data relating to the settlement and settlers of New York and New Jersey*, Vol. 3, Genealogical Publishing Co. pp. 41-43. For John Smith's Correspondence with Capt John Bowne, who died 13 March 1716, see: Edwin Salter and George Crawford Beekman 1887: "Old Times in Old Monmouth: Historical Reminiscences of Old Monmouth County", pp 287-288, Baltimore: Reprinted for Clearfield Co. by Genealogical Publishing Co., 2006.
21. Ibid.
22. *Journal of Lords of Trade and Plantations*, 'America and West Indies', Jan 1690, *Calendar of State Papers Colonial America & West Indies, 1689-1692*, Vol. 13 (1901), Item 710, pp. 115 and 200-215'.
23. Martha J. Lamb 1896, *History of the City of New York: its origin, rise and progress*, Vol. 1, Ch. XIX, '1689-1691; New York Under Leisler', New York, A. S. Barnes & Co., pp. 356-8.
24. *Journal of Lords of Trade and Plantations*, 'America and West Indies', Jan 1690, *Calendar of State Papers Colonial America & West Indies, 1689-1692*, Vol. 13 (1901), Item 710, pp. 115 and 200-215'.
25. Martha J. Lamb 1896, *History of the City of New York: its origin, rise and progress*, Vol. 1, Ch. XIX, '1689-1691; New York Under Leisler', New York, A. S. Barnes & Co., pp. 356-8.
26. James C. Brandow, 1983, *Genealogies of Barbados Families from Caribbeana and the Journal of the Barbados Museum and Historical Society*, Genealogical Publishing Co. Inc., p. 656; Abstracts of Wills on file in the Surrogate's Office, City of New York. William Smith Pelletreau, Vol. I, 1665-1707 p. 187. The correction is to be found in *Collections of the New York Historical Society*, 1907, vol. 40, p. 20. In fact, Elizabeth Bowne was an only child; Andrew Bowne could only have been her father. Thomas Wenham, John's executor, continued to have a turbulent life in New York including a period of exile; he became one of the first churchwardens of Trinity Church.
27. Martha J. Lamb, 1877, *History of the City of New York: its origin, rise and progress*, Vol. 1, Ch. XX, '1689-1691; New York Under Leisler', pp. 359-97; New York, A. S. Barnes & Co. Reprinted by Cosimo Classics, New York, 2005.
28. *Journal of Lords of Trade and Plantations*, 'America and West Indies', Jan 1690, *Calendar of State Papers Colonial America & West Indies, 1689-1692*, Vol. 13 (1901), Item 710, pp. 115 and 200-215'.
29. John Edwin Stillwell, 1970, *Historical and genealogical miscellany: data relating to the settlement and settlers of New York and New Jersey*, Vol. 3, Genealogical Publishing Co., pp. 41-43.
30. James C. Brandow, 1983, *Genealogies of Barbados Families from Caribbeana and the Journal of the Barbados Museum and Historical Society*, Genealogical Publishing Co. Inc., p. 656; Abstracts of Wills on file in the Surrogate's Office, City of New York. William Smith Pelletreau, Vol.

I, 1665-1707 p. 187. The correction is to be found in *Collections of the New York Historical Society*, 1907, vol. 40, p. 20.

31. Despite the imprisonment of many of the city's powerful merchants who opposed him, Leisler was unable to control them, and opposition to his authority increased. In early 1691 Colonel Henry Sloughter, King William's newly appointed Governor, sent in English troops under Major Robert Ingoldsby and, having arrested Leisler and the ringleaders, secured the city. On the advice of prominent merchants, probably those who had suffered at his hand, Leisler was charged with treason, found guilty, and was hanged and beheaded on 16 May. He was posthumously exonerated by the British Parliament in 1695.

32. New York City Wills 1665-1707, p. 104. The wider Bowne family had settled in Flushing, where later is to be found a farmer, Benjamin Haynes. See: Edith K. Wilson, 1987, *Bowne Family of Flushing, Long Island*, New York: Bowne & Co., Inc.

33. Esther Singleton, 1909, *Dutch New York*, New York: Benjamin Blom, Inc., Ch. V, pp. 110-113; Abstracts of Wills on file in the Surrogate's Office; City of New York, 1893. (V. 25), New York Surrogate's Court and William Smith Pelletreau, pp. 100-101 (248) and 105 (328). Interestingly John Haines was one of the overseers of the will of Colonel Lewis Morris of New York and New Jersey (p. 97).

34. Martha J. Lamb, 1877, *History of the City of New York: its origin, rise and progress*, Vol. 1, Ch. XX, '1689-1691; New York Under Leisler', pp. 359-97; New York, A. S. Barnes & Co. Reprinted by Cosimo Classics, New York, 2005.

35. New York City Wills 1665-1707, p. 104. The wider Bowne family had settled in Flushing, where later is to be found a farmer, Benjamin Haynes. See: Edith K. Wilson, 1987, *Bowne Family of Flushing, Long Island*, New York: Bowne & Co. Inc.

36. Hiram E. Deats, 1901, *The Jerseyman,* vol. 7, pp. 3, 20; Miller K. Reading, 1903, 'William Bowne of Yorkshire, England, and His Descendants', repr. 2009, *BiblioLife*, LLC., p. 17.

37. Elizabeth Donnan, 1965, *Documents Illustrative of the Slave Trade to America*, Vol. II. 'The eighteenth century', Octagon Books, p. 29. It may or may not be relevant to John Haynes's occupation that by her will dated 20 March 1720, Elizabeth Bond, widow of St George's Parish, gave her mulatto manservant called *John Haynes*, his freedom (Barbados Department of Archives: RB6/6/549).

38. Will of Robert Haynes, signed 18 September 1734; Executed 4 February 1736/7: Barbados Department of Archives: Wills, Vol. 27, p. 257; Vol. 30, p. 95.

39. Yellow fever, in 1740 known as la maladie de Siam, the black vomiting, or malignant fever: Henry Warren, 1740, *A treatise concerning the malignant fever in Barbados, and the neighbouring islands: with an account of the seasons there, from the year 1734 to 1738. In a letter to Dr Mead*, London: Fletcher Gyles. Warren observed (pp. 13-14) that the Africans rarely contracted yellow fever (a fact confirmed many times since): *'How*

comes it that the Negroes, whose Food is mostly rancid Fish or Flesh, nay often the Flesh of Dogs, Cats, Asses, Horses, Rats, &c. who mostly lead very intemperate Lives, and who are always worse clad, and most exposed to Surfeits, Heats, Colds, and all the Injuries of the Air, are so little subject to this Danger?'; see L. H. Berry, 1964, 'Black men and malignant fevers', *Journal of the National Medical Association*, 56, 43-7.
40. M. D. Ogden, ed., 1917, *Memorial Cyclopedia of New Jersey,* Vol. III, New Jersey: Memorial History Co; Edwin Salter and George Crawford Beekman 1887: 'Old Times in Old Monmouth: Historical Reminiscences of Old Monmouth County', p 287, Baltimore: Reprinted for Clearfield Co. by Genealogical Publishing Co., 2006.
41. Barbados Department of Archives: RB6/6, p. 247.
42. Presumably a kinsman of Elvie Thorne, the second wife of the testator's cousin, Richard Haynes.
43. Richard Perryman is said to have fathered five children by Thomazin Carrington [née Waterland], the wife of Dr Paul Carrington, whose complex matrimonial affairs are documented by James C. Brandow, 1983, *Genealogies of Barbados Families from Caribbeana and the Journal of the Barbados Museum and Historical Society*, Genealogical Publishing Co. Inc., pp. 212-3. These children were born after those of his wife, Ann.

15

The Family Multiplies - the Plantocracy [1700-1800]

Richard Haynes [1694-1739] - Captain Robert Haynes [1720-1753] - Edmund Haynes [c. 1739-1811] - Major-General Richard Downes Haynes [1746-1793]

Robert Haynes, the husband of Elizabeth, died in November 1696 leaving seven young children, according to the provisions of his will [see Figure 15.1]. Although they were their father's heirs, the boys were probably considered too young at the time to be signatories to the address of congratulation to Queen Anne on her Accession in 1702, which was drawn up on 18 May and signed by the principal officers and inhabitants of the island, including most of the planters of St John, from which the name Haynes was singularly absent.[1] Nevertheless, they would doubtless have joined in the celebratory march through Bridgetown that day to mark the occasion. At 11 a.m. on the instruction of the President and Council, all soldiers, constables, officers of the courts, gentlemen, lawyers, justices of the peace, clergy, all ministers and officers of state, mustered in the main square and marched through the capital to the accompaniment of trumpets and drums, much flag-waving and merriment.

William Haynes, who was probably the first son of Robert and Elizabeth, inherited a half share of the residuary estate with Richard. He may have been the one who with four others, witnessed the will of William Jordan of St Michael's Parish in April 1702.[2] Unlike Richard, he wasn't recorded in the 1715 census and may have died before he reached his twenty-first birthday, a condition of the bequest, as Richard became the undisputed owner of Newcastle. Born in 1694, Richard succeeded to the Haynes property in St John after his twenty-first birthday in 1715. How the estate was managed between Robert's death and his son taking over is unclear. It is

THE FAMILY MULTIPLIES – THE PLANTOCRACY

known that substantial numbers of sugar estates were abandoned because of the labour shortage following the prolonged epidemic to which Robert most probably succumbed possibly to be followed by his son William.

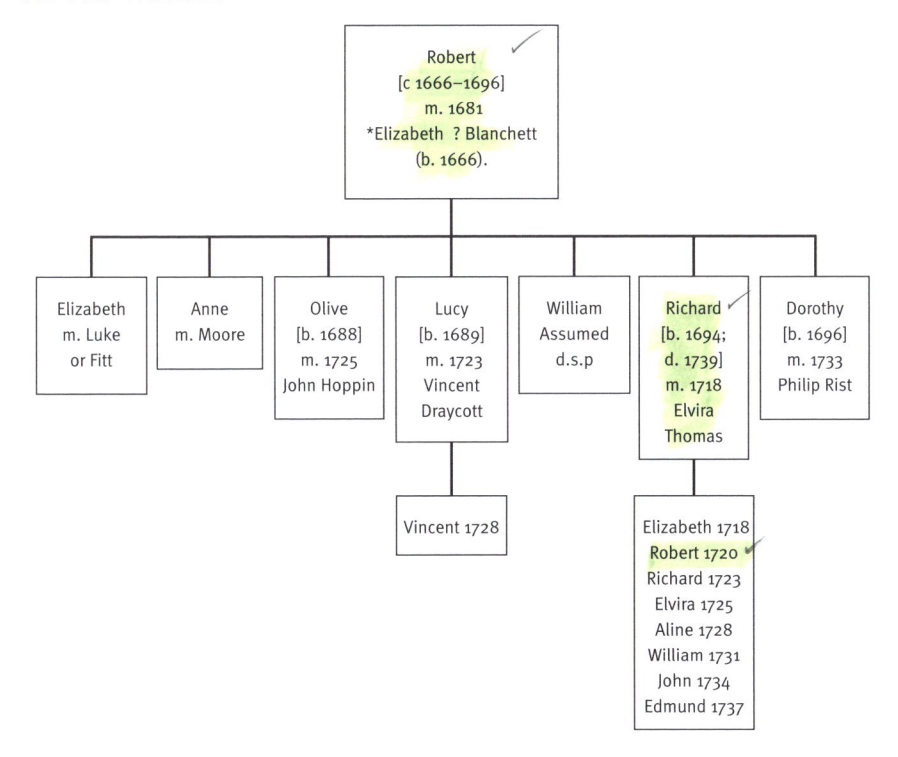

```
                        Robert
                     [c 1666–1696]
                        m. 1681
                  *Elizabeth ? Blanchett
                        (b. 1666).
```

| Elizabeth m. Luke or Fitt | Anne m. Moore | Olive [b. 1688] m. 1725 John Hoppin | Lucy [b. 1689] m. 1723 Vincent Draycott | William Assumed d.s.p | Richard [b. 1694; d. 1739] m. 1718 Elvira Thomas | Dorothy [b. 1696] m. 1733 Philip Rist |

| | | | Vincent 1728 | | Elizabeth 1718
Robert 1720
Richard 1723
Elvira 1725
Aline 1728
William 1731
John 1734
Edmund 1737 | |

*Elizabeth m.2 Scott and by him had one son William Scott b. 1702.

Figure 15.1 Family of Robert Haynes of Newcastle St John

Robert's first and second daughters, Elizabeth and Anne, were born about 1686 and 1687 respectively: nothing more is known of them with certainty except that they were no longer living at home in 1715 and may have married members of neighbouring properties. It was probably this Elizabeth who was the goddaughter of Tobias Swellivant, cooper of St Michael's, who mentioned her in his will dated 24 July 1693.[3] It is very likely that Elizabeth, married into the Luke or Fitt family, as Elizabeth Luke and Elizabeth Fitt were listed with her Haynes siblings, as the granddaughters of Anne Beachamp, formerly the wife of John Haynes who died in 1678. Likewise, Anne may have married John Moore who was witness to a deed, as an Anne Moore was also a granddaughter of Anne Beachamp.

Olive, Robert and Elizabeth's third-born, was unmarried and aged 27 in 1715, making her date of birth 1688. She subsequently married, on 14 May 1725, John Hoppin, who had been one of the witnesses of her father's will in 1696. John Hoppin had, with Henry Coombes, witnessed the signing and sealing of the deed of gift of George and Ann Summers to their daughter, widow Ann Crossing, of eleven named negroes on 28 April 1691.[4]

Robert and Elizabeth's fourth child, Lucy, was born in 1689. On 31 July 1723 she married Vincent Draycott, by whom she had a son of the same name who was baptised at Christ Church on 16 June 1728. The baby of Robert and Elizabeth's family was Dorothy, who was 19 years old in 1715. She married Philip Rist on 12 March 1733. It is of interest that Olive and Dorothy were both 37 at the time of their marriages. Lucy was 34 when she married Vincent Draycott and had only one child. Why the girls had to wait so long before marrying is again probably attributable to the epidemic in the 1690s which killed so many of their male peers. It is possible that they were all second wives, although there had also been a marked increase in the white population, from 12,528 in 1712 to 17,680 in 1736, so opportunities would have been increased by the influx of unattached men.

It appears that Dorothy Haynes had invested her inheritance wisely, as according to the will of William Jordan, planter, dated 17 September 1717, Dorothy had sold him land.[5]

Richard Haynes (1694–1739)

Richard Haynes succeeded to the Newcastle Estate on his twenty-first birthday in 1715. His mother, Elizabeth Scott, presumably widowed for the second time, was at that time living there with her son. Also living at Newcastle were his sisters Olive, Lucy and Dorothy, half-brother William Scott and Margaret King aged 17. The identity of this last person is not known, but a woman of this name was married to Daniel Hutchinson at Christ Church, Barbados on 12 June 1719.

Richard was 24 years old when he married Elvira Thomas at St John's on 30 July 1718. Elvira may have been the daughter of Edward

THE FAMILY MULTIPLIES – THE PLANTOCRACY

Thomas and Elizabeth Kent of St Philip, and it remains to be ascertained whether this Edward Thomas or, more likely, his father, was the indentured servant brought to Barbados by Robert Haynes in 1657. According to the birth date of their first child, Elizabeth, given in the IGI as 17 December 1718, Elvira must have been four months' pregnant at the time of their marriage. They had eight children in all. Their second child, Robert, was born on 20 November 1720, followed by Richard, 20 April 1723; Elvira, 29 January, 1725; Aline, 8 September 1728; William, 8 March 1731; John, 24 November 1734 and Edmund, 31 December 1737. This proved one baby too many for Elvira, who died soon after Edmund's birth, possibly from a combination of puerperal and yellow fever. Edmund also died and was buried on 9 March 1738.

Richard subscribed to the map of Barbados published in 1722 by William Mayo, which showed the position of 976 plantations with the name of the owners at the time of publication.[6]

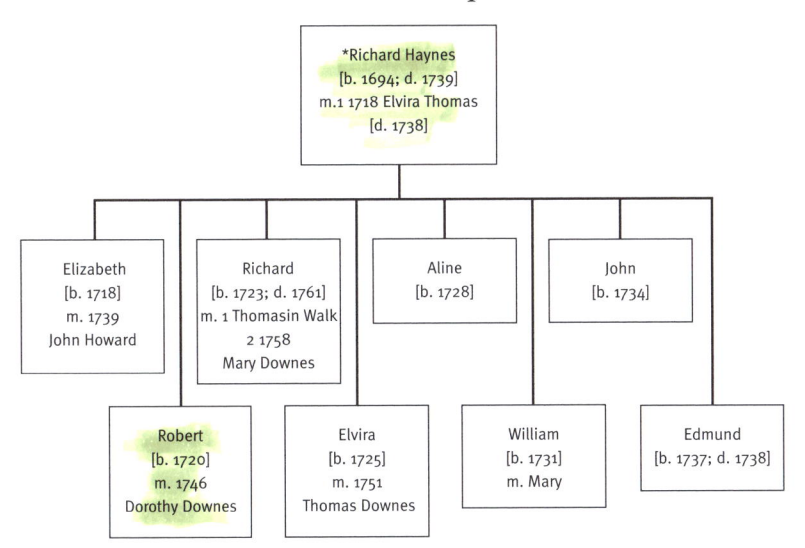

*m.2 1738 Elvie Thorne.

Figure 15.2 Family of Richard Haynes and Elvira Thomas of Newcastle Plantation

Within a year of Elvie's death Richard married another Elvie, Elvie (Elvira) Thorne of St Joseph, by whom he had one child, Edmund. Interestingly, a marriage settlement, drawn up in that year (1738) exists, which described the plantation of Richard Haynes of St John

275

to be 140 acres in area with 49 slaves.[7] This was about the same size as recorded for his father in 1685. The estate was bordered by the properties of Major Thomas Quintyne (Quintynes), Hon. John Colleton (Colleton), Thomas Dickenson, William Hoppin, John Coombes, Catherine Richards, John Beale, Edward Beale, John Hutchinson, Mary Welch, William Marshall, John Ince[8] and Elizabeth Study. Some of the neighbouring properties had changed hands, but the Colletons, Combes, Dickensons and Beales were still there. William Hoppin was a close relative of John Hoppin, Richard's brother-in-law. Captain Richard Haynes was mentioned in the will of John Townes, Richard's nearest neighbour in St John's Parish, dated 25 Jan 1725,[9] as owing the testator money, which was to be paid to *'Capt'* Thomas Quintyne and his wife Ann Quintyne (other neighbours).

Five of Richard's children by his first marriage are known to have been married: Elizabeth married John Howard; Robert married Dorothy Downes; Richard married first Mrs Thomasin Walk in 1746 and second Mary Downes, Dorothy's sister, twelve years later; Elvira married Thomas Downes, Dorothy and Mary Downes's brother; William married Mary by whom he had a daughter of that name in 1754. Thomas Downes and his sisters were members of a well-known large family that included Lt Col. Richard Downes, a prominent Barbadian who was Speaker of the Barbados House of Assembly for the years 1701–2 and 1708–12. He, along with Col. John Leslie[10] and Major Christopher Estwicke, were not only executors of Robert Haynes's will in 1696, in which they were described as *'his trusty and well beloved friends'*, but they were also named guardians of his children. Thus, Richard Haynes's upbringing was closely linked with that of the Downes family, and it is not surprising that two of his sons and a daughter should marry into it. The Downes were wealthy and could send their daughters to England, where one of them, Elizabeth, met and married William Franklin, the son of Benjamin, and the last Royal Governor of New Jersey.

Richard Haynes died at Newcastle Plantation on 31 January 1739, and was buried the following day in the old family tomb in St John's churchyard. It is possible that he died of the smallpox or yellow fever, both of which raged throughout the island in 1738.[11] He was survived by his second wife of one year, Elvy (Elvie) Thorne, but may not have lived to see the birth of his son Edmund. Newcastle then became the property of his eldest son Captain Robert Haynes.

For a significant part of Richard's life, Britain had been at war with France, which made the defence of Barbados essential. There were threats of invasion from Martinique and other islands. Thus many

Figure 15.3 Early Haynes Vault at St John's Church, Barbados

forts and gun batteries were constructed around the coastline. In fact 22 forts were built, chief among which was St Anne's and by 1736 there were 463 gun implacements.[12] All freemen were obliged to serve the militia regiment of their own parish, which obligation was to persist well into the nineteenth century and explains why successive heads of the Haynes household held military titles.

Captain Robert Haynes (1720–1753)

Robert Haynes the son of Richard married Dorothy Downes on Christmas Eve, 1746, at St John. The following 18 October saw the arrival of their firstborn son at Newcastle Estate whom they named Richard Downes Haynes. They may have had other children, but if they did, no records have come to the author's possession that could be corroborated. John Haynes, who died on Christmas Eve 1763 in St John, has been mentioned as their son.

By the 1730s all Barbadian freemen were obliged to enter

themselves into the regiment of their own district. Robert became a Captain in the St John militia.

In 1750 the Revd Griffith Hughes published his illustrated folio book: *Natural History of Barbadoes* in London. It includes a map listing the owners of the chief plantations on the island which

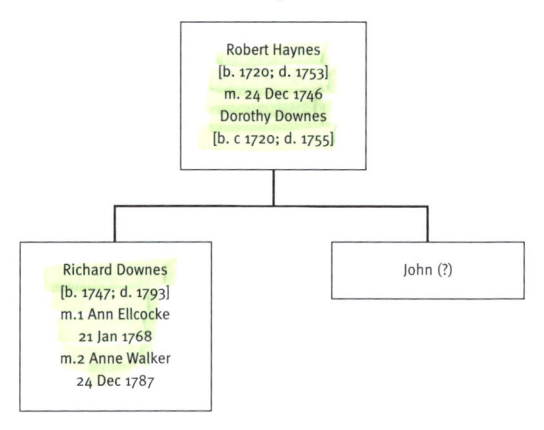

```
Robert Haynes
[b. 1720; d. 1753]
m. 24 Dec 1746
Dorothy Downes
[b. c 1720; d. 1755]
    │
    ├─── Richard Downes
    │    [b. 1747; d. 1793]
    │    m.1 Ann Ellcocke
    │    21 Jan 1768
    │    m.2 Anne Walker
    │    24 Dec 1787
    │
    └─── John (?)
```

Figure 15.4 Family of Captain Robert Haynes and Dorothy Downes

Toner[13] transcribed in alphabetical order. Of interest are the number and location of plantations owned by individuals with the name Haynes and its variants: Haines (this might have been a mistranscription of Harris): St Andrews; Hayns: St Peter; Haynes: St John. The relationship of these families, if any, is uncertain, but the proximity of the Hayns and Gibbs plantations in St Peter's, suggests these might have belonged to the descendants of Robert and Elizabeth Haynes of St Thomas's Parish, whose daughter married Joseph Gibbs. Also Captain Robert's brother, Richard, who married Mary Downes might have owned one of them. He died in September 1761 and two years later on 18 October 1763 his widow married William Cragg at St John's Church.

In 1750 there were about 125 landowners in St John's Parish, 55 of them listed as owners of 60 *'chief plantations'* by Hughes. This coincides with the estimated 52 owners of plantations of 10 acres and over in St John in 1840.[14] Thus Robert Haynes had inherited one of the largest plantations in the parish.

It is not known of what Robert died at the young age of 33, but smallpox was rife in early 1750s Barbados. This is evident from rather an unexpected source. The wealth of the planters and the

THE FAMILY MULTIPLIES – THE PLANTOCRACY

good life to be had on Barbados had not gone unnoticed abroad, and, even at this early date, the island was already attracting tourists. Paradoxically, in view of Robert's death, it became fashionable to go to Barbados for one's health. In 1751, at the age of 19, George Washington, the future first President of the United States, accompanied his elder half-brother, Lawrence, then aged 33 and suffering from tuberculosis, to Barbados for a change of air, which they hoped would ease Lawrence's symptoms. It was in fact the only place outside of the United States visited by George Washington, who kept a diary of his stay there.

Unfortunately for George, on 4 November 1751 he and his brother received an invitation to breakfast from Major Somers Clarke commander of James Fort, Carlisle Bay, and of the British forces on the Windward Islands, whose wife was ill with smallpox. Thirteen days later George went down with the disease, a serious attack from which he he took four weeks to recover. It marked him for life, but fortunately very little upon the face.[15] It has been speculated that the smallpox made George sterile. George and Lawrence rented Bush Hill House for their three-month stay and, while there, they saw some of the most extensive fortifications in British America and socialised with military men, experiences that probably stimulated George's interest in military service. Sadly Lawrence died in 1752, and shortly thereafter George inherited Mount Vernon and also obtained Lawrence's place in the Virginia militia and received the rank of major, which was the first step in his military career. Who knows how different the course of American history might have been, had the Washington brothers not visited Barbados?

Captain Robert Haynes would have had about a third of his land in sugar cane coming to harvest, having grown for 17 months; the rest would have been planted in young cane, with a significant amount of land set aside for food crops: guinea-corn (millet, a cereal used mostly by the slaves), yams, plantains, potatoes, rice, pineapple, guavas and other tropical fruits. He would undoubtedly have kept livestock: horses, mules, cattle, sheep, chickens, etc. on the remaining land. Fertiliser was recycled human and animal dung formed into a mound with soil and plant waste that cattle were allowed to trample for several months, before it was used to manure the crops. Chickens (appropriately named *'dung-hill fowls'*) would have been kept in large numbers. The grass growing in between the sugar cane would have been harvested as animal fodder.

For relaxation he doubtless attended the Beafsteak and Tripe Club in Bridgetown along with other military officers and members of the judiciary. He may even have met George Washington there in November 1751.

Edmund Haynes (c. 1739–1811)

Edmund was the only child of Richard Haynes by his second wife Elvie Thorne. He is believed to have been born in early 1739 around the time of his father's death. He married his mother's kinswoman, Ann Thorne, on 27 November 1766 in St Joseph's Parish Church, where their three children were subsequently baptised: Robert James (baptised 17 June 1768), Christian Ann (baptised 9 June 1771) and Mary Elvira (baptised 18 May 1775). Robert James Haynes married Elizabeth Battaley and they had three children, two daughters and a son. He, like his father, Edmund, played a full part in Barbadian politics. Christian Ann Haynes married William Francis Culpeper in 1799 by whom she had six children, two daughters and four sons. She died in 1817. Mary Elvira Haynes married Dr William Adamson in 1802 and had a daughter Elvira Haynes Adamson. The descent of Edmund's family has not been followed further.

Major-General Richard Downes Haynes (1746–1793)

Richard Downes Haynes was only six years old when his father died in 1753. He eventually inherited the Newcastle Estate. He was destined to become a Major General of the Barbadian Militia and Representative to the House of Assembly for the Parish of St John's, Barbados. Two great fires in 1766 and 1767 destroyed Bridgetown, and Richard would have participated in the decision which became law to control the planning and rebuilding of the town. The British parliament granted a total of £10,000 towards the redevelopment, largely through the efforts of Mr Walker, the colonial agent.[16]

On 31 January 1768 as Captain Richard Haynes, aged 21, he married Ann Ellcocke, also 21, at the Society Estate in the Parish of St John, the property of the Society for the Propagation of the Gospel. Ann, born on 14 March 1746, the fifth of ten children,[17] was the daughter of Grant Ellcocke, who at the time of his daughter's marriage, was the chief overseer of Society Estate.[18] After 13 years, mostly taken up with bearing nine children like her mother before her, Ann died on 25 February 1781, one week after giving birth to Edmund, her seventh son. She was not quite 35 years old. Her now barely legible memorial reads: *'Here rest till angels summon thee away to join in the choral symphony thy [light] shone tenderest sister parent friend & wife in death lamented as beloved in life a jewell [...] pattern of each heavenly grace from heaven receive the priceless crown'*. Six of their children including Edmund, survived to adulthood; Robert, the first-born, Richard and Edward, died in infancy.

Richard and Ann's nine children were: Robert (born 4 October 1768 and died the same night); Robert (born 27 September 1769, died 18 April 1851); Richard (born 18 December 1770, died 4 October 1771); Richard Downes (born 15 July 1772, died 25 May 1793); Edward (born 3 April 1775, died 15 April 1775); Henry (born 13 May 1776, died 17 January 1838); Anne Ellcock (born 8 August 1777, died 1857); Dorothy (born 4 January 1779, died after 1850); Edmund (born 19 February 1781, died 1846).

Richard stood for election to the House of Assembly in the Parish of St John in 1773 at which a most unusual chain of events occurred. The sitting Barbadian President, Samuel Rous, who was also the chief magistrate, not only appeared at the poll and voted himself, but influenced others to vote for J Miller, the candidate whom he supported. Richard, the unsuccessful candidate, had no alternative but to petition the Assembly, complaining of an *'undue return'*, and seeking the result be overturned. However, before the matter could be investigated, he was prevailed upon to withdraw his petition by the same president promising him his future support.[19] Nevertheless, the Solicitor-general, Henry Duke (St Michael's), a man of impeccable character and a constitutional expert, brought the matter before the house, in a strongly worded attack on the President and tabled the motion: *'That it is against the freedom of elections and the privileges of this house that a commander-in-chief should vote at, or interfere in, the election of representatives'*. The Speaker sidestepped the issue by postponing the matter for future consideration, which in the event never took place.[20]

Richard Downes Haynes was eventually elected to the House of Assembly as a representative for St John's and took his seat in January 1780. In the same election his half-uncle, Edmund Haynes, only seven years his senior, was elected representative for St Joseph's.[21]

It was surprising that the family survived unscathed the devastation of the major hurricane of Tuesday, 10 October 1780, which reached the island in the morning with violent winds and heavy rain. The wind at first north-easterly, gradually veered westward, increasing in ferocity, so that by noon many houses had been destroyed. The strength of the wind continued to increase for the next 12 hours. In the towns, Poyer described:

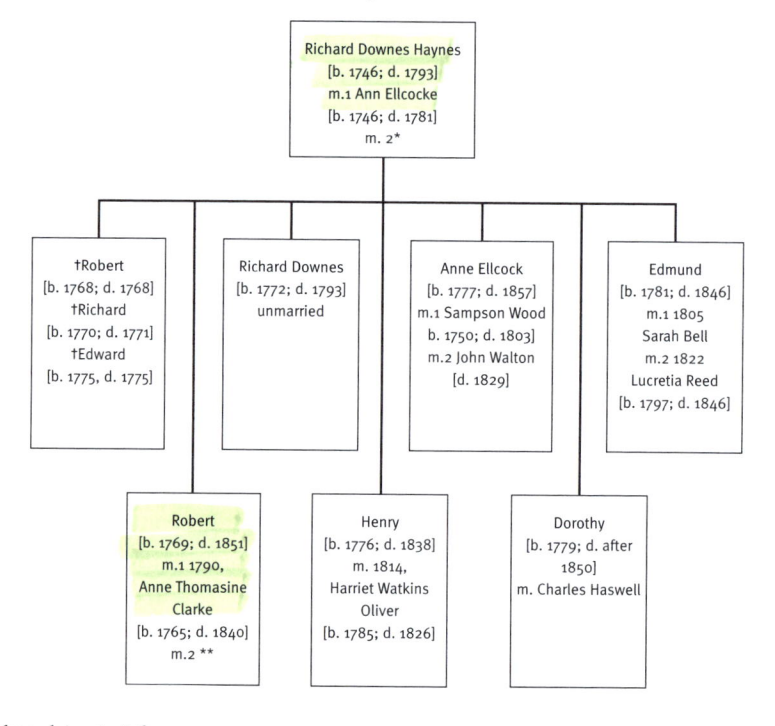

```
                    Richard Downes Haynes
                    [b. 1746; d. 1793]
                    m.1 Ann Ellcocke
                    [b. 1746; d. 1781]
                    m. 2*
```

| †Robert [b. 1768; d. 1768] †Richard [b. 1770; d. 1771] †Edward [b. 1775, d. 1775] | Richard Downes [b. 1772; d. 1793] unmarried | Anne Ellcock [b. 1777; d. 1857] m.1 Sampson Wood b. 1750; d. 1803] m.2 John Walton [d. 1829] | Edmund [b. 1781; d. 1846] m.1 1805 Sarah Bell m.2 1822 Lucretia Reed [b. 1797; d. 1846] |

| Robert [b. 1769; d. 1851] m.1 1790, Anne Thomasine Clarke [b. 1765; d. 1840] m.2 ** | Henry [b. 1776; d. 1838] m. 1814, Harriet Watkins Oliver [b. 1785; d. 1826] | Dorothy [b. 1779; d. after 1850] m. Charles Haswell |

†Children dying in infancy.
*Married secondly 1787 Anne Walker.
**Married secondly 1840 Anna Caitcheon.

Figure 15.5 The Family of Richard Downes Haynes and Ann Ellcocke

'... the havoc which met the eye contributed to subdue the firmest mind. The howling of the tempest; the noise of descending torrents; the incessant flashings of lightning; the roaring of the thunder; the continual crash of falling houses; the dismal groans of the wounded and dying; the shriek of despair; the lamentations of woe; the screams of women and children calling for help on those whose ears were now closed to the voice of complaint, – formed an accumulation of sorrow and of terror too great for human fortitude, too vast for human conception'.[22]

The force of the wind can be gauged from the fact that the heavy cannon were carried almost 100 feet from the forts and a 12-pounder was carried a distance of 140 yards.

The following morning revealed that not a single building on the island had escaped damage; Bridgetown was in ruins; St Michael's Church was no more; a ship had been driven into the Naval Hospital which was thereby destroyed and subsequently swept away; in St John's the church suffered severe damage. More than 4300 perished along with most of the livestock, but according to the Governor, Major General Cunninghame in his report, *'fortunately few people of consequence'* were among the dead: many had been washed out to sea. The situation was aggravated by slaves engaged in looting and the escape of 800 prisoners of war from the gaol.[23]

Richard Downes Haynes initially held the rank of Captain in the St John's Regiment of Militia and played an active part in the defence of Barbados, particularly in readiness for the threatened invasion by the French of 1782. He was promoted to Colonel and subsequently achieved the rank of Major General. In September 1781, Admiral Le Compte Francois Joseph Paul de Grasse had landed French reinforcements for General Washington's troops in Virginia, and then defeated the British fleet under Admiral Graves. This sealed the defeat of the British and insured the success of the American Revolution, which by 1782 was coming to an end. However, de Grasse, with Spanish help, was now intent on taking their Caribbean possessions from the British. By January 1782 his fleet began capturing British islands, Dominica, St Vincent, Grenada, and Tobago, forcing Montserrat to surrender on 22 February 1782. Within three months the only islands remaining under British rule were Jamaica, St Lucia, Antigua and Barbados. On 8 April de Grasse left Martinique with thirty-three ships-of-the-line for a rendezvous with twelve Spanish ships-of-line assembled in Cuba, for an attack on Jamaica. However, Admiral Sir George Bridges Rodney had meanwhile arrived in the Caribbean and immediately took command of the British Caribbean naval forces, including thirty-six ships-of-the-line. Through his intelligence network he knew exactly what the French were up to and, as de Grasse left Martinique, Rodney pursued him and after two days caught up with his fleet.

From 9 to 12 April the two fleets engaged at close range on parallel opposite courses near Les Iles des Saintes, a group of small islands off Dominica. Gaps appeared between the French ships and Rodney ordered his ships to turn ninety degrees and sail through the French line, effectively splitting it into four groups. Simultaneously the guns on either side of the British ships fired on the French who could do little in return. Several French ships were forced to surrender including de Grasse's 130-gun flagship, *Ville de Paris*, 400 of whose crew had been killed and more than 700 wounded. The French invasion ships were either sunk or captured; more than 6,000 men died. Admiral Rodney sailed the captured ships to Jamaica.[24]

With British naval supremacy restored, by late summer the Caribbean was safe enough for commerce to resume. Although the terms had been agreed, the preliminary articles of peace between Great Britain, France and Spain were not signed until 28 January 1783. On 13 September 1782, Richard Downes Haynes was confident enough to sail to England for the benefit of his health. He sailed in the brig *Venus*, on which he experienced significant hardships from violent storms that month. The journey between Barbados and England generally took thirty-one days in each direction when the weather was favourable; when very stormy it could last half as long again.[24] Although his journey was delayed, he made his passage safely. He probably visited Bath, although little is known of his whereabouts in England on this occasion. He would certainly have had the opportunity to undertake genealogical research on his family during his five months' stay. Richard returned to Barbados seven months later arriving on 16 April 1783.

Life was made easier in Barbados following the proclamation in December 1783 which permitted the importation of produce of the United States by British subjects in the West Indies. This greatly assisted reparation of the damage wrought by the great hurricane.

On 24 December 1787 Richard Downes married for the second time. His wife, Anne Walker, had been married twice before, first to George Foster Clark of the Ashford Estate, St John's,[26] secondly to Henry Walker, formerly owner of Walker's and Golden Grove plantations in the Parish of St George. She had one son by her first husband called Richard Foster Clark, who became an ordained clergyman.

Richard Downes Haynes was the Representative for St John's Parish in the Assembly and his cousin, Robert James Haynes (son of Edmund Haynes), was the representative for St Joseph's. In the 1780s there was growing disquiet over the fact that slaves or free men of colour could be killed by whites with little or no consequences to the perpetrators. This was clearly wrong in the

THE FAMILY MULTIPLIES – THE PLANTOCRACY

opinions of a growing number of whites, and in 1788 a bill was introduced to make such wilful murder a capital offence. It was supported by Sir John Gay Alleyne and others, but strongly opposed by Richard Downes Haynes and his cousin, and others in the assembly, and the bill was rejected.[27] It wasn't until 1805 that Governor Seaforth successfully introduced legislation which passed into statute making the murder of slaves and those of mixed race punishable by death.

Richard Downes Haynes seemed increasingly over-concerned about his health, although his son Robert believed his father to be healthier and stronger than most men of his age. Nonetheless, his father wouldn't believe it and in September 1792 he again undertook the hazardous journey to England, presumably to seek medical advice. He again sailed in the brigantine, *Venus*, a vessel owned by the firm Thompson & Rowlandson, *'merchants'* of Bridgetown, Barbados. It was to be his last journey.

In London he found accommodation in the handsome, and recently built, Warwick Court, a small cul-de-sac of Georgian terraced houses, off High Holborn, just north of Chancery Lane. Within four months Richard Downes Haynes, Major General of the Barbadian Militia and Representative of the House of Assembly for the Parish of St John, had died in London at his lodgings, aged 46.[28] His burial took place on 23 January 1793, four days after his death, at St Andrew's Church, Holborn, the entry in the parish register simply recording: *Richard Haynes Esqr, Warwick Court* (the term Esquire as used in Barbados signified membership of the House of Assembly).

The event which, according to family legend, indirectly led to the death of Richard Downes Haynes was an infection apparently made worse by his determination to watch a *'bruising match'*, a bare-knuckle prize fight, which bouts, though not always legal, were the most popular spectator sport in England at that time. Bare knuckle boxing, or pugilism, was a fixture of English society from the early 1700s up until the advent of gloves and 'The Marquis of Queensbury' rules in the nineteenth century. They were known to have involved women as well as men,[29] and the injuries received could be extremely serious. So, because of the damage inflicted, prize-fighters were only allowed to fight once per season. Fights were attended by members of all classes who were able to arrange travel to the often somewhat secret locations of the events. While gentlemen never participated in such fights, it being beneath their status, they did spar with each other and with teachers of the sport. His son, Robert, recorded in his notes:

> 'Whilst lying ill in London, he did insist on putting his head out of a window on a cold day in order the better to see a prize fight which was taking place in the street below. This induced a dangerous fever of which he died aged 46. Literally and truly an honest man. I knew well his worth! He cut short his life by many years.'[30]

The will of Richard Downes Haynes was dated 3 October 1792, presumably written on the journey to England. The estate passed to his eldest son, Robert, who was executor, along with significant debts. He also left his second son and namesake Richard Downes Haynes, £3000 for when he reached the age of 21 and £200 per annum thereafter.

Notes and references

1. 'Celebration of Queen Anne's Accession', *Journal of the Barbados Museum and Historical Society,* 1934, Vol. 1, pp. 36-8.
2. Will of a William Jordan, tin-plate worker of St Michael's Parish, dated 23 April 1700; Barbados Department of Archives: RB6/43/136.
3. Barbados Department of Archives: RB6/3/231.
4. Barbados Department of Archives: RB3/16/279.
5. Barbados Department of Archives: RB6/4, p. 407. There is an earlier will of a William Jordan, tin-plate worker of St Michael's Parish, dated 23 April 1700 (RB6/43, p. 136; Sanders II, p, 196). This was witnessed by a William Haynes and four others.
6. Ronald Tree, 1972, *The History of Barbados*, London, Hart-Davis; *The Genealogist* 17, March 1974.
7. Barbados Department of Archives: RB/88/74 [Written communication, Barbados Archives Department]. See also J. C. Brandow, 1983, *Genealogies of Barbados Families from Caribbeana and the Journal of the Barbados Museum and Historical Society*, Genealogical Publishing Co. Inc., p. 656.
8. John Ince married Mary Culpeper; her sister Margaret Culpeper married Edward Quintyne – *Barbados Wills and Administrations, Vol. III*, File Ref: 11 Article, Journal.
9. Barbados Department of Archives: RB6/22/551.
10. John Leslie was returned to the House of Assembly as representative for St John on 17 April 1700 – his name at the time was occasionally spelt Lesslye or Lesslie.
11. Smallpox: R. H. Schomburgk, 1848, *The History of Barbados*, Longman,

THE FAMILY MULTIPLIES - THE PLANTOCRACY

Brown, Green & Longmans, p. 85. Yellow fever, in 1740 known as la maladie de Siam, the black vomiting, or malignant fever: Henry Warren, 1740, *A treatise concerning the malignant fever in Barbados, and the neighbouring islands: with an account of the seasons there, from the year 1734 to 1738. In a letter to Dr Mead*, London: Fletcher Gyles.

12. Schomburgk, 1848, p. 193.
13. J. M. Toner 1892, *The daily journal of Major George Washington in 1751-2: kept while on a tour from Virginia to the Island of Barbadoes*, Albany, Joel Munsell's Sons.
14. Calculated from data provided Schomburgk, 1848.
15. J. M. Toner 1892, *The daily journal of Major George Washington in 1751-2: kept while on a tour from Virginia to the Island of Barbadoes*, Albany, Joel Munsell's Sons.
16. Schomburgk, 1848, p. 332.
17. Their details are given in E. C. Haynes, A. P. Haynes and E.S.P. Haynes, 1910, *Notes by General Robert Haynes of New Castle and Clifton Hall Plantations, Barbados and Other Documents of Family Interest*, Argus, London, p. 8.
18. The Ellcockes were descended from an old Barbadian plantation family. In 1642 Alexander Ellcocke had acquired 100 acres of land in St James, Barbados, from the London Merchant House, which had been leased ten thousand acres (about one tenth of the island) by the Earl of Carlisle. It was this land which was subdivided into large tenanted estates which were purchased and consolidated by individuals who emerged as sugar pioneers by the late 1640s (Hilary McD. Beckles, 1989). The London Merchants included Marmaduke Brandon, William Perkin, Alexander Bannister, Robert Wheatley, Edmond Forster, Robert Swinnerton, Henry Wheatley, John Charles and John Farringdon (Schomburgk, p. 262). Robert Wheatley became the fourth Governor of Barbados in 1629.
19. John Poyer, 1808, *The History of Barbados, from the First Discovery of the Island in the Year 1605, till the Accession of Lord Seaforth, 1801*, London: J. Mawman, p. 360.
20. Ibid. pp. 360-61.
21. Ibid. pp. 406-7.
22. Idid. pp. 448-9.
23. Ibid. p. 454
24. For details of Admiral Rodney's actions, see Schomburgk, 1848, pp. 343-6.
25. With the advent of steam crossings in 1842, ten days were saved in each direction.
26. Ashford was subsequently owned by Colonel William Matson Barrow.
27. *Journal of the Barbados Museum and Historical Society*, Vol. XLVI, 2000, pp. 13-51. See account of Governor Seaforth's subsequent legislation of 1805 in Schomburgk, 1848, pp. 370, 422.
28. Number 9, the house where he died, is now called Cobham House and is the address of The Guild of Air Pilots and Air Navigators, and other organisations and businesses.

29. George Rudé, 1971, *Hanoverian London 1714-1808*, London, ISBN 075093333X.
30. E.M.W. Cracknell, ed. (1934), *The Barbadian Diary of Gen. Robert Haynes 1787-1836*, The Azania Press, Medstead, Hampshire, UK, pp. 17-18. The reference to 'An Honest Man' might have been an allusion to the monument in Bath Abbey to George Gordon, Lord Byron's maternal grandfather: *'George Gordon Esq of Gight in the shire of Aberdeen. Died 9th January 1779 AN HONEST MAN'*.

16

Lieutenant-General Robert Haynes [1769-1851]: End of an Era

School - Stepmother - America - Marriage - Newcastle - Family - Expansion - Approaching Abolition - England - Back to Planting - Insurrection - Speaker - Resignation - Ann Lovell

Robert Haynes, the one who first penned the brief history of his family, was born at Newcastle Plantation on 27 September, 1769. He was the second and eldest surviving son of Richard Downes Haynes and Ann Elcocke.

Education

Robert's formal education began a month before his tenth birthday, when he went to school with the Revd Francis Fitchatt[1] at Thorpe Estate, St James. It was interrupted the following year when the great hurricane of 10 October 1780 destroyed the school and laid waste much of the island with considerable loss of life. Not only were 4326 people killed, but the damage finally ran to more than £1.3m.[2] Robert described this event as leaving him with his *'first sense of ruin and desolation'*. The school was moved to Joe's River Estate in St Joseph, then the property of Henry Evans Holder, where Mr Fitchatt had been given the living of St James the Vestry and Archdeaconry, and was allowed to build a parsonage with school dormitory attached. The Hayneses suspected that, although addressed as Reverend, Fitchatt appeared not have been a genuine clergyman and had not been ordained. However, this was untrue. Fitchatt, a Master of Arts, a gifted teacher and formerly a Fellow of St John's College Cambridge, was for 25 years successively the rector of St Joseph's, St James's and St Peter's where he had been installed only a few months before his death on 5 December 1802 aged 56.[3] It had been alleged that, in common with certain other schoolmasters and pseudo-clergy, Fitchatt was a propagandist for the anti-slavery campaign first started by the Society of Friends. This was probably because he was an occasional catechist at Codrington, a position he quickly resigned because the estate manager and Haynes kinsman, George Barrow, refused to give Fitchatt the use of the mansion house there. Had this been the case and his true sympathies exposed, he would surely have been expelled from the island, and certainly would never have become the Chaplain to the House of Assembly, which he did. Indeed, he married, had children and died on Barbados, where his descendants continued into the latter half of the nineteenth century. It is quite clear from Robert Haynes's correspondence that he had been well educated by the 'Reverend', and he himself in later life would in turn ensure the thorough education of all his children in Britain. He remained at the school until his eighteenth birthday in September, 1787.

Father's re-marriage

Robert seems to have been devastated when on Christmas Eve, 1787, his father remarried. He had lost his own mother six years earlier when he was 11½, and he had great difficulty in coming to terms with his new stepmother. She had been married twice before, first to George Foster Clark of the Ashford Estate, secondly to Henry Walker. Robert loved, admired and greatly respected his father, but took an instant dislike to his stepmother, and came to believe the marriage to be a *'disaster for the Haynes family'*, particularly for him personally.

America

It seems likely that his antipathy towards his stepmother galvanised Robert to leave Newcastle for a while. He went to Bridgetown where he boarded in the house in High Street owned by Mrs Sarah Bradford, and began to pen his diary. Here he met an American guest called Dickenson, a landed gentleman, said to be the Treasurer of Talbot County, on the coast of Maryland. Dickenson regaled the young Robert with his exploits and the active part he had played in the

American Revolution in September 1776, which intrigued the curious 18-year-old enough to make him determined to visit that continent. This would after all provide not only his last opportunity for respite from domestic confrontation, but also the only opportunity for adventure before he would begin the serious work in plantation management. The pair set sail for Baltimore from Carlisle Bay on 16 March 1788 in the brig *Lovely Lass*, arriving on 2 April. Robert then travelled widely in America, whether alone or in the company of Dickenson is not known. But he was there for six months, and it was not until 10 October that he began the return voyage via Cape Chesapeake on the same vessel, which made landfall in Barbados on 30 October.

The following day Robert began to assist his father in the management of the Newcastle Estate. He grew even closer to his father, but his time abroad had not helped his relationship with his stepmother; if anything his antipathy towards her was greater, bordering on hostility. Although she spent little on herself, and was neat, clean and tidy, she was accused by Robert of indolence and extravagance, neglect of all domestic concerns, and of allowing *'her favourite slaves to rob and plunder to an unlimited extent'*. He feared she would bring the estate to ruin. As for his father: *'Never did any man possess more of the milk of human kindness, a kind benevolent heart, an indulgent and affectionate husband and parent'*.[4]

Marriage

Nineteen months passed during which time he stood in for his father, who had become happy enough to leave Newcastle in his son's care. Meanwhile, Robert had met a 25-year-old mother of two, some four years his senior, who seemed to dispel the conflicts with his stepmother, for on Wednesday, 19 May 1790, Robert Haynes was married to Ann Thomasine Barrow, widow and relict of Nathaniel Barrow. The marriage took place in the cool of the evening at Mount Pleasant, the home of George Barrow, in the Parish of St Philip. George Barrow, the half-brother of Nathaniel Barrow, was the husband of Ann Thomasine's sister, Dorothy Clarke, which explains why Ann Thomasine's sister's husband, George Barrow, stood in as host at her marriage to Robert Haynes.[5] A family friend, the curate Revd Henry Husbands, officiated.

Ann Thomasine was born on 19 November, 1765, the daughter of William Clarke and Thomasine, his wife. William Clarke was Thomasine's second husband whom she had married as a widow on 4 February 1758 at Christ Church. Born Thomasine Bonnet in May 1738, she married first Robert John Meyers on 16 August 1753 at St Michael's Church. Her husband died five years later in 1758. William Clarke [b. 1735], her second husband, was overseer of the Thicket Plantation, a property which was to feature later in the spectacular slave revolt. He died in June 1766, his wife predeceasing him by seven months, being buried on 27 November 1765 at St Philip's.

Thomasine Bonnet had a famous, or rather infamous, ancestry. She was born in May 1738, the third child of Edward Bonnet and his wife Mary Thornhill. Edward, born in October 1713, was the eldest surviving son of Stede Bonnet and his wife Mary Allamby, whom he had married in Bridgetown on 21 November 1709. It was an early custom in Barbados to name children after the island's governor or speaker. Stede Bonnet, born in 1688, was duly named after Edwin Stede, then Lieutenant Governor (1685–90). Stede Bonnet had inherited his father's large plantation and was a major in the Barbadian militia. According to Johnson,[6] Bonnet's marriage to Mary was not a happy one and he was driven to desert his family by his wife's incessant nagging. Early in 1717 Stede became a pirate and, because of his sound education and family wealth and notwithstanding his plundering of the East Coast of America with Edward Teach (Blackbeard), became widely known as the *'gentleman pirate'*. However, within a year and a half he was captured off South Carolina, where he was tried and executed at Charleston on 10 December, 1718.

Ann Thomasine, the pirate's great granddaughter, had married Nathaniel Barrow on 19 December, 1782 at St Philip, and had a son, Samuel [baptised 4 March 1784] and a daughter, Sarah Ann [baptised 8 April 1785]. Samuel and Sarah Ann were to be brought up by their mother and stepfather alongside their younger half-siblings. Sadly Samuel Barrow died while at school at Codrington College.[7] Sarah Ann Barrow eventually married John Higginson.[8] From her portrait reproduced in Cracknell, Ann Thomasine appeared as a petite, round-faced lady wearing a bow in her hair.[9]

Newcastle

After their marriage the couple remained at Mount Pleasant for almost six months, George Barrow probably having returned to his English estate, before moving to Eastwood, on 5 November. Eastwood was a property of 30 acres owned by Robert's father. Their two eldest children, Dorothy, known as Dolly, and Richard, were born there. His father's death in England in January 1793 meant that Robert had to move at once from Eastwood to the plantation house at Newcastle. This enabled his two sisters, Ann Ellcock Wood and Dorothy Haynes, together with their stepmother, to decant to Eastwood, where their stepmother eventually died.

Robert catalogued the Newcastle Estate's acreage and value in his diary that year: 227^1/$_2$ acres, 111 slaves, 30 cattle, 3 horses; value £18,000. However, his father's legacies and debts amounted to £14,000. With his bride, two years earlier, he had acquired an interest in Mount Pleasant to the tune of £2100: he also possessed 37 slaves of his own, neither of which assets was he about to allow to be swallowed up by his father's debts. He, therefore, took on the Newcastle Estate as his father's executor rather than as his heir-at-law, because, at the time, he could not see how he could possibly repay his father's creditors.[10] In 1794 there was a deed signed[11] further to a Marriage Settlement of 1790 (Robert and Ann Thomasine) in which the Newcastle Plantation at St John was said to be 258 acres with 110 slaves. The estate was at that time bordered by the properties of John Adamson, Zachariah Howard, William Massiah, Christopher Massiah, James Edward Colleton, Edmund Haynes, William Shepherd, William Hollingshed, Thomas Coppin, James Johnson, Richard White, Robert Johnson Marshall deceased, and the sea.

On taking over the estate, he found that the Brazilian sugar cane that his father had planted was barely profitable, and so he managed to acquire from William Fernhall 1300 plants of the Bourbon cane – *'only one eye to a plant'* – for £32.10. 0d. This cane had been grown in Martinique and believed by Robert to have originated in Mauritius. In fact, Bourbon cane came originally from the French Isle of Bourbon, Réunion. This cane was unknown in the British West Indies until Admiral Sir John Jarvis, Commander in Chief of the Mediterranean fleet,[12] captured the island of Martinique, where the variety had been imported on the express instruction of King Louis XVI to be disseminated throughout the French Colonies. It thrived in the alkaline soil with just the right amount of rainfall during the growing period, but its prolific growth tended to impoverish the land which required frequent 'dunging'.

The death of Richard Downes Haynes had created a vacancy for a Representative of St John's Parish in the House of Assembly. On 15 July 1793 Robert was duly elected in his place, so becoming The Hon. Robert Haynes, the other member for St John being The Hon. Daniel Walcote. Robert was to be in the Assembly for the next 36 years. Two weeks later [29 July] he was appointed Colonel of the Centre Regiment of Horse by President William Bishop.

Family

Robert and Ann Thomasine were to have ten children. Their first child, a daughter, was born on 26 May 1791, while the couple were still at Eastwood. She was named Dorothy Ann: Dorothy after Robert's sister, with whom he remained in touch all his life, and who in turn had been named after their grandmother, Dorothy Downes; Ann was after her mother and grandmother Ellcock. Their first son, and heir, Richard, was born the following year and William a year later on 25 September 1793. Robert, busy as a Representative, was to face his second great loss. William died on 8 December, just two and a half months old. The diary entry states simply: *'My son William died (by inoculation), aged 2 months'*.

The practice of inoculation, or variolation, of infants against smallpox was widespread in the 1790s in the British Empire, where it had arrived from Turkey some 70 years earlier.[13] In Barbados, smallpox inoculation received great impetus following the arrival around 1750, of William Hillary [1697-1763], a Yorkshire physician, whose reputation had developed around his expertise on tropical diseases of the West Indies and on smallpox inoculation.[14] Smallpox afflicted about 60 per cent of the population at some time in their lives, and the mortality was one-third of those catching the disease. Inoculation by the removal of part of the crust of a scab of the *'more favourable'* form of the disease [Variola minor], and inserting it into the skin of child ('ingrafting') was what had presumably been done to William, but the child was overwhelmed by the ensuing reaction. Two months old was extremely young for this treatment. The orphanage of Thomas Coram used to inoculate all its children when they were three years old: the treatment had received royal support in Britain with the inoculation of the Prince of Wales.[15]

The 2 January 1795 saw the birth of their fourth child, called Robert after his father. Another son, George Barrow, followed 14 months later. The birth of a daughter, Fanny, in 1798, remains to be confirmed. Thomasine arrived next on 5 February 1799 and William Clarke on 9 January 1800. Henry Husbands followed in September 1801 and last of all John Aughterson on 27 June 1803, who was to survive only 18 months. Thomasine was named after her mother and maternal grandmother, Thomasine Bonnet; William Clarke after his maternal grandfather; Henry Husbands was named after the clergyman who married Robert and Ann Thomasine; John Aughterson was named after another close friend and plantation owner. Ann Thomasine was clearly a strong woman who bore twelve children in all.

The Bourbon sugar cane purchased by Robert turned out to be a godsend. Its yield was a vast improvement over the old Brazilian cane. The French revolution had a negative impact on sugar production in the French West Indies and because of the resulting diminished supply, the price of sugar rose rapidly, effectively more than doubling, so that by 1804 Robert had not only paid off his father's debts and legacies, but had upgraded the Newcastle plantation house and filled it with silver plate and other valuables and creature comforts fit for his growing family and heightened status.

The high level that the Haynes family had already achieved in their standard of living, and their generosity to visitors, were clearly evident from the account of a visit to Barbados in 1796 by Dr George Pinckard a London physician who took part in the expedition to the West Indies under General Sir Ralph Abercrombie against the French, who were instigating and helping the slaves to rebel against British rule.[16] Pinckard described the warm Barbadian welcome received in the home of *'Mr Haynes'* and his son, and his pleasure in being taken to the beautiful neighbouring 'Joe's River' estate, then in decay following the death of its owner.[17] This was followed by a trip to the Hayneses' ocean-side cottage called Bay-house, probably built by Edmund or Richard Downes Haynes, under a rock upon the open beach, to serve as a place of rest and refreshment for such of his friends who chanced to visit the windward coast. Haynes provided *'a rich and bountiful repast'*. All the good things of the island appeared to have been collected and staff allotted to attend on the

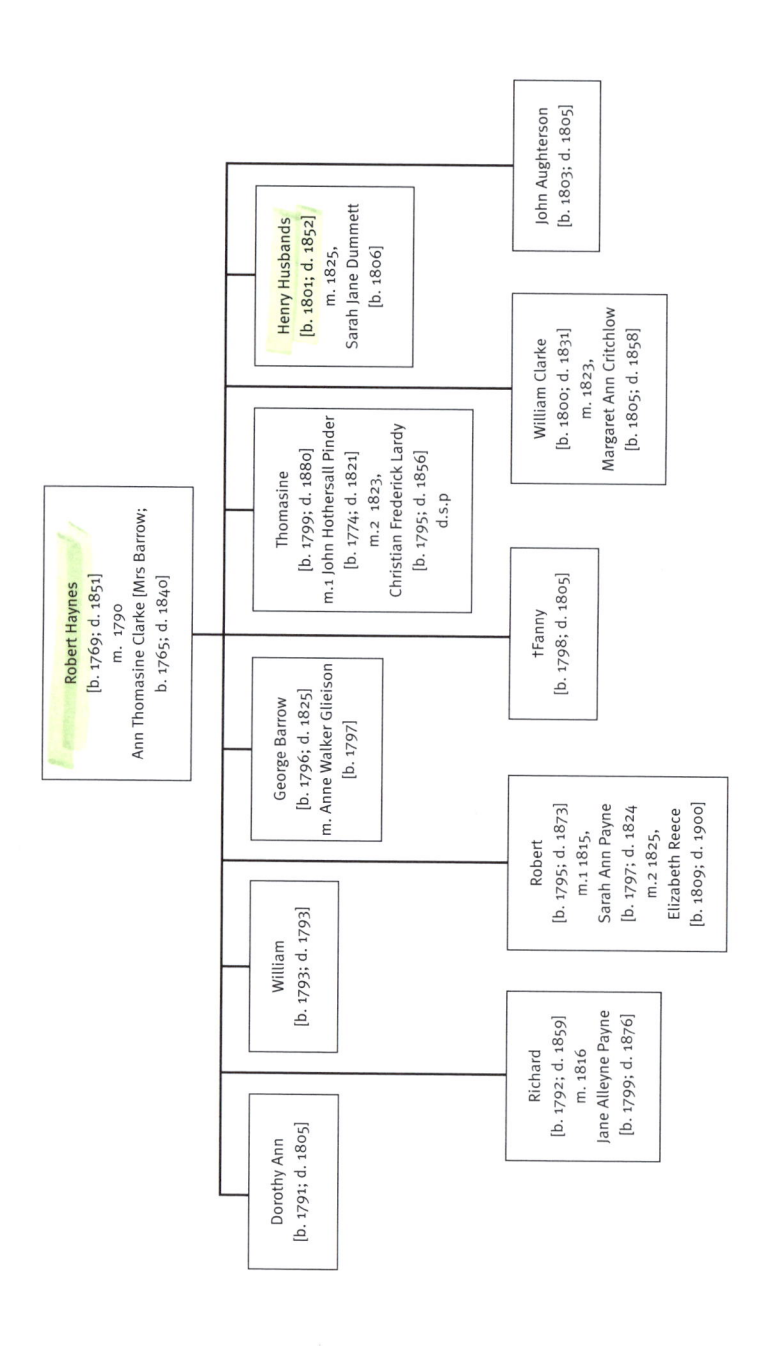

Figure 16.1 Family of Robert Haynes and Ann Thomasine Clarke

†Data to be confirmed

party included a cook, butler, fishermen [to net fresh fish], servants and slaves. The visitors were struck by how well Haynes treated his slaves, leading Pinckard to note: '*A supply of provisions was likewise collected for the refreshment of our slaves ... and these were regaled with no less liberality than their masters*' – the friendly hospitality of the Bay-house being extended equally to all. The meal on the beach that followed included: boiled, stewed and broiled fish, cold roast lamb, a cold turkey, fowls, tongues and crayfish accompanied by punch, a variety of excellent wines and tropical fruits. Just when they thought it all over there arrived a large bowl of grilled sprats and a copious supply of milk-punch. The continuous feasting lasted four hours in '*high banqueting and conviviality*'.

Expansion

Robert Haynes had not only paid all the debts and legacies but had also amassed £10,000 in local currency (about £7500 sterling; the pound sterling being fixed at 1.333 times the Barbadian pound currency) which was accruing interest, and he next embarked on an expansion of his land holdings. He acquired first the Baldocks (Baldricks or Baldwicks) Estate from the executors of Robert Trotman,[18,19] taking ownership on 9 August 1804 for which he paid £19,700. This estate comprised 168 acres with 96 slaves. His first action was to rename it Haynesfield.

It is by no means clear how these enormous sums were paid for in these property transactions, but private mortgages were highly likely sources of credit.[20] Robert Haynes mentioned that within a year of buying Haynesfield he had '*paid off every shilling except £2000 that I had borrowed of Thomas Sealy*', indicating that he had indeed taken on private loans. Clearly, the huge amounts that he had owed were probably paid for in sugar and rum. Local Barbadian gold currency had been unstable for over a hundred years. It was forbidden to export coins from Britain, and the English would not permit the colonists to mint their own. As a result, a mixture of British, Spanish, Dutch, French and Portuguese coins were used as Barbadian currency, along with a variety of Spanish and Portuguese colonial coins. In 1706 a bank was established to remedy the deficiency of gold and silver coin by means of a paper money currency. However, the population deeply resented being '*deprived*' of gold and silver and given *'worthless'* paper. In 1792 Governor David Parry established by proclamation the value of circulated gold coins at a standardised rate proportionate to the legal coin of Great Britain. Many commercial transactions at this time were conducted on a credit basis, with cash only being used occasionally to settle accounts. For a long time, cotton, tobacco and sugar were legal tender, and these were the 'currencies' that testators left by way of bequests in their wills. Presumably Robert Haynes was able to circumvent the currency difficulties through his sugar sales to England. Perhaps this was one of the reasons why his family eventually set up a shipping and trading business in Liverpool, where he could amass sterling.[21]

Some plantations were owned by absentee landlords living in Britain, a proportion of whom had inherited properties from relatives in Barbados. The Lane brothers, John and Thomas, inherited Seawells and Newton Plantations respectively. Thomas Lane, a London solicitor took on the management of both brothers' inheritance. However, in order to manage these complex businesses at the Barbados end, they needed someone in overall charge and until his death in 1803, Samson Wood, Robert Haynes's brother-in-law, managed the Lanes' plantations. On his death his widow, Ann Ellcock Wood, in a letter to Lane dated 19 March 1804, recommended John Farrell as plantation manager and her brother, Robert Haynes, as their Barbadian attorney[22] (also known as *'planting attorney'*, the colonial term for estate manager[23]). Judging from the fee notes he submitted, estate management was a lucrative source of additional revenue for Robert, and it continued up to the time of the death of both brothers in 1824. Attorneys generally charged about 5 per cent commission on the sale price of every crop under their care shipped to England. They also had the plantation houses at their disposal.[24] Dishonesty was rife amongst attorneys, who were able to spirit away sugar or rum, which they then sold and pocketed the proceeds. Attorneys would sometimes source plantation supplies from their own companies. There is little doubt that Robert Haynes's advice to Lane that he ship his sugar through Liverpool was to capture the business for Haynes, Lee & Co, rather than Lane's existing shippers.[25] Thomas Lane appeared to be wilier than most absentee owners. His legal practice was conducted from Goldsmiths' Hall, and it was there that Lane kept meticulous records relating to the brothers' Barbadian plantations, including the letters from his attorney, and copies of his correspondence to Haynes. This business and general correspondence, covering some 20 years, has survived.[26]

Approaching abolition

Unease began to grow among the heavily indebted planters at the increasing clamour of the abolitionists in Britain, on whom all the economic ills of the West Indies (including low property values and difficulty in obtaining credit) were blamed. The calls for compensation of the planters should abolition of slavery occur, intensified with powerful lobby groups emerging in Britain.[27] There was also the trouble stirred up on the island to this end by the Quakers. Robert Haynes loathed what he saw as the hypocrisy of the latter for their encouragement of the slaves to rise up, while simultaneously preaching their disapproval of violence and availing themselves of the protection of Barbadian law. He appreciated that slavery was open to abuse by some extremist owners and was not against some reform. He clearly considered that he and most owners treated their slaves with *'reasonable care and indulgence, save where such is neither reasonable nor warranted'*. Paradoxically, Haynes although a slave owner, was opposed to the slave trade and supported the Abolition Act, on the eve of which [16 September 1806] in a letter to Thomas Lane he wrote: *'I sincerely rejoiced at the abolition of the slave trade ...; I sincerely wish the trade had been abolished 20 years ago'*.[28] This was a view shared by many slave owners on Barbados, as their labour force was formed largely through the encouragement of marriage and natural increase in the population. Haynes and particularly his sister, Ann Ellcock Wood, appeared to form social relationships with their slaves, especially those of mixed blood, which led them to support requests for manumission. The later correspondence with Lane clearly shows such support for two sisters, Dolly Newton and Jenny Lane, daughters of the well-known *'Old Doll'*, and Jenny's two sons on the Newton Plantation. However, the Barbados Slave Manumission Records show that the grantings increased in the early 1830s before tailing off when compensation was on the horizon.

This should not be taken to imply that Haynes was anti-slavery – he was definitely not, but he was clearly opposed to the abusing of those in servitude and was convinced that slaves in his care were in better health and circumstances than, for example, the Irish in Britain. He was greatly perturbed by the recital of cruelties said to have been committed upon slaves in Barbados, which he knew to be totally false, and these, together with the hyperbole of the debates on the slave trade, were put down, as Schomburgk acknowledged, to the philanthropic fervour of the reporters, primarily the Quakers. However, Robert Haynes constantly feared the economic consequences of emancipation as evidenced by the increase in the numbers of people of mixed race acquiring important properties. So, in the autumn of 1802 he brought a bill before the House of Assembly *'to prevent the accumulation of real property by free Negroes and free persons descended from Negroes'*. This, in effect, limited the free black population's economic development and with it their impact on the wider community. For example, its provisions allowed freed individuals to own no more than five slaves.[29] Nine years later he led the opposition to a petition to the House of Assembly by free 'coloureds' for the right to testify in court against whites, again fully supported by his cousin Robert James Haynes. The Hayneses seemed consistently bent on neutralising any attempt at extending the rights of this group.

Most episodes of slave unrest were on a small scale and were quickly put down, but they nevertheless contributed to a growing anxiety. In January 1795 Haynes had been appointed Colonel of the St John's Regiment of Militia, by Governor George Poyntz Ricketts.[30] Ten years later an external threat, that of an invasion of the island by the French, led to his promotion to Lieutenant-General of Militia and Master General of Ordinance under Lord Seaforth as Governor.[31] He played an active part in preparing the people to defend their country, seconding John Pollard Mayers' martial law addition to the militia bill in the House of Assembly.[32] This gave the Governor the power, with the consent of his Council, to proclaim martial law for three days upon receiving information that the enemy's fleet was at sea and that it was heading for, and likely to attack, the island. On the 12 December 1804 war had been declared by Spain against Britain; a week later a general embargo was placed on Spanish shipping and England declared war against Spain on 11 January 1805. Napoleon sent a squadron of ships and 4000 troops to the Caribbean; they arrived at Martinique on 20 February. As soon as the news reached Barbados, the island was put on a state of defence, individuals pouring money, labour, horses and other draught animals into the war effort. The martial law bill which Haynes had seconded, had been enacted, and *'martial law was enforced in Barbadoes, and every preparation carried on with vigour to repel the blow, if that island should prove one of the objects of the enemy'.*[33] However, the legality of this particular proclamation on the part of the Governor, Lord Seaforth, was questioned in a motion to the Assembly by Robert's cousin, Robert James Haynes.[34]

There were several skirmishes at sea, but the French, under instruction from Napoleon to capture all the British West Indian colonies, never actually landed on Barbados: the French fleet had in fact sailed further north towards Martinique and then returned to Europe. On 4 June 1805[35] Nelson reassuringly called in at Carlisle Bay in HMS *Victory* with his fleet during his pursuit of both Admirals Villeneuve and Gravina who had joined forces, his fleet consisting of ten sail of the line and three frigates.[36] After taking on board 2000 troops under Lieutenant-General Sir William Myers, Nelson signalled to weigh anchor and departed the following day. Within three months the great man was dead, the news reaching Barbados on 20 December.[37]

The year 1805, a difficult one for Robert, must have been made infinitely worse by the untimely death of his youngest son, John Aughterson, in April, and that of his eldest daughter and first-born child, Dorothy, aged 14 on 26 June of consumption [tuberculosis], followed by Fanny on 4 July[38] from hydrocephalus.[39] Haynes was so concerned for the health and safety of the rest of his family that he made the decision that his eldest sons should leave Barbados immediately to complete their education in Britain. Thus, five days later (30 July 1805) saw the departure of Richard, Robert and George then aged respectively 13, 10 and 9, for England, accompanied by Mr Aughterson and their trusted and faithful manservant, the 39 year-old slave, Hamlet. Hamlet was born in Barbados in 1766 and was employed by Haynes as a sailor,[40] presumably on one of his company's ships. Just how much they were to be indebted to Hamlet became clear when, on disembarking at Liverpool, George fell overboard, and tragedy was averted only by Hamlet's quick thinking and swift action in jumping in the sea after him. It was most unusual for white boys to be able to swim at that time, but it was a common recreation among enslaved black Barbadians. Surprisingly, this courageous act on Hamlet's part doesn't appear to have been rewarded: at least, there is no record of any reward – certainly not by his freedom!

The Haynes boys went to school in Greenock, Scotland and the three older boys matriculated for Glasgow University: Richard and Robert matriculated in 1809, and George Barrow in 1810.[41] Why they were educated in Greenock remains a mystery, but it may have had to do with its accessibility by sea, as there was a frequent packet service between Liverpool and Greenock. Richard and Robert arrived back in Barbados on 17 July 1811.

England

With the defeat of the French, travel by sea became slightly less hazardous, and on 14 April 1808 his three remaining children, his only daughter, Thomasine, and sons William and Henry, sailed for Bristol accompanied by their mother, and Miss Elizabeth Clarke, presumably Ann Thomasine's unmarried sister, Dorothy Elizabeth Clarke, and were again joined on the journey by Hamlet. Their destination was Bath where they were to remain until joined ten months later by Robert. His presence in Barbados was required in the interim because of his intention to introduce a bill in September 1808 for extending the time of the interval between elections to the Assembly from one to three years. This raised much alarm and disquiet amongst the electorate, who were keen to retain their right to remove unsuitable representatives and also not to risk allowing the Assembly to become an *'instrument of oppression'*. His bill was nevertheless introduced on 1 November, but was defeated in the committee stage by one vote.

Two months later, on 30 January 1809, Robert was appointed Colonel and Commandant of the Batallions of St John's and St George's Militia. The following week, Robert put Thomas Francis, his overseer at Newcastle, in charge of both plantations before he himself set sail on 8 February 1809 in the *Irlam*, an armed merchantman, with George Keysor as master. The journey was delayed because of two events typical of that time. First, they encountered, attacked and captured two enemy prizes, which took up a great deal of time. Second, the weather became stormy and was, according to Robert, the most severe storm that he had ever encountered at sea. In all, the journey took 39 days and the *Irlam* docked at Liverpool on the 18 March where Robert was received with much hospitality by the ship's owners, Messrs Barton and Irlam. It seems that fellow

Barbadian, William Sealey, was also a passenger on the *Irlam*, for two days later the two travelled together to Bath, where they arrived on 28 April.[42] There was the happiest of reunions with his wife, Ann Thomasine, his daughter, Thomasine and Elizabeth Clarke, who were staying at number 12, Argyle Buildings[43] and, the biggest surprise of all, with his brother Henry, Captain Henry Haynes, RN, who stayed with the family for a long period.

Robert was also greeted on his arrival in Bath by a letter, dated 23 March 1809, from Thomas Lane, one of the two brothers for whom he acted as attorney in Barbados. In this letter Lane makes mention of Robert Haynes's ill-health (probably gout) and expresses the hope that his stay in England will help relieve it. Lane invited Robert and Mrs Haynes to visit him at Goldsmiths' Hall and urged him not to bother with any of his (Lane's) accounts until then: *'The accounts you may keep till then. I shall attend immediately to your Warlike Order'*. The attorney had clearly a magisterial approach to his client! The letter continued with a reference to Robert Haynes's *'experimental'* use of the Liverpool markets in which to sell his sugar and rum, and wanted none of it for his produce, saying: *'The experiment you have made can be of very little consequence however & so I don't regard it; but wo[ul]d rather confine ourselves to London and Bristol'*.[44] Robert had probably been trying to get the Lanes to agree to use the port of entry and shipping in which he himself was acquiring a commercial interest.

For the following six weeks, Robert was able to take the waters, visit doctors and renew acquaintances with Barbadians and absentee landlords who seemed to congregate in Bath.

On 6 May Henry accompanied Robert and family together with George Barrow, first to London, where they took lodgings in Great Russell Street and doubtless from where Robert met Thomas Lane, then to Hill Park, Kent to see his *'wife's kinsman'* [probably nephew], John Barrow. John Henry Barrow and his brother George had bought Branker's (formerly Chapman's) Plantation in St Philip in 1775, of interest as it was renamed Sunbury after the Barrows' home in Kent, although for quite some time it was known locally as *'The Brothers'*. John Henry Barrow was married to Frances Mayers of nearby Halton Plantation, St Philip, who was related to the first husband of Thomasine Mayers, the mother of Ann Thomasine. The Barrow family kept their house in Kent and John Barrow, John Henry's son, inherited both the English and Barbadian estates. It was he and his wife Mary [Mary Ward Senhouse] who were the Hayneses' hosts.[45] Ann Thomasine and Elizabeth Clarke remained with the Barrows for the rest of May, while on 1 June 1809 Henry took Robert to Portsmouth to visit the dockyard and to see the man o' war, launched on 25 June the previous year, which was then the pride of the Royal Navy.

HMS *Caledonia,* which would be painted by marine artist Thomas Butterworth, was then the largest ship of her class, a first-rate ship of the line, in the British Navy. She carried 120 guns, her burthen (weight fully laden) was 2505 tons, her gun deck was 197 feet in length and her extreme breadth was 53ft. She was bigger than her older sister ship, HMS *Victory* (100 guns; gun deck 187 feet; extreme breadth 51ft 10in; burthen 2162 tons), and was destined for distinguished service against the French in 1812, taking part in the bombardment of Algiers four years later. She was said to be *'faultless'* by her crew of more than 800 and became the design model for other three-deck first-raters. Renamed HMS *Dreadnought* in 1856, she became a hospital ship at Greenwich and was finally broken up in 1875.

On 7 June the Hayneses took their leave of Hill Park and returned to London, where they stayed in the Norfolk Hotel. Within two days, Robert experienced a crippling attack of gout, which kept him hotel-bound for almost three weeks. He was still in severe pain (*'torture'*) when they set off back to Bath, where he consulted a Dr Robinson, but to no obvious avail. He was so ill that he felt he couldn't undertake the long journey to Greenock, where his sons were at school and so his wife travelled there alone, having left their daughter, Thomasine, at her school in Bath. It distressed him to have to leave England without seeing his boys.

The coach journey back to Liverpool had to be made in very short stages, pain permitting, and on his arrival, he was prevailed upon by Mr (later Sir) William and Mrs Frances Barton to stay with them at their home, 114 Duke Street, while awaiting the arrival of the *Irlam* for the return journey to Barbados.[46]

Mrs Barton took good care of Robert Haynes and he became well enough to travel to Buxton in Derbyshire to take the waters, but this didn't seem to help. He left Liverpool on the *Irlam* on 1 September 1809. This time it was an enjoyable voyage with no storms or sea-battles to contend with, and he could hardly contain his excitement at the first sight of his *'own dear country'* on 3 October. This was the last time that Robert Haynes was to travel in the *Irlam* which he had come to know so well. It was in 1812 that after much pleading over many years and after the loss of some important ships, that it was decided to build a lighthouse on the Tuskar Rock off the coast of County Wexford, Ireland, near Rosalie Bay. It was while this work was in progress that the *Irlam*, sailing from Barbados to Liverpool with a full cargo, mainly coffee, a number of passengers and members of a regiment of the British Army, struck the Tuskar and began to break up in thick fog on the morning of Sunday, 10 May.[47]

All on board made desperate efforts to get on to the rock itself and with the help of the men who had been working on the building of the lighthouse, most of them succeeded. However, despite the best efforts of all concerned, a sailor, a drummer and four children were drowned. A passing ship, the brig *Sarah* of Workington, later hove to in response to the distress signals from the rock and took the *Irlam*'s master, Captain Keyser, seven army officers, all the women among the survivors and some others. Worsening weather conditions forced the *Sarah* to leave the area leaving behind three officers, sixteen privates, one woman and the mate and twelve seamen from the *Irlam*. The survivors were landed at Beaumaris, Anglesey on 12 May. Along with her other cargo the *Irlam* was carrying silver plate and a chest containing 200 gold guineas, all of which went to the bottom with the ship. However, by January 1815 the Liverpool owners had replaced the *Irlam* with another vessel of the same name for transporting sugar and coffee from Barbados to England.

Back to planting

All was well at Newcastle and Haynesfield on Robert's return, Thomas Francis having managed the plantations without difficulty. Robert was still unwell and feeling sorry for himself (*'alive but that is all'*, he wrote in his diary, dramatically; he seemed to have inherited his father's undue concern over health matters), but his health returned sufficiently for him to make further plans for expanding the business, and he looked around for suitable properties. On 10 February 1810 he purchased Clifton Hall for £35,350 cash. The property was in chancery and appears to have been sold at auction, as he blamed the under-bidder, James Maxwell, for making him pay £3000 more than he had intended.

Clifton Hall Plantation was part of the estate which first belonged to Prince Ferdinand Paleologus, a descendant of Greek royalty. The Prince was the last collateral descendant of Constantine VIII, the last Christian Emperor of Constantinople, who was murdered in 1453. He was born and raised in a house called Clifton in Cornwall, England, which was given to his father, Theodore, by King James I.[48] Being of royal blood, Ferdinand Paleologus inevitably found himself on the wrong side during the Civil War in England, and when Cromwell triumphed he, like many other royalists, fled the country. He moved to Barbados; was a member of St. John's Church Vestry for 12 years and church warden from 1655 to 1666. He died there in January 1678. His famous tombstone (he was buried in the Orthodox way in the churchyard) can be seen by all who visit St. John's Church, about a half mile from the house. He would have doubtless been personally acquainted with the other early settlers in St John, including Robert and William Haynes.

Clifton Hall needed a complete overhaul of its equipment; its fields were overgrown with grass and every road on the estate was impassable. Robert Haynes built on the more 'modern' part of the house in coral stone, which was completed in 1810. This called for strong management, and he installed Thomas Francis to sort things out, putting in his place Robert Reece, then living at Mount Wilton, as overseer of Haynesfield, which was by that time in excellent order. Reece also became a most successful overseer of the Newton Plantation, informing Lane that his '... *negroes [were] under good discipline, but happy, and rapidly increasing'.*[49] As an expression of his gratitude, Lane subsequently presented Reece with silver plate inscribed: *'Presented to Mr Robert Reece, for his particular attention and humanity towards the Negroes on Newton Plantation; and as a token of approval of his general good conduct by Thomas Lane Esq: London October 1815'.*[50] Thomas Lane would eventually write to Haynes, expressing surprise that the managers of Newton, one after the other, seemed able to buy their own plantations, the occasion being the purchase of Gibbons Plantation by Robert Reece. One cannot but wonder whether Robert Haynes was being entirely transparent in his dealings with Lane in view of his growing attachment, both in business and personally, to Reece, whose family was destined to be joined in matrimony through the marriage of Reece's young daughter, Elizabeth, and one of Haynes's sons.

Clifton Hall turned out to be more of a liability than the Lieutenant-

General had anticipated. The workers there were indolent and unhelpful, and the price of slaves had never been higher, since the transatlantic slave trade had been made illegal by the British Government in 1807. He took the decision to move 60 of his far superior workforce from Haynesfield, in families, to Clifton Hall. He then began to be dogged by a constant feeling of anxiety over owing so much money for Clifton Hall, especially with the prospect looming of the abolition of slavery altogether. Ever since his father's death, he had dreaded being in debt and he desperately wanted to reduce it. At the height of this mental turmoil, he discussed the matter with his family. Ann Thomasine and their son, Robert, back home from Scotland, no doubt seeing the toll the worry was having on him, urged him to reduce the debt by selling Haynesfield. Against his better judgment and to his perpetual regret, vehemently expressed in his diary, Robert allowed his own opinion, based on all his previous experience, not to sell, to be overruled, and the property was bought by his brother, Edmund, eleven years his junior, who had acquired part of the adjacent Belmont through his wife, Sarah Bell. Haynesfield, which had been improved and streamlined since its purchase, was sold for £18,000, £1000 less than he had paid for it six years earlier; but this was without the slave workforce which he had moved to Clifton Hall.

There was a transient respite in his anxiety at the prospect of now owing only £15,000, but the other stress never far in the background again rushed in to take its place. What would happen to their colonial trade if slavery were to be abolished? He inveighed against the government in Britain, yet seemed resigned to the inevitability of abolition and doubtless appreciated the abolitionist arguments, but the prospect of scarce, costly labour frightened him. To his diary he confided:[51]

'If as it may yet happen, that certain powerful ministers of Great Britain still further have their way in the ultimate passing of these proposals for the total abolition of slavery, it will be an ill day for many of us, for, in this event, nothing, so far as I can see, can possibly result otherwise than the most serious consequences to the trade of our Colonies hereabout from still further dearness and scarceness of labour'.

Yet, despite all his misgivings, within two years he was not only out of debt, but ready for his most expensive purchase yet.

In January 1812 he bought the Bath Estate with 400 acres of land and 212 slaves for £37,100. His old rival, James Maxwell, bid him up again, but this time John Thomas Lord of St Philips was the underbidder. Lord bid £37,000 local currency (about £27,820 sterling) to be paid in instalments; Robert Haynes topped that by £100 without any restrictions, which was preferred by the Court of Chancery.[52] He had no difficulty raising the money, but £32,000 of it was through loans. That same year, not only was the sugar crop poor, but, paradoxically, the price fell to a very low level, and despair once again set in. The United States declared war on Britain in Canada on 18 June 1812, and the ink was hardly dry on President Madison's signature when privateers left from almost every American port in droves, thirty from Baltimore alone, and swarmed over the Caribbean, taking a heavy toll on trade with England, many ships being lost to them.[53] Mail packets were particularly vulnerable. It wasn't until the 8 December 1812 that the Prince Regent's order of reprisals against America was proclaimed in Barbados. Robert did, however, have his brother, Captain Henry Haynes, RN, to thank for the capture of nine American privateers off Canada and Bermuda, during his command of HMS *Sapphire*.

On 1 May that year there was a major volcanic eruption on nearby St Vincent, preceded by earthquakes; the effects were felt throughout the Windwards. The 1230ft *La Soufrière* exploded, covering Barbados in darkness due to the 'rain' of choking volcanic ash. The volcano continued to erupt until July and the heavy explosions were loud even on Barbados *('like near cannonading')*. The atmosphere became more sulphurous, the wind hot and burning, and the whole population was terrified, believing that the Day of Judgment had come. The heat of the ash destroyed much of the vegetation, including many of the food crops, which shot up in price. The volcanic eruption also appears to have had an effect on the climate. There was little rain that year until October which prompted a note to the diary when rain started to fall again.

Along with the renewed hostilities, the eruption was the last straw! Haynes's anxiety rose to such a pitch that he came to fear being alone. His *'depression of spirits, constant agitation of mind and body'* made him question whether all his business effort had been worth while, with such little return from his investment in land. He owed £32,000 on the purchase price of Bath and Clifton Hall and this, along with plummeting sugar prices, caused him to become physically sick with worry. His tongue became sore, probably because of his diet, and the condition was wrongly perceived by him, ever the hypochondriac, to be life-threatening.

Ten months later (August 1813) his mental and financial pendulums had swung in the opposite direction. The war with France was going well; he was working well with the much admired Governor and forces commander, General Sir George Beckwith. The poor sugar crop he had feared turned out to be a great one, and the price was on the up as well. He made a cheery note that if this good fortune should continue, he would soon have the means to pay off all his debts. By mid-December his debts had been cleared.

It must have been with mixed feelings that Robert Haynes allowed his son, Henry Husbands, to enter the Royal Navy as a midshipman on HMS *Venerable* on 8 May 1814. The *Venerable* then in Carlisle Bay, flagship of Admiral Sir Philip Charles Durham, was on active service against the French. However, within a month he learned of Napoleon's abdication and the accession of King Louis XVIII and with it the cessation of hostilities between Britain and France. In January 1815 news arrived of the peace treaty signed on Christmas Eve between Britain and the United States, its ratification being received by Admiral Durham in March.

At last he felt able to concentrate on business matters. On 24 March 1815 after informing his client Thomas Lane of the usual list of expenses,[54] Robert Haynes went into great detail on the action he had taken over a troublesome, 20-year-old female slave of mixed-race called Polly Kitty Williams. She was the daughter of Kitty Thomas, a member of a difficult family, well known among white society, whose womenfolk, all of whom bore children of white fathers, refused to work ('*since her birth she has neversomuch as turned over one straw for you*'), and had become '*every day a regular nuisance to Newton Estate*'. This family had been of particular interest to Samson Wood, who had documented them in detail in his correspondence to Thomas Lane.[55] His wife, Anne Ellcock Wood, appears to have indulged them, as had the previous owner of the plantation, Elizabeth Newton; Mary Thomas, Kitty's sister, having been put to '*some little employment*' as her housemaid, Kitty being allowed to get away with doing nothing, as had her mulatto mother, Mary-Ann Saer, before her. Indeed they had their own slaves to wait on them. Doing little was by no means unusual for mixed-race female slaves, some of whom saw themselves a cut above their black sisters, seldom being called upon to do field labour, and Polly Kitty Williams was '*as white as either of us*', being in all probability at least three-quarters white. She was clearly causing trouble at Newton, and Haynes set about getting rid of her by finding a purchaser. In the event he approached a recently arrived naval captain, Thomas Percival, who was prepared to exchange her for a black girl of equal value;[56] Haynes jumped at the opportunity. Percival gave him a legally executed bill of sale, but Haynes then realised he had no authority to execute the counterpart himself, so requested Lane to do so, which in due course he did.[57] Haynes was also ready to dispose of another daughter of this family and indicated his intention of undertaking a similar transaction, when the opportunity presented itself. He was also intent on getting rid of the mother as well, but there is no record as to whether or not he was successful. Management of slaves was clearly not without its difficulties. Robert Haynes kept meticulous records listing the births and deaths of slaves in the Newton Plantation ledgers. These have survived and are in the collection of the Barbados Museum.

Peace with America had at last arrived, but in April news arrived that Napoleon had landed with his 'army' of guards in France and was making for Paris. Haynes anticipated that the price of sugar would fall more than was generally expected as a result of less harassment of shipping. As the Napoleonic Wars also wound down and finally ended in 1815, Britain found itself in a period of economic decline. Unemployment, reduced demand for manufactured goods, and a heavy war debt had all contributed, and what occurred in Britain would surely impact on Barbados. In addition, the political reforms that would affect the colonies were gathering momentum. Supplies needed to run the sugar plantations, such as timber, horses, cattle, flour, preserved fish and salt beef, which had previously come directly from the American colonies, now had to come from Britain. Consequently, there were increases in both material and shipping costs. Added to this was the advent of the extraction of sugar from beet grown in Europe, with the aid of machinery developed during the industrial revolution; and in Europe there was no need for slaves and no natural disasters to contend with.

Insurrection

The island's slave codes, laws regulating the behaviour of slaves, of 1661, 1676, 1682, and 1688, had created one of the most repressive regimes in the Caribbean. Attempts to use these laws to intimidate slaves and to discourage them from resisting did not succeed. Although they met with defeat, slaves undertook three major rebellions in Barbados in 1649, 1675, and 1692. No further slave insurrections occurred until 1816, possibly because of the strength of the colonial militia that was formed on the island, but also because the proportion of the population of African-European descent was on the increase, who, rightly or wrongly were believed by the planters to be less likely to foment trouble. Slave families had become encouraged by their owners, and the population grew to such an extent that there was little call for slave imports. The 1807 legislation abolishing the slave trade had, therefore, comparatively little impact on the Barbadian labour force, apart from the cost of 'indigenous' slaves, which was on the increase.

It was as Reviewing General of this militia that Robert Haynes found himself when, on Easter Day, 1816, all hell broke loose in the form of a long-anticipated slave insurrection, which has gone down in the annals of Barbadian history. The *'Easter Rebellion'* was led by an African-born slave named Bussa,[58] and began at 8 p.m. on the night of Sunday, April 14, in the Parish of St Philip and engulfed the southern half of the island for more than three days. The signal for revolt was the setting fire to cane and the ringing of bells. The insurrection was started at Bayley's Plantation (three miles south of Thicket Plantation), where Bussa was head ranger, but spread rapidly throughout the Parish and then into neighbouring Christ Church, St George and St John. It was the extremely careful planning of Bussa and his confederates, including Joseph Pitt Washington Franklin from the Franklin Estate (St Philip), King William, a friend of Bussa from the Sunbury Plantation (St Philip), and others[59] that created this element of surprise. The planning had been undertaken at a number of sugar estates, including Bayley's soon after the House of Assembly had discussed and rejected some of the provisions of the Imperial Registry Bill (a Bill raised in England by Wilberforce, requiring the compulsory registration of slaves to protect the Abolition Act and prevent the smuggling of Africans into the colonies) in November 1815, and by February 1816, the decision had been taken that the revolt should take place at Easter with Bussa as leader and field-commander. Franklin was a free man of mixed race, well educated, but described as *'a person of loose morals and debauched habits'*.[60] He was in the habit of reading to the slave population the inflammatory anti-slavery speeches emanating from England. This person was the source of most of the slaves' information.

As well as the 'scorched-earth' approach, the rebels turned the sails of all the cane mills into the wind to fly untended and become damaged. They ransacked a hardware store owned by Joseph Odwin Bayne and Samuel Clarke, and armed themselves with bill-hooks, cutlasses and some firearms. Bayne and Clarke subsequently claimed losses in excess of £10,000.[61] The rioting slaves even looted the militia's stores in St Philip and apparently carried the Battalion's flag into battle. The Christ Church militia were the first to fight off the rioters, caught in the act of looting the Golden Grove Estate. Most of the whites were tied up in defending their own lives and properties for much of Sunday night and Monday morning. By daylight, the regular troops were brought into action.

Bussa commanded some 400 fighters, but he was killed in the fighting. His men continued the fight until they were defeated by greater fire power, but it is reported that many went into battle shouting the name of Bussa. For this reason the insurrection has been known to generations of Barbadians as *Bussa's Rebellion*. There had been considerable coordination between the rebels on the various plantations in St. Philip. According to the Report of the Select Committee of the House of Assembly, Thickett was one of the first plantations mobilised into the rebellion. Thickett and Fortescue suffered more than £5000-worth of damage. Cane fields in nearly 40 estates were put to the torch, and plantation houses and buildings were pillaged and burned. Franklin Plantation and Oughterson were among the first to fall. The ground floor part of Oughterson house was destroyed and considerable damage in excess of £1200 was inflicted on the estate. Sunbury, owned by John Henry Barrow, sustained almost £4000 in damage. Paradoxically, it was Ruby, the plantation owned by a freed coloured, Jacob Belgrave, that suffered the greatest loss [£6720].[62]

The rebellion was described by Haynes in his diary as a *'hell-broth, which has been long in the brewing'*. Both the regular British troops of the First West India Regiment and Barbadian Militia were needed to put down the rebellion, several of whom were also killed.

Property and crops were looted and set on fire and Haynes described the ringleaders as a '*horde, inflamed with every vile passion, which committed every imaginable and filthy outrage in its path*', and those responsible were rounded up, tried and summarily executed. It was without doubt the first large-scale slave unrest in the British Caribbean. Between 500 and 1000 slaves were killed in the fighting; more than 140 slaves were executed; and 123 were forcibly removed from the colony. Joseph Pitt Washington Franklin was tried by court martial and convicted on 29 June for both inciting and aiding the slave rebellion. He was hanged at Enmore on 2 July 1816. Other ringleaders '*upon full evidence of their guilt*', were sentenced to be hanged upon their owners' plantations.[63]

One of the most vivid accounts of the insurrection was given by Haynes in reporting the event in a letter to his client Thomas Lane in England ten days after the event,[56] which expands on his diary entry and gives insight into his beliefs of the actual intentions of the slaves, based on intelligence extracted from the captives in the immediate aftermath of the Easter uprising, and his true feelings and forebodings for the future of the white planters in the wake of British government policy on emancipation:

'Dear Sir,

What I have [been] for some months dreading has at length come to pass. On Easter night the 14th of April an insurrection of the slaves commenced in the Parishes of St Phillip and Xt [Christ] Church, the slave population of which amounts to about 19,000. In the Parish of St Phillip and part of Xt Church most of the canes have been burnt, rum, sugar and provisions of various friends have been … destroyed to a considerable amount, and the neighbouring Parishes of St John and St George have partially suffered in a similar way. By a prompt movement of the Regular Troop and the Militia of the Island, tranquillity has been restored; only one white man has been killed. As well as I can calculate about two hundred [209] male slaves have been killed and executed and … 500 prisoners taken. You have been very fortunate at Newton, the Insurgents having never attempted that property you have escaped loss.

I wish I could say as much of Seawell, the floors doors and windows of the Dwelling House have been literally torn to pieces, the Building Doors broken open, seven hogsheads of sugar carried away, all the Rum which had been made let out of the butts and 23 acres of canes burnt. Three negro males shot, 2 legally hanged, 3 more imprisoned and several more I fear there will be evidence to convict … The Island has been under Martial Law for the last 10 days and is likely to continue so, much longer, and holding a Command I have never been able to put pen to paper before. Had the insurgents made use of the means they possess, it is most likely you never would have had this information from me, most of the Ringleaders have been taken and executed. Tranquillity I hope will be restored but I cannot promise for its duration ...

This is the result of Mr Wilberforce's and Mr Stephen's Registry Bill[64], the slaves had taken up the idea that the government had ordered them to be freed and that manumissions had been sent out for that purpose, but that their masters had with held them and they were determined to liberate themselves by a massacre of all the white men and their children and intended to take the white females as wives for themselves. I am fully persuaded that a general insurrection of all the slaves was intended throughout the Island but the active measures which were taken put a stop to the general conflagration which was the signal for revolt. I would advise you to give this the utmost publicity, if you think proper, without naming the Author, it is impossible to enter into a general detail of the losses of Individuals, which have been very great. No doubt you will have a variety of reports but the above you may rely on.

I am with much respect your humble servant.

Rob Haynes, Barbados, Bath.'

The active part played by Lieutenant General Robert Haynes in putting down the rebellion was not unnoticed by his white countrymen. At the Assembly meeting held on 6 August 1816, Haynes, along with Col John Rycroft Best, and a number of British officers received the thanks of the Council and Assembly and a *'grateful plantocracy'*. He was presented with a sword by the Officers of the Combined Corps of the Districts of St George and St John, in recognition of his skilful leadership.

The news of the insurrection slowly filtered through to England and the first questions were asked in Parliament on Tuesday 6 June.[65]

'Mr Pallmer begged to be informed ... whether government were in possession of any authentic documents on the subject of the awful event stated to have happened in the island of Barbadoes, in which, it was reported, the lives of a great number of persons were lost, and property destroyed to a great extent. Lord Castlereagh answered that government had not yet received the details of the melancholy event alluded to'.

A deputation of plantation proprietors arranged an appointment with Lords Bathurst and Castlereagh at the Colonial Office on 6 June, where they presented information received from Barbados in private letters like that from Robert Haynes and urged measures be adopted to prevent similar occurrences in Barbados or elsewhere.

After the rebellion, sugar prices fell sharply. This, along with the Abolition Act of 1807, restricted any further growth of British slave plantations, which caused Britain's share of sugar exports to start declining. There was an erosion of West Indian trade which, along with political uncertainty stemming from freed slaves, eventually would undermine the power of planters like Robert Haynes in the West Indies.

The planters made every effort to continue as normally as possible under the circumstances in which they found themselves. One month later, on 19 May 1816, the marriage took place of Robert Haynes's only surviving daughter, Thomasine, and John Hothersal Pinder, Robert's co-representative of St John's Parish in the House of Assembly, which must have been a happy interlude in an otherwise fraught time. Thomasine received £7000 from her father on her marriage. This was the value of the crop retained by him on the Newcastle Estate [416 acres, 180 slaves], which he had just passed on to his son, Richard, he and Thomasine having moved to the Bath estate in March 1815.

Two months after the uprising, Robert Haynes was in a reflective mood in his letter to Thomas Lane.[66] The island remained peaceful, but martial law was still in force, seemingly for the duration of the trials of the remaining slave ring leaders. But the situation had materially changed for the worse. No longer could plantation owners be confident that the same would not happen again. Yet Haynes looked '*forward for many years of peace and quiet*', but the cloud on the horizon was the part Britain would '*take in this unfortunate affair*'. He listed his own observations on the poverty he had seen in parts of the United Kingdom, particularly in Ireland, and how the slaves of Barbados were more enlightened and had more comfort than many of the poor in the mother country. Why couldn't the British government turn its attention and energy to putting this right, than '*meddle with the slave population of the West Indies*'? There were doubtless many acts of cruelty perpetrated against the slave population, but these may not have been as widespread as the abolitionists' hyperbole implied; undoubtedly many estate owners were accused of the cruelties committed by the few and it was left to periodicals such as *John Bull*, the *British Mercury*, the *Glasgow Courier* and the *Quarterly Review* and individuals such as John

Pollard Mayers, the agent of the island in England, to attempt to set the record straight from the planters' standpoint.

In September 1822 Robert Haynes again visited England (Liverpool and Bath). He arrived in Liverpool on 19 September, where a letter dated 5 September from Thomas Lane awaited his arrival. He was again concerned about his health, and travelled to Bath as quickly as possible to see his two brothers Henry and Edmund who had *'brought their respective families there'* to meet him, as he hadn't met any of them before. He was also looking forward to meeting old friends and relations who lived in Bath, some of whom he hadn't seen since his visit in 1809.

Robert always seemed to benefit from his visits to Bath and liked to stay there as long as possible. He liked to stay put and hated sleeping in different beds as it '... *does not agree with my health and I do so much feel better by my trip to this place*'.[67] He left Bath for Liverpool on Tuesday 22 October 1822 and on the way visited Thomas Lane's brother, John, at King's Bromley, Litchfield for two hours, no doubt to discuss Newton and Seawell.[68] He gave as his forwarding postal address: Thomas Lee Haynes, Liverpool. He set sail from Liverpool on the *Frances* accompanied by his daughter, Thomasine, and John Higginson, the shipping owner; the vessel docked in Barbados on 4 January 1823.[69] Thomasine had probably returned to attend the marriage of her brother, William Clarke Haynes, planned for two months later.

Both Thomas and John Lane died in 1824 within eighteen months of Robert's last visit, and from that time the owner of Newton and Seawell was probably John Newton Lane, who in 1834 was living at Brompton Hall; a member of the Lane family held the estates towards the end of the nineteenth century and so presumably they remained within that family throughout. However, we have come across no record as to whether or not Robert Haynes continued to help the Lane brothers' successor as their Barbadian attorney. Two events probably militated against him continuing after 1825: the tragic loss at sea of his son George Barrow and his election as Speaker of the House of Assembly.

The House of Assembly met on Thursday on the 3 April 1823, after a prorogation of four months. The main subject of discussion, was the choice of an Agent in England, in the place of the late Mr Jordan. John Pollard Mayers, Esq., was proposed as Agent by Robert Haynes, and seconded by John Hothersal Pinder; Thomas Howard Griffith,

Esq., was proposed by James S. Bascom and seconded by Robert J.Walcott. The majority favoured Griffith.[70]

Robert Haynes's military duties, usually battalion inspections, continued throughout his life in Barbados and these were generally called by notice in *The Barbadian* as on Saturday, 22 November 1823:

> 'His Excellency the Royal Governor having issued an Order for the Inspection of the Battalion of St. Philip's, by Lieut-Gen. Haynes, the Officers, Non-commissioned Officers, and Privates of this Battalion, are hereby required to meet in full uniform, with twelve rounds of cartridge, on their usual Parade-ground, by eight o'clock in the morning on the 28th instant, that being the day appointed by the Lieutenant-General for this purpose. By Order of Lieut.-Colonel Clarke, Wm. Murrell, Adjutant'.[71]

Another such inspection by the Lieutenant-General, this time of the First or Royal Regiment of Militia, intended to see the state of the Arms, clothing, etc. (deficiencies to *'be punished according to the Militia Law'*) took place on the morning of the 16 December 1823. Two days later he inspected the Life Guards. During this round of inspections, Robert Haynes doubtless found time to attend the marriage of his daughter, Thomasine, to Christian Frederick Lardy on 4 December. The troop and militia inspections and parades were dispensed with in 1833.

Lawlessness, never far below the surface, was getting so out of hand by late 1824 that on November 25 the members of the Barbadian Agricultural Society adopted measures to suppress the *'various and extensive depredations committed on the property of planters'*. The following resolutions were among those unanimously adopted:

> '1. That an Association be formed, and a fund raised by subscription, for the purpose of encouraging constables to take up slaves carrying sugar, large quantities of molasses, canes, and corn-meal to market, without a proper pass; each slave to be carried before a Justice of the Peace, and dealt with according to law.
> 2. That the funds raised be placed at the disposal of a committee for the purpose of granting monthly rewards to those constables who are active in carrying into effect the objects of the Association; and that Magistrates be respectfully solicited to give certificates to constables, whose activity and zeal merit reward.

John Higginson was appointed treasurer and Robert Haynes to the committee, whose role was to encourage all planters to subcribe, a list of subscribers appearing in *The Barbadian* the following week.[72]

Increasing lawlessness, exemplified by the wanton destruction of the Wesleyan chapel,[73] and indebtedness, brought the need for additional prison accommodation, and a committee was set up under Robert Haynes to make recommendations on the best way forward. At a meeting of the General Assembly on 8 March 1825 Robert reported the committee's recommendations to the Speaker, Cheeseman Moe, as follows:

> '... that if the Theatre of Bridge-Town can be purchased for the sum of £3,000 currency, that the same be so purchased and converted into a Parliamentry House, and a place for public Offices; that the vacant spot of land which is nearly opposite the Theatre be enclosed, and stabling erected for the shelter of the Horses of persons attending on public business. That the part of the Town-hall at present used for the meetings of the Legislature and as a Court House be converted into rooms for the better accommodation of imprisoned debtors,...'.

Speaker

About the time of George Barrow's death in 1825, after 32 years as Representative for St John, Robert Haynes was unanimously elected Speaker of the House of Assembly. This followed a few months in the same role in a temporary capacity following the sudden departure from Barbados of the Speaker, Cheeseman Moe. The circumstances of his elections were detailed in *The Barbadian*'s report of the meeting of the General Assembly at the Town Hall on Tuesday 3 May 1825. The members having taken their seats, the Clerk observed that he had just received a letter from the Speaker, which, with permission, he would read:

> 'My Dear Sir, Be so good as to state to the House, that as I am to embark this evening, it will be impossible for me to attend. Present the honorable members my last farewell, and my sincere wish that it may be the will of the Supreme Being that their exertions and energies may ever tend to the exaltation of the character, and true interests of their country. You will accept my warmest thanks for the able assistance I have received from you whilst I filled their dignified chair, and believe me to be, with the highest regard and esteem, My Dear Sir, Yours Most Faithfully, Cheeseman Moe'.[74]

Henry S. Cummins, Member for St Michael, then moved that the Honorable Robert Haynes, the senior member of the House, be called to the chair as Speaker *pro tempore*; the motion was seconded by Mr. Evelyn, and agreed unanimously; Robert was therefore called to the chair and said the following words:

> 'Gentlemen, Having done me the honour to appoint me your *pro tempore* Speaker, I cannot but confess that I am gratified that the house should unanimously consider me deserving of the situation. I am well aware of the difficulties I have to contend with in these perilous times, and the responsibility attached to this chair, and not least, my own inability to do justice to the arduous task you have imposed on me. Having anxiously anticipated that the day would arrive when all party spirit would be buried in oblivion, I hope I may venture to hail its commencement from the unanimity which has this day been shown in calling me to the chair; and for the good of my country, I trust I shall not be disappointed. I am plain and laconic – give me your support, and I will endeavour, to the best of my ability, to serve this house and my country, with assiduity, fidelity, and independence. Gentlemen, you have my best thanks.'

William Hinds (Member for St Peter) and Robert Haynes Jr (the Speaker-elect's son, Robert Haynes, the second Representative of St John's Parish) were then sent up to his Excellency the Governor (Sir Henry Warde) to inform him that the house had made choice of a Speaker *pro tempore*, in the absence of their Speaker, the Honorable Cheesman Moe; and to know the pleasure of his Excellency when he would receive the house. The two gentlemen withdrew for that purpose, and shortly after returned with an answer that his Excellency was ready to receive the house. On which, the Members attended the Governor in the Council Chamber, and presented their Speaker *pro tempore*, the choice of whom being approved by his Excellency, the Members returned to their house and took their seats.

The first piece of legislation over which he sat was a bill to appropriate a certain sum toward the building of a second place of public worship in Bridgetown. The Speaker seconded the motion to appropriate £5000 for the purpose and nominated six members of the house to act with the members of the Honorable Board of Council, to cooperate with the Lord Bishop in directing and superintending the building of another place of worship in Bridgetown, in such a situation, and in such a manner, as his Lordship may think best for the accommodation of the inhabitants of Bridgetown, and the Parish of St. Michael, whose population had greatly increased in recent years. This swelled the membership of the committee to oversee the work to 19 including Robert Haynes. Eighty-five individuals contributed £1867.10.0 to the Church-Building Fund.[75] The list included: Robert Haynes, £25; William M. Barrow, £10; Nathaniel Cave, £5; Edmund Haynes, £25; Richard Haynes, £25; Robert Haynes, Jun., £25; Henry H. Haynes,£1 5s; William C.Haynes, £1 5s; Miss Jane Payne, £5.

Up to this time the Church in Barbados had been under the distant jurisdiction of the Bishops of London. In 1825, under Royal Letters Patent of 1824 of King George IV, the first Bishoprics were established in the West Indies (Barbados and Jamaica). The first Bishop of Barbados was the famous Bishop Dr William Hart Coleridge, who arrived at the island on 28 January 1825 on HMS *Herald*. The Parish Church of Saint Michael now became the Cathedral of the new Diocese. Called to a meeting chaired by John Barrow at Lewis's Hotel on Thursday, 21 July 1825, the proprietors of estates throughout Barbados felt that to avoid discourtesy, they should present an address to the Bishop, including the following paragraph:

> 'We beg to assure your Lordship that we have the most sincere desire to afford the blessings of religious instruction to our slaves; and we promise that you shall always find us prompt and zealous in furthering every prudent measure, which may seem condusive to this objective.
>
> Weighted-down, however, as West India interests are by causes too notorious to require to be enumerated, we cannot, without injustice to others, consent to such a substraction of labour from the cultivation of the estates as would lead to a material reduction of income, but we very confidently believe that no such sacrifice is necessary to the accomplishment of the object contemplated, and we entertain no doubt that, under your Lordship's discreet and judicious direction, the spiritual interests of the slave will be shewn to be not incompatible with the temporal interests of the master'.

This address reflected the concerns of the planters on the loss of time that religious instruction of the slaves would incur. What was not apparently voiced at the meeting was the real reason why slaves had been forbidden membership of the Anglican Church by the plantation owners, their concern that the church might undermine their authority over the workers.These concerns were clear from the remarks of Hon. John Rycroft at the meeting, who was quoted as saying:

> 'The time generally allowed to slaves for rest from labour each day is three hours, and abundance of food is furnished to them without further care of labour, more than supplies their wants. To deprive the master of a portion of the time allotted to labour would be to rob him of the production of his property and the slave of that part of his allowance which he disposes of to his own profit. It would not only be unjust to the proprietor, but detrimental to the labourer'.

The Hon. James H. Alleyne then proposed that the Address be signed by the persons present, and be left at the counting house of Messrs. Higginson, Deane & Stott, for the signatures of such others who may wish to sign it, and had been prevented from attending the meeting (in the event 85 people signed the Address including: John Barrow, Robert Haynes (The Speaker) William Reece, Robert Reece, Joseph Mayers, Edmund Haynes, Robert Haynes Jr, John Higginson, W. Warre Barrow, John Henry Barrow and W.M. Barrow) and that a deputation be appointed to wait on the Lord Bishop with the Address. The deputation comprised: John Barrow, Hon. Renn Hamden, Hon. James H. Alleyne, Hon. R. Haynes, M. Coulthurst, Esq., Dr Maycock, and Forster Clarke, Esq. In the Bishop's reply he informed the members of the deputation:

> 'I propose then to impart religious instruction to every plantation thrown open to me, through the agency of catechists and teachers, licensed by the Bishop, after previous examination and subscription; acting under and directed by the minister of the parish within which they shall be appointed to act, paying every proper regard to the wishes of the master as to the time and frequency of instruction to the scriptures, the liturgy of our church, and such other religious works as are included in the catalogue of the Society for Promoting Christian Knowledge
>
> For the pecuniary support of these catechists and teachers, I look first to the monies placed at my disposal by His Majesty's government for the maintenance of a certain number of clergy and catechists within the Diocese, secondly, to the Society for the Conversion and Religious Instruction of Negroes, to whom I have ventured to propose such an exclusive application of their funds, and lastly, to yourselves, individually and collectively, through the formation of a branch association of the Society.

I trust that you will consider these measures to be prudential; and with your co-operation, under the Divine blessing, I cannot but anticipate from them the most beneficial results'.

On Tuesday, 25 October 1825 the newly elected members attended a Meeting of the General Assembly at the Town Hall. The new Members present on the returns of the elections for St John were the Hon. Robert Haynes, and Robert Haynes Jr, Esqrs. The members having severally taken the State Oaths before His Honour the President in Council, His Honour was pleased to direct them to return to the House, and make choice of a Speaker. The Members then withdrew to their House, and having taken their seats. The Honourable Robert Haynes was, on motion of Henry Cummins, seconded by Gabriel Jemmett, unanimously called to the Chair as Speaker. The Speaker addressed the House, thanking them for the honour thus conferred on him.[76]

Notwithstanding the time taken for transatlantic travel, Robert Haynes must by the time he was elected Speaker have known that the *Martha* had failed to reach Liverpool, assumed lost, with his son George Barrow on board. Possibly because of his grief, he threw himself even more actively into his civic work, and by December was discussing the question of a Petition for a Royal Charter for the Incorporation of Bridgetown at a public meeting of all interested residents held at the Town Hall.

The Petition was forwarded by the Governor, Sir Henry Warde, to be presented to His Majesty, praying his Royal Charter for Incorporating Bridgetown, which stated the wish of Mr Horton, the Under Secretary of State for the Colonies, that the petitioners should put His Majesty's Government in possession of the details of what they wished to be granted regarding such a Corporation. This was the purpose of calling the meeting, but it was poorly attended. Country gentlemen felt strongly about the parlous state of the policing of Bridgetown [all attempts to create a viable police force had failed] and the consequent lawlessness on the part of *'overbearing slaves, as well as free men in the shape of porters, carters, boatmen, &c'*. They often lamented the disorderly state of the town, and the general view was that, if anything could ensure the quiet and tranquility of the town, it was a Corporation; and as His Majesty's Ministers had asked for information on the principles upon which the Petitioners wished their Charter to be founded, it was felt at least they meant to consider the measure. At the meeting the Hon. John Alleyne Beckles was called to the Chair, and the following were appointed a Committee to prepare a plan for a Corporation for consideration by a General Meeting of the inhabitants: Hon. J.A. Beckles, Hon. Robert Haynes, Messrs. John Higginson, Richard Deane, Gabriel Jemmett, Henry S. Cummins, Benjamin Walrond, Benjamin Ifill, Thomas C. Spencer, Isaac Williamson Orderson, Henry G. Windsor, Francis Hodgkinson, Thomas Pierrepont, William Bovell, Alexander King, Abraham R. Brandon, Bezsin K. Reece, H. DeCastro, C.F. Sage, William Hawkesworth, and Abel Stuart. [77]

Other early Acts passed by the General Assembly of Barbados with Haynes as temporary Speaker included one on 13 September, 1825 to allocate £500 to purchase a piece of plate to be presented to James McQueen in recognition of his support of the West Indies and its inhabitants, including his publication of a pamphlet entitled 'The West India Colonies, the Calumnies and Misrepresentations circulated against them by the *Edinburgh Review*, Mr. Clarkson, Mr. Cropper, etc. etc. examined and reviewed'. Robert Haynes also called a *'Full and General Public Meeting of the Inhabitants of this Island'* at the Town Hall on Friday 23 September, at noon, for the purpose of taking into consideration the propriety of supporting a newspaper to be published in London by the same James McQueen, the present editor of the *Glasgow Courier*, who *'has upon all occasions, so ably defended the cause of the West Indians against the calumnies and machinations of their enemies'*.

At a meeting of the Assembly on 7 March 1826 Robert Haynes, as Speaker, brought up the question of the low stipends of the Barbadian clergy.[78] It was, therefore, decided to increase their stipends to £500 per annum, to be raised, if there were no funds left in the treasury, by a tax on slaves, mills and other items. The vestries were to be restrained from making their clergy gifts from the parish funds.[79]

On 23 October 1826 Robert Haynes, in his capacity as Speaker, wrote to the Governor, Sir Henry Warde, a summary of the House of Assembly's deliberations on the Consolidation Slave Bill which had passed the House of Assembly and the Honourable Board of Council. In particular, he wished to explain the reasons for not adopting certain of the recommendations contained in the dispatch of Lord Bathurst, Secretary for the Colonies, of 9 July 1823; especially, they could *'not yield to his Lordship's recommendations to prohibit the punishment of women by flogging, or the use of the whip in the field'.*[80] The main arguments appeared to be the widespread experience of the Assembly members that female slaves disregarded the authority of their owners to a greater extent than male slaves; therefore, to deprive an owner of this form of punishment would encourage a stronger disposition to disobedience. Imprisonment, which had been proposed as the alternative to flogging, was no answer, as the owners would be deprived of their slaves' services. In the case of owners possessing only one or two domestic female slaves, *'it is quite apparent how very objectionable must be the change from flogging to imprisonment'*. On whipping in the field, the Assembly disagreed to limit its use to a given number of *'stripes'* as advocated by Lord Bathurst. There was already provision to bring complaints against cruel owners before the courts. The Assembly also came down against compulsory manumission, whereby slaves who wished to could purchase their freedom against the will of their owners. They argued that if this were permitted, it would put an end to all security of mortgagees and other owners. They argued that the consequences of such manumission to slaves and their families could be disastrous, as the average number of effective, i.e. productive, slaves on a plantation was only about one-third of the total, the remaining two-thirds consisting of infants and the aged or infirm. These two-thirds were supported by the active one-third who would be the only ones with the wherewithal to purchase their freedom. With the departure of the effective slaves, the owner would not only be deprived of his income, but there would be no compulsion on the part of the freed slaves to support the remaining non-productive slaves. Other provisions suggested by Lord Bathurst had been incorporated in this or associated legislation.

The following year, on the 26 March 1827, the first steamship ever seen in Barbados entered Carlisle Bay. Called the *Liberatador*, it arrived after a 30 hours' journey from St Vincent. According to Schomburgk, all the quays and the shoreline were crowded with people trying to get a glimpse of the vessel, and marvelling that it could travel without sail against wind and current. It was followed that evening by another steamer, named the *Hamilton* after its owner. It had travelled from the Orinoco in Venezuela.

The religious education of the slaves continued to generate considerable anxiety in the minds of the white population. This came to a head in St Lucy's where the rector, the Revd W. M. Harte was an ardent protagonist of such education. The inhabitants were up in arms and passed a resolution on 17 April 1827 accusing the rector of attempting *'to destroy the distinctions which they deem so necessary to their safety'*, of preaching an offensive Easter sermon and of his *'disgraceful conduct whilst administering the most holy Sacrament of the Lord's Supper'*, which had the alleged effect on the slaves of *'inculcating doctrines of equality inconsistent with the obedience to their masters and the policy of the island'*. The Bishop of Barbados supported the beleaguered rector, but his accusers escalated the issue, and in October 1827 Harte appeared before two magistrates, who advised that the complaint should be abandoned. Harte was again brought before three other magistrates and his spirited protest against the legality of the proceedings ended with him being remanded on bail pending an appearance at the Court of Grand Sessions. He appeared on 14 December when, found guilty of misdemeanour, he was fined one shilling. Harte appealed against the verdict to the King, who granted an unconditional pardon which was proclaimed on 11 February 1828. The House of Assembly, with Robert Haynes as Speaker, then became embroiled in considerable acrimonious debate regarding the procedural aspects of the documents respecting this case passed between Barbados and London and eventually a highly critical reproach was sent to the Secretary for the Colonies, Sir George Murray, relating to the Revd Harte's pardon. Murray flatly refused to bring the matter before the King and on 7 May 1929 informed Governor Lyon that the inhabitants could not complain *'that their judicial rights were invaded by an act of the royal authority'*. He criticised the way the pardon had been proclaimed, but continued:

> 'By this unfortunate prosecution of a clergyman of the Church of England, distinguished, as I learn from the report of his diocesan, by personal and professional merits, the parties concerned have needlessly, but very seriously enhanced the difficulties of those whose endeavour it is to defend the character of the colonists against all unmerited reproach'.[81]

The Assembly had indeed scored an own goal.

Lieutenant-General Sir Henry Warde, the Governor, retired because of ill-health on 21 June 1827. It was Robert Haynes's duty, as Speaker of the House of Assembly, to present the address on behalf of the Representatives which included the passage:

'Whether we look to your Excellency's impartial, kind, and uniformly upright conduct as our chief magistrate for nearly six years, or whether we regard those social and domestic virtues displayed by your Excellency during your residence amongst us, we alike have cause to lament that we shall so soon be deprived of your Excellency's superintending care and valuable example; and when we reflect on your Excellency's condescending kindness and attention to us on all occasions, and the ready access afforded us at all times to your Excellency's presence, we should indeed be restraining our feeling, did we not frankly declare, as we now do to your Excellency, that your loss to us in particular will be long and severely felt'.[82]

The nature of the Governor's illness is not known, but an infectious fever rapidly spread through the Caribbean islands in 1828. It was called dandy-fever at the time because the musculoskeletal rigidity that was a symptom of the disease gave rise to a stiffness of gait which was likened to that of a dandy.[83] It is now known to be the *Aedes* mosquito-borne, viral disease called 'dengue' or 'dengue fever'. The illness is of sudden onset that usually follows a benign course with headache, fever, prostration, severe joint and muscle pain, swollen glands (lymphadenopathy) and, after a few days, a generalised rash. It goes by other names including 'breakbone' or or 'bone-crusher fever'. Victims of dengue often have contortions resulting from the intense joint and muscle pain.

The position of Speaker kept Robert Haynes in Bridgetown while the Assembly was in session, but this was seldom more than three consecutive days in any month. There was, however, a great deal of preparatory work to be undertaken, and there were ceremonial occasions which he was obliged to attend. He undertook this role for four full years until, in January 1829, he convened the Freehold of St John in the Parish Church, and there tendered his resignation both as Representative and Speaker of the House of Assembly. The Barbadian carried the news of his intended retirement on 13 January, 1829, noting that Robert Haynes Jr had '*recommended John Henry Nurse Esq., to the notice of the freeholders*'. He was undoubtedly a much admired Speaker, and when he undertook his last duty, the resignation of his Commission in the Militia into the hands of the island's Governor, Sir James Lyon, the latter in recognition of the General's services issued an order expressing his gratitude together with the unusual privilege to Robert Haynes of retaining in his retirement the local militia rank of Lieutenant-General.

A portrait was commissioned [1839] of Robert Haynes in his Speaker's robes from the artist, Sir John Watson Gordon, R.A. [1788-1864] of 123 George Street Edinburgh.[84] It may have ended its days in a London Gallery. Its whereabouts is unknown, although a number of copies are believed to exist in Barbados. The National Portrait Gallery has two lithographs (see p.xv) by Richard James Lane, printed by Michael and Nicholas Hanhart, after this painting, produced in 1851.

Family tragedy struck again in 1831 when William Clarke Haynes fell ill. His will dated 1 May, named his father, his brother Richard, his son Robert Haynes and brother-in-law Henry Crichlow as executors. He must have died soon after signing, as the will was proved on 16 May 1831. His father was to provide for William Clarke's children in his will ten years later.

Robert Haynes made another, this time longer, trip to England in 1831. He may even have been away when William Clarke died. In his absence, Barbados was visited by the hurricane of 10-11 August 1831, which ruined many buildings, including seven of the eleven churches on the island. St. Michael's Cathedral, east of Trafalgar Square, Bridgetown, was badly damaged. Parts of Bridgetown were levelled. Robert Haynes arrived back in January 1832 to find most of the houses still without roofs, many with flattened walls and St John's Church collapsed. More damage appeared to have been done in one and a half hours than in the 15 hours of the 1780 hurricane. It was reported that this hurricane was accompanied by earthquakes, as there were many ground fissures indicative of violent upward heaving.[85] Consistent with this there was also a 17-foot storm surge and the wind speed reached 130 mph, making it a Category 4 hurricane in today's classification. Most of the island's trees had been uprooted. Some 2000 people were killed including some troops.[86] The damage exceeded £1.6 million.

Robert again railed in his diary against the mistaken policy and broken promises of British ministers, which had added greatly to the plight of the planters, many of whom had been reduced to penury. Some relief was gained by the King's suspension of all import duties.

In 1835 a meeting took place in the rectory of St John's, which Robert Haynes chaired, and at which it was resolved to rebuild the destroyed Parish Church, the new one to be 80ft long and 50ft wide. A committee was elected and £765 was subsequently raised in subscriptions. It is situated near the cliff, more than 800ft above sea level.

Figure 16.2 St John's Church, Barbados (Photograph Meriel Johnson)

Robert became increasingly annoyed by what he perceived as a growing number of petty and, to him, wrong decisions by the legislature and British government since his departure from office, typical of which were those such as the embargo on American timber imports, more than 15 years after the peace, which now held up the rebuilding programme after the hurricane.[87] Joseph Sealy, the overseer at Newcastle, had squirreled away materials over several years of use for the maintenance of the Haynes properties, so that they were repaired faster and cheaper than any on the island. Clifton Hall was one of the few buildings which survived the hurricane relatively unscathed. Robert was also extremely grateful to his attorney, Nathaniel Cave, who worked incredibly hard in order to get his client's estate back in good order. The two men had become close friends over the recent years, and Nathaniel and his wife, Isabella Carr Wharton, had had their first-born son in 1830, whom they had named Robert Haynes (baptised Christ Church on 26 December 1830) after their friend and the island's most prominent citizen.[88] In 1832 the last major Haynes' acquisition was made within a year of the hurricane. Clifden, with debts against it of around £12,500, was sold by the Court of Chancery to Robert Haynes for the knock-down price of £12,823.[89] Having done all he could to deal with all his problems in Barbados following the hurricane, he determined to return to England, and he sailed from Carlisle Bay to Bristol on Sunday 13 May 1832, landing at Lamplighters near Bristol 39 days later, from where he set out for Bath.[90]

Infectious disease was the biggest cause of a shortened life in the 1830s, and in 1832 cholera was widespread throughout Europe. The Governor issued a proclamation on 27 March that year forbidding contact with any vessel arriving from Europe until it had been inspected by a medical team.[91] Measures adopted by the Legislature prevented the accumulation of rubbish in the streets, lanes and alleyways throughout the island. These thoroughfares had to be cleaned constantly. It is unlikely that these measures had to do with the fact that cholera did not cross to Barbados, which was more likely due to the fact that drinking water on the vessels was not polluted. Cholera, always endemic in Britain, was treated with naphtha from the Barbadian naphtha 'springs' and soon after the hurricane, naphtha became freed from a heavy import duty and made available in Britain.

Resignation

What prompted Robert Haynes's *'sudden'* resignation from office was not revealed although there had been much procedural and factional fighting within the Assembly and, following his resignation at the meeting of the General Assembly on the following 17 February, with the new Speaker in the chair, it was moved by William Eversley and seconded by William Oxley, and agreed without opposition that thereafter the Speaker should not make or second a motion unless the House should be in committee. It suggests that a number of motions might hitherto have been carried or otherwise by the Speaker's introduction or vote. He must have been embroiled in the protracted fracas over the Harte case and the highly critical reaction of the British Government to the involvement of the Assembly which arrived at this time, which, he may have concluded, reflected badly on him as Speaker. Added to this, he had reached his sixtieth birthday, and doubtless felt it was time to hand on the baton. But other matters were beginning to weigh on his complex mind.

Ann Lovell

He, or more plausibly, his son, Robert, had formed a liaison with a woman in Bridgetown called Ann Lovell, who may have been a single lady or the widow of a much older man Robert Lovell.[92] Born in 1801, she was suggested by her descendant the late Anthony Haynes Lovell to have been the daughter of Philip Lovell of Durant's Plantation, Christ Church, although evidence for this has not been found. However, assuming she was born of the Lovells of Christ Church, the parish where her son Robert Haynes Lovell was born in 1825, she could possibly have been the daughter of:

1. Edward and Mary Lovell and therefore younger sister of Edward (b.1789), Philip (b. 1791) and Mary (b.1792). This is unlikely as around the time of Ann Lovell's birth, Mary Lovell, with Anne Grassett, acquired the 350 acre Chancery Lane Estate in Christ Church.[93] Mary later moved to England where she died in 1843 leaving the property to her children, Philip and Mary [Hamden]. No Ann was mentioned.
2. Jarrett Lovell (b. 1764).
3. Robert Lovell who married Elizabeth Osborne in 1784.
4. Thomas and Mary Ann (Chaplain) Lovell of St Philip, who married in 1794.
5. Phillip Lovell born in 1774 at St Michael. This is unlikely unless she was a twin, as his son Philip was born on 3 April 1801 in the year of Ann's birth. However, her descendant Tony Haynes-Lovell speculated that she *'may have been a Mulatto and could have had a brother Johannes and a sister Mary,'*[94] so if this were to be the case, Phillip Lovell cannot be ruled out as her father. Philip Lovell of Barbados died on 6 August 1823 at Frenchay, near Bristol.[95]

Ann Lovell was from Bridgetown [St Michael], and on 5 February 1834 she owned 11 slaves, 3 males and 8 females (including 2 male and 5 female children), registered in her name, making it unlikely, though not impossible, that she was of mixed race, as from 1802 those of black descent were restricted to the ownership of no more than 5 slaves. On the same day a Mary Francis Lovell registered 4 slaves [3 males] in her name. Ann and Mary Francis may have been sisters.

Robert Haynes is said to have fathered two children by Ann Lovell, born in 1825 and 1829; they were both mentioned in his will. As Robert and Anne Thomasine were married in 1790, the Lovell children couldn't possibly have been the issue of an earlier marriage of Robert Haynes, the Speaker. We have not so far been able to find a record of the marriage of a Robert Haynes and Ann Lovell and have no reason to believe that one occurred; it is perhaps relevant that her children bore her supposed maiden name or a combination of Haynes and Lovell or Cave and Lovell.

Ann Lovell moved to England and came to live in Bath to be near her daughter, who was to become Mrs Ann Cave Dummett. Her first home was with her daughter in Argyle Street where they were captured in the 1861 census as lodgers of Mrs Lucy Bryant. She was clearly still an accepted part of the wider Haynes family, as Mary Frederica, the 19-year-old daughter of Henry Husbands Haynes and one of General Robert's granddaughters, was staying with Ann at that time. She died on 3 May 1876 at 4 Henrietta Street, Bathwick, aged 75 years. Her death was registered as the widow of Robert Lovell, Gentleman, information being given by her son-in-law, Joshua Rowley Dummett who was present at death.[96] The cause of her death was erysipelas[97] of about seven days' duration. Was there a Robert Lovell whom Ann married, or was this a pseudonym for Robert Haynes, after whom one of her two children was named? If the latter were true, then the registrar was deliberately given wrong information. The Robert Lovell who married Elizabeth Osborne, had he been still alive, would have been 72 in 1825, so is most unlikely to have been Ann's husband.

What is beyond reasonable doubt, however, is that at least one of Ann Lovell's children was fathered by a Robert Haynes, and Lieutenant-General Robert Haynes's final years were to be inextricably linked with those of this child, none of which appears to have been known to, or revealed by, the editors of Robert Haynes's 'diary'. Ann Lovell's first child, a boy, was baptised Robert Haynes Lovell on 28 December 1825 at Christ Church. Her second child, a daughter was named Ann Cave Haynes Lovell, after her mother, and someone

Figure 16.3 No 4 Henrietta Street, Bathwick, Bath; home of Ann Lovell

called Cave, who might possibly have been the child's father.[98] Illegitimacy was extremely common in Barbados in the first half of the nineteenth century and didn't carry the stigma that it then did in Europe. In 1835, for example, the number of illegitimate Anglican baptisms in Bridgetown was five and a half times that of children born in wedlock. In other words it was the norm. Likewise, there were more illegitimate than legitimate children baptised in St John and St Joseph in that year, although they were much closer in proportion.[99]

Notes and references

1. Not Flitchett as reported in: E.M.W. Cracknell, ed. (1934), *The Barbadian Diary of Gen. Robert Haynes 1787-1836*, The Azania Press, Medstead, Hampshire, UK.
2. Annual Register, 1781; quoted by R. H. Schomburgk, 1848, *The History of Barbados*, Longman, Brown, Green & Longmans, p. 47.
3. Vere Langford Oliver 1989, 'The monumental inscriptions in the churches and churchyards of the island of Barbados, British West Indies', *Stokvis Studies in Historical Chronology*, p. 137.
4. E.M.W. Cracknell, ed. (1934), *The Barbadian Diary of Gen. Robert Haynes 1787-1836*, The Azania Press, Medstead, Hampshire, UK.
5. IGI; George Barrow-Dorothy Clarke entry.
6. Charles Johnson, 1724, *A General History of the Pyrates, from their First Rise and Settlement in the Island of Providence, to the Present Time*, 2nd edn, London, p. 95.
7. E. C. Haynes, A. P. Haynes and E.S.P. Haynes, 1910, *Notes by General Robert Haynes of New Castle and Clifton Hall Plantations, Barbados and Other Documents of Family Interest*, Argus, London, p. 12.
8. Jonathan was clearly to make a marked impression on his half-brothers-in-law, as both Richard and Robert Haynes gave Higginson as a middle name to one or more of their children.
9. E.M.W. Cracknell, ed. (1934), *The Barbadian Diary of Gen. Robert Haynes 1787-1836*, The Azania Press, Medstead, Hampshire, UK.
10. Taking on the estate as executor rather than heir-at-law at that time freed the land from liability to ordinary debts of the predecessor in title (note by E.S.P. Haynes, p. 15 of Haynes Haynes and Haynes, 1910).
11. Barbados Department of Archives 202/41.
12. On the flagship *Victory* and of the Battle of St Vincent fame.
13. Schomburgk, 1848, p. 85.
14. J. M. Toner, 1892, *The daily journal of Major George Washington in 1751-2: kept while on a tour from Virginia to the Island of Barbadoes*, Albany, Joel Munsell's Sons, p. 43.
15. After inoculation, a child generally remained well for about a week, then on the eighth day the high fever set in and a few pustules occurred, which seldom left marks, and if the child survived another week, he became well again, apart from the site of inoculation which took some time to heal fully. It wasn't until three years after William's death that Edward Jenner in England successfully used material from a pustule of vaccinia (cow-pox) instead of smallpox, to induce cross-immunity against the variola virus.
16. G. Pinckard, 1806, *Notes on the West Indies. Vol. I written during the expedition under the command of the late General Sir Ralph Abercrombie: including observations on the island of Barbadoes, etc*, London: Longman, Hurst, Rees and Orme; Letter XXVII pp. 340-47; Abercrombie set sail on 28 November 1795 with the fleet under Admiral

Sir Hugh Christian carrying 15,000 men. They were to encounter one of the severest Atlantic storms ever experienced off the British coast, with the loss of ships and men. Pinckard did not mention Newcastle by name; 'Mr Haynes' could therefore have been Edmund.

17. Partly re-afforested, its 85 acres is now a tourist attraction known as Joe's River Tropical Rain Forest.
18. E.M.W. Cracknell, ed. (1934), *The Barbadian Diary of Gen. Robert Haynes 1787-1836*, The Azania Press, Medstead, Hampshire, UK.
19. J. C. Brandow, 1983, *Genealogies of Barbados Families from Caribbeana and the Journal of the Barbados Museum and Historical Society*, Genealogical Publishing Co. Inc.
20. Kathleen M. Butler, 1995, *The Economics of Emancipation: Jamaica and Barbados, 1823-1843*, University of North Carolina Press.
21. It wasn't until 1834 that the British Government introduced the colonial Banking Regulations, and two years later the Colonial Bank (later to become Barclays' Bank DCO) received its charter. It may have been the establishment of this bank that enabled many of the planters to take their money and leave Barbados, right at the time of emancipation.
22. Senate House Library, London University: 'Papers of the Newton Family (Newton Papers)', MS 523/567, 19 March 1804. Farrell became the subject of a number of accusations including theft, although his family and the Hayneses became joined in marriage.
23. Attorneys of multiple estates earned on average about £500 per estate per annum.
24. Planter indebtedness was more common amongst the absentee landlords, some of whom had to sell their estates; sale of such an estate netted the attorney 5% of the sale price.
25. Probably T & J Daniel of Bristol.
26. Senate House Library, London University: 'Newton Papers', MS 523.
27. John Pollard Mayers of Barbados worked tirelessly to obtain the highest possible compensation for Barbadians when emancipation was on the horizon (Butler, 1995, p.9).
28. Robert Haynes to Thomas Lane; Senate House Library, London University: 'Newton Papers', MS 523/630; 9 March 1807: the slave-trade abolition act received Royal Assent on 25 March 1807 and abolished the traffic and purchase of slaves from Africa after 1 May that year.
29. David Lambert, 2005, *White Creole culture, politics and identity during the age of abolition*, C.U.P.
30. Both Robert and his father served in the St John's Regiment of Militia also known as the First or Royal Regiment.
31. Francis Humberstone Mackenzie, Lord Seaforth: Governor 1801-3 and 1804-6.
32. Schomburgk, 1848, p. 361.
33. *The Times*, 3 July 1805.
34. Schomburgk, 1848, pp. 362-5.
35. Not 25 July as stated by the diarist. Nelson was back in Gibralter on 20 July; C. Hibbert, *Nelson: A Personal History*, London, 1994, p, 338.

36. *The Times*, 3 July 1805.
37. The victory at Trafalgar was celebrated in Bridgetown, three days later by a *'brilliant illumination'* [Schomburgk, p. 366]; a funeral sermon on the death of a hero was preached in St Michael's Church on the 5 January 1806.
38. *Barbados Mercury*, 4 July 1805.
39. Excess 'water (cerebrospinal fluid) on the brain', which in the early nineteenth century carried a high mortality through extensive brain damage.
40. Slave register for Robert Haynes, 1817.
41. *Caribbeana*, Vol. IV, p. 80.
42. William Sealey's kinsman, Joseph, became Robert Haynes's overseer in 1830.
43. Argyle Buildings designed by Thomas Baldwin, had been built in 1789.
44. Senate House Library, London University: 'Newton Papers', MS 523/967, 23 March 1809.
45. In 1835 with emancipation imminent, John Barrow, a Representative in the House of Assembly, migrated to Newfoundland.
46. Between 1790 and 1830 about 480 ships called at Bridgetown annually, each carrying about 100 passengers and crew, some carrying slaves. The Bartons' shipping company was founded by the brothers Thomas and William Barton in the 1790s and plied the Liverpool-Barbados route. After Thomas's death in 1799 it became part owned by George Irlam and John Higginson, two employees of the firm, as Barton, Irlam and Higginson. The owners had ships named after them. After the death of his partners John Higginson introduced his son, Jonathan Higginson to the company, which in due course he inherited on his father's death in 1834. Sir William Barton was Mayor of Liverpool in 1815-1816. Jonathan Higginson became the husband of Charlotte Haynes (Robert Haynes's granddaughter by his son, Richard). One of the company's ships, *Barton*, had as its captain master mariner John Gillespie (or Gilespy), who was involved in more than 40 transactions involving manumissions of Barbadian slaves between 1806 and 1818 [The deeds of manumission recording these transactions are in the Barbados Department of Archives: RB1/251-269 and RB1/229-248]. On 16 February 1816 a boat carrying people to the *Barton* moored in the Mersey overturned and three men drowned. Another captain of the *Barton*, Londoner Thomas Forest aged 60, died on 3 July 1819 at the Cape of Good Hope. After Jonathan Higginson had taken over the Liverpool end of the company, the Barbados subsidiary became Higginson, Deane and Stott. Richard Deane was Higginson's attorney and executor and Thomas Stott, a Bridgetown Merchant. The company acted on behalf of 30 estates and, as well as shipping services, provided credit secured on the properties, being the second largest company on the island in this respect around the time of emancipation. The company frequently bought estates for which they were creditors when the plantations fell into receivership (see Kathleen M. Butler, 1995, *The Economics of Emancipation: Jamaica*

and Barbados, 1823-1843, University of North Carolina Press, pp. 67-8 and 81-2).
47. *The Carlow Nationalist*, Monday, January 28 2002; *The Edinburgh Annual Register*, 1812, vol. 5, p. 2, John Ballantyne & Co, Edinburgh, p. 61.
48. Theodore Paleologus was a descendant of Thomas, brother of Constantine Paleologus, Emperor of the Byzantine Christian Empire. Constantine was killed by the Turks in 1453, whilst defending Constantinople. Thomas fled to Italy and settled in the Adriatic town of Pesaro, where his great-great-grandson Theodore, was born. Theodore married Eudoxia Comnena in 1593, but she died in 1596. After fighting as a mercenary for the Dutch against Spain, Theodore travelled to England around 1600, and became rider to the Earl of Tatershall at Lincoln Castle. He married Mary Balls, daughter of William Balls of Suffolk, and they had five children: Theodore, John, Ferdinand, Dorothy and Mary. By 1628 Theodore was living in Plymouth and the last years of his life were spent at Clifton, with Sir Nicholas and Lady Lower. Theodore died in 1636 and was interred within the church, with his memorial fixed near the vestry door.
49. A sign then accepted to occur only if the slaves were *'fed plentifully, worked moderately and treated kindly'*. See B. W. Higman, 1995. Slave Populations of the British Carribean 1807-1834. U.W.I.P. p. 304.
50. B W. Higman, 1995, *Slave Populations of the British Carribean 1807-1834*, UWIP.
51. E.M.W. Cracknell, ed. (1934), *The Barbadian Diary of Gen. Robert Haynes 1787-1836*, The Azania Press, Medstead, Hampshire, UK.
52. Chancery estate sales tended to achieve lower prices than private sales as many of them were initiated by creditors interested only in recouping their loans. Such sales increased in number as values plummeted within 15 years of emancipation.
53. Schomburgk, 1848, p. 383.
54. Senate House Library, London University: 'Newton Papers', MS 523/745, 24 March 1815. Kitty Thomas and her family are detailed by Hilary Beckles, 'Black female slaves and white households in Barbados', in D. B. Gaspar and D. C. Hine, eds, 1996, *More than Chattel: Black Women and Slavery in the Americas*, Bloomington: Indiana University Press, Ch. 6, pp. 111-125.
55. Senate House Library, London University: Newton Papers: M523/288; 1796; Sampson Wood's 'report' on the slaves.
56. Thomas Percival RN was Captain of HMS *Echo* from 1812 to 1817. Between 1814 and 1815 he was on service in the West Indies, including Bermuda and Jamaica.
57. Senate House Library, London University: MS 523/980, 31 Aug 1815, MS 523/745, 24 Mar 1815.
58. Busso or Bussoe; there were 32 male slaves of this name in 1834, all but one of them born before the insurrection.
59. *Journal of the Barbados Museum and Historical Society*, 2000, vol. XLVI, pp. 13-51.

60. Schomburgk, 1848, p. 395.
61. Senate House Library, London University: 'Newton Papers', MS 523/766, 25 April 1816.
62. *Journal of the Barbados Museum and Historical Society*, 2000, vol. XLVI, pp. 13-51.
63. E.M.W. Cracknell, ed. (1934), *The Barbadian Diary of Gen. Robert Haynes 1787-1836*, The Azania Press, Medstead, Hampshire, UK.
64. James Stephen Jr was William Wilberforce's nephew, a brilliant lawyer employed in drafting legislation at the Colonial Office. Wilberforce started work on the Slave Registration Bill in 1812, which he saw as necessary to ensure compliance with the 1807 Act abolishing the slave trade. If slaves were registered, it could be proved whether or not they had been recently transported from Africa. In 1815 the government had actually blocked the Bill's progress. In November 1815 a committee of the House of Assembly was elected to meet the Legislative Council in a joint session to petition the Crown against plans to impose the Registry Bill. Robert Haynes and John Hothersal Pinder were two of the members of that committee. Haynes's view of the cause of the rebellion is widely held. Professor Hilary Beckles has described it as 'an attempt to influence the general abolitionist politics of the time'.
65. *The Times*, 7 June 1816.
66. Senate House Library, London University: MS 523/769/1, 10 June 1816.
67. Letter to Thomas Lane; Senate House Library, London University: 'Newton Papers', MS 523/882, 20 October 1822.
68. John Lane and his brother had also inherited Bromley Manor with an estate of more than 3200 acres from their cousin Sarah Newton of Barbados. John Lane lived in Bromley Hall, a handsome mansion surrounded by an extensive park. All Saint's Church contains monuments to the Newton and Lane families.
69. *The Barbadian*, 4 Jan 1823; see *Journal of the Barbados Museum and Historical Society*, Vol. 1, pp. 64-72.
70. *The Barbadian*, Saturday, 5 April, 1823, p. 2.
71. Under a series of militia bills from 1812, all white and coloured men from 16 to 60 years of age were required to enrol themselves for service in the militia; Schomburgk, 1848, pp.190-1. After 13 August 1831 there were wealth and land-holding requirements for militia membership.
72. *The Barbadian*, Tuesday, November 30 1824, p. 2. Subscriptions: Hon. John R. Best for 12 estates £7.10; Hon. Sir R.A. Alleyne, Bart., for 4 estates £2.10; Hon. R. Hamden for 3 estates £1.17.6; Hon. J.A. Holder for one estate £0.12.6; Hon. John Braithwaite for one estate £0.12.6; Hon. & Rev. John H. Gittens for 2 estate £1.5; John Barrow, Esq. for 3 estates £1.17.6; Forster Clarke, Esq. for 22 estates £13.15; Hon. R. Haynes for 2 estates £1.5.0; Edward H. Senhouse, Esq. for one estate £0.12.6; J.D. Maycock, Esq. for one estate £0.12.6; John Wood, Jun., Esq. for 12 estates £8.2.6; Robert Reece, Esq. for 2 estates £1.5; Samuel Drayton, Esq. for 3 estates £1.17.6; William Sharpe, Esq. for 3 estates £1.17.6; Jonas Wilkinson for one estate £0.12.6.

73. Methodists were perceived as troublemakers by the white population – ready to stir up foment in the black community.
74. *The Barbadian*, Tuesday 3 May 1825. A man of this unusual name appears in Canada from 1830. His obituary appeared in the *The Bytown Gazette*, Wednesday, November 14 1838: 'At Peterborough, Newcastle District, U.C., on the 29 October, Cheeseman Moe, Esq., one of the oldest Lieutenants in the Royal Navy. During a residence of eight years in Upper Canada, his talents and inflexibility of moral character, secured to him, the esteem of a numerous and respectable circle of friends, who will long regret his death and respect his memory'.
75. *The Barbadian*, Friday, 15 July 1825, p. 2.
76. *The Barbadian*, Friday, 25 November 1825, p. 2.
77. *The Barbadian*, Tuesday, 6 December 1825.
78. £300 per annum, plus presents from their own vestries.
79. Schomburgk, 1848, p. 424.
80. State Papers (Presented by Command of His Majesty), Two Volumes: Vol. XXV, 'Relating to the Slave Population in the West Indies, etc.', Session: 21 November 1826-2 July 1827, pp. 271-4
81. Schomburgk, 1848, pp. 427-9.
82. Ibid. p. 425.
83. Ibid.
84. It became the possession of the subject's son, Robert, and Elizabeth his wife. It was purchased by their granddaughter, Caroline Haynes Whittle, after Elizabeth's death, but was later sold to members of the Haynes family who were lawyers in London [See: Brenda [Eleanor Brenda Violet Johnston née Whittle], *Memories and Confessions of a Grandmother*, published privately, 1949], probably Edmund Child Haynes or his son, Edmund Sydney Pollock Haynes.
85. Schomburgk, 1848, p. 41.
86. *The Times*, 10 October 1831.
87. The US President issued a proclamation on 5 October 1830 allowing vessels from British North American islands and colonies to enter US ports.
88. Robert Haynes Cave entered Rugby School aged 14 in November 1845; his parents were then living in Clifton, Bristol. He became a clergyman in Linbridge, England and married Elizabeth Yonge of Charnes Hall, Staffordshire. His sisters Frances Cave and Isabella Christian Cave married Elizabeth's brothers, Revd George Vernon Yonge and Norman Bond Yonge. All had issue. *Burke's Landed Gentry*, 1863, p. 1720.
89. Butler, 1995, p. 47.
90. Haynes, Haynes and Haynes, 1910, p. 23.
91. Schomburgk, 1848, p. 444.
92. Joanne M. Sanders 1984, *Barbados Records: Baptisms, 1637-1800*, Baltimore: Genealogical Publications, p. 235; Ann Lovell was listed as a widow in the UK 1861 census and was said to be the widow of Robert Lovell on her death certificate.
93. Butler, 1995, pp. 103-4.

94. Caribbean-L Archives, 21 August 2001.
95. *The Barbadian*, 18 October 1823; that there were Lovell's of mixed race at that time is beyond doubt. On Friday, 21 October 1825 *The Barbadian* carried the following announcement on page 2: *'Notice is hereby given that Kitty Lovell, a free coloured woman, hath petitioned for Letters of Guardianship of the Person and Estate of her Infant Daughter Eliza; and that Henrietta Mason, a free coloured girl, above the age of fourteen years, hath petitioned to appoint Kitty Lovell her Guardian'.*
96. The Dummetts lived at 31 Henrietta Street, Bathwick, a large triple-aspect house, opposite their mother.
97. A skin infection due to *Steptococcus pyogenes* involving the lymphatics which today is rarely fatal of about seven days' duration. The death was registered on 5 May 1871 by William Hayward, Registrar. She was buried in Bath Abbey Cemetery, just outside the City of Bath. Details of the plots are to be found in the Bath Records Office.
98. It is of interest that one of Ann Lovell's slaves in 1834 was a one year-old called James Cave.
99. Schomburgk, 1848, p. 90.

Part III
Back in England

17

Emancipation and the Move to England
[1835-1851]

Cutting the ties - Distribution of wealth - Bath - Death of Ann Thomasine - Marriage to Anna Caitcheon - Reading - Major Robert Haynes Lovell - Ann Cave Haynes Lovell - Will - Obituaries

In 1835 Lieutenant General Robert Haynes set about putting his Barbadian affairs in good order, beginning by transferring the plantations to his sons. His decisions were recorded in notes that were eventually added to the 'diary' and were later recapitulated in his will. The transfers reflected his intention or rather his determination to break his ties with Barbados. Why he came to this momentous decision, impacting as it did on the whole Haynes dynasty, is not known, but he had already severed his links with the Assembly, with which he had become increasingly disillusioned; the fact that the first persons of colour had been given the vote in 1831 (although it was to be another nine years before one sat in the legislature) would no longer have concerned him - indeed the act involved was as a result of a Bill introduced to the House of Assembly on 22 February that year, by his son, Robert;[1] the slave emancipation act had come into being, and, although initially out of sympathy, he would have adjusted to it; all his former slaves were to become free men in 1836 after their two year 'apprenticeship' - although slavery was actually abolished in Barbados in 1834, the slaves were not truly given freedom for another two years, and during this time were referred to as slave apprentices - and he probably feared that there would be a labour shortage after this period; he may have feared a backlash from the majority population for his part in putting down the 1816 insurrection - he should have had little to fear from his own workforce which by all accounts he had treated well; then there was the case of the slave, Robert James, who in December 1832, although found guilty of the rape of a poor white widow, had been reprieved of the death sentence,[2] an occurrence which shocked the white population and confirmed their deepest fears as to what might be in store – and the provision of police had to await December 1835; having a tendency to heightened anxiety bordering on hypochondiasis, he possibly feared old age and the likelihood of ill-health and wished to be within reach of Bath with all its doctors. Although unlikely, given the illegitimacy rates in Barbados, it is conceivable that Robert Haynes, feared embarrassment from possible revelations of his family's private life becoming public; having provided for all his legitimate children, he may have wanted to do the right thing by the newer members of his family.

One thing is certain; Robert Haynes wasn't the only planter who decided to leave the island around the time of emancipation, and the two were certainly linked. Many sold up and left for America, Canada, England and Australia. Some, for example Thomas Henry Dummett, settled in the southern states of America, where they set up new plantations run on slave labour. A large proportion of the Haynes family ended up in England; some settled, but others left for Australia and elsewhere.

England and the death of Ann Thomasine

Robert Haynes left Barbados on 13 May 1836, in the year that slavery was finally abolished. He and Ann Thomasine set out for England together, eventually to be joined by most of her surviving children, who became absentee landlords of their Barbadian estates. Mrs Haynes was 70 at the time of their departure from Barbados. One of their children and a few grandchildren remained on the island to look after the plantations. For the family remaining on Barbados little seemed to change. Although Britain repealed the apprenticeship laws in 1838, in their place came the Masters and Servant Act, or 'Contract Law', which bound former slaves as tenants on their former masters' estates, in effect enabling discrimination against black and mixed-race workers on the island and racial inequality to continue entrenched, disguised under the cloak of legality. The government of Barbados remained in the control of the whites, and the lot of the former slaves was arguably little changed throughout the rest of the nineteenth century, as few could afford to buy land and the majority continued their hard grind, working in sugar production.

On 10 February 1835 he had recorded the gift to his son, Henry Husbands, of the Bath estate. Although purchased for £37,100, the gift was reduced by £13,000, which his son had to find from the standing crops, compensation from the British Government for 243 apprenticed labourers (former slaves) and a mortgage Henry held on the Sterling Plantation.[3] As already mentioned, some years earlier he had given his son, Richard, the original Haynes home of Newcastle together with the estate of 416 acres and 180 slaves and standing crops, less a charge of £7000.[4] At the time of the gift he worked out that he had lived there for 45 years. The slave registration undertaken on 24 March 1834 revealed that the Newcastle Plantation had 220 slaves of whom 119 were males. It also gave Richard Haynes as the owner with his brother Robert acting as his Attorney at that time. Robert Haynes had also given Richard the Newcastle silver, furniture and wine, keeping back for himself a *mere* 40 dozen bottles which he had transferred to his Bath plantation for his own use. Richard also became the owner of Bissex Hill Plantation. His marriage settlement with Jane Alleyne Payne daughter of Joseph Alleyne Payne of the Bissex Plantation, St Joseph, dated 1816, is in the Barbados Department of Archives.[5] Robert's second son, Robert, was given Clifton Hall, at the time worth about £50,000, liable to certain annuities and subject to Robert paying his father's debt of £10,000 to his brother Henry. By these various manoeuvres it seems that each of the three sons had received goods and chattels of similar value.

The General also recorded earlier gifts of more than £15,000 to his son George Barrow, and £9,500 to his son, William Clarke, both of whom had already died. William Clarke's children would also be remembered later in their grandfather's will.

After giving Bath to Henry Husbands, he and Thomasine spent the rest of their time in Barbados from 16 February 1835 until their departure for England, in a property known as *Passage* in St Michael.[6]

In 1837 he and his sons received more than £35,000 compensation for their 1293 slaves; this was some £18,000 more than the amount borrowed against their Newcastle and Clifton Hall estates.[7] They clearly decided against further expansion over their 2000-plus acres, the third largest family landholding on the island comprising about 2% of the whole island, and bought only one new estate, Mount Harmony,[8] to add to their seven major estates and three smaller properties each of fewer than 50 acres. Their last major purchase had been in 1832 when Robert Haynes Jr, had bought Clifden Estate for under £13,000 in a chancery sale. This cash surplus and no debt enabled those of them who so desired, to be absentee landlords in England. They also acted as mortgagors to other plantation owners in difficulty.

In England he and Ann Thomasine settled in Bath. Ann Thomasine died of influenza on 11 March 1840 at Egham, Surrey, while visiting Mrs Elizabeth Sophia Lardy, her daughter Thomasine's widowed mother-in-law [d. 1845]. She was aged 74 years and 4 months.[9] She was buried in Egham parish churchyard, where her grave is numbered A74. Ann Thomasine's death certificate describes her as the *'wife of Robert Haynes Gentleman of Barbadoes'* which suggests that Robert might have been in Barbados when she died. Whatever his circumstances at the time of her death, Robert lost no time in marrying for the second time, for on 4 June 1840 he married Anna Caitcheon, and his new wife is mentioned in his will, drawn up on 23 December, 1841.

How long Robert had known his new wife I haven't been able to ascertain, but they may have met through the shipping business of Thomas Lee at Liverpool. Anna's father, the late Robert Hunter Caitcheon, was a master mariner of Scots descent, who may have carried Robert as a passenger on one of his vessels. Captain Caitcheon, in charge of a letter-of-marque, was the first Liverpool privateer to take a prize, the fine Bermuda-built brig, *Harriet*, which was apparently successful against the French in the West Indies in the mid-1790s. The *Harriet* was of 200 tons burthen and valued at £6000.[10] Caitcheon's *'Letter of Marque and Reprizals'* was dated 8 February 1813.[11] It was for the *'apprehending, seizing, and taking the ships, vessels and goods belonging to the United States of America or to any person, being citizen of or inhabiting within any of the Territories of the United States of America'*. Caitcheon's ship was a square-stern, 186- ton, two-masted brig called *Cockatrice*, of Liverpool, with the figurehead of a woman at the prow. The vessel was owned by Hamlet Mullion and John Goldie, merchants, of Liverpool.[12] She was mounted with 12 carronades, carrying shot of 9lb weight and had a crew of 36. On board the ship carried 36 small arms, 40 cutlasses, 10 barrels of powder, 30 rounds of great shot, and about 10 hundredweight of small shot. It was victualled for six months at sea.[13]

EMANCIPATION AND THE MOVE TO ENGLAND

Anna Caitcheon's mother was Jane Douglas, of Edinburgh. Anna was born on 7 August 1793 and baptised Hannah the following October at the Paradise Street Unitarian Chapel, Liverpool. She was one of seven children, five of them boys. We know something of her sister Jane and of her younger brother, William, both of whom were mentioned in Robert Haynes's will. At the time of the 1841 Census Robert and Anna lived at 1 Bathwick Street, Bath, with Robert Lovell then aged 15. Anna Haynes was given as aged 45 (age to the nearest 5 years) and Robert as 70.

Figure 17.1 No 1 Bathwick Street, Bath; home of Robert and Anna Haynes

They eventually settled at 1 Albion Place, Reading, Berkshire in St Giles Parish, an elegant, end of terrace, white-stuccoed, regency town house, which is still standing. At the time of the 1851 Census,[14] Anna Haynes was aged 57, born in Liverpool; Robert was aged 81, described as a *'fundholder'*, born in Barbados, British Subject. Robert A[H] Lovell, aged 25, described as Robert Haynes's grandson, an annuitant, born Barbados, British Subject, was there along with Jane Caitcheon aged 47, sister-in-law, annuitant, born in Liverpool, and Harriet Inglis aged 16, niece, scholar *'at home'* (i.e. educated at home), born in Seacombe, Cheshire. There were two resident house servants: Harriet Jones 26, and Mary Rush, 23. Within three weeks of the 1851 Census, Robert Haynes was dead.

Major Robert Haynes Lovell (1825–1903)

It is perhaps appropriate at this point to consider in more detail the likely parentage of Robert Haynes Lovell, which has been the subject of much speculation in internet correspondence. He was born of Ann Lovell and baptised, Robert Haynes, on 28 December 1825. Ann Lovell was then in her twenty-fifth year. Her antecedence is unknown, although the Lovells were well known in Christ Church – indeed, Lovell wills go back to 1699 in that parish. She may or may not have been of mixed race. No father was named in the baptismal registry of Robert, but given the convention of the time, he was possibly called Robert Haynes.[15]

Lieutenant General Haynes would have been 55 years old at the time of Ann's first pregnancy; 1825 was the year of his election as Speaker and the death of his son, George. He was late middle-aged, highly respected, married and had much to lose by a dalliance with a young woman less than half his age. Yet, on his departure from Barbados, he took upon himself the care, maintenance and education of young Robert Haynes Lovell, who joined him in England, living with him first in Bath and later in Reading, and was acknowledged as his *grandson*. Indeed, Haynes introduced Robert Lovell to his future bride, Harriet Kendal Inglis, a niece of his second wife, Anna. Lastly, he made generous provision for Robert Lovell in his will by way of annuities and capital.

It was not in Robert Haynes's nature to keep his true relationship from Robert Lovell, who was still living under his roof at the age of 25. On the contrary, there appears to be no need to discount his declaration in the 1851 census that Robert Lovell was his grandson. This being the case, the Robert Haynes who was the father of Robert Lovell can only have been the general's son, Robert, and it is therefore relevant to look at the circumstances of Robert Haynes Jr, in 1825.

Born in 1795, Robert Haynes Jr was 30 that year. He had married Sarah Ann Payne on 25 May 1815 and they had a son, Robert and two daughters. Robert is believed to have died in infancy. Sarah Ann died suddenly at Clifton Estate, St John's, after an illness lasting only eight hours[16] and was buried the same day, Friday, 30 July 1824, her three infants being mentioned on her epitaph. Her husband was therefore a widower until Monday, 25 September 1825, when he married, secondly, the 16 year-old daughter of Robert Reece, his father's overseer, who had just come of marriageable age. We conclude, therefore, that Robert Haynes, Jr, frequently in Bridgetown by virtue of his position as Representative of St John's Parish in the Assembly, was the more likely of the two Roberts to have fathered Robert Haynes Lovell, being young, fertile and unmarried. Why he didn't marry Ann Lovell may have had to do either with her race, station, or, more likely, his betrothal to Elizabeth Reece, who had returned from school in England. Given the prejudices of the time, and of Robert Haynes Sr, the balance of probabilities favours Robert Haynes Lovell not being of mixed race, as, while it was not uncommon for members of the 'plantocracy' to have two families, one of mixed race and the other white, seldom did members of the mixed race family leave the island. The final piece of the jigsaw is the entry for 'Haynes of Thimbleby Lodge' in *Burke's Landed Gentry* for 1848, three years before the death of Robert Haynes Sr, in which Robert Haynes (of Thimbleby, Yorkshire) had *'by Sarah-Anne daughter of Rev Mr Payne of Barbadoes, issue surviving:- Robert, Sarah-Anne, Jane-Alleyne'*. Clearly a living Robert, born after his first wife's death, had been substituted for the infant who had perished in Barbados, possibly the one for whom the simple epitaph, *'Robert Haynes Clifton Hall'*, was affixed to the large vault in St John's churchyard. It is possible that the Burke entry was a reference to Robert Haynes-Lovell, whose name and birth had been 'sanitised' for the purpose.

Robert Haynes Jr, and Elizabeth set about having their family at once and had in all eight children. By 1834 he owned Guinea and Clifden Plantations with 237 and 114 registered slaves respectively. He had also been given Clifton Hall Plantation by his father.[17] But after emancipation, Robert decided that living in England was a better option, and in the autumn of 1838, he bought the Thimbleby Estate in North Yorkshire, where, once he could safely hand over management of the plantations to his sons and attorneys, he and his family were to live. It has always been a puzzle why Robert, Elizabeth and family moved to Yorkshire, 250 miles from Bath or Reading – could it have had to do with the possible embarrassment in English Victorian society of Robert's illegitimate son?

So what of Ann Lovell's daughter, Ann Cave Lovell? The most likely father would be someone called Cave, the obvious candidate being Nathaniel Cave. At the time that Ann Lovell conceived for the second time, Nathaniel Cave was unmarried (he married Isabella Wharton on 17 November 1828). Like Ann Lovell he was from Christ Church – Ann Cave Lovell was baptised there on 19 February 1829. He was Robert Haynes Senior's trusted Attorney and would therefore undoubtedly have known Ann Lovell and her son. In 1829 Nathaniel Cave bought the 218-acre Sterling Estate from Jacob Belgrave's widow. Eight years later he bought Lightfoots Estate from the Hayneses' friend and business partner Thomas Lee [who had acquired it in 1835 and pocketed the £2700 slave compensation in the interim] for £14,285.[18]

The General mentioned Ann Cave Lovell in his will along with his grandchildren, and although he referred to her as Robert Lovell's sister, he did not identify her with himself in a family sense as closely as he did Robert. It was as though she did not seem to have been his granddaughter and he never wrote of her as such. However, in the 1861 UK census she is recorded as Ann C. H. Lovell, the H being taken to stand for Haynes, which therefore suggests she acknowledged a Haynes connection.

Robert Haynes's will,[19] written when he lived in Bath, was dated 23 December 1841 and proved at London with five codicils on 9 August 1851. It begins by reiterating what he had previously given his children: Richard had been given the Newcastle Estate; Robert had received Clifton Hall; George Barrow, deceased, property estimated to be worth £15,000; Thomasine, a marriage portion worth £7000; William Clarke, deceased, property to the value of more than £5000; Henry Husbands had been given the Bath Estate. He then returned to Robert Reece, his son's father-in-law, land forming part of the Hope Estate in Christ Church, which he had been looking after on Reece's behalf. He then put right the discrepancy between what William Clarke had received in his lifetime and what the other sons had received, by setting aside £4000 capital he had on interest in

Maynard's Estate in Christ Church[20] for the children of William Clarke in equal shares.

To his new wife, Anna (née Caitcheon), he left all the household furniture and personal effects and an immediate legacy of £500. Stephen Gordon, his young manservant was left £50.

He next gave Robert Haynes Lovell [his grandson], son of Ann Lovell of Bridgetown, an annuity of £200 from either Lovell reaching the age of 21, or the testator's death, whichever first, to be paid for the joint lifetime of himself and Anna Haynes. In fact Robert Haynes Lovell was 25 years old when the General died and was already in receipt of his annuity.

He left the residue of his personal estate, less funeral expenses, to his wife, unless she were to renounce it, in which case it was to go to her brother, William Caitcheon a grocer of Seacombe, Cheshire. However, he specifically excluded from his personal estate any money or financial securities, which were to be invested by his Trustee to produce a fund for the annuity already bequeathed in *'public stocks or funds or other government securities of the United Kingdom or on mortgage of any real securities in England or Wales, but not in Ireland'*, and the dividends and interest thereon to be paid to his wife for her own use and to provide for the maintenance and education of Robert Haynes Lovell.

After his wife's decease, the same trust funds were to be used to provide £500 to Anna's sister, Jane Caitcheon, the residue to be put in trust for Robert Haynes Lovell. He then covered the eventuality of Robert Haynes Lovell not reaching the age of 21 or leaving lawful heirs at the time of Anna Haynes's death, in which case the money was to be put in trust for the surviving children of his son, William Clarke Haynes, and for Ann Cave Lovell, sister of Robert Haynes Lovell, equally, provided they achieved the age of 21 years. In the event Robert Haynes Lovell received his inheritance in full.

His wife was nominated executrix in trust, to be followed after her decease by her brother, William Caitcheon, as executor in trust, but the latter would not be released from paying the estate any sums he had borrowed from the testator.

In the codicils Anna was left the lease on 1, Albion Place, Reading, an additional £1000 and the loan he had made to her brother, William Caitcheon. He left William an annuity of £40 and life annuities of £20 to his niece Harriet Kendall Ingles, his sister, Dorothy Haslam in Washington and his servant Stephen Gordon. He appointed his old trusted business partner, Thomas Lee of Liverpool, merchant, to be an additional executor.

The will was proved at London with five codicils on 9 August 1851 before the Judge by the oaths of Anna Haynes, widow, the relict, the executrix during life named in the will, and Thomas Lee the executor named in the fourth codicil to whom admon was granted having been first sworn by commission only to administer.

Anna Haynes and her sister, Jane Caitcheon, were to stay together after Robert's death, but not in Reading. They decided to make their home in Lancaster where they were recorded in the 1861 and 1871 Censuses at 5 East Road in St Anne's Ward. Anna was described as a *'fundholder'* and Jane as a *'landed proprietor'*. They had two women servants, a cook and a housemaid, who in 1871 were middle-aged sisters Betsy and Martha Hargreaves. Anna remained in Lancaster until her death at the age of 78 in December 1871. She was interred with her husband in St Giles' churchyard, Reading, on 23 December 1871.

Anna and Jane's brother, William Caitcheon and his wife Annabella Davies, emigrated from London to New Zealand in 1858, with their six children: Janet Douglas, Sarah Maria, Anna, Robert Hunter, William Davies, and Charlotte Haynes, on the 723-ton ship *Mary Ann* (Wm Ashby, Captain), which arrived at Auckland on 24 September 1858. The seventh and youngest of their children, Charles Haynes, was born in New Zealand.

Obituaries

Robert Haynes died at his home in Reading, Berkshire, on Friday, 18 April 1851, and was buried, following his funeral, in St Giles' churchyard one week later. On 26 April the following obituary appeared in the *Illustrated London News*:[21]

> 'This gentleman, for several years Speaker of the House of Assembly at Barbados, died on the 18th April in his 82nd year. He was a descendant of an ancient English family that was forced by their loyalty to emigrate to the West Indies in the time of Cromwell. Called at an early age to fill offices of trust in the House of Assembly as well as in the Militia of his native Island, Mr Haynes was promoted successively to the highest dignities in both the departments of public services; and in two emergencies – the threatened French invasion of 1805 and the insurrection of the Negroes in 1816 – he

gained universal approbation for his wisdom and skill. On the latter occasion, the Council and the Assembly tendered him their thanks and the Officers of the combined Corps of the Midland District of St John and St George, Barbados, presented him with the appropriate testimonial of a sword.

Some few years later Mr Haynes was unanimously elected Speaker of the House of Assembly.

Distinguished by great energy of character, singular aptitude for business, untiring industry, loyalty to his Sovereign and devotion to the interests of his native Island, he carried to the discharge of his public duties the integrity which marked his private conduct. Qualities such as these won him favour of successive representatives of the Crown in Barbados and on his final retirement the Governor, Sir James Lyon, was pleased to make recapitulation of Mr Haynes' services the subject of a general order in which was sent forth the high sense entertained by His Excellency of their importance and the unusual privilege to Mr Haynes of retaining in his retirement the local rank of Lieutenant General.

Mr Haynes leaves three sons: Richard Haynes of Clifton, Glos, Robert Haynes of Thimbleby Lodge, North Allerton, Yorks and Henry Haynes of Barbados'.

The Gentleman's Magazine[22] also published an obituary in the June issue which expanded on some of the points mentioned in the *Illustrated London News*:

'April 18th. At Reading in his 82 year Robert Haynes Esq. late of Barbados. This gentleman at an early period of life filled various offices of trust in his native Island. During the impending fears of French invasion in 1805 he was eminently useful to Lord Seaforth at the time Governor of the Windward Islands by whom and by the Admiral in Command he was held in high esteem for the sagacity and skill which he manifested in that emergency. For his military services also on the insurrection of the Negroes in 1816, Mr Haynes received the thanks of the Council and Assembly which was followed by the testimonial of a sword from the Officers of the Combined Corps of the Districts of St George and St John.

A few years later Mr Haynes was elected to the office of Speaker by the unanimous voice of the House of Assembly. Distinguished by the energy of character which led him to rely on none other than himself by singular aptitude for business, untiring industry, loyalty to his Sovereign and devoted patriotism, Mr Haynes carried with him the discharge of his public duties, the integrity which marked his conduct in the private relations of life. These qualities obtained for him the personal regard of a long series of representatives of the Crown in Barbados, among whom were Lord Seaforth, Sir George Beckwith, Sir James Leith, Lord Combermere, Sir Henry Ward[e] and Sir James Lyon.

On resigning his Commission into the hands of the latter Governor, his Excellency was pleased to make recapitulation of Mr Haynes services the subject of a general order in which document set forth showing the strong sense entertained by the Governor of those services together with the unusual privilege to Mr Haynes of retaining in his retirement the local rank of Lieutenant General.

Mr Haynes leaves three sons, Richard Haynes of Clifton, Glos, Robert Haynes of Thimbleby Lodge, North Allerton, Yorks and Henry Haynes of Barbados and numerous grandsons'.

His death did not go unnoticed in Barbados, and in due course a splendid marble memorial tablet surmounted by the Haynes family Arms was erected in the Parish Church of St John, Barbados:

'Sacred to the memory of Robert Haynes late of this parish, Esquire, Lieut. General of Militia and sometime Speaker of the House of Assembly, Barbadoes. His long and useful life was devoted to the service of his family and of his native island and it was his chief ambition to be accounted an honest man. Born 27th Sept. 1769, died 18th April 1851. Seest thou a man diligent in his business he shall stand before kings. He shall not stand before mean men. Prov. xxii 29th verse'.

Figure 17.2 Detail of the marble memorial to Lt-General Robert Haynes with the motto: *velis et remis* – St John's Church, Barbados

Notes and references

1. R. H. Schomburgk, 1848, *The History of Barbados*, Longman, Brown, Green & Longmans, pp. 431–2.
2. Ibid. p. 445.
3. There were 242 slaves, 115 males and 127 females registered in Robert Haynes's name on 24 March 1834.
4. E.M.W Cracknell, ed. (1934), *The Barbadian Diary of Gen. Robert Haynes 1787-1836*, The Azania Press, Medstead, Hampshire, UK.

5. Barbados Department of Archives: RB 260/222.
6. E. C. Haynes, A. P. Haynes and E.S.P. Haynes, 1910, *Notes by General Robert Haynes of New Castle and Clifton Hall Plantations, Barbados and Other Documents of Family Interest*, Argus, London, p. 23.
7. Kathleen M. Butler, 1995, *The Economics of Emancipation: Jamaica and Barbados, 1823-1843*, University of North Carolina Press, pp. 85-6.
8. Fifty-nine acres; purchase price £5035; Barbados Department of Archives:RB1/274, f. 296.
9. Her obituary appeared in *The Gentleman's Magazine* in April 1840, p. 445.
10. Charles Wye Kendall, 1932, *Private Men-of-War*, R. M. McBride & Company, p. 192.
11. ADM7/319.
12. Mullion is listed as the owner of the 270-ton Liverpool slaver, *John*, which on 28 July, 1807, transported 408 slaves from Angola, under Captain D. Phillips.
13. The *Cockatrice* had two suits of sails, four anchors, three cables, and about 10cwt of spare cordage. Dennison Murray was First Mate, Robert Parkinson, gunner, Robert Lowdia, boatswain, Millegan Procter, carpenter, Hugh Meredith, cook, and John Bannister, surgeon.
14. UK census 1851: HO107/1692 Reading St Giles.
15. One caveat is worthy of consideration. But for the fact that Robert Haynes Sr subsequently acknowledged Robert Haynes Lovell as his grandson, it was common practice in Barbados for children to be named after a senior political figure, such as the Speaker, to which office Haynes had just been appointed around the time of Robert Haynes Lovell's birth. A number of children had been named after the previous Speaker, Cheeseman Moe: for example, Cheeseman Moe Ryan, son of Michael and Mary Ryan, born in 1821 in St Michael.
16. *The Barbadian*, Tuesday, 3 August 1824, p. 1.
17. Clifton Hall had 261 slaves registered on 24 March 1834, 108 of them males.
18. Butler, 1995, p. 90.
19. Will of Robert Haynes of Bath, Somerset; probate 18 June 1851; TNA: PROB 11, 2137.
20. Proprietor: Joseph Briggs.
21. *Illustrated London News*, 26 April 1851.
22. *The Gentleman's Magazine*, June 1851, p. 669.

18

The General's Siblings [1772-1860]

Richard Downes Haynes [1772-1793] - Captain Henry Haynes RN [1776-1838] - Ann Ellcock Haynes [1777-1857] - Dorothy Haynes [1779-aft. 1850] - Edmund Haynes [1781-1846]

Richard Downes Haynes

Richard Downes Haynes was born on 15 July 1772 at Newcastle Plantation. When botanist Sir Joseph Banks visited Barbados he became acquainted with Richard Downes and his father in December 1787 and intended to use him as a courier as Richard was bent on travelling to England. Banks had been sending specimens of camphor and other trees he had found on Barbados.[1] He died on 25 May 1793 aged 20 years 10 months and 10 days and was buried the next day in the family vault in St John's. He was left £3000 by his father for when he reached the age of 21 and an annuity of £200. He thus failed by two months to receive his inheritance.

Captain Henry Haynes, RN

Henry was born on 13 May 1776. He entered the Royal Navy as a midshipman in 1793 and was made Master, then Commander, by Admiral of the Fleet Lord Gambier at Copenhagen. He later became Post-Captain, an important rank on the promotion ladder to admiral. He and his brother Robert met up unexpectedly in Bath in 1809.[2] He did not marry until he was 38; his bride was the 34-year-old Antigua-born Harriet Watkins Oliver. They were married in January 1815 in Bristol, and their Antiguan marriage settlement, including property on the island of Antigua and slaves specifically noted as forming part of this, still exists.[3] Harriet died in 1826 in Bristol where her monumental inscription was erected in St Paul's; Henry died aged 61 on 17 January 1838 in The Grove, Bath, Somerset. His monumental inscription placed in the entrance porch of the west door of Bath Abbey reads: *'Sacred to the memory of Captain Henry Haynes, Royal Navy, who died in Bath January 17 1838 in the 62nd year of his age'*.

Their two children Thomasine Oliver and Henry Freeman Oliver[4] were born in 1817 and 1818 respectively. Both benefited handsomely from Henry's will.[5] In 1843 Thomasine married Commander Sir William Sidney Thomas RN, fifth Baronet of Yapton and Dale Park, whose family was also originally from Antigua. They had four children: daughters Lucy Elizabeth and Isabella Montague Jane and two sons, George Sidney Meade and Frederick Louis Charles. Thomasine died in Melcombe Regis, Weymouth on 6 March 1853[6] after the birth of her second son and was buried in Radipole, Weymouth, Dorset, five days later. Freeman was educated at Isleworth School, Middlesex, then in Paris and Dresden, before entering Caius College, Cambridge on 11 April 1836 aged 17. After a distinguished academic career (MA Cantab., Fellow of Caius College), he was admitted to Lincoln's Inn on 12 January 1842 and called to the Bar on 6 May 1845. He practised on the South-Eastern Circuit, specialising in Equity Drafts and Conveyances. He wrote: *Outlines of Equity* and other works. His lecture notes and notebook on this subject are in the Shropshire Archives.[7]

Freeman married Emily Wilkes Child, daughter of Robert Child of Russell Square, London on 6 July 1843. They had sixteen children: Harriet Augusta [b. 1844]; Henry Sidney Freeman [b. 1845], Edmund Child [b. 1846]; Emily [b. 1848]; Lucy H. [b 1850]; Ellen Louisa [b. 1851]; Thomasine F. [b. 1852]; Adelaide M. [b. 1856]; Fanny L. [b.1857]; Herbert Pollock Gilbert [b. 1859]; Mary E. [b. 1861]; twins Edith and Rose [b. 1862]; Constance M. [b. 1865]; Freeman A. G. [b. 1867]; Florence [b. 1868]. The family lived at Edge Hill, Wimbledon and finally at Donhead Lodge, Wimbledon, Surrey, where Freeman Oliver died on 12 July 1880.[8] It was Edmund Child Haynes who first decided to retrieve what he could of the 'Diary' of General Robert Haynes, his great uncle, which notes were published after his death in 1910.[9]

Ann Ellcock Haynes

Ann Ellcock was born at Newcastle on 8 August 1777. She married first Sampson Wood, son of Sampson Wood Sr and Judith Richards. They had three sons and a daughter. All of them were given Haynes as a middle name: Henry Haynes (b. 1798), Helen Haynes (b. 1799), Richard Haynes (b. 1801) and Robert Haynes (b.1803). They all died in childhood. Sampson Wood, a former Speaker of the House of Assembly and estate manager of Newton, died in 1803, and on 19 March 1804 Mrs Ann Ellcock Wood, his widow, advised John and Thomas Lane to appoint her brother, Robert Haynes, as their attorney with John Farrell as manager of their two Barbadian plantations, Newton and Seawell.[10]

Ann Ellcock Wood married, secondly, General John Walton on 28 January 1806, by whom she had four daughters and a son all born in Barbados: Haynes (b. 1806); Dorothy (b.1808); Ann Ellcock (b. 1812); Charlotte (b. 1813); Henry Haynes (b. 1816). Henry Haynes Walton became a well-known London surgeon. In 1846, he married Elizabeth Haffey Reed in St David's Church, Exeter, who, although no blood relation, was the favourite niece by marriage of his uncle Edmund Haynes, by whom he had two children, Elizabeth (Bessie) and George, also a doctor. Sisters Dorothy and Haynes Walton remained unmarried and lived together to advanced years in London. The descent of neither family has been followed further. General Walton was Deputy Provost Marshal of Barbados for 26 years; he died at 'Goslings', Barbados, on 2 January 1829 and on the same day the President appointed Benjamin Walrond in his place. On 16 January 1829 Ann Ellcock Walton petitioned for letters of administration to her husband's estate.

Ann Ellcock Walton died on 26 August 1857 at Dorset Terrace, Hyde Park, London.

Dorothy Haynes

Dorothy was the younger daughter of Major General Richard Downes Haynes, born on 4 January 1779 at Newcastle, Barbados. She was just 14 when her father died. She married Charles Haswell of Dublin who was a member of the British diplomatic service. The couple moved to the United States where they lived in New York

City. On the 22 June 1850 Dorothy was left an annuity of £20 a year for life by her brother, Robert, according to the fifth codicil to his will. It mentioned her as '*my sister Dorothy Haswell of the City of Washington in the United States of America*'.

Dorothy and Charles had a son, Charles Haynes Haswell, who was born on 22 May 1809 at their home in North Moore Street, New York City. He was educated at schools in Jamaica, Long Island and New York City, where he attended Joseph Nelson's Collegiate Institute, and worked in the company of James Allaire, on steam engines. In 1836 Charles Haynes Haswell was appointed engineer in the US Navy, designing and supervising the construction of the first steam engines for warships, and became engineer-in-chief in 1844. Indeed, he designed the complete machinery for at least ten warships. He was dismissed from the navy in 1852 following his opposition to a design plan. Immediately after this the Haswells moved to Washington to sit for his portrait as first and last chief engineer of the US Navy. He continued as an engineer in civilian life, first for the Russian government in the Department of Naval Engineering and then as assistant engineer to the Board of Estimate. In recognition of his services, he received a diamond ring from Csar Nicholas. For over 40 years he was the Surveyor of Steamships for the Marine Underwriters of New York. He served under General Burnside in the Civil War. Surviving correspondence shows that as late as 1884, according to his letterhead, which bore the image of a ship, he remained a 'Consulting and Superintending Engineer'. A New Yorker to the core, he wrote several technical works including the *Mechanic's and Engineer's Pocket Book* in 1842 which was republished until 1913, and published his reminiscences late in life.[11]

Charles Haynes Haswell married Ann Elizabeth Burns on 19 September 1829 and they had nine children [Sarah Haynes (b. 1831); Edmund Haynes (b. 1833); Charles Roe (b. 1835); Frances Roe (b. 1837); Jane (b. 1838); Hansen (b. 1840); Gouveneur Kemble (b. 1842); Charles Haynes (b. 1845); Elizabeth Bulwer (b. 1851)]. He died on 12 May 1907, following a fall at his New York home, 324, West Seventy-Eighth Street, on 13 May 1907, a few days short of his ninety-eighth birthday. At the time of his death, he was said to have been the '*oldest and most distinguished civil and marine engineer in the United States*'.

Edmund Haynes

Edmund, the last child of Major General Richard Downes Haynes, was born at Newcastle plantation on 19 February 1781. He married Sarah Bell [b. 6 December 1779] in 1805. She was the daughter of Francis Bell, who died in 1808 leaving Bellmount [Belmont], the family home and estate, to his children. Edmund Haynes bought the other shares and by 1816 owned the property outright. Edmund sold Bellmount to Francis Cheeseman in 1827 for £1000. Sarah played a leading part in the work of the Moravian Mission and taught at the Sunday School there. She died on 11 November 1820 while visiting New York City; her memorial tablet can be found in Trinity Church. Although it is said that Edmund and Sarah had no children, and certainly none was acknowledged in Edmund's will, The Argosy Newspaper, Demerara, of Saturday, 9 August 1884, reported the death at Charlestown of "Joshua Haynes, aged 69, son of the late Edmund Haynes Esq. of Haynes-field, Barbados". He was born in 1815. Why Joshua appears to have been 'airbrushed' from Edmund's family history remains to be revealed.

Edmund was known throughout Barbados for his munificence. He gave four acres of land to the Moravian Mission and built a church on the land. When he sold the Haynesfield Plantation in 1838, there was a perpetual charge on it of £1000 to pay £60 per annum for the upkeep of the Moravian Church, Mount Tabor. The great house on Haynesfield was renamed Villa Nova. It was built after the 1831 Hurricane, which destroyed the previous plantation house from where he managed his three sugar estates: Haynesfield (Wakefield), Bellmount (Belmont) and Claybury, totalling over 1000 acres. Villa Nova was blessed by the Moravian Minister, John Gottlieb Zippel, in 1834.

Edmund married for the second time on 11 April 1822. He had just moved to England, although he continued with his business and philanthropic activities in Barbados, which he and his new wife visited at intervals, before finally selling the Haynesfield Plantation in 1838. His new wife was Lucretia [Lucy] Reed, the daughter of George Reed of Dockfour, Demerara, and Johnstone Street, Bath, England[12]. The marriage, conducted by Revd F Festing, Vicar of Wingham, took place at St Mary's Church, Bathwick.

Edmund Haynes lived in Summerland Place, Exeter, but died aged 65 while in Cheltenham on 2 May 1846. A memorial to his munificence was erected by the Rector of St John's Church, Barbados inscribed:

> 'This tablet is erected to the memory of Edmund Haynes, Esquire, late of Haynes-field in this parish who died in Cheltenham in the kingdom of Great Britain, in the 66th year of his age. It is designed to commemorate his Christian exertions and munificent contributions, in the re-erection of this sanctuary after the destruction of the former one in the hurricane of 1831. And it is inscribed with grateful and affectionate feelings by the Rector and Vestry of his native parish as a memorial to succeeding generations, so long as this temple shall have a name and place amongst them, of the obligations they owe to him, for this work of love to God and to man'.

He and Lucy, who died on 24 May 1854, are buried in the churchyard of St Peter's, Leckhampton. Their coped limestone memorial reads:

> 'In memory of Edmund Haynes, of the Island of Barbados, afterwards of the City of Exeter, England. Born at Barbados, Feb. 19, 1781 died at Cheltenham, May 2. 1846. And of Lucretia Haynes, his relict, fourth daughter of George Reed, Esq. of Barbados, and Dockfour, Demerara. Born at Barbados, Nov. 19, 1797, died at Exeter, May 24, 1854. Whose earthly remains repose together beneath this spot'.

Edmund and Lucy had no children.[13]

By his will dated 30 December 1845, Edmund left a small pecuniary legacy to his brother and £3500 to Miss Elizabeth Haffey Reed. He left lifetime annuities of £10 to his sister Ann Ellcock Walton and one of £80 to his sister, Dorothy Haswell. He left his wife an annuity of £1000 during her life. He also made bequests to Ann Elvira Reed (Lucretia's widowed sister-in-law) and, after the death of Lucretia, he bequeathed interest and dividends of £5000 to Elizabeth Reed (Lucretia's sister). Out of the principal residue of his estate after his wife's death he left £2000 to his nephew Freeman Oliver Haynes, £9000 to Elizabeth Haffey Reed (Lucretia's niece, daughter of Ann Elvira Reed), £1000 to Jane Morris Reed (Lucretia's niece, daughter of Ann Elvira Reed), £2000 to Lucy Haynes Reed (Lucretia's niece, daughter of Ann Elvira Reed), £1000 to Frances Haynes Reed (Lucretia's niece, daughter of Ann Elvira Reed), then £1000 each to his sister's children: Haynes Walton, Dorothy Walton, Charlotte (Walton) Brown and Henry Haynes Walton. The residual legatee was

Elizabeth Haffey Reed. After the death of Lucretia, the provisions of Edmund's will were contested and became a celebrated case heard in the English Courts of Common Law and Equity, used as a precedent thereafter.[14]

Notes and references

1. W.R. Dawson, 1958, 'The Banks Letters; A Calendar of the Manuscript Correspondence of Sir Joseph Banks, Preserved in the British Museum', British Museum, p. 788.
2. E.M.W. Cracknell, ed. (1934), *The Barbadian Diary of Gen. Robert Haynes 1787-1836*, The Azania Press, Medstead, Hampshire, UK.
3. Donald H. Simpson, ed., *The manuscript catalogue of the library of the Royal Commonwealth Society* (London, 1975), p. 141. Indexed: Reference RCMS 240/20 (former reference: MSS 741).
4. Trans Colonial Soc., Massachusetts, 1935, Vol. 68, p. 63; known as Freeman.
5. PROB 11/1190; 16 February 1838.
6. *The Gentleman's Magazine*, April 1853, pp. 454-5.
7. Shropshire Archives: 6241/1/165/1-21.
8. Boase, 1, 1400; Venn, II, 236 Al. Oxon; Inns of Court, Law Lists; The *Guardian*, 21 July 1880 [Obit.].
9. E. C. Haynes, A. P. Haynes and E.S.P. Haynes, 1910, *Notes by General Robert Haynes of New Castle and Clifton Hall Plantations, Barbados and Other Documents of Family Interest*, Argus, London.
10. Senate House Library, London University: 'Newton Papers', MS 523/567, 19 Mar 1804.
11. Charles Haynes Haswell, 1896, *Reminiscences of an Octogenarian of the City of New York (1816-1860)*. Harper & Brothers Publishers, NY.
12. Full details of the marriage were reported by Edmund Burke (Edit.), 1823, Annual Register; Vol 64, p 246.
13. The land on which Villa Nova still stands was in 1680 a plantation called Baldricks [or Baldock's], owned by Thomas Baldrick. By 1743 the plantation was 163 acres with 72 slaves. In 1804 Henry and Thomas Trotman, the executors of Robert Trotman, sold Baldricks to Robert Haynes for £19,700. He renamed it Haynesfield. Edmund Haynes, who acquired part of the adjacent Belmont through his wife, bought the remainder and in 1811 acquired Haynesfield from his brother. In 1838, when Edmund Haynes sold Haynesfield to William Thomas Sharpe for £25,000, it was 317 acres. Villa Nova, the house built in the 1830s, replaced the earlier plantation house. It is unusually elegant with a curved south front, identical with that of Government House. In 1907 it was separated from the plantation and sold for £800 to the Parish of St John as the Parochial Medical Centre. In 1965 it was bought by the Earl of Avon who sold it in

1971. It stands on a hillside 900ft above sea level, with refreshing breezes and temperatures 5 degrees lower than at sea level.
14. F. Pollock, O.A. Saunders, J.G. Pease and W. Bowstead, eds, 1908: *The Revised Reports; being a Republication of Such Cases in the English Courts of Common Law and Equity, from the year 1785, as are of Practical Utility*', Vol. 98, 1853-4, London: Sweet and Maxwell Limited.

19

The Diarist's Family [1792-1860]

Richard Haynes [1792-1859] - Robert Haynes [1795-1873] - George Barrow Haynes [1796-1825] - Thomasine Haynes [1799-1880] - William Clark Haynes [1800-1831] - Henry Husbands Haynes [1801-1852]

Robert Haynes the diarist and Ann Thomasine Barrow had ten children. Richard, Robert, George Barrow, Thomasine, William Clarke and Henry Husbands were to reach adulthood and marry; the others died in infancy or childhood. Only four, Richard, Robert, Thomasine and Henry are known to have outlived their parents.

They formed the typical wealthy Barbadian planter family, providing representatives on the Council, House of Assembly [including a Speaker], officers of the militia, and intermarriage with other planter families of influence.

Richard Haynes (1792–1859)

Richard, the eldest child and first heir, was born on 9 July 1792 at Eastwood Plantation. He was given the Newcastle estate in Barbados by his father. He married Jane Alleyne Payne of the Bissex Plantation, St Joseph, the daughter of Joseph Alleyne Payne and Jane Cragg, on 25 January 1816 and his marriage settlement of that year, is in the Barbados Department of Archives.[1] In due course his descendants owned the Bissex Plantation, St Joseph, as well, which became the family's principal residence. The year before their marriage, Jane Alleyne Payne and her sister Sarah Ann became the residuary legatees of their wealthy great uncle Richard Sharp of Philadelphia, so they brought much wealth to their respective marriages.

The Newcastle Plantation was 415 acres in 1825 and 455 acres in 1861. When his father transferred the estate to him, he recorded: '*To my son Richard I have given Newcastle, where I resided for 45 years; 416 acres, 180 slaves, subject and liable to about £7,000; the crops I left on it should almost have made that'.* He also gave Richard his chest of plate, furniture, and wine (with the exception of about 40 dozen).[2] Clearly wine was a popular beverage in the Haynes household.

Richard was an innovator, keen to take advantage offered by new developments in engineering, which he no doubt saw as an advantage over the spiralling cost of manual labour after emancipation. In 1846, for example, he installed the first steam mill at Newcastle.

There are numerous records relating to Richard Haynes in the Barbados Department of Archives for 1825 and 1842-1861. In 1825, for example, Richard took out a mortgage on Springfield Estate in order to help the owner, Charlotte Pile, a widow, who was unable to find a buyer when her existing creditor was foreclosing.[3]

Richard and his wife were passengers to Liverpool on the *Frances* that left Barbados on 28 July 1825 and, again with his family, on the *Higginson* that set sail on 23 June 1830. Eventually he and several of his family moved to England around the same time as his brother, Robert, and settled in Clifton, Bristol at 33 Ashley Villas, Richmond Park. In the 1851 Census, he and his wife Jane had four servants, one of whom, Leah Redwar, was a 58-year-old widow from Barbados.

Richard died in Clifton on 17 November 1859, and both he and, subsequently, Jane were buried at St John's Church, Apsley Road. Interestingly it was at 32, Upper Park St, Clifton, where his nieces, Sarah Anne Haynes and Jane Alleyne Haynes, the unmarried daughters of Richard's brother, Robert by his first wife (Richard's sister-in-law, Sarah Ann Payne), were recorded as living in the UK 1881 census, then aged respectively 62 and 60.

Richard and Jane had five children: Robert Haynes, Joseph Alleyne Haynes, Ann Jane Haynes, Charlotte Haynes, and John Higginson Haynes.

Robert Haynes (1818–1853)

Robert Haynes of Newcastle, Barbados, the eldest son, was born in 1818. On 6 May 1847 he married Emily Anne Mairis, eldest daughter of Major Valentine Hale Mairis and Elizabeth Edwards, who was born 15 July 1825 in Clifton, Bristol[4] and baptised in Bishop's Lavington, Wiltshire. The couple had their home at Dawford, St Mary's,

Bridgnorth. They had one child, Richard, born 4 January 1850 at Danesford, Bridgnorth. They had a household that included a manservant, a nurse for Richard, a cook and two maids. Robert, described as a landed proprietor, drowned while swimming at Aberystwyth, Wales, on 14 July 1853, but his body was never recovered. His death possibly took place while he was visiting his first cousin, Henry Husbands Haynes, who was Chief of the Police in Wales. He was entered on his parents' memorial in St John's Church, Clifton, Bristol. His death made way for the Barbados estate to be passed on to Joseph Alleyne Haynes. Emily Anne died a few months later on 22 December 1853 at Clifton, where she was buried with her husband.[5] Infant Richard was brought up by his aunt Anne Walker Shepherd, named in Emily's will as one of Richard's guardians, and her daughter Thomasine Haynes. As an 11-year-old he was living with them at 14 Peterborough Place, Paddington, Marylebone where he was educated both privately and at Westbourne College, Bayswater. On 1 October 1869 he was admitted, pensioner, to Gonville and Caius College, Cambridge, where he remained only four terms. At the time of the 1871 census, he was single and still in Marylebone staying at the Great Northern Hotel, where he was described as living off dividends, but later that year, Richard married Catherine Fisher [b. 1844, Cambridgeshire] in Cambridge. They lived first in Clifton, Gloucestershire, where their son, Robert Mairis, was born in 1873, then in Swanage, Dorset, where daughter, Agnes Emily was born two years later. Richard died on 3 July 1884 aged 34. Catherine moved to Alverstoke, Hampshire, where she was living with her unmarried daughter in 1901. At that time her son, Robert, also unmarried, was serving as a Lieutenant in the Royal Navy on the ram, HMS *Hotspur*, then a guardship in Bermuda.

The Hon. Joseph Alleyne Haynes MCP (1820–1894)

Joseph Alleyne Haynes was born on 1 November 1820 and held the Newcastle Plantation. He also owned the West India Estate. On 5 July 1854, in the Cathedral, St Michael, he married Margaret Ann Howell, daughter of Colonial Treasurer Benjamin Howell and his wife Mary Simpson. They had eight children: Richard (1856-1937); Carleton (1858-1945); Alleyne (1859-1938); Mary Howell (1860-1957); Jane Charlotte 1862-1928); Arthur Percy (1864-1928); Agnes Payne (1865-1923) and Helen Margaret (1868-1937). Of these eight, five were married, but between them produced only two children, one of whom, Joseph Alleyne, son of Arthur Percy bore the Haynes name, which is now extinct in this branch.

Newcastle possessed one of the six fresh water springs in St John. In 1860 the Bridgetown Waterworks Company began to work on channelling water from springs in Newcastle, through pipes to the city. In 1864 additional sources to be tapped by the water company were the springs at Codrington College. Today some of this water still supplies Barbados.

Joseph Alleyne built St Margaret's Church, Parish of St John, Barbados in 1862 on the site of the ruins of the Glenburnie boiler house, much of the materials from which were used in its construction. The church, in a remote situation, was named for his wife Margaret Ann. Her memorial is to be found in the sanctuary of the church, Parish of St John, Barbados: *'In memoriam. Margaret Ann Haynes beloved wife of Joseph Alleyne Haynes of New Castle Estate. A worthy follower in piety benevolence and devotion as wife and mother, of St. Margaret of Scotland of glorious memory, after whom this Church is named. Died 13th Feb 1917. Aged 90 years. R.I.P.'*

A memorial to Joseph Alleyne Haynes, who died on 1 July 1894, his wife and five of their children, is in St John's Church, Barbados.

Ann Jane Haynes

Ann Jane Haynes, who may have been the first-born child, is recorded as receiving a slave, whom she owned outright, as a christening present.[6] Ann Jane ceased to be recorded as an owner around 1824 and most probably died in childhood.

Charlotte Haynes (1819–1906)

Charlotte Haynes, the second daughter, was born on 14 January 1819 in Barbados. She married her cousin, Liverpool banker Jonathan Higginson [b. 1807], on 2 January 1838 at St George's Church, Everton, Lancashire. They shared a common grandmother, Ann Thomasine Clarke, but different grandfathers. Charlotte was the granddaughter of Lieutenant General Robert Haynes; Jonathan was the grandson of Nathaniel Barrow.

Jonathan Higginson was the son of John Higginson, a partner in the shipping company, Barton, Irlam & Higginson, operating between Liverpool and Barbados. The shipping company was started by the Barton brothers, Thomas and William. After Thomas's death it became part owned by George Irlam, as Barton & Irlam. John Higginson became a third partner, who left Barbados bound for Liverpool in the *Irlam* on 17 July 1823 and, after his death in 1824, George Irlam and Sir William Barton having also died, Jonathan became the sole owner of Barton, Irlam & Higginson, based in Liverpool. Its Barbadian subsidiary was renamed Higginson, Deane and Stott to reflect its new partners. Barton, Irlam & Higginson was a large company, owning 1232 slaves in Barbados, which at the time of emancipation netted them £26,822 in compensation.[7] The company advanced mortgages to plantation owners, such loans generally being secured against the properties. Occasionally such loans appeared to be secured on gentlemen's agreements and were binding in honour only. In 1823 Sir William Barton, the senior partner, bought Sandy Lane estate for £11,142. The estate was in receivership, and appeared in Chancery again in 1829, when its liabilities to the partners alone had risen from £3000 to more than £23,000. John Higginson, the sole surviving partner, bought it again, for this figure, to clear the debt to the partnership. In 1832 Richard Deane, acting as John Higginson's executor, bought Grettons Estate out of chancery for £16,860. The estate was indebted to the partnership for more than £10,600. Just before his death, John Higginson had agreed to buy the estate, provided £5000 remained in trust for the women relatives of the late owner. The inventory of John Higginson's property taken in 1835 valued his seven estates[8] at £77,750, considerably less than their purchase price. The partners continued to buy property. Thus in 1837 Stott bought Mount Clapham, clearing its debts including 90 per cent of those owed John Higginson's executry. Within a year the partners had resold it for a handsome profit. In 1841 Jonathan Higginson bought Palmer's, another estate indebted to his father's executry, Barton, Irlam & Higginson, the Imperial Government (for the 1831 hurricane damage) and Barbados' largest merchant house, Thomas & John Daniel. The huge price Higginson paid cleared all these debts which totalled £23,541.[9] However, Barton, Irlam & Higginson went into receivership in November, 1847, along with dozens of other overseas traders caught up in the food and money panic of 1847-8,[10] and Jonathan became bankrupt, indebted to the Royal Bank of Liverpool [which also became insolvent] to the extent of £463,000,[11] the firm owing around £1m.

Jonathan and Charlotte Higginson had six children: John [b. 1838]; Jane [b. 1840]; Charlotte Elizabeth [b. 1841]; Emily Jane [b. 1843]; Arthur [b. 1845]; Jonathan [b. 1848]; all except Jane, who died just before her second birthday, reached adulthood. They were the godparents [sponsors] of Henry Higginson Robinson, born 11 May 1846, son of the Revd John Matthews Robinson and Margaret Sparrow his wife. Henry was baptised in Liverpool on 19 June 1846, by his father, who had been in England since attending Cambridge University.[12]

Jonathan died in 1859 in Edinburgh from *'softening of the brain'*, his entire estate being valued at just £50. Two years later John, the eldest child of Charlotte and Jonathan, married Helen Elizabeth Thurburn and the couple moved to New Zealand. In the same year, their daughter, Charlotte Elizabeth, married Revd Henry Brougham Bousfield ninth child of William Cheek Bousfield and Rebecca Richings. They moved to South Africa where in 1878 Henry became the first Bishop of Pretoria. His wife and daughters were inveterate diarists and many of their notes and diaries survive,[13] covering various aspects of their lives, family and the Boer War.

About ten years after the death of Jonathan Higginson, Charlotte left with their daughter, Emily Jane and sons, Arthur and Jonathan, for the United States. The boys travelled in 1869 and Charlotte and Emily Jane embarked on the ship *Idaho* at Liverpool, travelling saloon class the following year. They arrived in New York on 16 May 1870 and made for the Colorado Territory, where Jonathan had gone the previous year.[14] They became cattle ranchers. Charlotte died in Denver, Colorado, on 16 March, 1906 and is buried at Riverside Cemetery, Denver. Her estate included a marriage dowry of £8000 dating from 1838 (£5000 contributed by Richard Haynes, £3000 from Jonathan Higginson).[15] Her daughter Emily remained unmarried and lived with her mother until Charlotte died. She eventually returned to England where she died in 1928. Arthur died in 1881 and is buried in Riverside Cemetery. Jonathan married three times and has descendants in Colorado. He died on 9 May 1919 in Denver.

John Higginson Haynes (1829–1910)

John Higginson Haynes was born on 3 November 1829 at Everton, Liverpool, and died on Christmas Day 1910 at Bournemouth. He married Inez, born in Winchester 1832. They had two children: Edward [b. 1857] who became deaf as a result of scarlet fever contracted at the age of 12, and Lucy [b.1860], both born in Hursley, Hampshire. Ten years later, the family lived in Egham, Surrey, where Inez died in 1878. John subsequently married Rosalie Somer, a family friend. Rosalie was born in Southampton in 1830. She was staying with the family at the time of the 1861 census. In 1881, John and Rosalie lived at The Rectory, Ashprington, Devon, with Edward and Lucy. All four were still in Devon in 1891, but at Kenton, Tiverton. By 1901 the family consisted of John, Rosalie and 40 year-old Lucy; they had moved to Cavendish Road, Bournemouth. It was to be their last home together. John Higginson Haynes appears to have lived off his investments throughout his life and appeared to have no occupation.

Robert Haynes (1795–1873)

Robert's early life and his first marriage have already been discussed in Chapter 17. By Sarah Ann Payne, he had a son, Robert, who appears to have died in infancy, and two daughters, Sarah Anne and Jane Alleyne. Sarah Ann's monument in St John's Church reads: *'Three helpless infants, pledges of mutual affection the objects of her tenderest anxiety and most watchful care, alike deplore her untimely fate'.*

On Monday, 25 September 1825 Robert married for the second time. His new bride was Elizabeth, the 16-year-old daughter of his father's overseer, Robert Reece and his wife Sarah Knight. In March that year, Sarah Reece and Elizabeth had returned on the ship *Venus*, from Bristol[16] after the completion of Elizabeth's education in England. The marriage was solemnised in Christ Church, Barbados by the Rector, the Revd Dr Thomas Harrison Orderson.[17]

Robert Haynes was elected with his father to serve St John in the Assembly in October 1825 and took his State Oath before the President in Council [John Brathwaite Skeete]. His first duty with the other Assembly Members was to elect his father on the motion of Henry S. Cummins, seconded by Gabriel Jemmett, both members for St Michael. Robert Haynes senior was unanimously called to the Chair as Speaker.[18]

On 22 February 1831 Robert introduced a bill to the House of

Assembly, conferring upon 'coloured' inhabitants the same rights as those of the white inhabitants: to elect or be elected to the Assembly; to be vestrymen, to serve as jurors to try real actions. The proviso was that they must have the same qualifications of age and property holdings stipulated for the whites. The bill became law on 9 June 1831, only four members voting against.

On 23 June 1830, *The Barbadian* announced the departure of Robert Haynes Jr and his family on the *Higginson*, bound for Liverpool. He was accompanied his brother, Richard and his family, William Reece, his wife's kinsman, and the Reverend J. H. Gittens. The purpose of his visit to England was not mentioned. It seems likely that it was also this Robert that left Barbados for a visit to the USA in the Brig *D.H. Miller*, arriving in New York on 14 May 1833. His age was given as 35 and occupation, planter. With emancipation on the horizon, he may have been wondering whether to move his growing family (four by 1833) to England or America as some of his fellow planters had done.

Elizabeth was to have nine children: George Higginson; Elizabeth; William; Henry Higginson; Edmund Lee; Jonathan Wynyard; Caroline Ann; Frederick Hutchinson; Henry. Robert and Elizabeth Haynes were clearly remembering their Liverpool family and business contacts in their choice of names such as Lee and Higginson. Hutchinson, the name given to their second youngest child, was after the clergyman who had married their eldest daughter two and a half years earlier. All except George Higginson were known to have married and had families, several of them large ones.

In 1838 Robert decided to move his family to England, possibly because he feared for their safety after emancipation, possibly because his father and uncles were also set on doing the same. He purchased the Thimbleby Estate and 2600 acres at Osmotherley, Northallerton, Yorkshire, where he spent the rest of his days.[19] Thimbleby had previously been occupied by Richard William Christopher Peirse and his family, which may explain how Robert came to know of the property, for a James Peirse resided in Barbados. Robert and later his son William held the Lordship of the manor of Thimbleby until 1879, together with the advowson of the Rectory of Laceby, Lincolnshire, which latter position was held by successive clergymen who were members of the Haynes wider family.

His arrival at Thimbleby was trumpeted by the 13 October 1838 issue of *The Barbadian*:

'Thimbleby (Yorkshire). Robert Haynes Esq., the gentleman who purchased the estate of Thimbleby of R.W.C. Peirce Esq., made his first visit there on Saturday last.[20] He was met at Thimbleby by his tenants and many respectable tradesmen of North Allerton when his horses were immediately taken from his carriage and he was drawn by the villagers to Thimbleby Lodge amid the hearty cheering of those assembled. Refreshments were distributed. There appeared one wish – that the present proprietor might long live to enjoy this beautiful estate.'[21]

As was indicated in Chapter 17, his father had given Robert the Clifton Estate, prior to leaving Barbados. He also held the Guinea Estate. These he operated as an absentee landlord through his sons until his eldest son, William, inherited them.

According to his great granddaughter, Eleanor Brenda Violet Whittle, Robert was improvident and boasted that he had done with business, and eventually there was very little money left. That seems hard to believe as his son and heir William went on to buy The Hall at Patrick Brompton, North Yorkshire. However, English estate incomes were on the decline, and his massive programme of reconstruction at Thimbleby must have drained the resources of the wealthiest of planters. Family anecdotes relate him recalling small loans. This may have been an age-related foible rather than of necessity, as he saw his time out at Thimbleby and all his family seem to have been well provided for. In the 1851 census Thimbleby boasted eleven servants to look after five family members including Robert's daughter, Sarah Anne, by his first marriage. By 1771 there were seven live-in servants for five adult family members. However, many other staff occupied the Lodge and Home Farm.

Robert died and was buried in Yorkshire. The funeral was covered by the local newspaper. The cortège left the Hall at noon followed by family, friends and tenantry on horseback. The shops remained closed out of respect. It was universally felt that the Parish of Osmotherley had lost an upright man, a kind neighbour and an excellent landlord. There is a memorial tablet in St John's Church, Barbados which reads: *'May we like him depart in peace. In affectionate remembrance of Robert Haynes of Thimbleby Lodge, Yorkshire, who departed this life Feb 17 – 1873, aged 78 years. "Blessed are the dead which die in the Lord, that they may rest from their labours: and their works do follow them".'* It would be wonderful to learn that Robert Haynes-Lovell was at the funeral.

After Robert's death and after leaving Thimbleby in 1879,

THE DIARIST'S FAMILY

Elizabeth changed houses a number of times. In 1881 aged 72 she lived at 86 Bootham, Marygate, St Olave, York, with two maids, (Margaret Hodgson 25 and Jane Walker 24), both from Yorkshire. In the census she was described as a widow, an annuitant, born in Barbados. According to her great granddaughter, Eleanor

Figure 19.1 Tomb of Robert, Elizabeth and William Haynes, Osmotherley, Yorkshire (photograph: Trilby Johnson)

Brenda Violet Whittle, who at the age of seven, was taken to see Elizabeth in York, Elizabeth *'had to live very quietly'*. She was a lovely white-haired old lady who wore *'a lawn cap with streamers, lawn collar and cuffs and a dress made of rich black corded silk that could literally stand alone'*. Eventually she went to live at The Hall, Patrick Brompton, North Yorkshire, to be with her sons William and Jonathan Wynward for several years. She is captured there in the 1891 Census return at the age of 82, *'living on her own means'*.

Elizabeth died of influenza aged 90 at Patrick Brompton, during the Boer War. She was interred with her husband and son, William, and 5-month-old infant, Robert (d. 18 July 1862), the son of Edmund Lee Haynes, in St Peter's churchyard, Osmotherley (Figure 19.1).[22]

The family's arms [Haynes Thimbleby Lodge, Northallerton, Co. York] resemble closely those confirmed on Nicholas Haynes of Hackney in 1578: a stork, wings displayed, proper, in the beak a serpent (snake) of the last. Arms: Quarterly; first and fourth arg. three crescents, paly wavy, gules and azure; second and third, gules, two billets argent.[23] Interestingly, in *A History of Yorkshire North Riding*,[24] the Haynes Arms are given: *'Haynes of Thimbleby, argent*

373

three crescents paly wavy gules and azure', thus omitting the Foxley quartering.

The family's one-time presence in Osmotherley is recalled in the name of the inn at the junction of the A19 and the Thimbleby Road: Haynes Arms.

Children of Robert Haynes and Sarah Ann Payne

Sarah Anne Haynes (1819–1889) and Jane Alleyne Haynes (1821–1890)

Sarah Anne Haynes and Jane Alleyne Haynes, Robert's daughters by his first wife Sarah Ann Payne, remained unmarried and lived with their father and subsequently with their stepmother first in Barbados and then in Yorkshire. Although they made frequent visits to family, particularly in the Clifton area, Thimbleby remained their home where they had lived from the moment their father had bought it, along with their half-siblings as and when they arrived. Subsequently, as the others married and left home, they were left with only Henry, their youngest half-brother, the three being recorded there in the 1871 census. The two sisters had become Thimbleby 'fixtures' and frequent reference to them was made in correspondence addressed to other siblings. After their father's death and the subsequent disposal of the estate, Sarah and Jane moved to Clifton to be near their Aunt Thomasine, where they took lodgings with the Hawkes family at 32 Upper Park Street.

Sarah Anne died in Clifton (Barton Regis) in December 1889 aged 70. Jayne Alleyne returned to Barbados where she died in October 1890 and was buried in Westbury Cemetery.

Children of Robert Haynes and Elizabeth Reece

George Higginson Haynes (1826– ?)

George Higginson was the first child of Robert and Elizabeth Haynes. He was baptised on 26 September 1826 at St John's Church, Barbados. Thereafter nothing is known of him to the writer.

Elizabeth Haynes (1828–1900)

Elizabeth Haynes was born in St John, Barbados in 1828. She was the first daughter of Elizabeth, and the third daughter of Robert Haynes. She was ten years old when the family moved to Osmotherley. As the most prominent family in the village, they attended St Peter's Church every Sunday and where, in her late teens, Elizabeth made a huge impression on the young curate, who had just completed his practical training at Osmotherley in early 1848. The Reverend William Hilton Hutchinson was ordained priest by the Bishop of Chester on Sunday, 20 Feb 1848 and had moved on to a curacy at Weaverham, Cheshire, but kept in touch with the Hayneses by letter, from which it is clear that Elizabeth was on the point of departing for Barbados, possibly to find a husband. He appreciated he was not of her social standing and mentioned to Elizabeth that he *'might first obtain some preferment in the Church more worthy'* of her *'acceptance'*.[25] His proposal was accepted and the couple were married on 19 July 1848 at Osmotherley.

The marriage was announced in due course in *The Barbadian* on 2 September 1848:

> 'Married. At Osmotherley, York, by Revd Joseph Mitton, M.A. Vicar, Rev. William Hilton Hutchinson, of Weaverham, Cheshire, youngest son of the late Teasdale Hutchinson Esq. of Grassfield House, [Pately Bridge] Co. York, to Elizabeth, third daughter of Robert Haynes Esq. jur., of Thimbleby Lodge in same county and Proprietor of Clifton Hall and Guinea Estates in this Island and granddaughter of Robert Reece Esq'.

It is not known whether or not Elizabeth's grandfather, Robert Haynes Sr, or her half-brother, Robert Haynes Lovell attended.

Elizabeth's father, who held the advowson of Laceby in Lincolnshire, arranged for his son-in-law to have the living as soon as a vacancy permitted and, after a year or so at Thimbleby, where Elizabeth had her first baby, Haynes Hanley, in August 1849, the couple stayed at Laceby for 13 years, where they had four more children by 1856 [William Hilton (b. 1851; Henry Teasdale (b. 1852); Elizabeth Reece (b. 1854); Caroline Haynes (b. 1856)]. A sixth child, a son, Frederick William, was born 14 years later in Harrogate, perhaps while the couple were visiting family at Pateley Bridge or Osmotherley. After Laceby, William and Elizabeth moved to the Rectory at Welney, Norfolk, in 1862 where, on 20 September, ten years later,

William had a heart attack and died. He was interred in Welney churchyard.[26]

Elizabeth moved to Scarborough in the house next to her recently bereaved mother, Elizabeth Haynes. When the latter moved to York to be near her son, Edmund, Elizabeth moved to a smaller house on the South Cliff, where she remained for 12 years with her two teenage daughters and infant son.

She went to live with her daughter Caroline and her family (the Whittles) at Beverley where she began to suffer from what appears to have been Parkinson's disease. The end of terrace house in Beverley was crammed with her family portraits and furniture.[27] However, Caroline was herself very ill, and the family was advised to go abroad for her health, so Elizabeth had to return to Scarborough, where she died a paralysed little old lady at 4 West Street, and a stained glass window was put up in her memory behind the font in St Martin's Church.

William Haynes (1830–1889)

William Haynes was born in St John, Barbados on 10 May 1830 and baptised on 19 July that year. Just before his 31st birthday on 18 April 1861 he married 19-year-old Ellen Farrell Gittens at the Parish Church of St Philip. She was the daughter of John Gittens MCP and Ellen Hinds. William became MCP and determined that his future rested with the family plantations. It wasn't therefore until his father died, that he came to live in Yorkshire. As his father's eldest surviving legitimate son, William's arrival from Barbados to take possession of the Thimbleby estates was recorded: '*A deputation of the Tenantry waited upon him, presenting an address, beautifully illuminated, congratulating him upon his accession, to which he replied, and thanking them in a most feeling manner*'.

He held on to the Thimbleby Estate until 1879 when, with falling farm incomes, he sold it to John Storey Barwick JP, whose family held the estate until 1982.[28] He moved to Norwood House, Falsgrave, York, where he and his family resided at the time of the 1881 census. He then moved to the The Hall, Patrick Brompton where he was joined by his brother, Jonathan Wynyard, and then by his mother.

William inherited his father's plantations. Clifton Hall Plantation passed through William's family (Clarence Farrell, William Lyall) until the death of Mr. William [Will] Lyall Haynes in 1976. His heirs then separated the house and four acres of land from the plantation, and in 1979 it was sold.

William and Ellen had ten children: Robert (b. 1862); Ellen Louisa Wynyard (b. 1863); Henry Gittens (b. 1864); Edmund Lyall (b. 1866); Elizabeth Woodfield (b. 1867); William Clarke (b. 1868); Constance Farrell (b. 1869); Clarence Farrell (b. 1870); Charles Herbert (b. 1872); Emily R. (b. 1874).

Their son Robert seems to have departed from the life of a planter or country squire. On 19 July 1879, as *'Robert Haynes, son of William Haynes Esq., of Thimbleby Lodge, Northallerton, Co. Yorkshire'*, he became apprenticed to Charles Arthur Head and Thomas Wrightson both of South Stockton, Yorkshire, to become an ironfounder and engineer. His father was a party to the apprenticeship indenture.[29] In 1881, aged 19, he was lodging with Ann Lamb at 13 Woodland Street, Stockton-on-Tees, where he was working as a clerk in the iron works.

William died in 1889 and was interred with his father in the churchyard of St Peter, Osmotherley. His memorial in St John's Church, Barbados reads:

'Gloria Deo. Sacred to the memory of my beloved father William Haynes, son of Robert Haynes of Thimbleby Lodge, Yorkshire & Clifton Hall, Barbados who died in England January 1889 aged 59. "Father thou art no more on earth but thy goodness is remembered with regard and high esteem by those most dear to thee". Requiescat in pace'.

The inscription on his Osmotherley tomb gives his age correctly at 58 years.

Ellen Haynes had moved to St Martin's, Scarborough by the time of the 1891 census. She was the Head of the house, aged 48, born in Barbados.[30] She would be near her sister-in-law, Elizabeth.

Henry Higginson Haynes (1832–1853)

Henry Higginson was born in 1832 in Barbados. There is a memorial to him or his namesake in St John's Church, Barbados, which is taken to be this man:

'Sacred to the memory of Henry Higginson Haynes sent by his father from England to the Island of his birth for the fulfilment of a duty assigned him, he died, by a sudden providence January 15th 1853. Frankness of character a lively temper and a generous disposition, had won for him the affectionate regard of all his friends. "I must work the works of him that sent me while it is day: the night cometh when no man can work". Aetatis suae XXI.'

The *'sudden providence'* was acute peritonitis (abdominal inflammation) of three days' duration, he being taken ill on Thursday, 13 January, he died three days later on Saturday, at Guinea Estate.[31]

Edmund Lee Haynes (1837–1894)

Edmund Lee was apparently born while his mother was on the journey to England in 1837.[32] He married Susan Thomas Howell, daughter of Benjamin Carleton Howell and Mary Simpson, whose older sister had married his first cousin Joseph Alleyne Haynes. They had eleven children, all except one of whom were born in Barbados where Edmund Lee worked in his father's plantations. The exception was Mary Elizabeth, who was born in 1863 at Thimbleby Lodge and registered at Amotherby, a small village near Malton in North Yorkshire [See UK 1881 census]. Their children were: Emily Lyall (b. 1861), Robert Haynes (b. 1862; d. 1862), Mary Elizabeth (b. 1863), Louisa Howell (b. 1864), Caroline Annie (b. 1865), Robert (b. 1867), Charles Wynyard (b. 1869), twins Florence Agnes and Mary Louisa (b. 1870), Frederica Howell (b. 1873) and Alice Maud (b. 1874). Infant Robert, who died aged 5 months of a diarrhoeal disorder while at Thimbleby on 18 July 1862, was buried in Osmotherley churchyard, where he was subsequently joined by his grandparents.

Edmund Lee Haynes was Speaker of the Barbados House of Assembly in 1870. Some time after his father's death, he retired as a sugar planter and he and his family came to England where they lived in York at 29 St Mary's, Marygate. He was joined there for a time by his mother. In the 1881 census they were recorded at this address with all their surviving children: Emily Lyall (20), Mary Elizabeth (17), Caroline Annie (15), Robert (14), Charles Maynard (11), Florence Agnes (10), Mary Louisa (10), Frederica Howell (8) and Alice Maud (6). He returned to live in Barbados where, according to his obituary,[33] he died as the result of an accident on 25 October, 1894, aged 58.

Jonathan Wynyard Haynes (1839–1914)

Jonathan Wynyard was the first of Robert and Elizabeth Haynes's children to be born in England, a year after their arrival at Thimbleby. He first joined the army in 1855 in the East Yorkshire Regiment. He saw service in India during the Mutiny. He afterwards transferred to the 42nd Highlanders (Black Watch). In order to take service in the Ashanti war of 1873-4 in Gold Coast (now Ghana), he transferred again and served under Sir Garnet Wolseley in the march to Kumasi. He served in the engagements of 27 October 1874, in which he was slightly wounded, and of 3 November near Dunquah (brevet of Major; medal with clasp). In 1878 he was in the 2nd West India Regiment.

Jonathan married an Osmotherley girl, Anne (Annie) Elizabeth Yeoman, daughter of William and Hannah Yeoman. Yeoman was a family long associated with Osmotherley village and their names are recalled on memorial tablets in the tower, nave and chancel of St Peter's Church. The East window, a fine example by Kempe, is a memorial to John Yeoman, 1828–1890. In 1890 the Yeoman family and the Misses Yeoman, were among the principal landowners in the Osmotherley district.[34]

Jonathan, accompanied by his wife, was posted with his regiment to British Guiana where in 1878, his daughter Alice was born. He retired from the army in 1879 with the rank of Lieutenant-Colonel, and for a time resided at Ripon and Patrick Brompton with his mother, where, after the death of his brother, William, both were recorded in the 1891 census. He moved to Sowerby and took an active interest in local affairs, being a churchwarden. He died there and is interred in St Oswald's churchyard.

Jonathan and Annie had four children: Cecilia Elizabeth (b. 1873), William H. (b. 1875), Mary Wynyard (b. 1876) and Alice Yeoman (b. 1878).

Caroline Anne Haynes (1844–1904)

Caroline was born at Thimbleby in 1844. On the 26 September 1861 at the age of 16, she married the Barbadian clergyman, Edmund Hinds Knight, son of Edmund Knight of Barbados, at Osmotherley.[35] He was distantly related to Caroline, as her eldest brother's [William's] mother-in-law, was Ellen Hinds of Barbados and Caroline's grandmother was Sarah Knight. Again, her father came up with the living at Laceby, which Edmund Knight filled. They were still living at The Rectory, Laceby, Lincolnshire, in 1881 after Thimbleby had been sold. Edmund eventually went blind.

The Knights had five children: Edmund Haynes (b. 1863); Caroline Ann (b. 1871); Kathleen Maud (b. 1874); Robert Haynes (b. 1876); Emily (b. 1881).

Frederick Hutchinson Haynes (1851–1927)

Frederick was a cavalryman in 21st Hussars and the Royal Army Pay Department (serving with the Royal Scots Greys and later the 4th Hussars). He served in India [Indian Army cavalry] and Gibraltar, Egypt and elsewhere as Chief Paymaster. His first commission was in 1869 and he rose to the rank of Hon. Major in November 1885. While serving in Barbados he formed a polo team of himself and three nephews. He left the army in 1911 with the rank of Colonel.

Frederick married Kathleen Kavanagh the daughter of Thomas Henry Kavanagh, one of a small number of civilians to be awarded the Victoria Cross. Thomas Kavanagh lived out his final years in Gibraltar, where he died and is buried. His daughter and son-in-law probably saw much of him during Frederick's posting in Gibraltar.

Frederick was a talented violinist and qualified ARCM. He died aged 78 at Cheltenham. Kathleen outlived her husband.

Frederick and Kathleen had two daughters, Kathleen ['Kitty'] and Maud Agnes, and a son, Henry [Harry].

Henry Haynes (1853–c.1931)

Henry, known as Harry, was the youngest child of Robert and Elizabeth Haynes. Born at Thimbleby in 1853, Henry studied land agency under John Coleman at Riccall Hall, Yorkshire, where he was staying at the time of the 1881 census. He became the Agent for Lord Clifton, and lived on the Clifton Estate near Nottingham, next to the Hall. Henry's estate papers (1926–1931) have survived and are now held in the Manuscripts and Special Collections Department of Nottingham University. On 21 April 1892, at the age of 39, he married 25-year-old Annie May Pilkington of Roby Hall Lancashire, the daughter of William Pilkington and Elizabeth Lee Watson. The ceremony at St Bartholomew's Church, Roby, was witnessed by Henry's brother, Jonathan Wynyard Haynes. They had two children, Norman Pilkington [b.1893] and Evylyne Elsie [b.1899]. Norman was recorded as being present at the funeral of Colonel Frederick Hutchinson Haynes, [his great uncle] at Cheltenham in 1927. Evylyne married Sir Ronald Weeks; she died in 1932.

Robert Haynes Lovell (1826–1903)

The reasons for considering Robert Haynes Lovell to be the son of Robert Haynes Jr [of Thimbleby], were set out in Chapter 17. He married Harriet Kendall Inglis on 4 April 1854 at Brydekirk, Dumfries, Scotland.[36] Harriett and Robert met when they both lived with Robert's grandfather in Reading, where they were captured by the 1851 census. Born in Seacombe, Cheshire, the daughter of Peter Inglis and Ellen Gordon, she appears to have been the niece of Anna [Caitcheon] Haynes. Robert was sometimes referred to as Robert Lovell-Haynes. He was remembered handsomely in his grandfather's will. In the 1861 census Robert Haynes Lovell aged 35, described as an annuitant, was living with his 26-year-old wife, Harriet Kendall, born in Seacombe, Cheshire, at Underwood House, Bootle, Lancashire,[37] with their 56-year-old general servant, Nurse Ann Eccles. His occupation in the 1881 census was given as a retired Major of Militia. This was The Haddington Artillery Militia, formed in 1855 and one of 39 brigades attached to the Territorial Army, in this case the Eastern Division. He and his wife lived at Fremington in 1891 and at 166 Boutport Street, Barnstaple, for a time before 1881.

Robert and Harriett had 11 children: Robert Haynes (b. 1857); Mary Lee Haynes (b. 1857); Nellie Haynes; Ellen Gordon Haynes (b. 1859); Annie Douglas K. Haynes (b. 1862); Henry Hunter Haynes (b. 1863); Charles Booker Haynes (b. 1865); Jessie Haynes (b. 1867); Avice Norah Haynes (b. 1869); Edith Haynes (b. 1871); Nina Marianne Haynes (b. 1876). The fact that every child bore Haynes as its last forename reflected Robert's wish to carry forward his parentage in perpetuity. Their son, Robert became a well-known London surgeon whose practice was at 3A Hans Crescent, London SW1. Charles Booker Haynes Lovell moved to Australia in the mid-1880s, where he founded a large Haynes-Lovell dynasty. Most of their daughters married and had issue.

Robert Haynes Lovell died on 15 June 1903 at Newport, Barnstaple, Devon, following a two months' bout of influenza complicated by dementia. His own will is of interest as he had obviously inherited all the plate marked with the Haynes crest, which he passed on to his son, Robert. His estate was valued at almost £18,000.

George Barrow Haynes (1796–1825)

George was born on 16 March 1796, the fifth child of Robert and Ann Thomasine. He was sent to school at Greenock, Scotland in 1805 at the age of nine. George fell overboard while landing at Liverpool in 1805 and his life was saved only by the prompt action of their slave Hamlet, who accompanied them, along with Mr Arthur Oughterson. George Barrow matriculated in 1810.[38]

He married Anne Walker Glieison (daughter of the late Arthur Oughterson) in September 1819 and they had two daughters: Thomasine (b. 1821) and Mary Ann (b. 1823). Mary Ann Haynes married a London postmaster, William Slade, and they lived at 94 Sutherland Farm, London, where they were captured in the 1881 census. Also living with the Slades was Mary Ann's unmarried elder sister, Thomasine. Ten years later, the two sisters, Mary Ann by then a widow, were living at 74 Clyde Bank, Bath. Thomasine died there in June 1897, Mary Ann nine months later.

George Barrow was a prominent businessman in Bridgetown and there are numerous references dealing with his activities in *The Barbadian*. For example, on 13 September 1823 he placed an advertisement for an overseer who was wanted on a sugar plantation on a neighbouring island, *'a person who perfectly understands the management of such a property'*. He was also co-opted to serve on numerous committees, including the one dealing with repairs to the pier head, confirmation of which post-dated his fateful voyage. The Haynes family had set up a shipping business with Thomas Lee of Liverpool and their company Haynes, Lee & Co (known as *Thomas Lee Haynes & Co* in Liverpool) sailed merchant vessels between Barbados and Liverpool. The company's vessels included the brigs: *Amity*, N. Vaughan, Master; *Pandora*, Joseph Corry, Master; *Betty*, John Wedgewood, Master; *Atlantic*, Joseph James, Master; *Highlander*, Helear Vibert, Master; *John and Mary*, Samuel Shaw, Master; *Marques Wellington*, William Procter, Commander; *Martha*, Thomas Dixon or Charles Hays [or Hayne], Master; *Fame*, Matthew Jacobson, Master; *Robert*, William Cannon, Master; and *Thomas*. Other ships bore the Haynes name. Thus, the schooner, *Robert*

Haynes, plied between Nickerie [Dutch Guiana, now Surinam] and Barbados; the brig, *Robert James Haynes* [owned by Bovells & Heyes], was on the Liverpool run. Apart from the post, a few passengers and occasional troops, vessels leaving Barbados carried mainly agricultural products: sugars, molasses, ginger, arrowroot, coffee, etc. Those arriving from England brought grain [generally oats], potatoes, cheeses, hams, salt, hay, beer, coal, bricks, building lime, hoops, soap, glass and crystal, earthenware, pitch-pine and specific freight ordered by individuals. Hardwoods were generally imported from South America. The returns of changes in the slave holdings of plantation owners and others in the British colonies had to be made following the abolition of the slave trade in 1807. Such returns mention *'Haynes, Lee & Co'* and *'Barton, Irlam & Higginson'*, the slaves in question having occupations related to commercial warehousing and port work such as cooper, porter, etc. The company's address in 1825 was 9 Church Lane [74 Church St] Liverpool. The address also housed John Lee & Co. George Barrow Haynes was effectively the Haynes half of the business and made a number of transatlantic crossings. The slaves belonging to the Lee-Haynes partnership are listed in the Barbados slave registers. All were porters who worked in the company store or on the docks. George and his family set sail for Liverpool on 24 July 1824, in the *Martha*, with Thomas Dixon, Master, but the ship appears to have suffered damage and had to return to port for repairs arriving on Friday 30; they set sail again after the damage had been repaired on Monday, 2 August.[39] Their crossing appears to have been uneventful and the *Martha* made several more retun trips over the next 12 months. On Tuesday, 23 August 1825 George set sail with fellow passengers J. Walsh and James F. Thompson, again on the *Martha*, Charles Hays this time being the Master. This was to be George's last journey. The *Martha*, laden with sugar, arrowroot, sweetmeats, molasses, rum and cotton,[40] sank without trace, with the loss of all on board. It is possible that the repairs had not been adequate to keep the vessel seaworthy. Numerous merchantmen, transports and packets had foundered on the south-eastern reef. The loss of the *Martha* was followed on 5 June 1827 by that of HM packet-ship *Cynthia* on the Cobbler's rocks. Although the precise fate of the *Martha* is not known, these disasters contributed to protracted discussions on the need for protective lighthouses. George Barrow Haynes, John Higginson and other Bridgetown merchants were appointed to a sub-committee to assist the Commissioners in superintending the pierhead repairs and improvements in the Mole-Head [area of the port]. On 2 September 1825 the establishment of the sub-committee received the assent of John Braithwaite Skeete, President of His Majesty's Council and Commander in Chief of Barbados in George's absence.[41] An executorial notice by Ann W. Haynes of her claim to the estate of her husband, George Barrow Haynes, was published in *The Barbadian* of 26 May 1826. The same newspaper also published the following notice on 4 July 1826:

'George Barrow Haynes, Merchant, sailed for Liverpool on 23 last August in Ship Martha, which has not been heard of. Partnership Thomas Lee Haynes Co. in Liverpool and Haynes Lee Co. Barbados dissolved so far as said G.B. Haynes is concerned. Signed: Ann W. Haynes Exx., Thomas Lee'.

The firm was in fact carried on by Thomas Lee and Nicholas R. Garner as Lee & Garner in Barbados, and Thos. Lee & Co. in the company's headquarters: 4, Water Street, Liverpool. Thomas Lee, a frequent transatlantic voyager, remained a friend and business associate of Robert Haynes for the rest of the General's life. Lee & Garner owned 112 slaves for whom the company eventually received £2239 compensation. Although Lee & Garner was a comparatively small company, such merchants in effect controlled the sugar market in Britain. They took consignments of sugar and sold it on for further processing. They received commission, brokerage and loan interest from the proceeds of sugar and rum sales and credited the planters with the residue. On the return trip they transported supplies ordered by their client estates and some passengers. They provided their customers with credit and mortgages, held against the estates. When these were obliged to foreclose, they became estate owners. Thomas Lee made a profit from the the quick turnround of its Mangrove Pond and Lightfoots estate holdings. It was entirely logical for the Haynes family to enter into this side of the business and endeavour to reduce the impact of the middleman.

Anne Walker Haynes married again on 13 May 1828 at St Michael's Church, Bridgetown, to become Mrs George Frederick Shepherd, but was widowed for a second time by December 1853, when she was living in Carlisle, Cumberland. In 1861 she and her daughter, Thomasine, had moved to 14 Peterborough Place, Paddington, where she brought up her great nephew, Richard Haynes, the orphaned son of Robert and Emily (Emily had named Anne Shepherd as one of her son's guardians in her will of 16 December 1853). She ended her days in England and was buried along with her in-laws, Richard, Jane, Robert and Emily Haynes, and Thomasine and Frederick Lardy, at Clifton, Somerset.

Thomasine Haynes (1799–1880)

Thomasine was born on 5 February 1799. On 19 May 1816, a month after the serious Easter Sunday slave uprising, Thomasine was married at St John to John Hothersal Pinder, a man more than 25 years her senior. He was born the son of William Pinder of Ashford (Ayshford) Plantation, St John, and his wife, Ann Isabella Hothersal daughter of John Hothersal of Hothersal, or Hogstye, Plantation.[42]

The Hothersals were an old Barbadian family. In 1639 Capt Thomas Hothersall acquired 100 acres of land in Christ Church from the London Merchant House, which had been allocated 10,000 acres (about a tenth of the island). The land was subdivided into large tenanted estates which were purchased and consolidated by individuals who emerged as sugar pioneers by the late 1640s.[43] By 1641 Hothersall had acquired 1000 acres at Pool, Claybury, St John.[44] Likewise, the name Pinder is recorded in the census of Barbados for 1679, when Richard Pinder was mentioned in Colonel Lyne's Regiment of Militia. The Pinders also settled in Christ Church Parish.

Thomasine received an outright gift of £7000 from her father as her marriage settlement. Her husband was a Major in the Barbados Militia. He was also a representative of the Assembly for St John from 1803 to 1808 and at the time of the slave uprising in 1816. For some unrecorded reason he decided to live in England, and, like so many in this narrative, settled in Bath, where his wife had attended school and where his brother, Francis Ford Pinder, and sister-in-law, Elizabeth (Senhouse) Pinder, then lived in Gay Street with their family. While attending divine service at All Saints Chapel, in the Parish of St Swithin, Bath, on Sunday 11 March 1821, John Hothersal Pinder dropped dead. He was 46. A monument to him was erected by Thomasine on the right hand side of the upper gallery of St Swithin's, the Parish Church, Walcot, Bath. She may have had the help in this task of her father when he visited Bath the following year. It reads: *'To the memory of John Hathersall Pinder Esquire, of the Island of*

Barbados in the West Indies, who died suddenly while attending the performance of Divine Service in the Chapel of All Saints in this Parish on Sunday 11th Day of March 1821. This monument is erected by his afflicted widow Thomasin Pinder'. The spelling of Thomasine without the 'e' and 'Hathersall' are of interest. In the same church is a memorial to Francis Ford Pinder, and his second wife, Elizabeth Senhouse. Francis died on 27 January, 1843.

It is likely, although not certain, that Thomasine accompanied her husband to England. She knew Bath well and had family and friends there. Her husband had written his will on 13 January 1821, only two months before he died, which suggests that he knew his health was impaired. He left Thomasine £150 a year for life, on top of his marriage settlement of £600 per annum, together with all his silver except his large two-handled cup, which he left to his brother, the Revd William Lake Pinder, at her death. He had previously sold Ashford Plantation to his friend, William Matson Barrow, in December 1817, but he gave his wife the right to appoint by will £8000 out of £10,000 secured against Ashford. He also left Thomasine his slaves for life, thereafter to the Revd William. There were numerous other bequests to family members and one to his god-daughter, Ann Campbell Higginson, daughter of John Higginson Esq. of Everton, Liverpool. This would appear to have been the sister of Jonathan Higginson who married Thomasine's niece, Charlotte Haynes, daughter of her eldest brother, Richard.

Two and a half years later, on Thursday, 4 December 1823, Thomasine married a soldier, Lieutenant Colonel Christian Frederick Lardy, at St George's Church, Barbados. The marriage ceremony, reported in *The Barbadian* on 6 December, was conducted by the Revd William Lake Pinder, the brother of Thomasine's first husband. Lardy was then an officer in the 4th or King's Own Regiment and was later in Her Majesty's 53rd [Shropshire] Regiment of Foot [Infantry]. He was born in Ceylon [Sri Lanka] in 1795 into a Swiss military family that came to settle in Egham, Surrey, the son of Peter [Pièrre] Lardy [d. Egham 1827] and his wife Elizabeth Sophia. Peter Lardy was formerly an officer of the de Meuron Regiment [eventual rank Lieutenant-Colonel, 1803] along with Lieutenant F. Lardy both of whom were listed on the day it was disbanded in 1816.[45] Christian probably met Thomasine during a tour of duty in Barbados. She and her new husband returned to England, on what must have been her fifth transatlantic crossing, where they settled at 2, West Park, Westbury-on-Trym, Bristol. Thomasine was mentioned in her father's will of 23 December 1841, as Thomasine Haynes, not Lardy. Christian died in the winter of 1856 and his will dated 14 April 1848 was proved on 12 February 1856. Thomasine was the principal beneficiary and executrix, the only other beneficiary being Christian's sister, Louisa Georgiana Stevenson, widow of the Reverend Thomas Stevenson of Winchester. His principal bequests were in the form of shares in the London and South Western Railway and Southwestern Steam Packet Company and his consolidated stock in the Midland Railway. British railways and canals that linked London, the Midlands and the industrial north were the investments favoured by sugar planters and their descendants who had moved their capital away from the West Indies.[46] Thomasine was his residuary beneficiary. Thomasine died in June 1880 in Barton Regis Gloucestershire aged 81 and was interred with her husband at Clifton, Bristol. In the same cemetery rested her sister-in-law Ann Walker Shepherd, wife of George Barrow Haynes, and her brother, Richard Haynes and his wife Jane Alleyne Payne, together with their eldest son, Robert Haynes and his wife Emily.

William Clarke Haynes (1800–1831)

William Clarke was Robert and Ann Thomasine Haynes's fifth son. He was born on the 9 January 1800 at the Newcastle Estate. On 17 March 1823 William married Margaret Ann Critchlow, daughter of Henry Critchlow and Eleanor Ann Butler, of St Michael. The marriage, which took place at Highgate, was solemnised by the Revd William Garnett.[47] The Critchlows were descended from the earliest Haynes to arrive at Barbados through Anne Haynes who married Richard Perryman in 1678. William Clarke Haynes acquired Bannatynes estate in 1830, partly by means of a private mortgage from Sarah Bispham. He also received large monetary gifts from his father, presumably towards this purchase. But his ownership of his family's new home was to be short-lived.

William and Margaret Ann's first child, a son, born in 1824 was named Robert after his grandfather Haynes. Three others followed: Henry Critchlow (b. 1826), named after his maternal grandfather; Anne Thomasin (b. 1828), named after her grandmother Haynes; William Clarke (b. 1831), named after his father. William Clarke Haynes lived just long enough to see the birth of his last son and namesake, as he died in May 1831. Baby William Clarke, who was not baptised until 7 July 1831 at Christ Church, was mentioned in his father's will.

William had fallen ill around April and he signed his will on 1 May. He died soon afterwards, as the will was proved on sixteenth of that month. A few years later his father recorded that he had given his son at various times sums totalling £9500. This was less than the General ultimately gave his other sons, and intent on doing the right thing, he remembered William Clarke's children in his own will.

William's will is deposited in the Barbados Archives.[48] He left his wife his three men-slaves, Sandy, John Francis and Henry. Bannatynes was to be sold and the proceeds divided equally between his wife and four children when they attained the age of 21 years. In the interim the interest on the children's share was to be put to their maintenance, support and education.

The executors were named as his father Robert Haynes, who may have been abroad at the time of his son's death; his brother Richard Haynes; Robert Haynes, and his brother-in-law, Henry Crichlow.

After his death, Bannatynes estate was sold in Chancery in 1832, the purchaser being Robert Reece, the father of William Clarke's sister-in-law, Elizabeth. Sarah Bispham attempted to gain the compensation due for the slaves belonging to Bannatynes, because of her mortgage held on the property during William's lifetime,[49] but she was unsuccessful, apparently because the ownership of a property purchased through Chancery took precedence over any prior mortgage on it.

Robert Haynes (1824–1905)

Robert Haynes was born in 1824 in Christ Church, Barbados. He married Mary Harbourne Straughan on 7 August 1849 and they had six children all born in Barbados: Robert Harbourne (b. 1852); Mary Harbourne (b. 1853); William Lindsay (b. 1855); Helena Harbourne (b. 1859); Eugene Harbourne (b. 1863); Edgar Vere [Harbourne] (b. 1864).

Robert and his siblings received an equal share of the £4000 mortgage that their grandfather, the General, held on Maynard's Estate in the Parish of Christ Church, Barbados under the terms of his will.

In 1861 Robert reported to the Acting Colonial Secretary,

J. Hampden King, on a survey of Barbados taken on the night of 7 April 1861, in which a large number of statistical analyses were made on the demographics of the island's population. In his report he noted the population to comprise 152,727 persons: 10.86 per cent white, 23.66 per cent coloured and 65.48 per cent black. There were 70,070 children (45.88 per cent of the population) under 15 years of age.[50]

Robert died when he was 81 in 1905. In his will dated 8 June 1905,[51] which was proved in December that year, he left his entire estate to his unmarried daughter, Helena Harbourne Haynes and his son, William Lindsay Haynes, his executor. His eldest son, Robert, was still alive at the date his father died, but apparently received nothing.

Henry Critchlow (1826–1854)

Henry was born in 1826. On 21 February, 1850 he married his first cousin, 20-year-old Sarah Dummett Haynes, daughter of Henry Husbands Haynes and Sarah Jane Dummett, at St Michael, Barbados. They lived at The Farm, Barbados. He had shared in his grandfather's bequest of £4000. Henry died on 2 July 1854 aged 28. His monument in St John's Church reads: *'Henry Crichlow Haynes Esqr. Born 1826. Died July 2nd 1854 aged 28 years. Universally regretted for his amiability of disposition. This Tablet is erected by his affectionate Brother'.*

Ann Thomasin (1828–1911)

Born in 1828, Ann Thomasin Haynes was baptised at Christ Church, Barbados, on 23 August that year. She moved to England and lived with her widowed aunt Jayne Alleyne Haynes in Clifton. She died in 1911 aged 83 having never married.

William Clarke (1831–1875)

William Clarke Haynes was born on 7 July 1831 and married Charlotte Maria Dalby at St George's Church, Guelph, Ontario, Canada on 12 August, 1852, by whom he had four children: William Clarke (b. 1853) named after his father and grandfather; Charlotte Maria (b. 1855) named after her mother; Anne Thomasine (b. 1857) named after her aunt; Margaret Anne (b. 1858) named after her grandmother. William Clarke Haynes died aged 45 in Fontabelle, St Michael, Barbados, on 27 May 1875; his wife died in 1908.

Henry Husbands Haynes (1801–1852)

Born on 10 September 1801, Henry Husbands Haynes decided to follow his uncle Henry to a career in the Royal Navy, and on 8 May 1814 he entered as midshipman on HMS *Venerable*, the 74-gun flagship of its new commander Rear-Admiral Sir Philip Charles Henderson Durham (not Dereham as mentioned in the Diary).[52] On 16 and 20 January 1814, the Venerable[53] captured two French frigates, the *Alcmène* and *Iphigénie*, off the Canaries, for which the Admiral and Captain James Andrew Worth of the *Venerable* were decorated.[54] On 8 February 1814 Admiral Durham's squadron was alerted to the presence of the American Warship, USS *Constitution*, by HM brig *Columbine*, to the windward of Barbados. HMS *Venerable* called at Barbados at the end of December and remained there until January 1815, so presumably Henry was reunited with his family. The *Venerable* again set in chase of American privateers to the windward of Barbados on 13 February 1815. On May 26 the vessel appeared off Falmouth[55] accompanied by a frigate. After telegraphing HMS *Scamander*, laying in Carrick Roads, the latter proceeded to join them. Their destination was said to be Lisbon. In fact the *Venerable* crossed the Atlantic again, where on 31 July 1815, in Carlisle Bay, it received on board the Governor, Sir James Leith, and a large number of troops, and then weighed anchor and sailed to Guadaloupe, which the British forces restored to the French Emperor, following Napoleon's final defeat at Waterloo. So Midshipman Haynes would seem to have had a very exciting introduction to naval life!

Henry married Sarah Jane Dummett, daughter of Dr Edward James Dummett M.D., and Rebecca McGrath in St Michael's Cathedral, Bridgetown, on 28 June 1825[56] and they had 12 children: Ann Thomasin (b. 1826); Robert (b. 1829); Sara Dummett (b. 1830); Robert James (b. 1832); Henry Husbands (b. 1834); John Torrance (b. 1836); Christina Lardy (b. 1839); Mary Frederica (b. 1841); George Barrow (b. 1842); Elizabeth Reece (b. 1844); William Clarke (1847); Joshua Rowley (b. 1850). Sarah Jane Dummett was the sister of Joshua Rowley Dummett who married Anna Cave Lovell. She and Henry named their last child after her brother.

In 1834 Henry is listed as a 35 year-old farmer, sailing from Barbados to Virginia in the brig *Amulet*, which arrived in Alexandria, VA, on 9 September. His father recorded the transfer of the Bath Plantation to him on 10 February 1835; he also owned the Bush Hall plantation.

He died on 14 August 1852. His memorial in St John's Church, Barbados, reads: *'Sacred to the memory of Henry Husbands Haynes Esqre sixth Son of Robert and Ann Thomasine Haynes of this Island, who died August 14th 1852 aged 51 years. This tablet is erected to his memory by his affectionate Widow and Children. 'Our days on the earth are as a shadow and there is none abiding'. 1. Chron. 29. 13'.*

Ann Thomasin

Ann Thomasin was born in 1826 and baptised on 27 May that year at St Michael, Barbados. Nothing more is known of her.

Robert

Robert was baptised on 3 January 1829 and is believed to have died in infancy.

Sara Dummett

Sarah, the eldest daughter of Henry Husbands Haynes, was born in 1830 at St Michael, Barbados. She married her first cousin Henry Crichlow Haynes on 21 February 1850 in St Michael, Barbados. They lived at The Farm, St Michael. Henry Crichlow died on 2 July 1854 and his widow left for England, where she settled in Bath near to her Uncle Joshua Rowley Dummett in Bathwick, Bath.

After almost three years she married, secondly, a 49-year-old widower, Robert William Elton on 10 March 1857 in Bathwick Church, Bath. The marriage was reported in *The Barbadian* on 8 April. The Reverend Caddle Holder, the brother of the groom's first wife, Ashley Holder, the Vicar of Avenbury, Herefordshire, was assisted by the Rector of Bathwick. The groom, a Lt- Colonel in the Bengal Army, was the second son of Jacob Elton of Little Burstead, Essex. Sara Elton appears as a widow of 5 Nelson Place, Bath in the

1861 census where she was being visited by her mother. She married for a third time at Marylebone, London in 1875. Her husband was John William Christian, a Mancunian wholesale goldsmith. The 1881 census gives her address as 28 Somerleyton Rd., Lambeth; John William's age as 31 and hers as 48, when she was actually 50 years old.

Robert James

Robert James was born at the Bath Estate, Barbados on 28 June 1832. He married Elizabeth Stewart Gill on 29 May 1854 in St. John. She was born at St John in 1830. Robert James Haynes was admitted pensioner at Caius College, Cambridge on 27 June 1851. He was educated at Summer Hill School, St George, Bristol with four others from Barbados [1851 Census] and at King's Bruton School, Somerset. He matriculated in 1851, was an Exhibitioner, and gained his BA degree in 1855 and MA in 1859. He clearly interrupted his studies in order to return home to be married. While at Caius he witnessed the will of Emily Anne Haynes, the widow of his first cousin, Robert, on 16 December 1853.

He was ordained Deacon at Peterborough Cathedral in 1855, became priest at Gloucester on 30 May 1858, was appointed Rector of Ashton, near Rotherham and was Canon at St George's, Bloomsbury, from 1860 to 1863. After the death of his first wife, and without issue, he married Annabel Selina Gambier, the daughter of William Gambier and Annabella Frances Garth-Colleton, at Dover in 1859. They had five children. Robert James died about 1899 at 22, Overcliff, Gravesend, Kent.

Henry Husbands

Henry was born at the Bath Estate in 1834. He appears in the Bruton Register as attending King's School; he is mentioned there as the former proprietor of Senior's Estate, Barbadoes.[57] He was a first-class cricketer, having played (batting and fielding) on one occasion for Barbados internationally in the 1864/65 season when he scored six runs. He was Chief Constable of the Welsh Police [Head of Wales Police] according to the 1881 census return, when he was staying with his widowed sister, Elizabeth Reece Beckles and his sister, Mary Frederica, at 71 Shirland Rd, London. He died in Bridgetown, Barbados, in February, 1892.

John Torrance

John was born in 1836. He was a sugar planter. He married Catherine Garrett Reece, thus firming the long-held links between the two families, and they had eight children most of whom married and had issue. He died in Barbados.

Christina Lardy

Christina was named after her uncle, Christian Frederick Lardy, the husband of Aunt Thomasine, at her baptism in January 1839 at St Michael, Barbados. She married Manoel da Silva Ferreira on 6 January 1859 in St James's Church, Paddington, London and the couple moved to Rio de Janiero, where Christina died on 11 October 1913.

Mary Frederica

Mary Frederica was born on 30 September 1840 at Spooner's Hill, St Michael, Barbados. She was unmarried at the time of the 1881 UK census when she was recorded as living with her widowed sister Elizabeth Beckles and brother, Henry Husband Haynes. She lived at The Ferns, Strand on the Green, Chiswick, and was involved in the development of the Haynes family genealogy with Andrew Mack Haines, an American kinsman.[58] It was she who acquired Lieutenant-General Robert Haynes's original notes, which she copied and then returned the original to a friend of the general's second wife, Anna Caitcheon. Their whereabouts are unknown. Mary Frederica died aged 59 at Brentford, Middlesex on 15 November 1899.

George Barrow

George Barrow was born in 1842 and baptised at St John's Church, Barbados on 25 October that year. On moving to England he lived first in Hackney, then with his mother and sister Mary, at 16 Cavendish Road, Marylebone, where he met and in 1875 married Agnes Louise Punch [b. 1857, Barbados]. Agnes may have already died by the time of the 1881 Census when George was captured living with his mother-in-law Agnes S. Punch, a widow aged 44, at Abbey Road, London. George was described as a merchant in East India, as was his 28 year-old brother-in-law, John J. Punch. In late 1881, George married his cousin Ellen Gordon Haynes Lovell at Barnstaple, Devon. Ten years later, still married, he was boarding in a lodging house in Prittlewell, Essex. He died in Brighton, Sussex, in 1906.

Elizabeth Reece

Elizabeth Reece was born in 1844 and was married at the age of 16 to Edward Hyndman Beckles who was born in 1838 at the Bay Estate in St Michael, Barbados, the son of Bishop Edward Beckles and Margaret Walcott. The marriage took place on 14 July 1862 at All Souls, St Marylebone, London. Edward Hyndman Beckles was educated at London House, London and also at Morden Hall, Surrey. He served in the Commissariat and HM Civil Service, Control Staff in Sierra Leone, where he died on 23 July 1872, his thirty-fourth birthday. Elizabeth and Edward had three children, Edward Hyndman, who lived for only 10 weeks in 1863, Elizabeth Reece Beckles born 1864 in Barbados and Edward Hyndman Haynes Hunte Beckles born 1867 in England. In 1881 she was already widowed, living at 71 Shirland Rd, London, and is recorded together with her sister, Mary Frederica, and brother, Henry Husbands. All of their ages as appearing in the census were less than their actual ages by six years. Her surviving children both married and had issue. Elizabeth Reece Beckles eventually returned to Barbados where she died on 21 February 1920.

William Clarke

William Clarke Haynes was born in 1847. He joined the Royal Navy and rose to the rank of lieutenant. He married Barbara Strachan of Turriff, Aberdeenshire in 1870 and they lived first in Bayswater, London, then in Bray, Berkshire, and subsequently in Old Machar, Aberdeen. They had five children. William Clarke died in 1893.

Joshua Rowley

Joshua Rowley Haynes was baptised on 28 August 1850 at St Michael, Barbados. He was named after his maternal uncle, Joshua Rowley Dummett, who married Anna Cave Lovell and lived in Bath, England. Joshua was in Bath, England at the age of eleven and visited England again at the age of nineteen *en route* for Sydney, Australia. He was one of only three passengers that embarked on the 937-ton clipper, *Walter Hood* of the White Star Line, which left London on 22 January 1870. The vessel encountered a south-easterly gale on 24

THE DIARIST'S FAMILY

April off the coast of New South Wales and, having lost many sails and booms, at 7.30 p.m. on Monday 25 April the mountainous seas drove the vessel onto a reef 200m off the shore. Eleven souls were lost including Joshua, who drowned on Wednesday 27 April as he attempted to swim for the shore. At the inquest held on 30 April Joshua was identified as a native of Barbadoes in the West Indies who had been living for a time in London.[59] His body was interred in a common grave with the others washed ashore. A memorial to those lost is to be found at Cudmirrah Nature Reserve, Wreck Bay, New South Wales.

Notes and references

1. RB260/222.
2. E.M.W. Cracknell, ed. (1934), *The Barbadian Diary of Gen. Robert Haynes 1787-1836*, The Azania Press, Medstead, Hampshire, UK.
3. Kathleen M. Butler, 1995, *The Economics of Emancipation: Jamaica and Barbados, 1823-1843*, University of North Carolina Press, p. 51 for references.
4. B. Burke, (1855), 'A Visitation of the Seats of Arms of the Noblemen and Gentlemen of Great Britain', vol. 1, p. 31, Hurst and Blackett; the marriage was reported in *The Patrician*, vol. 3, 1847, p. 599. The Valentine H. Mairis Papers are in a collection held by Georgetown University: http://library.georgetown.edu/dept/speccoll/cl107.htm.
5. All three of Emily's younger sisters died at Clifton between 1847 and 1853, one of whom, Maria Adelaide dying on 3 November 1853, only six weeks before Emily.
6. Barbados Slave Register, 1823.
7. Butler, 1995, Table 3.2.
8. Joe's River; Foul Bay; Rowans; Congo Road; Sandy Lane; Cane Garden; Castle Grant.
9. Butler, 1995, pp. 82-3.
10. David Morier Evans, 1849, *The Commercial Crisis of 1847-1848*, London: Letts Son, and Steer, pp. 81, 92.
11. William B. Dana, 1848, 'Advances Upon Bills of Lading', in *Merchants' Magazine and Commercial Review*, pp. 631-2.
12. The Robinsons were an old Barbadian family. Henry Higginson was the grandson of Dr Robert Robinson who may have assisted Robert Haynes when he was unwell in England in 1809. He became a barrister and emigrated to Australia where he was involved in the sugar industry in Queensland. His descendants still own sugar plantations there. He had twin brothers, Robert Alleyne and Haynes Sparrow Robinson. It is

apparent from the interchange of names how closely the Haynes, Higginson and Robinson families were connected.
13. Archives of the Church of Southern Africa, deposited at the University of the Witwatersrand; Lambeth Palace; United Society for the Propagation of the Gospel; The National Archives.
14. 'Declarations of Intention, Territory of Colorado, First Judicial District, Arapahoe County', in *The Colorado Genealogist* [1984], vol. 45, pp. 3-11, 60-62; 250-252; [1985] vol. 46, pp. 39-63, 105-118, 139-152; [1986] vol. 47, 22-49.
15. Personal Communication from Grant Houston, descendant.
16. *The Barbadian*, Tuesday 22 March p. 2; See also note 27.
17. *The Barbadian* 25 September 1825.
18. *The Barbadian*, Tuesday 29 November 1825, p. 2.
19. Thimbleby Lodge, the main estate house, is a large Georgian building on two floors with a columned portico surmounted by two lions. It has an impressive entrance hall with five reception rooms. The house is set in 63 acres of parkland. It was remodelled by the removal of one floor in the early 20th century. The grouse moor extended to 1334 acres and the Home Farm with pheasant shoot to 1231 acres (*Country Life*, 2 June 2005, advertisement).
20. Thimbleby was bought by Richard Pierce Esq., of Hutton Bonville, in 1694, remaining in his family until purchased by Robert Haynes in 1838; Bulmers 1890, Osmotherley Parish.
21. The Thimbleby Estate Deeds (3) for 1838 and 1862 are lodged with the North Yorkshire County Record Office, Cat Ref: TD 83 [MIC 2568]; see also, *A History of Yorkshire North Riding*, Vol. 1, 'Osmotherley', p. 437.
22. Infant Robert was Robert and Elizabeth's grandson, son of Edmund Lee and Susan Haynes, who was born in Barbados and baptised at St John in March 1862. It is known that Edmund and Susan were in England that year as their next child, Mary Elizabeth, was born at Thimbleby in 1863.
23. Sir Bernard Burke, 1884, *The General Armory of England, Scotland, Ireland and Wales*.
24. *A History of Yorkshire North Riding*, Vol. 1, 'Osmotherley', p. 437.
25. Letter dated 1 April 1848, Weaverham, in the possession of the writer's grandson, the late Peter Hanley Hutchinson.
26. Frederick William Hutchinson: Reminiscences of a Lincolnshire Parson. Published privately.
27. Brenda [Eleanor Brenda Violet Johnston née Whittle], *Memories and Confessions of a Grandmother*, published privately, 1949.
28. Bulmers informs that the estate was still in the possession of the executors of William Haynes in 1890; *'Thimbleby Lodge now let for a term of years to a gentleman as a shooting box'*; Thimbleby remains in private ownership.
29. Teesside Archives: U.HW/2/48, 21 July, 1879.
30. UK Census 1891: RG12/3967, f. 23, p. 39.
31. 'Obituary', *Journal of the Barbados Museum & Historical Society*, Vols 20-21, p. 97.

32. UK Census 1881.
33. *Yorkshire Gazette,* 24 November 1894, p. 6.
34. 'Bulmers' 1890, Osmotherley Parish.
35. *The Gentleman's Magazine,* November 1861, pp. 558-9.
36. He used the name Robert Haynes when he married.
37. UK Census 1861: Barbados Department of Archives: RG9/3955, f. 8, p. 18.
38. *Caribbeana,* Vol. IV, p. 80.
39. *The Barbadian,* Tuesday, 3 August 1824.
40. *The Barbadian,* Friday, 26 August 1825.
41. *The Barbadian,* Friday, 9 September 1825, p. 2
42. J. C. Brandow, 1983, *Genealogies of Barbados Families from Caribbeana and the Journal of the Barbados Museum and Historical Society,* Genealogical Publishing Co. Inc., p. 447.
43. Hilary McD. Beckles, 1989, *White servitude and black slavery in Barbados,* Univ. Tennessee Press, Knoxville.
44. *Journal of the Barbados Museum and Historical Society,* 1983, vol. 37 p. 233.
45. De Meuron's Regiment was a Swiss regiment, formed locally in Neuchâtel in 1781 for service with the Dutch East India Company. During its time with the Dutch, it was stationed in South Africa and Ceylon as well as serving as marines aboard the French fleet of Admiral Suffren. In 1794 the Dutch Governor, Johan van Angelbeek, permitted a detachment of troops under Captain Peter Lardy of the regiment de Meuron to be sent from Ceylon to Negapatam to assist the English against the French in Mauritius and Bourbon [Archives of the Dutch Central Government of Coastal Ceylon, 1640-1876, 3173] and in 1795 [the year of Christian Lardy's birth] it made a secret contract with the British and served in India until 1806. Governor van Angelbeek permitted the regiment to leave for India on condition that it took no part in the siege of Colombo. During the Third Mysore War of 1799 it served under Arthur Wellesley [later Duke of Wellington] at the siege of Seringapatam on 4 May, where Peter Lardy was wounded in the left arm. On its return to Europe, it saw garrison duty in various posts in England and the Mediterranean until 1813, when it was shipped to North America for the War of 1812. Here it saw more garrison duty in Lower Canada with the only action being at Plattsburg in September 1814 where it was assigned the rearguard for the withdrawal of the British. In 1816, its contract expired and the regiment was disbanded. The soldiers were given either land grants in Canada or transportation back to Europe. Peter Lardy returned to England where he and his wife settled in Egham, where he was to write his will on 4 July 1816.
46. Butler, 1995, p. 71.
47. *The Barbadian,* 19 March 1823, p. 2.
48. Will of William Clarke Haynes 1 May 1831; proved 16 May 1831. Barbados Archives Department.
49. Butler, 1995, pp 99-100.
50. State Papers, Session 5 February 1863-28, July 1863, Vol. XXXIX, 1863, p. 32.

51. Will of Robert Haynes 8 June 1905; personal communication Angela Lloyd-Roberts, UK; Barbados Department of Archives: Wills, Vol. 95, p. 419.
52. Until December the previous year the ship had been under the command of Admiral Sir Home Popham RN, but on 23 December 1813, HMS *Venerable*, under Admiral Durham, is recorded as bound for the Leeward Islands. Admiral Durham is commemorated in Largo Parish Church, Fife
53. Captain James Andrew Worth, bearing the flag of Rear-Admiral Philip Durham, on his way to take chief command of the Leeward Isles. The encounter was the subject of a painting by Thomas Whitcombe (1763-1824): *'H.M.S. 'Venerable' Engaging the French Frigate 'Alcmène' Off the Canary Islands on 16th January 1814, with Another French Frigate 'Iphigenie Bearing Away to Escape the Fight'*.
54. The Navy List, 1850, p. 300, Publ., The Admiralty; *James' Naval History*, 'Epitomised in one volume by Robert O'Byrne', by William James, Adamany Media Corporation, p. 485.
55. Admiral Durham remained in the Leewards on HMS *Barossa*.
56. According to the report in *The Barbadian* of 28 June 1825, Henry married Mary Dummett, but this was followed by a correction.
57. Bruton Register, 1826-1890, Bruton (England) King's School, Thomas Augustus Strong, 1892, H. Frowde.
58. E. C. Haynes, A. P. Haynes and E.S.P. Haynes, 1910, *Notes by General Robert Haynes of New Castle and Clifton Hall Plantations, Barbados and Other Documents of Family Interest*, Argus, London, pp. 24-8.
59. *The Australian Town and Country Journal,* 14 May 1870; for the aftermath see: 'Six Weeks at Wreck Bay', *The Sydney Mail*, Saturday 20 August 1870; for full details and summary of contemporary accounts see: http://www.ozhistorymine.com/html/walter_hood.html.

Epilogue – A Family of High Respectability

'A celebrated people lose dignity upon a closer view' – Napoleon Bonaparte

This then is the story of the family that Lieutenant-General Robert Haynes considered to be one of high respectability and ancient lineage. It was of course no more ancient than any other family, but one cannot fail to be impressed by how much of the family's folklore on the Tudor and Stuart Hayneses can be corroborated by documents still in existence; it is equally impressive for a family whose descendants left so many footprints in England to have left so little evidence to link the same family with the Barbadian migrants. It is difficult to gauge how much of what the General knew of this link was based on knowledge passed on from generation to generation on the island, or conceivably derived as a result of genealogical research undertaken by his father or other members of the wider family, during trips to England in the eighteenth century. Correspondence with the College of Arms or other depositories, for example, would have revealed much of what the General mentioned, including the 1578 coat of arms of Nicholas and William; its existence was certainly known to the young diarist as early as 1787, five years after his father's first journey to England. However, as with most family lore there were inevitably some embellishments such as the Foxleys *living* at Foxley Grange or the Hayneses *owning* Haines Hill; knowledge of the existence of the former could only have been by someone familiar with or having toured the Bray area of Berkshire, and the latter might have been assumed by someone familiar with the last will of Richard Haynes of Wargrave. Arms based on those of Nicholas but with less detail and an esquire's helm added (Nicholas was not styled 'esquire'), were subsequently used by the

General and some of his descendants, but their earliest use in Barbados is difficult to ascertain.

Figure 20.1 Arms of Haynes of Barbados in the twentieth century taken from a bookplate

Tantalising questions remain as to whether the Barbadian Hayneses were indeed descended from Nicholas of Hackney, his brother, William, their kinsmen Henry of Hackney or John of Hoxton, or from an entirely different family such as that of Nicholas of Winterbourne, near Salisbury (Ch. 13, note 7). As yet there is no *proof* that there were male descendants of Nicholas of Hackney other than his son Richard, living at the time of the settlement of the island, although there is circumstantial evidence for a Robert Haynes in the East London area. If the Robert Haynes and his family on Barbados in the 1640s really were descendants of the Hayneses of Reading, the notion of this being a family of 'high respectability' is difficult to square with what we now know of the allegations of corruption, misappropriation and fraud, which ultimately led to the ignominious departure of John and William from royal service. The discovery of one's ancestors being in Queen Elizabeth's employ is understandably a cause for satisfaction, but service at their 'below stairs' levels cannot be taken as a criterion of respectability *per se* – far from it. The allegations against Christopher included embezzlement, smuggling and piracy, and he narrowly escaped arrest and worse over his involvement with Shelley and the escape of Arundell and Paget to France. Nicholas, the only brother with children that we can be certain survived him, seems to have kept his job and presumably his reputation in the Royal Household largely intact, although he was no better than other purveyors in delaying payment for his purchases, and he died before the spate of dismissals of the mid-1590s. Perhaps all this self-serving behaviour was the norm in Elizabethan England, with the Hayneses merely taking leaves from their superiors' books, but given the changes in legislation to remedy and prevent the corruption of royal purveyors, possibly not! Yet as I write, the newspapers are daily brimming with scandals of Honourable and Right Honourable members of the British parliament making claims for apparently unsupportable or non-existent expenses, so it would appear that obtaining money by deception and theft from the state continue to be attractive to its servants suggesting that human behaviour hasn't really changed over the past four centuries.

Turning to the move to Barbados, what sort of people were these early colonisers? In the case of the Hayneses, they, like many other early settlers, are generally assumed to have been the sons of minor gentry; Robert was certainly styled *'gentleman'*. Above all they were English and free men with financial resources – an important distinction (the Scots and Irish migrants being mainly indentured servants). The first colonial settlers, it was said, lacked morals and virtue, were ruthless and unprincipled and it was *'the desire of enriching themselves by any means that led them to distant climes'*.[1] In Barbados they apparently soon became dissolute in their manners and profligate in conduct. Some of the earliest accounts of the planters on Barbados give them as quarrelsome, idle and heavy drinkers.[2] Barbados was described by Henry Whistler in 1655[3] as England's dung-hill, a dumping-ground for '*its rubidge: rogues and whores and suchlike*', the latter of whom '*if handsome*' made a wife for some rich planter.

Nothing is known of the family's royalist credentials, but there must be doubt that this was entirely the reason for the very earliest Hayneses migrating there, as they were on the island and trading by 1641, years before Cromwell became Lord Protector. It is possible that they took the royalist side at the time that Cromwell sent General George Ayscue to secure the island,[4] but they don't appear to have left any documentary evidence to this effect, and a royalist claim for one's ancestors was highly fashionable in the eighteenth and nineteenth centuries. It would be an entirely different matter if they were of the family of the royalist John Haynes, *'barbadoed'* in 1656, and released in 1660.

The early acquisition of wealth from sugar was the foundation on which the family's reputation and status would develop. By the end of the eighteenth century and no longer the nouveaux riches, the Hayneses amassed the visible trappings of wealth and success: table silver, fine porcelain, furniture, portraits, huge quantities of the finest wines and most important of all, the symbols of lineage, for it was then that the use of the coat of arms adopted by Nicholas and William in 1578 really became evident. They took to Freemasonry: they enjoyed the dynastic trappings of power within the legislature: their children were sent to Britain to be educated: they remained staunch royalists: they were honourable, correct, decent, reputable, industrious, estimable – and respectable – highly respectable – as the diarist himself claimed; the paradox of the means whereby all this success had been acquired failing to register in the colonial conscience.

The clamour for emancipation in the first decade of the nineteenth century led to new challenges and fears. The slave uprising of 1816 did much to stoke the belief that the old order would disappear, and yet respectability still did not wholeheartedly embrace propriety, but the General's responsible behaviour towards his illegitimate grandson would reveal a decency born of duty and fairness that had been recognised by his peers in other aspects of his life.

Emancipation signalled it was time for many of the family to leave the island while their wealth, including the generous slave compensation, permitted, but they kept the plantations in the family until they saw how the dust would settle on the new Barbados. In England, country estates, large city houses and railway stock were acquired and plantation wealth was gradually diluted amongst the children of large Victorian families. The census returns repeatedly give the occupations of the family's descendants as, fundholders, annuitants, and *'living off own means'*. However, public schooling and university graduation soon led to the emergence of distinguished doctors, surgeons, lawyers and writers, with their sanitised island past echoing from time to time in the confines of the drawing room. Is it possible for any of us, even now, easily to face the past realities of respectability, in case perhaps more than we realise is transferred in our genes?

And what happened to ... ?

Newcastle, in the Parish of St John, Barbados, was the Haynes family seat from the late 1640s. Built of coral, the large rectangular plantation house withstood the great hurricanes of 1731, 1780, 1831 and 1898. It was handed from father to son in an unbroken line through more than three centuries until Joseph (Joss) Alleyne Haynes sold the plantation lands in 1960. The last descendant to live in the Newcastle plantation house was Mary Haynes who died in 1954. After changing hands several times the magnificent house became a government children's home, which was closed in 1985.[5] In 2004 it stood derelict, stripped of the verandahs and all its contents, reclaimed by the wilderness, and beyond restoration; a silent metaphor of a family of high respectability.

Figure 20.2 Newcastle Great House St John, Barbados 2004

Notes and references

1. R. H. Schomburgk, 1848, *The History of Barbados*, Longman, Brown, Green & Longmans, p. 93.
2. Sir Henry Colt, 1631, quoted in R.S. Dunn, 1972, *Sugar and Slaves: the Rise of the Planter Class in the English West Indies*, University of North Carolina Press.

3. Ibid.
4. E.M.W. Cracknell, ed. (1934), *The Barbadian Diary of Gen. Robert Haynes 1787-1836*, The Azania Press, Medstead, Hampshire, UK.
5. H. Fraser and R. Hughes, 1986, *Historic Houses of Barbados*, The Barbados National Trust and Art Heritage Publications, p. 32.

Index

A'Beare, William 35
Abercrombie, General Sir Ralph 295, 333
Aberystwyth, Wales 366
   death of Robert Haynes at 366
absentee landlords 298, 303, 334, 344, 346, 372
Abshawe, John 53
Abston 182
Acatry, the [achatry, acatery, chaterie, catre, catery, le kathery] 39, 41, 42, 43, 45, 46, 48, 50, 52, 58, 59, 61, 65, 66, 72, 131, 140, 145, 163, 164, 165, 174
   abolition of 66
   clerk of 164
   household reforms, January 1779 66
   pastures of at Tottenham, Sayes Court and Creslow 41, 42, 45, 59
   Sergeant [Sargeant] of 39, 41, 42, 43, 45, 46, 48, 50, 52, 58, 61, 65, 131, 145, 163, 165,
   work of 42, 45, 165, 174
   yearly food consumption 165
Adam, William 112
Adams, Lawrence 86
Adamson, Elvira Haynes, daughter of Dr William Adamson 280
Adamson, John 293
Adamson, Richard, plantation of 232, 233
Adamson, Dr William, husband of Mary Elvira Haynes 280
Addy, Henry, Commissioner of the Exchequer 53
Addye, Ellen, formerly Mrs Arthur Kay 213; *see also* Welshe, Ellin
A Dene, William 8
Admiral, Lord High 80, 83, 133
   after pasture 43, 44
Ailleward [Hailleward, Ailward, Aylward, Aylleward] family 24, 25, 26
   connection with More 24
Ailleward, Joan 24, 25
Ailleward, Nicholas 26
Ailleward, William 25
Albany, New York 261
   Peter Schuyler, mayor of 261

Alcmène, frigate 390, 398
Aldercarr 123, 202
Alderne, Captain Thomas 248
Aldworth, Andrew 215
Alexandria, Virginia 391
Allamby, Mary, wife of Stede Bonnet 292
Allen, Thomas 203
Alleyne, Hon. James H. 321
Alleyne, Sir John Gay 285
Alleyne, Hon. Sir. R.A., Bart. 337
Allington Castle 112
Alsop, Joseph 263
Alton, Hampshire 83
Alverstoke, Hampshire 366
America, southern states of 344
American Revolution 283, 291
American timber imports, embargo on 328
Amity 382
Amotherby, Yorkshire 378
Amulet, brig 391
Anderson, Mr Justice 107
Andros, Governor Sir Edmund 257
Angell, William, purveyor of sea-fish 59, 82, 145, 152
Angola, slaves from 355
Anne of Cleves 158
Anne (Stuart), accession of 272
Anthony, Viscount Montague 56, 71
Antigua 283, 356-357
anti-slavery campaign/movement 290, 312
Antwerp 83
Anyon, Thomas, President of Christ Church College, Oxford 199
Appleton, William, Town Clerk of Rye 141, 142
apprenticeship(s) 13, 135, 163,
   indentures 13, 377
Arborfield 25
Archer, Mary, daughter of Nicholas Enos 245
Argosy Newspaper, Demerara 360
Armiger 162
arms, coat of xiv, 19, 172-182
   earliest use by Barbadian Haynes family 402
   dormant 27, 119
      modification and reassignment of 27, 119

405

for sale by heralds 20, 117, 118
Foxley 23
Haynes xiv, 19, 172-182, 203, 228, 399
   brisures 173-182
   crest 175, 178
   motto 63, 174, 181, 183, 354
   torse 175, 178
arms and crest, exemplification of xvi, 3, 9, 19, 20, 21, 22, 117, 118, 153, 175, 176, 178, 180
Arran, a slave 256
Arrion, Henry 245
Arscott, Arthur, husband of Sarah Pooler 267
Arthur's Close, *see* Carr's Close
Arundel, 14, 74-108, 135, 147, 149, 153, 161, 167, 168, 173, 188, 189, 198, 204, 210, 215, 219, 221, 222
   Borough Minute Book 76-77, 102, 103
   Borough Seal 76
   Bridgewarden's Accounts 76, 102, 106
   Burgess chest 76
   Burgess(es) 76, 77, 78, 83, 99, 102, 104
   Brooks 76
   Castle 82, 168
   Corporation 102, 222
   Crown House 76
   Council 75-77, 82, 95, 103
   Custom House 78, 80, 81, 88, 95
   customer 74, 77, 78, 79, 80, 91, 95, 97, 102, 107, 138, 161
   Earl of 52-53, 76, 77, 80, 85, 89, 93, 95, 99, 104, 221
   exports from 75, 81, 108
   George Inn, 99, 104
   Haven 86, 91
   High Street, 76, 99, 104
   Little Park 104,
   market 81
   Mayor of 74, 76-77, 88, 91, 93, 94, 102, 104, 222
      appointment of 102
   Rape of 83, 102-103, 108
   searcher 78, 80, 92, 94, 97, 106
   Town House 76
Arundell, Charles 85, 87, 89, 91, 93, 94, 95, 96, 400
   death in Paris 96
   flight to France 85, 87, 89, 94
   guilty of treason 96
Arundell, Robert 246
Ashanti War 379
Ashby, Captain William, of the *Mary Ann* 352
Ashford Estate, 284, 287, 290, 385, 386
Ashprington, Devon 370
Ashton, Richard 231
Ashton, Yorkshire 392
Ashurst, John 233

Assinegoes 231-2
Atlantic 382
Attorney 27, 77, 78, 199, 298, 303, 316, 328, 334, 335, 345, 350, 358
   Robert Haynes as 298, 303, 316, 345, 358
Austen, Lawrence, Yeoman of the Larder 163
Australia 344, 381, 394, 395
Avon, Earl of [Sir Anthony Eden] 363
Awdyence, Richard, 71
Ayscue, General George 401
Ayshford Plantation, St John, *see* Ashford Estate

Bacon, Sir Francis 73, 158
Bagendon, Gloucestershire 23, 28, 179, 184
   church 23
Bakehouse, the [Great] 58, 59, 109, 113, 161, 163
   Privy 113, 116
   sergeant of 109, 113
Baldrick [Baldricks, Baldwicks or Baldocks] Estate 297, 362
Baldrick, Thomas 362
Baleere [Balier or Blyer], Sarah, daughter of Henry Timberlake 193-194
   suit against Sir William Balfoure 193
Baleere, Timothy, vicar of Tichfield, husband of Sarah 193
Balfoure, Sir William, Lieutenant of the Tower of London 193-194
Ballard, Thomas 108, 216-217, 221, 222-224
   marriage to Susanna Meade, St. Dionysius Backchurch, London 222
   Mayor of Arundel 222
   will of 113, 222, 223, 224
Balls, Mary 336
Balls, William 336
Baltimore 291, 308
Banckes, Richard, exiled catholic brother of Thomas 95
   letter to brother, Thomas 95
Banckes, Thomas, recusant 87, 89, 95
   barque owned by 87, 89, 95
   signing of the cocket 95
Banister, Henry 200, 214
Banks, Sir Joseph, botanist 356, 362
Bannatynes Estate 387, 388
   acquisition by William Clarke Haynes 387
   purchase by Robert Reece 388
Bannister, Alexander 287
Bannister, John, surgeon on the *Cockatrice* 355
Banyster, John, esquire 40
Barbadian Agricultural Society 317-318
Barbadian Militia [*see* militia]
   Corps of the Midland District of St John and St George 353
   St John's Regiment 283, 300, 334
Barbadian, The xii, 317-318, 326, 337, 338,

INDEX

339, 355, 371, 375, 382, 384, 386, 391, 396, 397, 398
  executorial notice in 384,
Barbados [Barbadoes]
  acquisition of land 227, 230-235
  Archives, Friends of xii
  as a health resort 279
  Bridgetown (formerly Indian Bridge) xiv, 250, 272, 280, 283, 285, 290, 292, 319, 320, 326, 330, 332, 335, 349, 351, 382, 383, 384, 392
    fires 235, 280
    hurricane damage to 283, 327
    petition for a Royal Charter for the Incorporation of 322-323
    policing of 322
    second church 319-320
    St Michael's Cathedral 327, 384, 390
    Theatre 318
    Town Hall 318, 322-323
    Trafalgar Square 327, 335
    Waterworks Company 367
  Carlisle Bay 246, 255, 256, 279, 291, 301, 309, 324, 329, 390
  census of 1679/80 240, 257, 385
  distemper; *see* Yellow Fever
  early settlers of xiv, 227, 306, 401
  England's dung-hill 401
  exodus from by planters 344, 371
  export tax of 4½ per cent 235
  fortification of 277
    St Anne's Fort 277
  Government House 362
  Governor of,
    Beckwith, General Sir George 309, 353
    Combermere, Lord (Sir Stapleton Cotton) 353
    Cunninghame, Major General James 283
    Granville, Sir Bevill 268
    Leith, Sir James 353, 390
    Lyon, Sir James 325, 326, 353
    Parry, Major David 298
    Warde, Sir Henry 319, 322, 323, 325-326
  history of 227, 267-268, 286, 287, 354
  Holetown Village 227
  House of Assembly xv-xvi, 249, 268, 276, 280, 281, 285, 286, 291, 294, 300, 311, 312, 315, 316, 318, 323, 325, 326, 335, 337, 343, 352-354, 358, 364, 378
    Agent in England 316
    Speaker of 276, 316, 318, 325-326, 352-354, 358, 364, 378
  Indian Bridge (*see* Bridgetown)
  labour shortage 256, 273, 343
  lawlessness of 317-318, 322
  Legislative Council 337
  legislature 227, 318, 328, 329, 343, 402

Mayo's map of 275
Moll's map of 243
parishes (Barbados)
  Christ Church 230, 231, 251, 263, 264, 274, 292, 311, 312, 313, 329, 330, 331, 348, 350, 370, 385, 388, 389
  St Andrew's 243,
  St George's 230, 246, 270, 284, 302, 311, 313, 314, 353, 367, 386,
  St James's 229, 243, 287, 289
  St John's xv, 218, 228, 229, 230, 231, 232, 233, 234, 235, 236, 237, 238, 239, 240, 242, 244, 247, 249, 250, 251, 254, 255, 256, 259, 264, 267, 272, 273, 274, 275, 276, 277, 278, 280, 281, 283, 284, 285, 286, 293, 294, 300, 302, 306, 311, 313, 314, 315, 318, 319, 322, 326, 332, 334, 349, 353, 354, 363, 367, 370, 375, 376, 385, 392, 396, 403
  St Joseph's 275, 280, 289, 332, 345, 364
  St Michael's 244, 245, 252, 257, 258, 264, 265, 268, 272, 286, 319, 320, 330, 345, 355, 266, 370, 387, 389, 390, 391, 393, 394
  St Philip's 247, 249, 250, 255, 275, 291, 292, 303, 311, 312, 330, 376
plague of locusts 235
police 322, 344,
settlement of xiv-xv, 217, 227-235, 400
threatened French invasion of, 1782 283
threatened French invasion of, 1805 300
violent distemper in 255
Baret, John, Clerk of the King's Kitchen 162, 170
Barford, William, Burgess of Arundel 103
  dismissal of 103
  living off proceeds of piracy 103
Barker, Jasper 100, 101
Barker, John, of Hurst 163
Barker, Mary, wife of Christopher Haynes of Billingshurst 100, 101
Barker, Robert 101, 108
Barlee, John, husband of Mary, sister of Governor John Haynes 180
Barnard, Thomas, of Wargrave 38
Barnard, Thomas, resident of Ferring 90
Barnes, Thomas 231
Barnstaple, Devon 381, 393
Barrey, Richard, Lieutenant of Dover Castle 141, 143
Barrow, Ann Thomasine [née Clarke] 291-292, 296, 364
Barrow, George, husband of Dorothy Clarke 290-291, 293, 303
  host of marriage of Lt Gen Robert Haynes 291

407

Barrow, John, son of John Henry 303, 320, 321, 335, 337
Barrow, John Henry 303, 321
Barrow, Mary, née Mary Ward Senhouse, wife of John 304
Barrow, Nathaniel 291, 292, 367
  Ann Thomasine, relict of 291
Barrow, Samuel, son of Nathaniel and Ann Thomasine 292
Barrow, Sarah Ann, daughter of Nathaniel and Ann Thomasine 292
  wife of Jonathan Higginson 292
Barrow, W. Warre 321
Barrow, William Matson, 287, 386
Bartlett, Martha 245
Barton 335
Barton & Irlam, Messrs 302, 368
Barton, Irlam & Higginson 335-336, 367-368, 383
Barton, Mrs Frances 304
Barton, Sir William 304, 335, 368
Barton, Thomas 335, 368
Barton Regis, Clifton 374, 387
Barwick Hall, Essex, 202
Barwick, John Storey, JP 376
Baschurch Rectory, tithes from 133, 147
Bascom, James S 317
Basing, Lord St John of 164
Bassingshaw Ward, London 162
Bath
  Abbey 288
    cemetery 339
    monuments 357
  All Saints Chapel, St Swithin 385
  Argyle Buildings 303, 335
  Argyle Street 331
  Clyde Bank 382
  Estate, Barbados 308, 315, 345, 350, 392
  Gay Street 385
  Henrietta Street 331, 332, 339
  Johnstone Street 360
  St Mary's Church, Bathwick 360
  taking the waters at 303,
  The Grove 357
  Thomasine's school 304
Batt, Thomas 230
Battaley, Elizabeth, wife of Robert James Haynes 280
Bayard, Colonel Nicholas 260, 261, 262,
Bayley's Plantation 311
Bayliffe, John, lawyer of Hackney 177
Bayne, Joseph Odwin, hardware store owner 312,
Beachamp, Ann(e) [d. 1722], formerly wife of John Haynes 240-242, 251, 264, 266, 267 273-274
  will of 240, 251
  grandchildren mentioned in 240, 267, 273-274
  great-grandchildren mentioned in 240
  key to Haynes family relationships in 241
Beachamp, Thomas 242
Beachamp, William 58
Beafsteak and Tripe Club, Bridgetown 280
Beale, Accuna 255
Beale, Edward 254, 276
Beale, John 276
Beard, William, fisherman of Rye 143
Becke Street, Wokingham 173
Beckles, Bishop Edward, father of Edward Hyndman 394
Beckles, Edward Hyndman, husband of Elizabeth Reece Haynes 394
  children:Elizabeth Reece [1864] 392, 394
  Edward Hyndman Haynes Hunte [1867] 394
Beckles, Hon. John Alleyne 323, 336, 337
Beckles, Professor Hilary McD 287, 397
Beckwith, General Sir George 309, 353
Bedford House, Bloomsbury Square 43
beer, export of 40
Belgrave, Jacob, free coloured plantation owner 312
  widow of, 350
  sale of Sterling estate by 350
Bell, Francis, 360
Bell, Great, of St John At Hackney 125, 129
Bell, Major [Philip], former Governor of Providence 239
  Governor-in-chief 1645 239
  Lieutenant Governor of Barbados 1641 239
Bell, Sarah, wife of Edmund Haynes, daughter of Francis 282, 360
  death in New York 360
  inheritance of share of Bellmount Estate, Barbados 360
Bell Tavern, King Street, Westminster 196
bell-penny 125
Bellamy, one, sale of tithes by William Haynes to 148
Belmont [Bellmount] Plantation 307, 360, 362
Bennett, Joan 105
Bennett, Judge John 64,
Benson, Captain Robert [d.1676] 230, 232, 242, 247, 249-252
Benson, Ensign Robert [d.1674] 232, 249
Benson, Katherine 232
Benson, Mary [Mrs Downes] 232
Berde, Richard 89, 93
Berden, Essex 221, 224
Berden [Bearden] Hall 221, 222
Bermuda 308, 336, 346, 366
Best, Hon. Col. John Rycroft 314, 227
Bestbitche [Besbytche], Mathew 51, 58

INDEX

Bestbitche, Richard, father of Mathew 51
Bett, Robert, fishing master and Jurat of Rye 141
Bettes [or Battes], Richard, shipmaster 81
Betty, brig 382
Beverley, East Yorkshire 376
Bigge, Alexander, son of Thomas, 64, 207
Bigge, Sara [née Gunnell] 64, 207
Bigge, Thomas, fishmonger, husband of Sara Gunnell 49, 62, 207
Bigge, Walter, fishmonger 62
Billings, Lt Roger 146
Billingsgate (Byllingesgate) Street 40, 41, 48
Billingshurst 74, 83, 100-101, 104, 108, 187, 188, 210
Binfield, Berkshire 26, 32, 37
Birchet, Thomas, Mayor of Rye 144
Bishop, James, Esq., Deputy Governor of Connecticut 263
Bishop of Chester 375
Bishop of London 122, 201, 224, 247
Bishop, William, President of Barbados 294
Bishop's Egney, Stepney 122
Bispham, Sarah 387-388
Bissex Hill Plantation, St Joseph, 345
Black Friars, house and land of, Winchelsea 132, 148
Blacke, Richard 139
Blackler, Peter, master of *The Expedition* 217
Blanchett, Elizabeth, supposed wife of Robert Haynes [d.1696] 241, 255, 273
Blisworth, Northamptonshire 133
Blockeswiche, Agnes, of Old Fish Street 48, 68
Blotheman, Robert, Clerk 208
Bludder [or Bludde] Henry, yeoman of the counting house 54, 70
Boade, purveyor of grain 113
Boer War 369, 373
Boiling (Bayling) House 254,
Bond, Elizabeth 270
Bonne (Bonner), Ellen, wife of William Haynes 10, 15, 69, 134
Bonne (Bonner), Harmon, of Wolverhampton, father of Ellen 10, 134, 157
Bonner, Edmond, Bishop of London 69, 134
Bonnet, Edward 292
Bonnet, Stede, the *gentleman* pirate 292
   execution at Charleston 292
   association with Edward Teach alias *Blackbeard* 292
   unhappy marriage of 292
Bonnet, Thomasine, daughter of Edward Bonnet 292, 295
Bootle, Lancashire 381
Bores Head, The, Old Fish Street 48
Botyler, Robert 18
Bourbon 293, 397
Bourbon sugar cane 293, 295

Bournemouth 369-370
Bousfield, Revd Henry Brougham, First Bishop of Pretoria 369
Bousfield, William Cheek, father of Revd Henry Brougham 369
Bovell, William 323
Bovells and Heyes 383
Bowles, T & J, Map of Barbadoes printed by 243
Bowman, Robert 242
Bowne, Andrew, 258, 259, 262, 264, 269
   Alderman of New York City 258
   executor of John Haynes, his son-in-law 263
   Governor of East New Jersey
   quietus granted 263
   settlement in Chingaroras, Monmouth County, East Jersey
   will of 264
Bowne, Andrew Junior, born Andrew Haynes, grandson of Governor Bowne 259, 264
Bowne, Elizabeth, daughter of Andrew, wife of John Haynes of New York 241, 255, 258, 259, 264, 269
Bowne, Elizabeth, wife of Andrew 258
Bowne family 258, 264, 270
   coat of arms 268
Bowne, John, Captain, merchant of Monmouth County, East Jersey, 259, 262, 269
   inventory of John Haynes's possessions taken by 262
Bowne, Obadiah, second husband of Elizabeth (Bowne) Haynes 263, 264
   children: John, Ann, Lydia 263, 264
Bowne, William, of Yorkshire 270
Bradford, Mrs Sarah xiv, 290
Braithwaite, Hon. John 337
Bramshill 23, 26, 211
Branch, Elizabeth, daughter of John Haynes [d.1678] 241, 264, 265
Branch, John, son of Samuel, husband of Elizabeth Seaward 241, 264, 265
   children: James, Olive and Anne; triplets [b. 1715] 265
   John [b. 1710] 265
   Samuel [b. 1713] 265
   death 265
   will of 265
Branch, Samuel, first husband of Elizabeth Haynes 241, 264, 265
   children: Elizabeth 241, 264, 265, 266
      possible marriage 266
   John; *see individual entry*
   Olive 241, 264, 265, 266
      possible marriage 266
   Samuel; *see* individual entry
   co-executor of the Will of John Haynes of New York 262
   early death of 264

409

Branch, Samuel, son of Samuel, husband of
    Barbara Price 241
  children: Elizabeth b. 1723 265
    Richard b.1719 265
    Richard b. 1725 265
    Samuel b. 1721 265
    Stephen b. 1722 265
Brandenburg, Marquis of 191
Brandon, Abraham R. 323
Brandon, Marmaduke 287
Brandt, Elizabeth, widow 238, 243
Branker's (formerly Chapman's) Plantation, St
    Philip 303
  renamed 'Sunbury' 303
Bray xvi, 21, 23-6, 28, 29, 119, 394, 399
  Church - see St Michael's Church
Bray, John 264
Bray, Susanna, wife of John 264
Brenda (Eleanor Brenda Violet Whittle; Mrs
    Johnston) xii, 338, 372, 373, 396,
Brett, Thomas 240
Bridewell; see prison
Bridge, John 123
Bridger (Brydger) family 78
Bridger, Thomas, 78, 79, 94, 102
  birth in Easebourne 102
  burgess of Arundel 78, 102
  controller of Arundel 78, 102
  mariner of Felpham 78, 102
Bridgnorth, Danesford 366
Bridgnorth, Dawford, St Mary's, 365-366
Bridgetown; see Barbados, Bridgetown
Bridgewarden 76, 91
  Account Book 76
Briggs, Joseph 355
Briggs, Mary 178
Briggs, Robert 178
British Government, compensation from to
    planters 299, 334, 345, 350, 368, 384, 388,
    402
British Guiana (see Demerara) 379
British Mercury 315
British Parliament 57, 114, 136, 248, 270,
    280, 315, 401
  Members of 401
broadcloth, uncoloured 82, 161
Broadwater, Sussex 222
Brodehinton [Broad Hinton; see Hinton] 163
Brook, John 137
Brooke, Chidley 261
Brooke, Sir William 57, 121
Brookesbie, Bartholomew, scrivener of London
    56
Bromley Hall 337
Bromley Manor 337
Brothers, The, [see Sunbury Plantation]
Brown, Charlotte, née Walton 358, 362

Browne, Anthony, of Cowdray Park, Sussex,
    Viscount Montague 71
Browne, Elizabeth, wife of William Harvey 214
Browne, John 265
Browne family, Clerks of the Green Cloth 42
Bruce, Sir John, of Airth, Stirlingshire 191
  coal, sold by 191
Bryant, Mrs Lucy, of Bath 331
Brydekirk, Dumfriesshire 381
Buckhurst, Thomas, Lord, Lord Lieutenant of
    Sussex 56, 83, 104, 114
Bull of Pope Pius V [1570] 97
Bulverhide 222
Burghley, Lord - see Cecil, William
Burns, Ann Elizabeth 359
Burrowes, James, of St James 243, 252
Burrowes, Elizabeth, wife of James, sister of
    Robert Haynes of St Thomas 244, 252
Burston, John 255
Bush Hill House 279
bushel measures 114, 115
Bussa [Busso or Bussoe] 311-312, 336
Butchers, Master and Wardens and Commonalty
    of 192
Butler, Deborah (aka Eleanor Ann), Mrs Henry
    Critchlow 387
Butterworth, Thomas, artist 304
Buxton, Derbyshire 125, 304
  taking the waters at 304
Byack, Grace 238
Byack, William 229, 238
Bynyon, Owen, customer of wine 79
Byrch 252
Byshe, Edward, herald 180
Byshley, Bennett, purveyor of grain 113
Bytown Gazette, The 338

Cadiz 255, 268
Cage, Anthony, husband of Anne Haynes of
    Hoxton 178-179
Caitcheon, Anna [Hannah], second wife of
    Robert Haynes 282, 343, 346, 351, 352,
    381, 393
  executrix in trust 352
Caitcheon, Jane 348, 351, 352
Caitcheon, Robert Hunter, master mariner 346
  Captain of the brig *Cockatrice* of Liverpool
    346
  privateer under Letter-of-Marque and
    Reprizals 346
Caitcheon, William, brother of Ann, grocer of
    Seacombe, Cheshire 351, 352
  children: Anna, Charlotte Haynes, Janet
    Douglas, Robert Hunter, Sarah Maria,
    William Davies 352
  migration to New Zealand 352
Caius College, Cambridge 357, 366, 392

INDEX

Caliver 98, 107, 168
Calverly, Richard, Jurat of Hastings 56
Cambridge Heath, Hackney 125, 199, 222, 224
Cammelland, Allen, will of 239
Campion, Saint Edmund, S.J. 84
Cannon, William, Master of the brig *Robert* 382
Canterbury, Archbishop of 65
Cape Chesapeake 291
Cape of Good Hope 335
carbuncular fever 159
Carell, Sir John, *see* Caryll
Carlisle Bay, Barbados, *see* Barbados, Carlisle Bay
Carlisle, Cumberland 384
Carmichael, Archibald 254
Carolina, Lords Proprietors of 257
   recruitment of settlers for 257
Carr, Robert 53
Carres Close 53
Carrington, Elizabeth 265
Carrington, Paul 271
Carrington, Thomazin, née Waterland, wife of Paul 271
Carrick Roads 390
carrier routes 86, 99
Carter, Robert, glazier 45
Carye, Edward, Esquire, Keeper of Marylebone Park 41, 44, 66
Carye family 44
Caryll, Edward, esquire of Benton's Place, Shipley, Sussex, recusant 88, 89, 93, 105
   accusation by Christopher Haynes of harbouring fugitives at Shipley 88, 93
   estate steward and executor of Philip Howard, earl of Arundel 93
   examination at own request 88, 89
   imprisonment of 93
Caryll, Sir John, Shelley's nephew 105
Caryll, John, of Warneham 89
cash for honours 20, 119
Castle, Le, St John's Street, London 178
Castleman, Thomas 206
Castlemans, Hare Hatch 206
Castlereagh, Lord 314
Catcher, John, lessee of the Stocks Market 153
catechists 290, 321
Catelsgrove, Whitley 6
Cater, Catre – *see* Acatry
Catholic League 191
Catholicism 85
   possible return of England to 260
Cavare, Sarah 245
Cave, Frances 338
Cave, Isabella Christian 338
Cave, James 339
Cave, Nathaniel, attorney 320, 328, 350
Cave, Robert Haynes 338

Cawet, Margaret, wife of Stephen 5
Cawet, Stephen 5, 16
Cawet, Thomas, son of Stephen 5
Cecil, Sir Robert 163, 170
Cecil, William, Lord Burghley 41, 45, 47, 55, 58, 59, 95, 107, 114, 141, 142, 143, 159, 165, 247, 250
Ceylon [Sri Lanka] 386, 397,
Chalcottes and Wyldes, Manor of, 52
Chalk, Kent 51, 69, 111, 112
Chancery, 170, 199,
   Barbados Court of Chancery 308, 329,
     estate sales in 305, 308, 329, 336, 346, 368, 388
   High Court of 61, 65, 162, 211-212, 219-220, 224
   Chancery Lane 285
Chander, Thomas 243
Chaper, Agnes, of Lodesworth, second wife of William Dartnall sen 188-189
Chaplain, Mary Ann, wife of Thomas Lovell 330
Charles I 125, 230
Charles II xv, xvi
Charles, John 287
Charleston, South Carolina 292
Charter of Elizabeth 1559-60 4, 5, 9
Cheeseman, Francis 360
Cheeswright, Elizabeth 362
Cheeswright, John 230, 232
Cheeswright, John, son of John 232
Cheeswright, Mary (née Crossing), wife of John 232, 249
Cheltenham 361, 380-381
Chessington 176-179, 191, 208
Chesswell, Thomas 141, 143
Chete [Chet] *alias* Foxley 22, 23
Chevenesse 18
Chichester 8, 68, 94, 102, 106, 107, 108, 168
   Bishop of 81, 107, 210
   Captain 80
   customer of 68, 91, 95, 97-98, 102, 106-107, 131, 137
   Rape of 83
Chickell (or Clinkett), Thomas 255-256
Child, Emily Wilkes, Mrs Freeman Oliver Haynes 357
Child, Robert 357
Chingaroras, Monmouth County, New Jersey 258
cholera 159, 329
Christian, Admiral Sir Hugh 333-334
Christian, John William, goldsmith 392
Chrychton, James, BA, schoolmaster of Hackney 201
churches
   St Andrew's, Holborn 285

411

St Augustine's, Hackney – see St John At Hackney
St David's Exeter 358
St. Dionysius Backchurch, London 222
St George's, Everton 367
St George's, Guelph 389
St Giles in the Fields 53, 60, 247, 253
St Giles, Reading 109, 125, 352
St John at Hackney 124, 128-129, 160, 183, 186, 189, 190, 195, 196, 199, 200, 204, 213, 214, 216, 218, 9, 221, 223, 224, 228
St John's, Barbados xii, 228, 235, 237, 249, 255, 277, 278, 306, 327, 328, 354, 361, 365-367, 370, 372, 374, 377, 389, 391, 393
St John's, Clifton, Bristol 366
St Margaret's, Parish of St John, Barbados 367
St Martin's, Scarborough 376-377
St Mary Aldermary, London 209
St Mary's, Bathwick 360, 391
St Michael's, Bray 21
St Nicholas, Hurst 163, 170
St Oswald's, Sowerby 379
St Peter's, Osmotherley 373, 375, 379
St Stephen Walbrook and St Benet Sherehog, London 209
St Swithin's, Bath 385
churchwardens
  George Willson of Hackney 200
  John Foxley of Bagendon 23
  Richard Haynes of Reading 7-8, 109, 155
  Richard Haynes of Hackney 196, 200, 201, 202, 214
  Thomas Haynes of Whitley 36
  Walter Halliley of Hackney 201, 214
Civil War 181, 196, 223, 230, 306
Civil War, American 359
Clamelane Farm, Westchalke [West Chalk], Kent 112,
Clapton House and Pond 123
Clarenceux King of Arms 9, 19, 117, 147, 174
Clark, George Foster 284, 290
Clark, Richard Foster, clergyman, son of George Foster Clark 284
Clarke, Ann Thomasine, daughter of William, wife of Lt-Gen Robert Haynes 291-296, 303, 304, 307, 343-346, 364, 367, 382, 387, 391
  burial at Egham 346
Clarke, Daniel, Queen's cook, 52
Clarke, Dorothy, daughter of William, wife of George Barrow 291, 333
Clarke, Edward 50
Clarke, Elizabeth [Dorothy Elizabeth] 304
Clarke, Forster 321, 337
Clarke, Lt.-Colonel William Murrell

Clarke, Major Somers 279
Clarke, Mrs Thomasine, née Bonnet, wife of William Clarke 292
Clarke, Samuel 312
Clarke, William, Overseer of Thicket Plantation 292
Clarke, William of St Michael's Parish, London, deputy to William Haynes 142, 143
Clarke, William, Secretary to the Council of the Army, Papers of, 248
Clarkson, Mr. 323
Clavis Apocalyptica 221
Claybury Estate 360, 385
clergy, stipends of 323
Cleveland, John, royalist poet 181
Cleveland, Thomas, Earl of 122
Cley, Norfolk 115,
Clifden Plantation 329, 346, 349
Clifton, Bristol, 336, 338, 353, 365, 366, 374, 385, 387, 389, 395,
Clifton Hall Plantation, 305-307, 308, 328, 333, 345, 349-350, 355, 362, 372, 375, 376, 377, 398
Clifton, Lord, Clifton Estate, Nottingham 380
  Henry Haynes, agent for 380
Climping (Clymping), Sussex 74, 79, 82, 99, 105, 107, 132
  Christopher Haynes's lands at 79, 82, 105, 132
  St Mary's Church 82
    incumbent of 83
    Norman tower of 82
Clynsall (Clinsall, Climsoll, Climsoldes), Edmund, wife of Joan Bennett, Barfham 105
Clynsall, Thomas, mariner 87, 89, 90-91, 94-95, 105
  meeting with Shelley at Haynes's behest 90
  release from the Marshalsea 90, 94
  sailing to France 87, 91, 95
  tenant of Christopher Haynes 90, 94, 105
Cluniac Abbey of St John of Pontefract, Yorkshire 40
Cloth, Green 42, 46-48, 53, 70-71, 116, 123, 136, 139, 144, 162-166, 196
clothiers, 36, 79, 82, 120
Cobham House 287
Cobham, William, Lord, warden of the Cinque Ports 56, 57, 121
Cockatrice, 346, 355
Cockman, Richard 243
Codrington College 290, 292, 367
Codrington, Col. Christopher 233, 250
Codrington, Sarah 265
Codrington, William 265
Cofferer 78, 112, 163, 364, 365, 170
Coggeshall, Essex 120

412

INDEX

Cokk, John 122, 128
Colbroke, Elizabeth 10, 15, 74, 75, 78, 102, 157, 168
Colbroke, John, father of Elizabeth 10, 75
Colbroke, Richard 82
Colbroke, Thomas, Searcher of Arundel, brother of Elizabeth 77-79, 102
Cole, John 247
Colecroft [Cotecroft] 123
Coleman, John, of Riccall Hall 380
Coleridge, William Hart, first Bishop of Barbados 320
   address by planters to 320-1
   address to planters by 321-2
   religious instruction of slaves 321-2
Colford [Colneford] Hill 123
College of Arms 19, 172, 399
Colleton, Col. James Edward 256, 293
Colleton, Col. Thomas 255
Colleton, Hon John, of Colleton Plantation 276
Colleton, Sir John, Lord Proprietor of Carolina
Colleton, Sir Peter, Lord Proprietor of Carolina 254, 257, 267
Collumbell, Francis 202
Colne Engaine, Essex 123, 202
Colne Priory Manor Court Rolls 123, 203, 214
Colombo, siege of 397
Colonel Lyne's Regiment of Militia 385
Colonial Bank (Barclays Bank DCO) 334
Colorado Territory 369, 396
Coltman, William 211
Columbine, HM brig 390
Combermere, Lord - *see* Barbados, Governor of
Combes [Coombes], 276
   Henry 274
   John 276
Commission for Household Causes 57, 114, 145
Commonalty of Butchers, Master and Wardens and 192
Commonwealth, The xvii, 1, 44, 127, 180, 222, 223, 244, 249, 253
Comnena, Eudoxia 336
Company of Royal Adventurers Trading to Africa: *see* Royal African Company
Connecticut 180, 181, 184, 261, 263
Consolidation Slave Bill, deliberations of the House of Assembly on 323-324
   permission for compulsory manumission, arguments against 324
   provisions for punishment of women slaves in, arguments for 323-324
   punishment of field workers, arguments for 324
Constable, 51, 83, 122, 162, 170, 239, 392
   purveyance in the presence of 68-69
   receipt given to 68
   statement of purchases signed by 68-69

Constantine VIII, Emperor of Constantinople 306, 336
Controller [comptroller] of Arundel 78, 88, 91, 92, 102
Cooke, Robert, Clarenceux King of Arms 9, 19, 117, 147, 174
Cooper, Thomas 231
Cope, Lady Elizabeth, wife of Sir Anthony 193
Cope, Sir Anthony 193
Cope, Sir William, of Hauwell, Oxfordshire 191
   indebtedness to Henry Timberlake 211
Copenhagen 356
Copford Hall and Manor 180-181
Coppin, Thomas 293
Copyn [Coppin], Alice 119
Copyn, John 119
Copyn, Miss [mother of Agnes Enowes] 119, 120
Copyn, William of Earls Colne, will of 127
Coram, Thomas 295
Corde, John, mariner 81
Cordell, Sir William, Master of the Rolls 61
   godfather of William, son of John Haynes 61
   inquisition before 71
Cornelis, Christian 47
Cornelis, Sebrant 47
corruption 110, 113, 114, 151, 400, 401
Corry, Joseph, Master of the brig *Pandora* 382
Cortlandt, Stephanus van, Mayor of New York 260-261
Cote Crofte (Cotecroft) 123, 202-203
Cotrington - *see* Codrington
Cottesmore, Margaret, wife of Thomas More of Lancelevey, Sherfield on Loddon 101
cotton 227, 231, 232, 249, 298, 383
Cotton, Mary (Shelley), wife of 105
Cotton, Mary, daughter of Sir George, wife of Sir John Caryll and niece of William Shelley 105
Cotton, Sir George 105
Coulthurst, M. 321
Council of New York 261, 268
Countinghouse 163, 165
Couper, Robert 17, 169
Court of Chancery; *see* Chancery
Court of Grand Sessions 325
Court of Great Ward 192
Court of Requests 50, 58, 63, 71, 199
Court of Wards and Liveries 56, 108, 220
Cowdall (Kowdale), Thomas, Merchant Taylor 48
   family of 68
Cox, Joseph 258
Coxe, Leonard, Head Master of Reading Grammar School 17, 74
Coxson, William, jurat of Rye and fishing master 141, 142, 143

Coytemore, Roland, of Wapping, a mariner, husband of Christian Haynes 121
Cracknell, Everil Marion Winifred (née Thomas) xviii, 21, 83, 110, 125, 129, 185, 292
Cradle, Francis, 99, 104
Cragg, Jane, wife of Joseph Alleyne Payne 364
Cragg, William, husband of Mary Haynes, widow 278
Crane, Anthony 151
Creslow Brook 42
  Manor and Pastures, Buckinghamshire 39, 41-43, 61, 185
  'Hayne Houses ground and pasture' 43
Cressell, Margaret, wife of Thomas 32
Cressell, Thomas 32
Cressell, William, son of Thomas 32
crest [of Haynes arms] 10, 63, 73, 81, 147, 151, 172-175, 177-178
Critchlow, Ann, née Perryman 240-241
  guardian of three children of her cousin, John Haynes 263, 266
Critchlow, Henry, husband of Eleanor Ann Butler, father of Margaret Ann 387
Critchlow, Margaret Ann, wife of William Clarke Haynes 296, 387
Critchlow, Olive, née Branch 266
Crofte, Sir James, Comptroller 151,
  petty graft and dishonesty tolerated by 67
Crompton [or Crumpton], Anthony, 94
Cromwell, Oliver, Lord Protector xv, 181, 203, 218, 248, 306, 352, 401
  'Arch-tyrant' xv
Cromwell, Richard 248
Crooke, Paul 214
Cropper, Mr (James) 323
Crossing, Ann, née Summers, wife of William 255, 274
Crossing, Mary, sister of William and wife of John Cheeswright 267
Crossing, Richard 232
Crossing, William 254-255, 267
Cuba 283
Cudmirrah Nature Reserve, Wreck Bay, New South Wales 395
Cull, Edward 219
Culpeper, Margaret [Mrs Edward Quintyne] 286
Culpeper, Mary [Mrs John Ince] 286
Culpeper, William Francis 280
Cummins, Hon. Henry Stephen 319, 322, 323, 370
Cunninghame, Major General James, Governor of Barbados 283
currency 231, 297, 298
  coins, mixture of 297
  colonial Banking Regulations and
  cotton as 231, 298
  indigo as 231
  local (Barbadian pound) 297, 308
  paper 297-298
  sugar as 231, 298
  tobacco as 231, 298
Custom House Quay, Barking 191, 211
Custom House, New York 260-261
  expulsion of commissioners from 260-261
custom of money, farmer of 177
customer (s)
  fees and rewards of 102
  of Arundel *see* Arundel
  of Chichester 91, 95, 97, 98, 102, 106, 107
  of Rye 50, 80, 97
Cynthia, H M Packet-ship, sinking of 383

Dagges, case of 107
Daglingworth, Gloucestershire 23, 179, 184
Dalby, Charlotte Maria 389
Dale, Valentine, Commissioner of oyer and terminer for Middlesex 133
  acting Lord High Admiral 133
Danyell, William, esquire 56
Darcy, Sir Thomas, Lord Sheffield 128
Dartnall [Dartnoll, Dartnell, Darkenall,Darknall], Christopher, brother of William 210
Dartnall, Dorothy, sister of William 210
Dartnall, Edward, brother of William 210
Dartnall Edward of Slinfold, father of William 187, 192
Dartnall, Elizabeth Penfold, wife of Edward Dartnall of Slinfold 187-188,
  will of 188, 210
Dartnall family, origins of 187
Dartnall, Hester, daughter of Sebastian, wife of Francis Phillips 212
Dartnall, Jane, widow of Sebastian 190
Dartnall, John, brother of William, yeoman of Kirdford 188
  will of 188, 189, 194
Dartnall, Mary, sister of William 210
Dartnall, Sebastian 190, 192, 210-212
Dartnall, Thomasine, wife of William sen 100, 101, 186-189, 195
  death of 160
Dartnall, William, husband of Thomasine Haynes 62, 87-88, 91, 92, 95, 186, 187, 189, 195
  bequest from brother John 188
  careless conversation with Pellet 62, 87-88, 91
  children by Thomasine 189
  Jasper 64, 100, 187-194
    children by Rose:
      Ann 190
      Jasper 189-190
      Thomasine 190

414

INDEX

appointed executrix by Sebastian 190
Faith 190
William, godson of Henry Timberlake 190-191, 193
citizen and fishmonger of London 191
coal agreement 191
contests will of John Haynes of Marylebone 64, 198
death in Southwark 1659 194
farmer of Leadenhall Flesh Market 192-193
marriage to Rose Merrick 190
Robert Peere, sued by 192
suit against Lady Elizabeth Cope 193
suit against Sarah Baleere 193-194
wharfinger at Custom House 191-192
  damage to goods during transit by lighter 192
  case brought by Simonds 192
  petitions to East India Company 191-192
Nicholas 194-195
  trial and execution for piracy 195
Thomasine 194
William, M A, godson of Christopher Haynes of Arundel, 187, 188, 189
child by Agnes Chaper 188
  Elizabeth 188
domicile: Drungewick, Wisborough Green 91, 187
marriages, Wisborough Green 187, 188
Dartnoll's Cause, a petition to the House of Lords 212
Davenant 49
David, Richard, purveyor of grain 113
Davies, Annabella, wife of William Caitcheon 352
Davies (Davyes) William, farmer of Patching, Sussex, servant of William Shelley 84, 85, 86, 89, 94
Davison, Robert, groom of the chamber 199
Dawford, St Mary's, Bridgenorth 365-366
Daye, John, doctor of laws 56
Deacon, Richard, skinner 50
Deane, Richard 323, 335, 368
death in the Haynes family 154, 157-159
DeCastro H. 323
deception 58, 401
Defoe, Daniel 121
Demerara, British Guiana 360, 361
De Meuron Regiment 386, 397
dengue, or dandy, fever 326
De Vere, John, 16th Earl of Oxford 120, 127
D H Miller, brig 371
Dickenson, Capt Francis [Frances] 231, 232, 233, 234, 235, 249, 254

Dickenson, Francis, son of Capt Francis 255, 256, 259, 264, 267
Dickenson of Talbot County 290, 291
Dickenson, Thomas 276
Dieppe 85, 86, 90, 93
diet 155, 161, 166, 167
differencing marks (cadences) 17, 117
  annulet (ring), fifth son 17
  crescent, second son 17
  fleur de lys, sixth son 17
  martlet, fourth son 117
  mullet, third son 174
  rose, seventh son 17
Dillon, James 230
Dissolution of the Monasteries xvi, 3, 4, 6, 16, 29, 42, 48, 68, 74, 132
Dixon, Thomas, master of the brig *Martha* 382, 383
Dockfour, Demerara 360, 361
Dolphin, The, Old Fish Street 48
Dominica 283
Domus Providencie 45, 152, 161-165
dossers 138, 142, 143
Douglas, Jane, Mrs Robert Hunter Caitcheon 347,
Dove, John, High Sheriff of Wiltshire 248
Dowglas, Thomas 38
Downer, a carrier of London 87
Downes, Captain Thomas 232, 267
Downes, Dorothy 276, 277, 278
Downes, Elizabeth, wife of William Franklin, son of Benjamin 276
Downes, Lt Col Richard, Speaker of the House of Assembly 256, 276
Downes, Mary 232, 249, 275, 276, 278
Downes, Richard, Barbados Treasurer 267, 268
Downes, Thomas 275, 276
Downinge, Edmund 54
Dragon, The 81, 86, 90
Drax, Anne, sister of James, wife of Nicholas Enos 245, 246
Drax, Captain [subsequently Lt Col] James 230, 245, 246, 253
  Introduction of Dutch milling machinery 246
  Removal of sugar from the *Samuel* 246
  Return to England 246
  Slave ships of 246
Drax, Dorothy (Lovelace) second wife of Henry 246
Drax, Frances, daughter of Henry 247
Drax Hall Estate, Jamaica 246
Drax Hall and Plantation, Barbados 246 247
Drax, Henry, son of James 245, 246, 247, 253
  Indenture with Nicholas Enos 245
Drax, James, son of Henry 246
Drax, Margaret, wife of Lt Col James 252
Drax, Meliora, wife of Robert Pye 253

415

Drax, Phalatia wife of Thomas Gomeldon (Gumbleton) 253
Drax, William 246
Draycott, Vincent 273, 274
Drayton, Samuel 337
Druez, Jane 29
Drungewick 87, 108, 187, 210
Dudley, John 53, 58, 130
Dudley, Robert, 1st Earl of Leicester 70, 121, 130, 148, 152, 165
   Amye, wife of 152
      death of, 152
   death of 152
   William Haynes, servant of, 152
Duelling 55
Duke, Henry, Solicitor-General of Barbados 281
Duke of Bedford's Road, London 43
Dummett, Dr Edward James father of Sarah Jane 390
Dummett, Joshua Rowley, husband of Ann Cave Haynes Lovell 331, 391, 394
Dummett, Mary 398
Dummett, Sarah Jane, wife of Henry Husbands Haynes 296, 389, 390
Dummett, Thomas Henry 344
Dung-hill fowls 279
Dunkerton 218
Dunning, John 50, 68, 131 137-138, 139
   customer of Chichester 68, 131
   former Sergeant of the Acatry 50, 131
   freeman of Rye 131
   jurat of Rye 137
   mayor of Rye 131, 138
Dunninge, William, Sheriff of Berkshire 14
Duntisborne, Gloucestershire 23, 179
Duppa, Geoffrey, brewer of London 169
Durant's Plantation 330
Durham, RN, Rear Admiral Sir Philip Charles Henderson 209, 390, 398
   commemoration in Largo Parish Church, Fife 398
Durham [Deron], Thomas, of Hurst, Sergeant of the Larder 163
Dutch East India Company 397

Earls Colne, Essex 119, 120, 127, 129, 180, 185, 202, 214, 221, 245
   Manor Court Rolls 127, 128
   Project xiii, 129, 214
earthquakes 308, 327
East India Company, *see* India Company
East Yorkshire Regiment 379
Eastwood Plantation 293, 294, 364
ecclesiastical court 202
Edinburgh Review 323
Edney, Captain (Peter) 231
Edward IV 45, 162

Edward VI 16, 19, 44,
Edwards, Elizabeth, wife of Major Valentine Hale Mairis 365
Egham, Surrey 346, 370, 386, 397
Eliott, Blithe, wife of George Haynes of Much Hadham, Hertfordshire 180
Elizabeth I xv, xvi, 4, 16, 37, 42, 44, 45, 46, 47, 48, 51, 60, 62, 79, 80, 84, 96, 97, 106, 110, 111, 112, 113, 114, 122, 131, 132, 133, 137, 138, 143, 146, 155, 158, 161, 163, 164, 165, 167
Ellcocke, Alexander 287
Ellcocke, Ann, first wife of Richard Downes Haynes 278, 280-282, 289
Ellcocke, Grant, father of Ann 280
Ellis, Richard, financial officer to Robert Dudley, Earl of Leicester 130
Elmer, Edward 122
Eltham Ordinances 166, 171
Elton, Jacob, father of Robert, of Little Burstead, Essex 391
Elton, Lt. Colonel Robert William, second husband of Sara [Dummett] Haynes 391
emancipation xv, xvii, 300, 313, 334, 335, 336, 343, 344, 349, 355, 365, 368, 371 402
embezzlement 400
Englefield, Sir Francis 4, 84
   estates of 4
Enowes [*see also* Enos], Agnes [Annis, Anne] (Mrs Nicholas Haynes) 10, 15, 20, 109, 119, 120, 123, 124-125, 157, 168, 187, 221, 245
   death of 15, 124-125, 187
Enowes, Edith, third wife of Richard 123
Enowes, Edmund 123
Enowes, Richard, yeoman, father of Agnes, 10, 119, 120
   attempt to reconcile the Countess and Earl of Oxford 127-128
   servant of the Earl of Oxford 119, 120
   will of 120, 128
Enowes, Dutch ancestry 120, 245
Enos, Edward 245
Enos, Margaret, sister of Nicholas, wife of James Drax 245, 246
Enos, Mary, daughter of Nicholas 245
Enos, Nicholas 221, 245-246
   Anne Drax, wife of, 245, 246
   bequest from Edward Hopkinson 221, 245
   indenture with Henry Drax 245
   will of 252
entertainments, Elizabethan 167-168
Epton, Henry, husbandman 60
erysipelas 159, 331
escheator 56, 75, 219, 220
esquire's helm 63, 174, 399
Essex, Robert, Earl of 145

INDEX

Estcourte, Thomas, citizen and leatherseller of London 183, 208, 209
Estsmythefeelde 32
Estwicke, Major Christopher 256, 268, 276
Estwicke, Richard 230
Eton College, lands of 44
Euston Road, London 43
Evelyn, John 66, 177
Ewery 163
Exeter 66, 69, 248, 358, 361
exemplification of arms and crest xvi, 3, 9, 19, 20, 21, 22, 117, 118, 153, 175, 176, 178, 180
Expedition, The 217, 229
Eversley, William 329
Eynon, John, vicar of St Giles, Reading [1520-1533] 8

Falcon 74, 81, 82, 218, 229, 233, 243
Fame 382
Faringdon, Abbot Hugh Cook 8, 17
Farm, The, Barbados 389, 391
Farrell, John 298, 334, 358
Farringdon, John 287
Fawcett, Edward, Commissioner of the Exchequer 53
Feason, John, of Cawson, Cornwall 217
Felpham, Sussex 78, 89, 94, 102
Fenne, John, Burgess of Arundel 104
Fenner, Thomas, merchant 97, 106
Fernhall, William 293
Ferreira, Manoel da Silva 393
　marriage to Christina Lardy Haynes 393
Ferring, Sussex 89, 90, 94
Festing, Revd F, Vicar of Wingham 360
Filborowe [Filborow; Filborough Farmhouse], Eastchalke, Kent 111
Finchley 52, 169, 178, 183-184, 205
First or Royal Regiment of Militia 317
First West India Regiment 312
fish days, compulsory 138, 144
Fisher, Catheryn, wife of Richard Haynes 366
Fisher, Thomas, former Sergeant of the Bakehouse 113
fishermen of Rye 51, 134, 137, 139, 140-141, 144-145, 150-151, 165
　alleged threats and abuse towards William Haynes by 141-142
　trouble caused by 165
　Burghley's intervention in 141, 165
Fishmongers, Mystery or Company of, London 13, 49, 138
Fitchatt, Revd Francis 289-290
Fitt, Elizabeth 266, 273
Fitt family 273
Fitt, Robert 265
Flemish hoy 83

Fletcher, John, former purveyor of sea-fish 144
Flitchett, *see* Fitchatt
Flora, a slave 256
Flushing, New York 270,
Flying Horse, The, Hackney 122
Fokes, Sir Robert 21
Folkes [ffolkes, Folks, Folk, Fokes, Folx, Faulks, Faux, Fawkes, Fulks] 20, 21, 24, 27
Folkes, John 10, 11, 19, 21, 24, 25, 26, 27, 102, 119, 176
　grave of
　payment of seat in St Giles 11
Folkes, John, father of John 25-27, 29
　identification with John Fulks, woodward of Windsor Forest 29
　owner of land at Hartley 27, 37
　woodward of the Abbot of Abingdon 29
Folkes, Maude (Mawd), assumed wife of John, 27
Folkes, Thomasine, daughter of John and wife of Richard Haynes of Reading 3, 10, 11, 15, 19, 20, 26, 119, 176, 187
　children of 15
　maternal descent of 176
　pregnancies of 11-12
Folkes, William, of Hartley 27
Fontabelle, St Michael, Barbados 390
Ford, Sussex 74, 79, 82, 132
Ford [Forde], Anne, wife of Richard Haynes of Wargrave 15, 31, 32, 33
　family, Wargrave and Hare Hatch 32
Ford, William [Wylliam] 33
Forest, Captain Thomas 335
Forster, Edmond 287
Forte, Nathaniel 239
Fortescue, John 114
Fortescue Plantation 312
Foster, George, planter of St John's Parish 239
Foster, Hester 239
Fothergill, John, woodward for Middlesex 44
Fowler, a servant of William Shelley 94
Fowler, Nicholas, fishing master of Rye 141, 142, 143
Foxley [(de)Foxle, Foxele, Foxlee; Voxley], 19, 20, 21, 22, 25, 174, 184
Foxley, arms of xv, 21, 23, 24, 27, 63, 117, 119, 147, 174, 175
　quartering of 9, 21-22, 27, 63, 73, 117, 119, 147, 151, 174, 175, 176, 197, 374
Foxley, Chete, alias 22, 23
Foxley, coheir of 24, 26, 28, 176
Foxley, daughter of Thomas [I] 26
Foxley, family of 19, 21, 22, 23, 27, 119
Foxley, Grange xv, xvi, 21, 22, 399
Foxley, heraldic heiress of 24
Foxley, John, Esq xv, 9, 11, 19, 20, 22, 28

417

Foxley, John, of Bagendon, Gloucestershire 23, 28, 184
Foxley, John, Sir [I] [c.1258-1324] 23, 25, 26
  collateral descent from 22, 23
Foxley, John, Sir [II] [c. 1316-1378] 23, 24
Foxley, John of Blakesley, Northamptonshire 22, 23, 28
  John, son of 23
Foxley, John, of Purton
Foxley, John, Esq. [d. 1419] of Rumbaldswyke, Sussex, son of Sir John [II] [d.1419] 23, 24, 25, 26, 28, 119
  Alice, daughter of 25
Foxley Manor, xvi, 21, 24,
Foxley, maternal descent from 22, 24, 26, 176
Foxley, Richard of St Lawrence, Reading 22
Foxley, Thomas, of Bagendon, son of John 23
  Alice, wife of 23
  Jane, daughter of 23
  John, father of, churchwarden of Bagendon 23
  Thomas, son of 23
Foxley, Thomas [I], of Bray [1290-1360], Constable of Windsor Castle, father of Sir John [II] 24, 26
Foxley, Thomas, of Bray, son of Sir John [II] 24, 118
Foxley, Thomasin(e), Thomasen, 3, 19, 20, 27, 184
Foxleys of Bagendon, Gloucestershire 23
Foxleys of Highworth, Wiltshire 23
Foxleys of Hull, Yorkshire 23
Foxleys of Kent 23
Foxleys of Norfolk 23
Foxleys of Northamptonshire 23
Foxleys of Sussex 23
Frances, The 316, 365
Francis, Thomas, overseer 302, 305, 306
Frankfurt 191
Franklin Estate, St Philip 311-312
Franklin, Joseph Pitt Washington 311-312
  Court Martial 313
  execution of 313
Franklin, William, son of Benjamin 276
fraud, allegations of 400
Freehold of St John, Barbados 326
Freemasonry 402
Fremington 381
French Revolution 295
Freshstore, yeoman purveyor of 66, 132
Friern, Manor of 178
Fulconis (or Fulco; *see* Fokes), Sir Robert 21
Fulk, Master, king's gardener at Windsor 30
Fulks *see* Folkes
Fynch, Francis, husbandman 60
Fynnett, Robert, mayor of Dover 50

Gage, Sir John, Comptrollor 65
Gaines, Mrs Ann, widow 238, 242
Gambier, Annabel Selina, second wife of Robert James Haynes 392
Gambier, Admiral of the Fleet Lord (formerly Sir James) 356
Gambier, William 392
Garatt, Nathaniel 233
Garetson, Peter 47
Garner, Nicholas R. 384
Garnett, Revd William 387
Garth-Colleton, Annabella Frances, wife of William Gambier 392
Garton, Francis 99, 104
Gates, George, Purveyor of the Queen's Mouth 113
Gawyn, Richard 105
Gaymer, Henry, 139, 141, 143
  deputy fish purveyor 149
  detention of Julio Marino by 151
Gayneford, Richard, husband of Dartnall 210
General Session of the Peace 55
gentleman, title or style of 12, 35, 75, 117, 118, 127, 169, 228, 231, 290, 401
  cause for ridicule 117
  of the Bedchamber 152
  usher 163
George Inn, Arundel 99, 104
George of Amsterdam 47
George of Purmerend 47
Gerard, Richard 27
German Imperial army 191
Ghana, *see* Gold Coast
Gibb, Henry, Groom of the King's Bedchamber 191
Gibbons Plantation 306
Gibbs, Alice, daughter of Joseph 244
Gibbs, Elizabeth, wife of Joseph 244
Gibbs, Haynes, grandson of Robert and Elizabeth Haynes of St Thomas's Parish 244
Gibbs, Henry 243, 244
Gibbs, Jane, daughter of Joseph 244
Gibbs, Joseph, husband of Elizabeth Haynes of St Thomas's Parish 244, 278
Gibbs, Katharine, daughter of Joseph 244
Gibbs, Mary, daughter of Joseph 244
Gibbs, Joseph, son of Joseph 244
Gibbs, Thomas, son of Joseph 244
Gibbs (Gibbes), plantations of 244, 278
Gibbson, Robert 242, 252
Gill, Elizabeth Stewart, Mrs Robert James Haynes son of Henry Husbands 392
Gillespie [Gilespy], John 335
Gittens, Ellen Farrell, Mrs William Haynes of Thimbleby 376, 377
Gittens, Revd Hon. John H. 337, 371

INDEX

Gittens, John, MCP 376
Glasgow Courier 315, 323
Glasgow University, matriculation for 302
Glaston, Somerset, Monastery of 56
Glenburnie boiler house, site of 367
Glieison, Anne Walker, Mrs George Barrow Haynes 296, 366, 382, 384, 385, 387
Glover, Alexander 52, 54
Glover, A. daughter of Alexander 69
Glover, Blanche, wife of Alexander 52
Goare, Cyprian 232
Godsmarke, Henry, feter 142, 143, 144
Gold Coast (Ghana) 379
    Dunquah 379
    Kumasi, march to 379
Golden Grove Estate 284, 312
    looting of 312
Goldie, John, merchant and ship owner 346
Goldsmiths' Hall 298, 303
Gomelton [Gumbleton], Thomas 253
Gonell, Gonnell, *see* Gunnell
Gordon, Ellen 381,
Gordon, George, Esq of Gight 288
Gordon, Sir John Watson, R.A. [1788-1864] 327
Gordon, Stephen 351
Goringe, Henry 86
Gossfenn [Gosse (Gosses) Fenn] 123, 202, 203
Gounde - *see* Gunnell
gout 303, 304
grain, prices of 114, 115, 116
    spoiling by mould 114
Grange, The, Frampton on Severn 179
Granville, Governor Sir Bevill 268
Grasse, Admiral Le Compte, Francois Joseph Paul de 283
Graves, Admiral Samuel 283
Gravina, Admiral (Don Frederico Carlos Gravina y Napoli) 301
Great Northern Hotel, Marylebone 366
Great Russell Street, London 303
Green Cloth, Board of 46, 47, 48, 53, 139, 163, 164, 165, 166
    Clerk of 42, 46, 71, 123, 136, 162, 164, 196
    shortage of money of 116, 144
green fish, 68, 137, 149
Greene, Robert 118, 127
Greenock, school at 302, 304, 382
Greenwich 66, 304
Greenwich Palace 39, 42, 112, 113, 131
Gregory, William, husbandman 60
Grenada 283
Grettons Estate 368
Grey Friars, land of, in Winchelsea 132, 148
Griffith, Thomas Howard 316, 317
Griland, Martin 243
Groom Hayleman William Haines, 146
Groom of the Bedchamber xv, 161, 191, 199

Groom Porter 166
Grove, Hugh 248
Grove, John 32, 37
Grove, William, yeoman of Wargrave 32
    Proctor for Anne Haynes, Widow 32
Guadaloupe 390
Guinea Estate 372, 378
guinea-corn (millet) 279
Gunnell [Gonnell, Gonall, Gonnele], Abraham, son of William 34
Gunnell, Christian 35, 205, 209
Gunnell, Christofer 35, 205, 207
Gunnell, Danyell 34, 205, 206
Gunnell, Dorytie 34, 205, 209
Gunnell family 205-209
    influence on John Haynes by 64
Gunnell, John, brother of Thomas 35, 206, 207
Gunnell, John, son of William, 34
Gunnell, Lydia 209
    will of 215
Gunnell, Moyses [Moses] 34, 49, 59, 98, 99, 193, 205, 206
    bequest from Christopher Haynes 98, 99 206
    burial in St Nicholas Cole Abbey 206
    children: Edward, Hellen, John, Margaret, Moses, Sara 206
    death of, from plague 193, 206
    fishmonger 49, 59, 206
    Katherine, wife of 60
    burial 206
Gunnell, Raphe (Ralph), brother of William, son of Thomas of Hare Hatch 209
    children: Anne, Elizabeth, Hannah, Lydia, Mary, Matthew, Parnell, Thomas 209
    citizen and skinner of London
    grandchildren 209
    marriage to Lydia Halley 209, 215
Gunnell, Richard 34, 205, 209
Gunnell, Sara, wife of Thomas Bigg 35, 49, 62, 64, 205, 207, 208
    Alexander Bigge, son of 64
    bequest of Uncle John Haynes 64, 207
Gunnell, Susanna 35, 205, 209
Gunnell, Thomas, husband of Christian Haynes 14, 15, 20, 31, 33, 34, 35, 37, 38, 156, 205, 209, 215
    witness of inventory 35
Gunnell, Thomas, son of Thomas 35, 205, 209
Gunnell, Thomas, son of Thomas of Hare Hatch 34
Gunnell, Thomas, the elder of Hare Hatch 34, 38, 209, 215
Gunnell, William, son of Thomas, 35, 50, 63, 64, 65, 135, 183, 193, 197, 205, 207, 208
    children: Thomasine 64 208
    John 64, 65, 208
    marriage to Elizabeth 208

Gunnell, William, brother of Thomas 35, 206
Gunnell, William, junior, citizen and skinner of London, 50, 65, 183, 208, 209
  putative son of Thomas Gunnell of Hare Hatch 35, 209
Gunnell, William of Hare Hatch 35
Gunnell, William, of Wargrave 35
Gustavus Adolfus, King of Sweden 191
Gyboncokes 25, 29

Hackett, Richard, plantation of 232
Hackney
  Cambridge Heath, 124, 199, 222, 224
  collectors for the poor 200, 201
  fashionable schools in 121
  Flying Horse Inn, 122
  Grove Street, 177
  Hasting's Mead [Jerusalem Close], 122
  House of Correction, 200, 202
  Mare [Mayre; Meer] Street, 122, 224
    Haynes home in 122, 124, 128, 199
  Manor of Grumbolds, 122
  Manor of Kingshold, 121
  Manor of Wick [Wyke], 177
  overseers of the poor, 200,
  physic garden at 121
  St John At Hackney [St Augustine] 124, 128, 186, 189, 190, 195, 196, 204, 216, 218, 219, 221, 223, 228
    Great Bell 124, 125
    Parish Records, state of
      plague entries in 124, 129, 159, 216, 218, 219
    Vestry 124, 200, 202, 204, 213, 214, 219
  St John's Mead, 122
  surveyors of the highways 200
  Well Street, 122
Hagar, Cinaine, a slave 256
Haines, Andrew Mack 393
Haines (Heynes), George, groom of the Buttery 184-185
Haines [Haynes] Hill, Nr Reading, Berkshire xv, xvi, 31, 163, 170, 197, 399
Hains, Captain, a pirate 147
Hales, Marvin 257, 268
Haler, John, master mariner and ship owner 85, 86, 90, 91, 106
Hall, The, Patrick Brompton, North Yorkshire 372, 373, 376
Hallage 79, 103
Halley, Lydia - see Lydia Gunnell
Hallilie (Halliley), Walter 201, 202, 203, 214
Halton Plantation, St Philip 303
Hamden, Hon. Renn 321, 337
Hames, Richard 218, 224, 229
Hamilton, second steamship to arrive in Barbados 324

Hamlet, slave mariner 301, 302, 382
Hanaper 40, 65
Hancock, John 18
hangings, painted cloth 6,
Hannis, John, will of 218, 224, 230
Hannis, Richard 217, 218, 227, 229, 230
Hanson, Robert 215
harbinger 109, 110, 111, 112, 125, 162, 167
Hare Hatch, Berkshire 32, 33, 34, 35, 38, 205, 209, 215
Hargreaves, Betsy and Martha 352
Harlakenden, Mabel 180
Harman, Thomas, fisherman of Rye 143,
Harnes [Harness] 8, 17
Harold Wharf, Deptford 42
Harris, Dorothy 121
Harrison, William 118, 127
Hart, Sir John 179
Harte, Revd W. M, rector of St Lucy 324, 325, 330
Hartley [Hertleigh] 26, 27, 29, 37
Harvey (Harvie), Thomas, father of William 214
Harvey, William 178, 214
  Elizabeth Browne, wife of 214
Hastings, harbour project 56, 57,
Hastings of Durham, Buckinghamshire 179
Haswell, Charles, of Dublin 282, 358, 359
Haswell, Charles Haynes son of Charles and Dorothy 359, 362
  author: Mechanic's and Engineer's Pocket Book
  children: Charles Haynes, Charles Roe, Edmund Haynes, Elizabeth Bulwer, Frances Roe, Gouveneur Kemble, Hansen, Jane, Sarah Haynes 359
  marriage: Ann Elizabeth Burns 359
Haswell, Mrs Dorothy, wife of Charles, sister of Lt-Gen Robert Haynes 181, 182, 358, 359, 361
Hatsill, Captain Henry 248
Hatton, Christopher, Chancellor 114
Havering 113
Hawkes family of Clifton 374
Hawkesworth, William 323
Hawley, Captain William 230
hay, composition agreement for 202, 203
Haynes Arms, Thimbleby 374
Haynes [Hawes, Hawnes, Hawenys], Bedfordshire 3, 17
Haynes [Hayne, Hayn, Heaynes, Hain[n][e][s], Hane[s], Han[n]ys, Hannis, Hawenys, Hawnes, Heyn[e]s], Hein[e][s], Aines etc], Alan 70, 152
Haynes, Alice [Alis, Alles] [b. 1568] 99, 186, 189, 204
Haynes, Andrew, aka Bowne; *see* Andrew Bowne, junior 259, 264

420

INDEX

Haynes, Ann, of Hackney [1601-1607],
  daughter of Richard 159, 199, 216, 217
Haynes, Ann Jane, daughter of Richard 265, 267
Haynes, Anne, [Mrs Stewart], daughter of
  Elizabeth 230, 235, 236, 238, 239, 242
Haynes, Anne, daughter of John [d.1678] 236,
  238, 240, 241, 242, 254, 263, 266-267
Haynes, Anne, wife of William Byack, Barbados
  229, 238, 244-245
Haynes, Arms of; *see* Arms, coat of
Haynes, Arthur Percy, CMG [1864-1928] 234,
  366, 367
Haynes, Bartholomew, husband of Sarah Cavare
  245
  Bartholomew, son of 245
Haynes, Benedict [Benedicke] of Hackney
  father of Henry 176, 177, 180, 200, 209
Haynes, Caroline Anne, Mrs Edmund Hinds
  Knight, daughter of Robert Haynes of
  Thimbleby 379
  children: Caroline Ann Knight, Edmund
    Haynes Knight, Emily Knight, Kathleen
    Maud Knight, Robert Haynes Knight
    380
  marriage at Osmotherley 379
  move to Laceby, Lincolnshire 380
Haynes, Charlotte, Mrs Jonathan Higginson 335,
  365, 367-369, 386
  burial in Riverside Cemetery, Denver 369
  children:
    John Higginson [b. 1838] 369
    Jane Higginson [b. 1840] 369
    Charlotte Elizabeth Higginson [b. 1841]
      wife of Revd Henry Brougham
      Bousfield, first Bishop of Pretoria 369
    Emily Jane Higginson [b. 1843] 369
    Arthur Higginson [b. 1845, d. 1881] 369
    Jonathan Higginson [b. 1848] 369
  departure for the United States 369
  marriage to second cousin, Jonathan
    Higginson 367
Haynes, Christian, Mrs Thomas Gunnell [*see*
  Gunnell]
  children of [*see* Gunnell]
Haynes, Christopher [d.1586] 74-101
  allegations against 74, 79-83, 400
  alleged piracy 74, 103
  alleged smuggling 74, 105, 400
  and the escape of Charles Arundel and Lord
    Paget 85, 87, 89, 91, 94
  bridgewarden of Arundel 76, 91
  burgess of Arundel 74, 76-77
  death in London of 100
  examination of 83, 85, 88, 92
  false cocketting 95
  godfather of William Dartnoll 100, 189
  house of, in Arundel 86, 104

insulting the mayor 76
introduction of Clynsall to Shelley by 90, 94
letter from brother, John 62, 91, 92, 95
Mayor of Arundel 74, 76
purveyor of sea-fish
Reading cloth, assistance with sale of
tavern and tabling house 86, 99, 135
transport of grain by 81
sentence of 100
vessel [hoy] owned by 74, 81, 82
will of 99-100
Haynes, Christopher of Billingshurst, son of
  Robert Haynes of Wiggenholt 100, 101
  children: Christopher, Joan, John, Thomas
    101
  marriage 101
  William Tredcrofte's bequest 101
Haynes, Edmund, son of Elizabeth, 227, 235,
  238, 239, 240, 254
  inheritance of one third share of the
    Newcastle Plantation 235, 239
Haynes, Edmund, son of Richard and Elvira
  Thomas 273, 275
Haynes, Edmund, son of Richard and Elvira
  Thorne 275, 276, 280, 284, 293
  election to House of Assembly 281
  marriage to Anne Thorne 280
  children: Christian Ann, Mary Elvira, Robert
    James 280
Haynes, Edmund, son of Richard Downes
  Haynes 281, 282, 307, 316, 320, 321, 356,
  358, 360-362
  acquisition and sale of Belmont [Bellmount]
    360, 362
  burial at St Peter's Leckhampton
  Claybury Estate 360
  death in Cheltenham 361
  donation of land to Moravian Mission 360
  home in Exeter 361
  marriage to Sarah Bell 360
  marriage to Lucretia [Lucy] Reed 360
  memorial in Barbados 361
  purchase of Haynesfield 360, 362
  son, Joshua 360
  will of 360, 361, 362
Haynes, Edmund Child 338, 357,
Haynes, Edmund Lee 371, 373, 378, 396
  birth at sea 378
  children:
    Emily Lyall [1861] 378
    Robert Haynes [1862] 378
    Mary Elizabeth [1863] 378
    Louisa Howell [1864] 378
    Caroline Annie [1865] 378
    Robert [1867] 373, 378, 396
    Charles Wynyard [1869] 378

421

Florence Agnes and Mary Louisa [1870] 378
Frederica Howell [1873] 378
Alice Maud [1874] 378
death of 378
retirement and move to England 378
Speaker of the House of Assembly, Barbados 378
York 378
Haynes, Edmund Sydney Pollock 338
Haynes, Elena, widow of Arundel 107
Haynes, Elias and Agnes of Berkshire 173
Haynes, Elizabeth, daughter of John [d.1678] and Anne 236, 238, 240, 241, 262, 264-266
Haynes, Elizabeth, daughter of John and Elizabeth Haynes 236, 237, 238
Haynes, Elizabeth, daughter of Robert and Elizabeth of St Thomas 244, 278
Haynes, Elizabeth [b 1707], great granddaughter of John [d.1677/8] and Anne 259, 263, 265
Haynes, Elizabeth, of Hackney [1606-1607] 216, 217
Haynes, Elizabeth of London, sister of Benedict of Hackney 176, 177
Haynes, Elizabeth, of St John, Barbados [d. 1674] 229, 230, 234, 235-238, 239, 242
  children: Anne, Edmund, John see *individual entries*
  grandchildren of 236
  identification of husband of 237
  Robert Stewart, son-in law, wife of Anne
  step-mother of William 236, 237
  will of 236, 237, 239
Haynes, Elizabeth of Thimbleby, Mrs William Hilton Hutchinson 371, 375-6
  birth in Barbados 375
  children of: Caroline Haynes, Elizabeth Reece, Frederick William, Haynes Hanley, Henry Teesdale, William Hilton 375
  death of 376
  period in Beverley with daughter Caroline Whittle 376
  marriage at Osmotherley 375
  memorial window 376
Haynes, Elizabeth, wife of John of Barbados and New York [see Elizabeth Bowne]
Haynes, Emanuel, Lord of Copford 180
Haynes Estate, Barbados, burial from, 1647 181
Haynes family, 248, 401, 402
Haynes, Fanny 295, 296, 301
Haynes, Fanny L. 357
Haynes, Francis, passenger bound for Barbados 245

Haynes, Frederick Hutchinson, Col., ARCM, 371, 380, 381
  children: Henry [Harry], Kathleen [Kitty], Maud Agnes 380
  death at Cheltenham 380, 381
  military career 380
  marriage to Kathleen Kavanagh 380
Haynes, George Barrow 295, 296, 301, 302, 345, 350, 364, 382-385, 387
  children: Mary Ann [1823], Mrs William Slade, Thomasine [1821] 382
  fall overboard 301, 382
  loss at sea 316, 322, 383, 384
  marriage to Anne Walker Glieison 382
  matriculation 302, 382
  Mole-head Committee 384
  partnership in Haynes, Lee & Co 382, 383
  slaves owned by 383
  school at Greenock 302
Haynes, George Barrow, son of Henry and Sarah Jane Dummett 390, 393-394
Haynes, George of Much Haddam, miller 180, 181, 185
  Agnes Allis, wife of 185
  Blithe Eliot, wife of, 180,
Haynes, Henry, Captain, RN 281, 282, 303, 308, 356-357
  children:
    Henry Freeman Oliver, Fellow of Caius College 357
    barrister 357
    children: Adelaide M, Constance M, Edith, Edmund Child, Ellen Louisa, Emily, Fanny L, Florence, Freeman A G, Harriet Augusta, Henry Sidney Freeman, Herbert Pollock Gilbert, Lucy H, Mary E, Rose, Thomasine F
    Lincoln's Inn 357
    marriage to: Emily Wilkes Child 357
    Thomasine Oliver, Lady Thomas 357
    marriage to Commander Sir William Sidney Thomas RN 357
  marriage to Harriet Watkins Oliver
  midshipman and commander 356
  meeting with Robert in Bath 303
  will of 357
Haynes, Henry Husbands, RN, son of Lt-Gen Robert Haynes 295, 296, 331, 364, 366, 389, 390-395
  acquisition of Bath Plantation 391
  Bush Hall Plantation 391
  children of:
    Ann Thomasin [1826] 390, 391
    Robert [1829] 390, 391
    Sara Dummett [1830] 390, 391-392

422

INDEX

marriage to Henry Crichlow Haynes 391
marriage to Robert William Elton 391
marriage to John William Christian 392
Robert James [1832] 390, 392
  Canon, St George's, Bloomsbury 392
  Deacon, Peterborough Cathedral 392
  marriage to Annabel Selina Gambier 392
  marriage to Elizabeth Stewart Gill 392
  pensioner at Caius College 392
  witness to will of Emily Anne Haynes 392
Henry Husbands [1834] 390, 392, 394
  Chief Constable, Welsh Police 392
  first-class cricketer 392
  Senior's Estate, Barbados 392
John Torrance [1836] 390, 393
  marriage to Catherine Garrett Reece 393
Christian Lardy [1839] 390, 393
  marriage to Manoel da Silva Ferreira 393
Mary Frederica [1841] 331, 390, 392, 393, 394
  acquired Lt-Gen Robert Haynes's original notes 393
George Barrow [1842] 390, 393-394
  marriage to Agnes Louise Punch 393
  marriage to Ellen Gordon Haynes Lovell 393
  merchant, East India 393
Elizabeth Reece [1844] 390, 392, 394
  children: *see under* Edward Hyndman Beckles
  marriage to Edward Hyndman Beckles 394
William Clarke [1847] 390, 394
Joshua Rowley [1850] 390, 394-395
  drowned in the sinking of the *Walter Hood* 395
  memorial at Cudmirrah Nature Reserve NSW 395
marriage of 390
midshipman on HMS *Venerable* 390
Haynes, Henry of Hackney 190, 201, 203
Haynes, Henry, son of Robert of Thimbleby 371, 380-381
children:
  Evylyne Elsie [b. 1899], wife of Sir Ronald Weeks 380, 381
  Norman [b. 1893] 380-381
marriage 380
Haynes, Hezekiah, Colonel, later Major General, Cromwell's Army 180, 181, 184
Haynes, Hezekiah, grandson of Major General Hezekiah 180, 181
Haynes, James, of Barbados 245

Haynes, James, Major General, killed at Santo Domingo 244
Elizabeth, wife of 244
Haynes, Jane, daughter of John and Elizabeth Haynes of Barbados 236, 237, 238
Haynes, Joan, daughter of Christopher of Billingshurst 101
Haynes, Joan, daughter of Nicholas 15, 119, 120, 123, 154, 186, 187, 195
Haynes, Joan, wife of John of Marylebone and London 15, 62, 65
burial of 62
Haynes, John, clothworker and Merchant Tailor 48, 68, 135
John, son of 135
Thomas Kowdale (Cowdale), business partner of 48
Haynes, John, Constable of Arundel Castle 168
Haynes, John, grandson of John of Barbados [d.1678] 240, 266
Haynes, John, great grandson of John of Barbados [d.1678] 240, 266
Haynes, John of Barbados [d.1678], son of Elizabeth 227, 234, 235, 236, 237, 238, 239-241, 251, 254, 257, 262, 264, 265
children:
  Anne, wife of Richard Perryman jnr *see individual entry*
  Robert, husband of Elizabeth Blanchett *see individual entry*
  John of Barbados and New York *see individual entry*
  Richard 238, 240, 241
  Elizabeth, wife of Samuel Branch *see individual entry*
  married secondly John Seaward 241, 264, 265
wife Anne 238, 240, 241, 242, 243, 254, 257, 262, 264, 265, 266, 267
  married secondly Beachamp 240, 241, 242
will of 240, 241, 251, 264, 266, 267
Haynes, John of Barbados and New York, 238, 239, 241, 243, 254, 255, 257-264
appointment to Council of New York 261-2
appointment as Commissioner of Revenues, New York 260
ejection from Custom House 261
husband of: 1. Hannah Walcott, Barbados 241, 259
  child: John 241, 258, 259, 263
husband of 2. Elizabeth Bowne, New York 241, 255, 258, 259
  child: Andrew 241, 258, 259, 264
will of, Pelletreau transcript of 262-263
witness of will of Marvin Hales 257, 268
witness of will of Nicholas Walcott 257, 268

Haynes, John, of Barbados, son of John of
    Barbados and New York 241, 258, 259,
    263
  Catharine, wife of 259
  children: Robert 263
      death possibly from Yellow Fever 264
      marriage to daughter of John Hein 263
      will of 263-264
    John 263
    Elizabeth 263
  children's guardian: Ann Critchlow 240, 263,
    266
  death of 263
  importation of slaves by 263
  links with the Bownes 263
Haynes, John, son of Richard of Hackney
    [1600-1607] 216, 217
Haynes, John of Marylebone and London 10, 36,
    38, 39-65, 71, 66, 72, 112, 131, 145, 163-
    165, 166, 199
  action against Edward and William Page 62-
    63
  action brought against by William Page 71,
    199
  Agnes Thomson, wife of 10, 36
  allowance of 46, 67, 166
  birth year 39
  binding over by Privy Council 39
  burial in the vault of St Nicholas Cole Abbey
    62, 65
  cash bonuses, receipt of 164
  commissions 56, 57
  deception of Her Majesty by 58-59
  discharge from office 58
  encroachment by 40, 65
  gift for Queen Mary 40
  house in Queenhithe 61, 72
  keeper of the Pastures 41-43, 44, 54, 58
  lack of scruples of 58
  lodge in Marylebone Park 43, 44, 45, 55, 61
    alterations to 44
  Joan, wife of, burial of 62
  merchant of London 39, 47
  messuages in Byllyngesgate Strete 40
  messuages in Old Fish Street 46,
    granted by Royal Patent 46
      confiscation from the London
        Minoresses (Poor Clares) 48, 68
  Overseer of the Queen's Meadows 66
  perks of
    grazing at Tottenham Court 44
    lease of a cottage in Hampstead 42
    lease on the Manor of Raynehurst, Kent
      51-52, 112
    lease on Rectory of Kirkdale, Yorkshire 51
    lease on mansion house of Sturminster
      Newton, Dorset 51

Queen's Servant 41, 51
responsibilities of 39, 45-46
retail fish business of 40, 47, 48, 49, 59, 64,
    65
sentence of 73
Sergeant of the Acatry 39, 41-3, 45-6, 48, 50,
    52, 58, 61, 65, 131, 145, 163-5,
shop in Parish of St Gabriel Fenchurch
taxation assessments of 59, 72
testamentary capacity of 64
theft from 55
under-keeper, Marylebone Park 43
will of 38
  contest of by Jasper Dartnall 64
  possible influence of Gunnells on 64
William, son of 61
  baptism of 61
  godparents of 61, 62
  presumed death of 61
William Gunnell, faithful servant of 63
yeoman of the store 40
Haynes, John of Newcastle, Barbados, son of
    Captain Robert 277
Haynes, John of Shoreditch, father of Robert
    205, 230
Haynes, John of Wokingham 170, 173
Haynes, John, Sheriff of Wiltshire's Trumpeter
    229, 247-248, 401
  conviction for high treason 248
  transportation to Barbados 229, 247-248
Haynes, John, son of Christopher of
    Billingshurst 101
Haynes, John, son of Robert, mariner, and
    Elizabeth, of Shadwell 230, 234
Haynes, John Higginson, son of Richard Haynes
    of Clifton 365, 369-370
  marriage to Inez 370
    children: Edward, Lucy 370
  marriage to Rosalie Somer 370
Haynes, Jonathan Wynyard 371, 376, 379, 380
  army career 379
    Ashanti War 379
    Indian Mutiny 379
    march to Kumasi 379
    under Sir Garnet Wolseley 379
  children: Alice Yeoman, Cecilia Elizabeth,
    Mary Wynyard, William H 379
  marriage to Anne Yeoman 379
Haynes, Joseph Alleyne, The Hon, of Newcastle
    Estate, son of Richard 365, 366-367, 378
  builder of St Margaret's Church, St John,
    Barbados, 367
  children of
    Agnes Payne (1865->1923) 366
    Alleyne (1859-1938) 366
    Arthur Percy (1864-1928) 366
    Carleton (1858-1945) 366

424

INDEX

Helen Margaret (1868-1937) 366
Jane Charlotte 1862-1928) 366
Mary Howell (1860-1957) 366
Richard (1856-1937) 366
   son, Joseph Alleyne 367, 403
marriage to Margaret Ann Howell 366
memorial to 367
West India Estate 366
Haynes, Joseph (Joss) Alleyne, son of Arthur Percy 367, 403
Haynes, Joshua, son of Edmund 360
   death at Charlestown, Demerara 360
Haynes, Lee & Co, Barbados 298, 382-383, 384
Haynes, Mary, last Haynes to occupy Newcastle 403
Haynes, Mary, sister of Governor John Haynes, wife of John Barlee 180
Haynes, Nicholas [d. 1641] of Winterbourne, near Salisbury, Wiltshire 248, 400
   children: Richard, Susanna 248
Haynes, Nicholas, painter and stainer, Aldgate Ward, St Katherine Crechurch parish 204, 247
   marriage to Elizabeth Dawes 247
Haynes, Nicholas, Purveyor of Her Majesty's Grain xv, xvi, 8, 9, 10, 11, 14, 15, 19, 20, 24, 25, 27, 33, 35, 36, 52, 62, 63, 71, 87, 99, 100, 109-125, 129, 130, 131, 135, 151, 153, 154, 155, 156, 157, 159, 161, 162, 167, 168, 169, 171, 173, 174, 175, 176, 177, 178, 180, 181, 184, 186, 187, 193, 195, 196, 197, 198, 204, 205, 219, 220, 228, 233, 245, 273, 399, 400, 402
   arms of, original 118, 174, 175, 181
   birth year 123, 154
   burial entry in the Parish Record of St John at Hackney 124, 161
   children of 186:
     Thomasine, wife of William Dartnall 10, 15, 36, 119, 120, 186-195
     Joan (Jane), wife of John Kay 10, 15, 36, 119, 120, 187, 195-196
       bequest from Thomas Haynes 36, 195
       daughter: Thomasine Kay 195
     Richard 10, 15, 36, 119, 120, 187, 196-204
     Alice 10, 15, 36, 119, 120, 176, 187
     Robart, son of Nicholas (?)15, 119, 187, 204-205, 233
     John (?) 233
     Nicholas (?) 15, 119, 233
     Christian (?), wife of Rowland Coytemore 15, 119,
   dockets signed by 114, 115
   exemplification of Arms quartered with Foxley, *see* Exemplification
   harbinger, *see* harbinger

   lease of Filborow [Filborough; Felborough], Estchalke [East Chalk], Kent 52, 111
   marriage, *see* Enowes [See Enos], Agnes [Annis, Anne] (Mrs Nicholas Haynes)
   purveyor of wheat in Norfolk 114-116
   theft of his clothing 121
Haynes of Barbados, respectability of xv, xvii, 399, 402, 403
   lineage xv, xvi, 28, 399, 402
   trappings of power 402
   trappings of wealth 402
Haynes of Chessington, Surrey, London and Hackney 176-178, 179, 191, 208
   Alice Maria [Hatton] 178
   Benedict; *see* Haynes, Benedict [Benedicke] of Hackney
   Henry; *see* Haynes, Henry of Hackney
   Matthew 177
   Thomasine [Evelyn], 177
   William 177, 178, 208
Haynes of Daglingworth, Duntisborne and Bagendon 23, 179
Haynes of Finchley, London and Hoxton 169, 178-179
   of Walbrook Ward [City of London] 178
   John, father of Robert 178, 205
     Margaret Man, wife of, 184
   Lyonell 178
   Richard 178
   Robert 178, 205
   Thomas 178, 205
Haynes of Frampton on Severn 176, 179
   Anne 179
   Giles 179
   Richard 179
   Richard son of Anne 179
   Thomas 176
Haynes of Much Hadham, Hertfordshire and Essex 179-182, 185
   Emanuel, brother of John, Governor of Massachusetts Bay 180
   George of Much Hadham, husband of Blithe Eliott, father of John [d. 1606] 180, 185
   Hezekiah; *see* Hezekiah, Colonel, later Major General, James, Cromwell's Army
   Hezekiah; *see* grandson of Hezekiah
   John [d.1606], husband of Mary Mitchell, 179
   John [1594-1654] 180
     Governor of Connecticut 180
     Governor of Massachusetts Bay 180
     marriage to Mabel Harlakenden 180
     marriage to Mary Thornton 180
     migration to New England 180
   John, *see* grandson of Hezekiah
   John, *see* son of Hezekiah

Robert, eldest son of John, Governor of Massachusetts Bay 180, 181, 182
  death of only son of 181
William 185
  children: George, William 185
Haynes of Reading 3-16, 173-176, 179, 182
Haynes of Sussex 101, 168
Haynes of Westbury on Trym, Wick, Abton and elsewhere in Gloucestershire and Dorset 182, 185
  Thomas 182
    Agness Wall of Westbury, wife of 182
Haynes of Winterbourne 248, 400
Haynes, Richard, Constable of Salisbury Castle 162, 170
Haynes, Richard, fruiterer to Henry VIII 7
Haynes, Richard (1792-1859) of Bissex Hill, Newcastle Barbados and Clifton, England 320, 345, 353, 354, 364-365, 387, 388
  burial at St John's Church, Clifton 365, 387
  Bissex Hill Plantation 345
  children: Ann Jane 367
    death in childhood 367
    gift of a slave at her baptism 367
  Charlotte (Mrs Jonathan Higginson) 367-369
    John Higginson 369-370
  Joseph Alleyne [1820-1894] 365, 366-367, 378
  Robert [1818-1853] 365-366
  Eastwood, birth at 364
  marriage settlement, 1816 364
  migration to England 365
  mortgagor of Springfield estate 365
  Newcastle, father's gift of 364-365
  silver plate, father's gift of 365
  steam mill, installation of 365
  wine, father's gift of 365
Haynes, Richard of Hackney [1565-1634] 36, 62-63, 64, 72, 123, 128, 147, 190, 196-204, 216-223, 228, 233, 251
  bequests received by 36, 62-63, 100, 147 196-197
  burial 204
  children 216-223: Ann 159, 199, 216, 217
    Elizabeth 216, 217
    John 216, 217
    Susanna 100, 216, 217, 219-223
    William 216, 217, 218
  churchwarden 200-201, 202, 214
  Clerk of Her Majesty's ordnance 199
  composition agreement for hay, role in fulfilling 171, 202, 203
  death 204
  groom of the chamber 199
  gunner 199
  heir of Christopher of Arundel 100, 147

inheritance of property at Earls Colne, Essex 123, 202-203, 214
  surrender of 203
plumber's bill 200
possible first wife of 203, 216-217
possible second wife of 217
residence in Allhallowes Parish, Bread Street Ward 147
residence in Mare Street, Hackney 128, 200
shop 203
solicitor to John Haynes 62-63, 72, 199
subsidy assessments 199
vestry member, St John At Hackney 200, 201, 202, 214
vexatious suit against 197-198, 213
Haynes, Richard of Newcastle, Barbados [1694-1739] 256, 271, 272, 273, 274-277
  children: Aline 275
    Edmund 275
    Elizabeth, wife of John Howard 275, 276
    Elvira, wife of Thomas Downes 275, 276
    John 275
    Richard, husband of Mary Downes 275, 276, 278
    Robert [Captain], husband of Dorothy Downes 275, 276, 277-280
    William 275, 276
  death 276
  marriage settlement of 275-276
  marriage to Elvie Thorne 271, 275, 276, 280
    child: Edmund 276, 280
  marriage to Elvira Thomas 274
Haynes, Richard, son (alleged) of Richard of Hackney 217-218
Haynes, Richard, son of John [d.1678] 238, 240, 241
Haynes, Richard, Queen's Gunner, Tower of London 32, 161
Haynes, Richard, of Reading xv, xvi, xvii, 3-16, 11-15, 19, 22, 23, 24, 26, 31-148, 162, 173, 178, 179, 181
  children of 11-15, 31-148
    order of birth of 14, 154
  churchwarden of St Giles Reading 7-8
Haynes, Richard, of Wargrave [Waregrove] and Hinton 31-33, 34, 35, 37, 75, 98, 157, 159, 168, 399
  death 32, 98, 159
  house in Hinton 32, 33
  marriage 32, 157; *see also*: Anne Ford
  will of [1566] 32, 34, 399
Haynes, Richard, of Wokingham 162, 170, 173
  enfeoffment of goods of Roger Landen 170
Haynes, Richard Downes of Newcastle, Major General, Barbados [1746-1793] 272, 277, 278, 280-286, 289, 294, 295, 358, 360

426

INDEX

children: Ann Ellcock 358; *see* Ann Ellcock Wood
   Dorothy 358-359; *see* Mrs Dorothy Haswell
   Edmund; *see* Edmund, son of Richard Downes Haynes
   Edward 281, 282
   Henry, Captain RN; *see* Captain Henry Haynes RN
   Lt-Gen Robert, the Diarist 281, 282; *see* Lieutenant-General Robert Haynes
   Richard 281, 282
   Richard Downes, son of Richard Downes 281, 282, 286, 356
     meeting with Sir Joseph Banks 356
     premature death 356
   Robert 281, 282
concern over health 284, 285
death 286, 294
election for House of Assembly 281
journey to England 284, 285
marriage to Ann Ellcocke 278, 280
marriage to Anne Walker 284, 290
Representative to the House of Assembly [St John's] 281, 284, 285
will of 286

Haynes, Robert, Capt (1720-1753), husband of Dorothy Downes 275, 276, 277-280
Captain of St John Militia 278
children: John 277, 278
   Richard Downes; *see* Richard Downes Haynes of Newcastle
death aged 33 278
marriage 277

Haynes, Robert, Lieutenant-General xii, xiv-xvii, 21, 32, 161, 181, 203, 229, 232, 234, 281, 282, 286, 289-332, 343-354, 355, 357, 358, 359, 360, 362, 364, 367, 370, 375, 381, 384, 388, 395, 399-400
attorney to the Lane Brothers 298, 303, 316, 358
backlash, possible fear of 343
children: Dorothy Ann 294, 296, 301
     death from consumption 301
   Fanny 295, 296, 301
     death from hydrocephalus 301
   George Barrow; *see* George Barrow Haynes
   Henry Husbands; *see* Henry Husbands, RN
   John Aughterson 295, 296, 301
     death of 301
   Richard; *see* Richard Haynes (1792-1859), of Bissex Hill
   Robert; *see* Robert Haynes of Thimbleby
   Thomasine; *see* Thomasine, daughter of Lt-Gen Robert Haynes
   William 294, 296
     death by inoculation 294
   William Clarke; *see* William Clarke Haynes, son of Lt Gen Robert Haynes
concern over health 301, 303, 305, 316, 344,
death 352, 354
depression and anxiety of 308
diary of: *see* Cracknell
education of 289-290
education of the children of 290, 301, 348, 351
father's remarriage, his reaction to 290, 291
inspections, military, by 317
integrity of 353
journey to America 290-291
journeys to England 302, 303, 316, 327, 329, 344-345
last home in Reading 347, 351
marriage to Ann Thomasine Barrow 282, 291
marriage to Anna Caitcheon 282, 343, 346
marriage settlement [1790] 293
memorial to, in St John's Church, Barbados 354
meeting with brothers in Bath 303, 316
obituaries 352-354
presentation of sword to 314, 353
Representative of St John's Parish xv, 294, 315, 318, 319
   resignation as 326
Reviewing General of Militia 311
sale of Haynesfield 307
sense of ruin and desolation 289
Speaker of House of Assembly xv, xvi, 316, 318-326
   resignation as 326, 329-330
temporary residence in *Passage*, St Michael 345
will of 350-352
   codicils to 350, 351-2

Haynes, Robert, of Barbados [Fl 1640s] 228, 229, 230
father of Edmund, Anne and John by Elizabeth 237, 238
father of William and John by former wife 233, 238
husband of Elizabeth 237
indentured servant of 230
land transactions 231-233
mariner 230, 234
putative son or grandson of Nicholas of Hackney 233
transfer of property to William Haynes 232

Haynes, Robert of Barbados [c.1666-1696], son of John [d. 1678], 238, 241, 254-256, 272-274
children: Anne, wife of Moore 256, 273, 274
   Dorothy, wife of Philip Rist 256, 273
   Elizabeth, wife of Luke or Fitt 256, 273

427

Lucy, wife of Vincent Draycott 256, 273
    mother of Vincent Draycott 273
Olive, wife of John Hoppin 256, 273
Richard; *see* Richard of Newcastle,
    Barbados [1694-1739]
William 256, 273
    died before reaching majority 272
    eldest son 256, 272
co-executor of his brother, John Haynes of
    New York 262
death 255, 272
education of children 256
lease of land to Francis Dickenson, gent 254
halving the plantation 254
marriage to Elizabeth Blanchett 255, 256
remarriage of widow 256-257
St John's Plantation of 254, 255
surveyor of roads and Vestry 255
will of 255, 256
Haynes, Robert of Clifton (1818-1853) 365-
    366, 385
death by drowning 366
marriage to Emily Anne Mairis 365, 385
Richard, son of 366, 384-385
    children: Agnes Emily, Richard Mairis,
        Lieutenant RN 366
    marriage to Catherine Fisher 366
Haynes, Robert of Finchley 178, 205
Haynes, Robert of Shadwell, mariner 230, 234
Elizabeth, wife of 230, 234
John, son of 230, 234
    baptism of 230, 234
Haynes, Robert of Thimbleby, 295, 296, 319,
    320, 321, 322, 326, 327, 329, 345, 349,
    353, 354, 364, 370-381
arms of Haynes of Thimbleby 373-374
attorney for Richard Haynes, Bissex Hill 345
children of, with Ann Lovell: 330, 349
    Ann Cave Haynes Lovell (?) 331, 350
    Robert Haynes Lovell 330, 331, 348-350,
        381-2
children of, with Elizabeth Reece: 371
    Caroline Ann 371, 379-380
    Edmund Lee 371, 378
    Elizabeth; *see* Elizabeth of Thimbleby, Mrs
        William Hilton Hutchinson
    Frederick Hutchinson 371, 380
    George Higginson 371, 374
    Henry 380-381
    Henry Higginson 371, 377-378
        death and memorial [aged 21] 378
    Jonathan Wynyard 371, 379
    William 371, 376-377
children of, with Sarah Ann Payne 349, 370,
    372
    Jane Alleyne 349, 370
    Robert 349, 370

Sarah Anne 349, 370
Clifden Plantation 329, 346, 349
Clifton Hall Plantation 345, 349, 350, 372
death and funeral 372-373,
Guinea Plantation 349, 372
improvidence of 372
introduction of bill granting rights to
    'coloureds' 370-371
memorial in St John's Church, Barbados 372
move to England 371
purchase of Thimbleby Estate 349, 371-372
    advowson of the Rectory of Laceby,
        Lincolnshire 371, 375
    programme of reconstruction 372
recall of small loans 372
Representative of St John 319, 322, 349
tomb of 372, 373
Haynes, Robert, putative son of Nicholas of
    Hackney; *see* Robert Haynes of Barbados
    [Fl. 1640]
Haynes, Robert, son of John of Shoreditch 205,
    230
    baptism at St Leonard Shoreditch 205, 230
Haynes, Robert, son of William, possibly of St
    Thomas's Parish, Barbados 238, 243-244
burial of, St Philip 244
Elizabeth Brandt, wife of 328
Elizabeth daughter of, wife of Joseph Gibbs
    244
property transactions of 243
will 244
witness to will of Robert Chander 243
Haynes, Robert James, son of Edmund 280
marriage to Elizabeth Battaley 280
Representative of St Joseph 284, 300, 301
Haynes, Robert James, son of Henry Husbands
    390, 392
Haynes, Simon, Dean of Exeter 66, 69
Haynes, Susanna, daughter of John and
    Elizabeth Haynes of Barbados 236, 237,
    238, 251
Haynes, Susanna [1612-1686] daughter of
    Richard Haynes of Hackney 100, 187, 199,
    203, 216, 217, 219-223
baptism 203, 216
hospital lease 222-223
inheritance 204, 217, 219
inquisition at the Questhouse 219
marriage to Edward Hopkinson, chapman, of
    Halifax 219, 220, 221
    children: Anne, Edward, Margaret, 221
marriage to Nicholas Meade 221, 222
    children: Joseph, Nicholas, Susanna 222
marriage to Thomas Ballard; *see* Thomas
    Ballard
mother 217, 219
death and burial 223

INDEX

Haynes, Thomas, of Frampton on Severn 176, 179
Haynes, Thomas, of Hackney, son of Henry 177
Haynes, Thomas, of Westbury on Trym 182,
Haynes, Thomas, of Whitley 7, 10, 16, 31, 33, 34, 35-37, 63, 82, 98, 109, 120, 125, 147, 197, 205
   Agnes Wheatley, wife of 36, 37
   best gold ring 36, 63, 197
   churchwarden of St Giles, Reading 36
   removal of statues and altar 36
   death 37
   servants 7,
   will of 38, 98, 120, 147, 205
Haynes, Thomas, son of Christopher of Billingshurst 101
Haynes, Thomasine, daughter of Lt-Gen Robert Haynes 295, 296, 350, 364, 374, 385-387, 393
   marriage settlement 350, 385, 386
   marriage to Frederick Christian Lardy 386-387
   marriage to John Hothersal Pinder 385-386
Haynes, Thomasin[e], of Chessington-at-Hoke, wife of John Evelyn, daughter of William 177
Haynes, Thomasine, of Hackney, daughter of Nicholas, wife of William Dartnall 10, 15, 36, 119, 120, 186-195
Haynes, William, '*comfot maker*' of London 135
Haynes, William, Groom of Her Majesty's Stable 145, 146
   Yeoman of the Toyles (Toils) 145
Haynes, William, of Chessington-at-Hoke, 177, 178, 208-209
Haynes, William of Hackney [b. 1603], son of Richard 216, 217, 218
Haynes, William, of Hurst 31, 32
Haynes, William of St John, Barbados [d.1679] 230-233, 236, 237, 238, 240, 242-244, 247, 250 251
   acquisition of land 230-232
   brother John 237
   children: Robert 243-244
      Theodore 238, 242
   godfather of John Haynes
   marriage to Anne Gaines 238
   sister, wife of Richard Cockman 243
   son of Robert Haynes 233, 236, 238
   son-in-law (step-son) of Elizabeth 235, 236, 237, 247
   transfer of slave, Mary 242
   witness to sale of slaves 242
   witness to indenture 242
   witness to will of Robert Benson 250

Haynes, William, of Thimbleby 371, 376-377, 379
   birth 376
   children of: Charles Herbert 377
      Clarence Farrell 376, 377
      Constance Farrell 377
      Edmund Lyall 377
      Elizabeth Woodfield 377
      Ellen Louisa Wynyard 377
      Emily R 377
      Henry Gittens 377
      Robert 377
         apprenticeship of 377
      William Clarke 377
   Clifton Hall estate 376
   death 377, 379
   inheritance of Thimbleby 376
   interment at Osmotherley 376
   marriage to Ellen Farrell Gittens 376
   memorial in Barbados 377
   purchase of The Hall, Patrick Brompton 376
   sale of Thimbleby 376
Haynes, William, pensioner 146
Haynes, William, pirate 146-147
Haynes, William, purveyor of sea-fish 10, 13, 14, 15, 19, 21, 24, 27, 35, 36, 45, 47, 48, 50, 51, 56, 59, 62, 63, 66, 69, 71,73, 82, 99-101, 117, 118, 119, 122, 127, 130-148, 151, 154, 155, 157, 159, 160, 161, 164, 165, 169, 173, 175, 176, 179, 180, 198, 203, 207, 213
   allegations of abuses against 57, 114, 144, 145
   arms, coat of 19, 27, 117, 118, 119, 135, 147, 151, 173, 175-176, 179, 180
   death of 147, 213
   difficulties with Rye fishermen 134, 139-144, 145, 165
   executor and heir of Christopher, 159, 198, 203, 206
   exemplification of arms 19, 147, 151, 174-175; *see also* Exemplification of Arms and Crest
   fishmonger 130-131, 133, 147, 164, 206, 207
   John Haynes, servant to 135
   leases
      Holy Cross Chapel, Winchelsea 132
      house and land of the Black Friars, Winchelsea 132
      house and land of the Grey Friars, Winchelsea 132
      land in Blisworth, Northamptonshire 133
      manor and parsonage of Stanton Lacy, Shropshire 131
      mansion in Porchester [Moralles], Hampshire 132
      parsonage of Bisley, Gloucestershire 133

rectory of Porchester 132
rectory of Shipton, Hampshire 132
St Anne's Chapel, Winchelsea and associated land 132
Stocks Market in Poultry 122, 130, 133-134, 140, 147
letter to Mayor and Jurats of Rye 139, 140, 144, 147, 150-151, 165
licence to import fresh fish in barrels 136-138
marriage to Ellen Bonne 15, 69, 134
purveyor of the Freshstore 66, 132
replacement of 59, 145, 165
residence in Baddinges Warde, Rye 136, 169
seal of 73, 147-148, 151, 173
tithe rents from the rectory of Baschurch 147
use of the Royal Prerogative 136, 138, 141
William, son of 119, 127, 135, 159, 160
burial of 135
yeoman of the Guard 131
Haynes, William, yeoman of Much Hadham 185
children: George, William 185
Haynes, William Clarke, son of Lt Gen Robert Haynes 295, 296, 316, 327, 345, 350, 351, 364, 387-390
acquisition of Bannatynes estate 387
birth at Newcastle Estate 295, 296
children: Ann Thomasin [1828] 387, 389
Henry Critchlow [1826] 387, 389
marriage to Sarah Dummett Haynes 389
Robert [1824] 387, 388
children of: Edgar Vere [Harbourne] 388
Eugene Harbourne 388
Helena Harbourne 388
principal beneficiary 389
Mary Harbourne 388
Robert Harbourne 388
William Lindsay 388
father's executor 389
marriage to Mary Harbourne Straughan 388
William Clarke [1831] 387, 389
children of: Anne Thomasine, Charlotte Maria, Margaret Anne, William Clarke 389
marriage to Charlotte Maria Dalby 389
gifts from his father 345, 350
marriage to Margaret Ann Critchlow 316, 387
premature death of 327
will of 388, 389
Haynes, William Lyall, son of Clarence Farrell 376
last Haynes to own Clifton Hall estate 376
Haynesfield Plantation, formerly Baldricks 297, 305, 306, 307, 360, 362

Haynes Lovell, Ann Cave, Mrs Ann Cave Dummett 331, 343, 350, 351, 391, 394
controversy over father's identity; *see under* Robert Haynes of Thimbleby
Haynes Lovell, Anthony [Tony] 330
Haynes Lovell [Lovell-Haynes], Robert (1825-1903), grandson of Lt-Gen Robert Haynes 330, 331, 343, 347, 348-350, 351, 355, 372, 375, 381-382
baptism 330, 331
children of: Annie Douglas K. Haynes 381
Avice Norah Haynes 381
Charles Booker Haynes 381
emigration to Australia 381
Edith Haynes 381
Ellen Gordon Haynes, wife of George Barrow Haynes 381, 393
Henry Hunter Haynes 381
Jessie Haynes 381
Mary Lee Haynes 381
Nellie Haynes 381
Nina Marianne Haynes 381
Robert Haynes 381
surgeon of Hans Crescent, London 381
death, cause of 381
marriage to Harriet Kendall Inglis 348, 381
military career: Haddington Artillery Militia 381
parentage of 331, 381
provision for by Lt-Gen Robert Haynes 351, 381
will of 381-382
Haynes Hill [*see* Haines Hill],
Hays [or Hayne], Charles, Master of the brig *Martha* 382
Head, Charles Arthur, iron-founder 377
Hein, John 263
Hein, Charles, son of John 264
godson of Robert Haynes 264
Heir-at-law 293, 333
Henley, herbage at 110
Henri III of France 85
Henry III 48
Henry VI 104, 162
Henry VII xiv, 29, 162, 182
Henry VIII xiv, 7, 8, 44, 48, 49, 54, 60, 68, 69, 122, 162, 165
herald 10, 11, 20, 21, 22, 27, 117, 118, 119, 147, 176, 180, 248
bogus 118,
herd/heard [herdsman] 45, 52
herd, John Haynes's private 41
heretics 95
Her Majesty's 53rd [Shropshire] Regiment of Foot [Infantry] 386
Her Majesty's Ordnance 199
Hertford, Earl of 65

430

INDEX

Heynes, Nicholas, messenger to the Royal Household 162
   attendance at funeral of Henry VII 162
Higginson, Ann Campbell, daughter of John Higginson 386
Higginson, Deane & Stott, Messrs 321, 335, 368
Higginson, John 292, 316, 318, 321, 323, 335, 367, 368, 383, 386
   partner in Barton, Irlam and Higginson 335, 367, 368
Higginson, Jonathan, son of John, husband of Charlotte Haynes, 335, 367-368, 386
   bancruptcy 368
   children (*see under* Charlotte Haynes)
   death from 'softening of the brain' 369
   purchase of Palmer's estate 368
Higginson, 365, 371
Highgate, Barbados 387
High Holborn 219, 285
high respectability, family of, xv, xvii, 399, 400, 403
Highlander 382
Highlanders, 42nd (Black Watch) 379
Hill Park, Kent 303, 304
Hillary, Dr William 294
Hinds, Ellen, Mrs John Gittens 376, 379
Hinds, William, representative for St Peter 319
Hinton [Hynton, Henton], Berkshire, formerly Wiltshire 31, 32, 33, 63, 163, 168, 173, 197, 208
Hirstenhalderst [see Hurst]
HMS *Barossa* 398
HMS *Caledonia* 304
   renamed HMS *Dreadnought* 304
HMS *Echo* 336
HMS *Herald* 320
HMS *Hotspur* 366
HMS *Sapphire* 308
HMS *Scamander* 390
HMS *Venerable* 309, 390, 398
HMS *Victory* 301, 304, 333
Hobbes, Francis, servant of Edward Caryll 89
Hodgkinson, Francis 323
Hodgson, Margaret
Holder, Ashley, Mrs Robarts William Elton, deceased
Holder, Henry Evans 289
Holder, Hon. J.A. 337
Holder, Revd Caddle 391
Holetown Village, Barbados 227
Holland, wars with 223
Hollingshed, William 293
Holy Cross Chapel, Winchelsea 132
Homerton 122
Honor 247
Honorable Board of Council 319
Hope Estate, Christ Church 350

Hopkinson, Edward [d.1640] 216, 217, 219, 220, 221
   children: Anne 221
      Edward [b.1636] 221
         bequeathed the Hackney and Arundel properties of the Haynes family 221
      Margarett [Mrs Watson] 221
   cousin, Nicholas Enos; *see* Nicholas Enos
   property in Halifax 221
Hopkinson, Edward of Halifax, father of Edward 219
Hoppin, John 255, 273, 274, 276
Hoppin, William 276
Hopton, John 8, 14
Horsham market 89
Horthorne, John 8
Horton, Mr (Sir Robert John Wilmot-Horton), Under Secretary of State for War and the Colonies 322
hospital leases 222-223
Hospital of St John of Jerusalem 54, 122,
Hospital, Naval, 283
Hothersal, Ann Isabella, wife of William Pinder 385
Hothersal, John, father of Ann Isabella 385
Hothersall, Captain Thomas 230
House of Correction, Hackney, 200, 202
Household, Royal xvi, xvii, 3, 4, 12, 16, 36, 37, 39, 40, 41, 42, 45, 47, 59, 75, 82, 87, 109, 112, 118, 122, 130, 136, 141, 144, 146, 161, 162, 164, 166, 168, 185, 196, 199, 206, 213, 401
Houston, Grant xii, 396
Howard, Bartholomew, of Ratcliffe, Middlesex, mariner 246
Howard, Catherine 85
Howard, Henry, Lord, uncle of the Earl of Arundel, nephew of Queen Catherine Howard 84, 85,
Howard, John 275, 276
Howard, Mary, wife of Edmund Perryman 266, 267
Howard, Philip, Earl of Arundel 93, 95-96
   conviction of treason – sentence commuted 96
   death in custody 96
Howard, Queen Catherine 85
Howard, Zachariah 293
Howell, Benjamin Carleton, Colonial Treasurer 366, 378
Howell, Margaret Ann, Mrs Joseph Alleyne Haynes 366
   memorial to 367
Howell, Susan Thomas, Mrs Edmund Lee Haynes 378

Hoxton [Hoggeston], Hackney 122, 169, 178, 179, 205, 233, 400
Hoy, 81, 83, 86, 91
Hudson Bay Company 257
Huett, Dr, of Hackney 124
Hughes, Revd Griffith 278
Humfrey, John. Arundel coroner, burgess and one-time customer 77, 91, 102
Humite (hummetty) 9, 27, 175
Hundred of Brothercross, Norfolk 115
Hundred of Gallowe, Norfolk 116
Hundred of Holt, Norfolk 115
Hundred of Osulston (Ossulston), Hackney, Middlesex 59, 109, 196, 213
Hundred of Smithdon, Norfolk 116
Hundred of South Erpingham, Norfolk 115
Hunsing, Robert, purveyor of sea-fish to Henry VIII 48
Hunt, William 50, 138
hurricane, damage from 235, 237, 239, 255, 294, 282, 284, 289, 327, 328, 329, 360, 361, 368
Hursley, Hampshire
Hurst (Hirstenhalderst),
  House 163
  Lodge 163
  poor of 32
  woods of the Abbot of Abingdon 29
Hurst, Anthony, of Eltham, Kent 192
Hurst, Mary, widow of Anthony 192
Husbands, Reverend Henry 291
Hutchinson, Daniel 274
Hutchinson, John 276
Hutchinson, Major Peter Hanley xii, 396
Hutchinson, Revd Frederick William 396
Hutchinson, Revd William Hilton 375
Hutchinson, Teasdale, father of Revd William Hilton 375
Hutton Bonville 396
Hyampsill, Robert 243
hydrocephalus 301
Hythe, Kent 68

Idaho 369
Ifill, Benjamin 323
IGI (International Genealogical Index) 107, 108, 210, 255, 267, 275, 333
Iles des Saintes, Les 283
Ilicke, Thomas, of Dunkerton, Wiltshire 218
illegitimacy in Barbados 332
Imperial Registry Bill 311
Ince, John 286
indentured servants 217, 223, 227, 249, 401
India [East India] Company 191-192, 397
Indian Mutiny 379
indigo 231, 232, 249

influenza (sweating disease; grip) 158, 188, 346, 373, 381
Inglestorpe Manor 203
Inglis, Harriet Kendal[l], wife of Robert Haynes Lovell 308, 348, 381
Inglis, Peter 381
Ingoldsby, Major Robert 270
  arrest of Leisler by 270
ingrafting 294
Inn, Flying Horse 122
inoculation 294, 295, 333
  of Prince of Wales 295
inquisition(s) 14, 16, 18, 56, 70, 71, 75, 90, 102, 132, 184, 219
Insurrection of 1816 289, 311-315, 336, 343, 352, 353, 385, 402
interrogatories 39, 53, 60, 72, 122, 127, 128, 141, 143, 148, 154, 197
Iphigénie, frigate 398
Ireland, poverty in 300, 315
Irish, Thomas, master of the *Falcon* 224
Irlam 302, 303, 304, 305
  capture of prizes by 302
  loss of 305
Irlam, George 335, 368
Isake, Rafe 51,
Isham, timber merchant of London 86, 90, 106
Isle of Bourbon, Réunion 293
Isleworth School 357
Ixodes ricinus 268

Jacobson, Matthew, Master of the brig, *Fame* 382
Jackson, Robert, Mayor of Rye 151
Jamaica 246, 283, 284, 320, 359
James I 96, 102, 113, 118, 306
James II 223, 260
James IV of Scotland 121
James V of Scotland 8
James VI of Scotland 146
James, Joseph, Master of the brig *Atlantic* 382
James, Robert, slave guilty of raping a white woman 344
Jane, a slave of Anne Haynes 240
  children: Crab and Will 240
Jarvis, Admiral Sir John, CIC Mediterranean Fleet 293
Jefferey, John, Jurat of Hastings 56, 57
Jemmett, Gabriel 322, 323, 370
Jenkins, Judith 243
Jenner, Edward 333
Jenyns, Thomas 27
Jerusalem, Hospital of St John of 54, 122
Jerusalem, Prior of St John of 44
Jesuit threat xvi, 84
Jewkes, Edward, Sergeant of the Bakehouse 113
Jockey, Joan, attack on 128

INDEX

Joe's River Estate 289, 295, 334, 395
John and Mary 382
John, Bishop of Winchester 132
John Bull 315
John Lee & Co 383
John, slaver of Liverpool 355
John of London 247
Johnson, Charles 292, 333
Johnson, Henrik 47
Johnson, Hugh, vicar of St John At Hackney 201, 214
Johnson, James 293
Johnson, John 231
Johnson, Meriel xii, 328
Johnson, Robert 293
Johnson, Trilby, Dr xiii, 373
Johnston, Eleanor Brenda Violet 338, 396
Johnston, Ivor, of Helensburgh xii, xiv
Johnston, William 267
Jones, Harriet 348
Jones, Robert, Richard Besbitche's deputy 69
Jordan, Mr, Agent of House of Assembly in England 316
Jordan, William, planter 272
   will of 274, 286
juniper 55
Jurat(s); *see under* Rye

Kavanagh, Kathleen 380
   Thomas Henry, VC father of Kathleen 380
Kay [Kaye; Key(s)], Arthur, son of John of Woodsom 123, 124, 195
Kay, Ellen Lynnet, wife of 123, 195
Kay, John, of Woodsom, Yorkshire 123
Kay, John son of John of Woodsom 123, 129, 162, 187, 195
   arms impaling Haynes [Heynes] of Hackney 196
   children by Bridget: Bridget 196
      Dorothy 196
      John 196
         Arthur, his son 196
         John, Bart., his son 196
      Lucy 196
   children by Joan Haynes: Thomasine 187, 195, 196
   Clerk Comptroller of the Green Cloth 123, 162
   Clerk of the Averye 196
   marriage to Bridget 196
   marriage to Joan Haynes 123, 187, 195
   scrivenor of Hackney 196
   will of 123, 129
Kay, Thomas, Avenor and *'accomptant'* 213
Kempe 379
Kent, Elizabeth, wife of Edward Thomas 250
Kenton, Tiverton, Devon 370

Keysor, George, Master of the *Irlam* 302
Kidney, Mary 243
King, Alexander 323
King, J Hampden, Acting Colonial Secretary 389
King, Margarett 256
King's Bromley, Litchfield 316
King's Garden, manor of, Colne Engaine/White Colne, Essex 202
King's Own Regiment 386
King's Shop, Rye 136, 138
Kingsland Leper Hospital 223
Kingsland Road 124, 223
Kingston 178
Kirdford, Sussex 189, 190, 194, 210
Kirkdale, Yorkshire [Monastery of Newborowe], Rectory of 51
kitchen 43, 46, 163, 185
Kitchin, Thomas 232
Knatchbull, Reynold 52, 112
Knight, Edmund, of Barbados, father of Edmund Hinds 379
Knight, Revd Edmund Hinds 379, 380
   Rector of Laceby 380
Knight, John, Commissioner of the Exchequer 53
Knight, Sarah 370, 379
Knights Hospitaller 177
Knollys [Knolles], Sir Francis, Treasurer of the Household 4, 45, 57, 142
Kowdale; *see* Cowdall
Kumasi 379

Laceby Rectory, Lincolnshire, advowson of 371, 375, 380
Ladd, John, of London, mariner 246
Lamb, Ann, Stockton-on-Tees 377
Lambwood Hill, Grazeley, Shinfield Parish 27
Lamplighters, near Bristol 329
Lancaster, East Road 352
Lancaster, William, Sargeant of the Acatry 58
Landen, Roger, of Wokingham 170
Lane, Jenny, daughter of *Old Doll* 299
Lane, John, absentee landlord 298, 337
Lane, John Newton 316
Lane, Richard James, lithographer 327
Lane, Thomas, solicitor, absentee landlord 298, 303, 306, 309, 310, 313, 315, 316, 337, 358
Langbourn Ward, London 40
Langley, Sir John, goldsmith and Alderman of London, godfather of William, son of John Haynes 61
Langley, Thomas, purveyor of grain 113
Larder, The 163
Lardy, Mrs Elizabeth Sophia 346
   death of Ann Thomasine Haynes, while her guest 346

433

Lardy, Lt-Col Christian Frederick, second husband of Thomasine Haynes 296, 317, 385, 386, 387, 393
  marriage of 317, 386
  military career 386
Lardy, Lieut F 386
Lardy, Peter [Pierre], father of Christian Frederick 386, 397
  Lt-Col, de Meuron Regiment 397
La Soufrière, St Vincent, eruption of, 308
Lathbury, John 25
laudanum 159
Laudian persecution 180
Laughton, Sussex 222
laws against piracy 97
laws for the preservation of the Queen 97
  impact on Christopher Haynes 97
Lay Subsidy Rolls for Wargrave 37,
Lay Subsidy Rolls for London, 1576, 1582 48, 61, 68, 72, 105, 112, 133, 145, 147, 149, 213
Leadenhall, Flesh Market in 191, 192, 194
Leadenhall Street 191
Leckhampton, St Peter's Church 361
Leddra, Nathaniell 245
Lee & Garner 384
  commission, brokerage and interest on sugar sales 384
  slave compensation 384
  slave ownership by 384
Lee, Thomas, of Liverpool 346, 350, 352, 382, 384
  Lightfoots 350, 384
  Mangrove Pond 384
Leeward Islands 398
Leicester, Earl of, Lord Steward, *see* Dudley, Robert
Leipzig 191
Leisler, Jacob 260, 269, 270
Leisler's Rebellion 260
Leith, Governor Sir James, Barbados 353, 390
Leith pier 195
Lennard, Elizabeth 178
Lennard, William, mercer, father of Elizabeth 178
Lennox, Margaret, Dowager Countess of 121
Leslie, Andrew 254
Leslie, Capt [later Colonel] John 240, 254, 256, 257, 267, 268, 286
Leslie, Elizabeth, wife of Sir Peter Colleton 257
Letter of Marque and Reprizals; *see under* Captain Robert Hunter Caitcheon 346
letters patent 46, 50, 65, 71, 111, 137, 170, 320
Lewis's Hotel, Bridgetown 320
Lewis, John, indentured servant 250
Lewkenor, Richard, Sergeant-at-Law 86, 104
Lewkenor, Thomas of Angmering 88, 93, 94

Leytonstone 113
Liberatador, the first steamship seen in Barbados 324
licence to sell raw hides
Life Guards, Inspection of 317
Lightfoots Estate 350, 384
Ligon, Richard 234
Linbridge 338
Lincoln 144, 336
Lincoln, Earl of, Lord Admiral of England 80, 103
Lincoln's Inn 357
Lindsay (Lindsey), Alexander 233, 249, 250
Ling (*Molva molva*)
Lingard (Lyngarde), William 132
Little Burstead, Essex 391
Little Sixpence 203
Little Wakering, Essex 211
Littlehampton 78, 94
Littmans and Giffords 120
Liverpool 298, 301, 302, 303, 304, 305, 316, 322, 335, 346, 347, 348, 352, 355, 365, 367, 368, 369, 371, 382, 383, 384, 386
  Royal Bank of, 368
Loftis, Richard, mercer, father of Blanche Glover 52
Lombley, Francis, farrier, of Hackney 203
London, Bishop of 122, 201, 224, 247
London, City of 10, 14, 39, 49, 56, 65, 68, 70, 72, 125, 148, 149, 178, 212
  Aldersgate Ward 213
  Aldgate Ward 204
  Bassingshaw Ward 162
  Bread Street Ward 48, 56, 134, 147, 149
  Cornhill Ward 133
  Langbourn Ward 40
  Queenhithe Ward 72, 105, 130
  Walbrook Ward 178
London Company of Fishmongers 49, 138
London, Great Fire of, 49, 65, 134, 223
London,
  Great Russell Street, 303
  Merchant House 246, 287, 385
  Norfolk Hotel 304
  Old Fish Street 46, 47, 48, 49, 56, 59, 61, 64, 65, 68, 140, 149, 152, 157, 190, 207, 208
  Middle Row 49
  Parish of St Botulph, Aldersgate 213
    St Gabriel Fenchurch 40, 65
    St Gregory by St Paul 251
    St Mary Antill [St Mary-at-Hill] 40
    St Mary Magdalene 134
    St Mary Somerset 61, 72, 191
    St Mary the Virgin, Aldermanbury 251
    St Mary Woolchurch 133
    St Michael, Cornhill 47,

INDEX

St Michael Crooked Lane 190, 210
St Nicholas Cole Abbey 48, 49, 59, 61, 62, 65, 68, 72, 73, 100, 108, 127, 134, 135, 145, 149, 206, 207, 215
St Peter 72
Port Book 67
Port of 47
Long Ditton 178
Long, John, Master of Reading School 17
Long Melford 61
Lord Admiral, The 80
Lord Chamberlain 110, 162
Lord Chamberlain's Accounts 146, 199
Lord, John Thomas 308
Lord Mayor's Court of London 248
Lord Steward 45, 52, 53, 152, 162, 165, 170
Lord Warden of the Cinque Ports 56, 144
Lords Lieutenant 167
Louis XVI of France 293
Louis XVIII of France 309
Lovell, Ann 289, 330-332, 338, 348, 349, 350, 351
  death from erysipelas 331
  move to England 331
  slaves registered by 330-331
  'widow' of Robert Lovell 331
Lovell, Johannes, possible brother of Ann 330
Lovell, Mary, possible sister of Ann 330
Lovell, Ann Cave [Mrs Dummett]; see Ann Cave Haynes Lovell and Ann Cave Dummett
Lovell, Edward and Mary 330
  children of: Edward, Mary, Philip 330
Lovell, Eliza, daughter of Kitty
Lovell, Jarrett 330
Lovell, Kitty 339
Lovell, Mary Francis 331
  slaves registered by 331
Lovell, Phillip 330
  Philip, son of 330
Lovell, Robert, alleged husband of Ann 331
Lovell, Robert, husband of Elizabeth Osborne 330
Lovell, Thomas and Mary Ann 330
Lovely Lass 291
Lowdia, Robert, boatswain of the *Cockatrice* 355
Lower, Sir Nicholas and Lady 336
Lowther, Christopher, servant of Henry Drax 247
Luke family 273
Luke, Elizabeth 266, 273
Lussher (Lusher), William, of Thakeham, Deputy Vice-admiral of Sussex 83
Luxborough 113
Lydesey, Robert 56
Lyon, Lieutenant General Sir James Frederick, Governor of Barbados 325, 326, 353

Lynnet [Lynet; Lin(n)et], Ellin, daughter of Hippolitus 195
Lynnet, George, Constable of Hackney 122
Lynnet, Hippolitus [Hippolite; Ippolite], son of George 110, 111, 112, 122, 123, 124, 162, 195, 203
  accusation of perjury 122
Lynnet, William, son of Hippolitus 201, 203, 214

Macfarlane, Professor Alan xiii
Mackenzie, Robert Humberstone, Lord Seaforth 268, 285, 300, 301, 334, 353
Madison, US President 308
Maidenhead Inn, Dyott Street, parish of St Giles in the Fields 53
Mairis, Emily Anne, wife of Robert Haynes of Clifton 365-366, 385, 387
  death and burial at 366
  son: Richard Haynes
  will of 366, 392, 395
Mairis, Major Valentine Hale, father of Emily Anne 365, 395
  wife: Elizabeth Edwards 265
Mairis, Maria Adelaide 395
Malbon, William 17, 169
malmsey 104, 158-159
Man, Margaret, wife of John Haynes of Finchley 184
Manorial court 120, 123, 202, 203
Manors of Sheriffes and Barwick Hall 202
Mansion House 133, 148,
Manuden, Essex 222
Manumission 299, 324, 335
Margaret [Tudor] of Scotland, wife of King James IV 121
Marino, Julio 151
market men 138
Marques Wellington 382
Marshall, Robert Johnson, deceased 293
Marshall, William 276
Marshalsea: *see* prison
Martha 382, 383, 384
  damage suffered by 383
  sinking of 383, 384
Martial Law 300, 301, 313, 315
Martinique 277, 283, 293, 300, 301
Martyn [Martin], Joan, second wife of Sir John de Foxley II 24, 26
Martlet 9, 17, 63, 174, 184
Mary, Queen 4, 40, 44, 80, 170
  Haynes's gift to 40
Mary, Queen of Scotland 84, 96, 125-126
  execution of 126
Mary Ann 352
Marylebone [Marybone] 39, 41, 50, 54, 56, 58,

435

59, 60, 61, 63, 67, 169, 173, 183, 208, 366, 392, 393, 394
  dangers in the vicinity of 55-56
  Great Northern Hotel 366
  magistrates 55
  Park 41, 43, 44, 46, 52, 56, 58, 59, 60
    improvements to buildings in 44-45, 55
    improvements to watercourses in 44
    keeper of 41, 43, 44, 54, 55
    Long (Lustie) Lane, 56
    poachers in 55
    removal of hinges from the gate of 55
    repairs to fences of 44, 45
    underkeeper of 39, 41, 46, 52, 55
    woods in 59, 60
Mason, Henrietta 339
Massachusetts Bay 180
Massam, William, transport of wheat by 82
Massiah, Christopher 293
Massiah, William B. 293
Master or Mr to signify a member of the gentry 127
Masters and Servant Act (Contract Law) 344
Mathews, Richard 8
Mauleverer, Dorothy, wife of John Kay of Woodsom 123, 195
Mauritius 293, 397
Maxwell, James 305, 308
Maycock, J.D. 321, 337
Mayers, Frances, wife of John Henry Barrow 303
  of Halton Plantation 303
Mayers, John Pollard 300, 315-316, 334
  amendment to the Militia Bill 300
  Barbadian Agent in England 315-316
Mayers, Joseph 321
Mayers, Thomasine 303
Mayhew, Graham 159, 160
Maylkyn, Francis, churchwarden, Hackney 200
Maynard, Jacob 265
Maynard's Estate, Christ Church 351, 388
Mayo, William 275
McGrath, Rebecca, wife of Dr Edward James Dummett 390
McQueen, James 323
Mead(e), Agnes 224
  will of 224
Meade, Edward, of Berden Hall, grandfather of Nicholas sen 224
  will of [1577] 224
Meade, Joseph, Fellow of Christ's College, Cambridge 221, 224
  author of *Clavis Apocalyptica* 221
  mentor of John Milton/Henry More 224
Meade, Mary, wife of Nicholas of Berden Hall, Essex 221

Meade, Nicholas, of Berden Hall, father of Nicholas 221
Meade, Nicholas, husband of Susanna Haynes 221, 222
  children: Joseph, Nicholas, Susanna 222
  will of 222
Melcombe Regis 357
Mendoza, Don Bernardino de, Spanish Ambassador 96
Meredith, Hugh, cook on the *Cockatrice* 355
Mermaid 256
Merman of Amsterdam 47
Mery, Richard, vicar of St Mary's Church, Climping 83
Merrick, Rose, wife of Jasper Dartnall 189, 190, 191, 193
Meyer, Nicholas de 261, 262
Meyers (Mayers), Robert John, first husband of Thomasine Bonnet 292
Michell, George B. 179
Middlestreet Ward, Rye 141
Midhurst, Sussex 75, 82, 102
Militia xv, 258, 260, 277, 278, 279, 280, 283, 285, 292, 300, 302, 311, 312, 313, 317, 326, 334, 337, 352, 354, 364, 381, 385
  Bill 300, 337
  Christ Church 312
  Haddington Artillery 381
  law 317
  New York
  obligation to join 277
  service in 337
  St George's Battalion 302
  St John's Regiment of, aka First or Royal Regiment 283, 300, 302, 317, 334
  St Philip, Battalion of 317
  Virginia 279
Millard, Nicholas 214
Millcroft, Sussex 222
Miller, Frederick 69
Miller, J. 281
Miller, Richard, Keeper of the Queen's grounds at Deptford 66
Milles, James, customer of Rye 80
Millis, Mathew, William Haynes' deputy in Rye 149
Mills, John 230, 231, 249
Milton, John 224
Minbah, a slave
Minoresses of London, known as the Poor Clares 48, 68
Minvielle, Gabriel 61
Mitchell, Mary, wife of John Haynes of Much Hadham 180
Mitton, Revd Joseph, MA 375
mixed race, people of 285, 300, 309, 344
  acquisition of property by 300

INDEX

right to testify against whites 300
slave ownership by 300
Moe, Cheeseman, Speaker of the House of Assembly 318, 319, 338
Mole-Head, Bridgetown harbour 384
Moll, Herman, geographer 243
monarch, New Year's gifts to 40
Monastery of St Peter and St Paul, Baschurch, Shrewsbury 147
Monmouth County, East New Jersey 258, 259, 261, 262, 263, 268
Montserrat 283
Montague, Anthony, Viscount 56, 71,
monumental inscription(s) xii, 228, 244, 252, 333, 357
  St John's Barbados 228
Moor Place, Hartley Battle 29
Moore, Anne, Barbados 274
Moore, John, Barbados 254, 274
Moralles, Porchester 132
Moravian Church, Mount Tabor 360
Moravian Mission 360
More, Christian, daughter of Nicholas 29
More (Moor, Moore, atte More), Cristyne, [Christian, Xrine] 29
More, arms of 11, 24, 25, 29
More, Dorothea, wife of Thomas Perkins, sister of Thomas More of Lancelevey 101
More, Giles, clerk to Edward Caryll's ironworks
More, Henry atte 29
  Christian, wife of 29
  Nicholas, son of 29
More, Jane, daughter of Nicholas 29
More, Joan, wife of John of Bray 25, 26, 28
More, John 24, 29
More, John, of Bray 26, 29
  William, heir of 29
More, John 25, witness
More, John, vicar of St Giles 17
More, links with Ailleward 24, 25, 26
More, Maude (Mawd) 11, 26, 27, 119
More, Nicholas of Barkham and Colemans More 19, 25, 26, 102
  Nicholas, son of, fuller 25
More, Nicholas, son of Henry atte 29
  Jane Druez, wife of 29
More, Nicholas, vicar of St Giles 17
More of Bray 26, 29
More of Wyford, Pamber, Hampshire 29
More, Richard atte
More, Richard, of Bray 29
  John, son of 29
  Katherine, widow of 29
More, Robert 27
More, Thomas 101
More, Thomas, father of William of Sherfield-on-Loddon 101

More, William atte, of Wyeford, Pamber 29
  Henry, son of 29
More, William of Loseley, Vice-admiral of Sussex 79, 83, 102
More, William, of Sherfield-on-Loddon 75, 101
More, William, possible brother of Nicholas 25
More and More Molyneux family of Loseley Park 79
Mores, Manor of, 25, 29
Morgan, Robert, merchant of Barbados 231, 251
Morgan, William, of Chilworth Manor, Surrey, Deputy Vice-admiral of Sussex 79, 83, 102
Morgin, Elizabeth, wife of Robert Haynes 251
Morish, Charles 231
Morris, Colonel Lewis, of New York 262, 270
Mould Meade (Broade Marsh) 66
Mount Clapham Estate 368
Mount Harmony plantation 345
Mount Pleasant 291, 293
Mount Vernon 279
Mount Wilton 306
Mullion, Hamlet 346, 355
Murray, Dennison, first mate of the *Cockatrice* 355
Murray, Sir George, Secretary for the Colonies 325
Murrell, William, adjutant 317
Murrell's Closes 43-44
Mychell of Amsterdam 47
Mychell, John, yeoman of Shipley, servant of Edward Caryll 88, 89, 93, 94, 105, 106
Myddleton, Edward, Mayor of Winchelsea 132, 148
Myers, Lieutenant-General Sir William 301
Mysore War, Third 397

Naldect (Naldret), John, husband of Mary Dartnall 210
Napoleon, Emperor 300, 301, 310, 399
Naphtha 329
Navigation Act of 1651 246
Neale, Margery, second wife of Walter Bigge 62
Neale, Mrs, godmother of William, son of John Haynes 61, 62
Necton, William, feodary of Middlesex 43, 52, 56
Negapatam 397
Nelson, Admiral Horatio 301, 334,
Nether Tote, Slinfold 210
Neuchâtel 397
Nevill, Dorothy, Countess of Oxford, wife of John de Vere 16th Earl 127-128
New England 180, 258
New Haven, Connecticut 261, 262, 263
New Year's gifts 40
New York 239, 240, 241, 254, 255, 257, 258,

437

259, 260, 261, 262, 263, 266, 268, 269, 270, 358, 359, 360, 362, 369, 371
Newbery, Ralf 38
Newburgh, Monastery of [Newburgh Priory], Yorkshire 51
Newcastle Great House and Plantation xv, 228, 229, 231, 234, 238, 240, 241, 272, 273, 274, 275, 276, 277, 280, 289, 290, 291, 293, 295, 302, 305, 315, 328, 334, 345, 350, 356, 358, 360, 364, 365, 366, 367, 387, 403
Newport, Barnstaple, Devon 381
Newport, William, Sheriff of London 49
Newton, Dolly, daughter of *Old Doll* 299, 309
Newton, Elizabeth 309
Newton Estate 298, 299, 306, 309, 310, 313, 316, 334, 337, 358
   plantation ledger
   record of births and deaths of slaves
Newton, John
Newton, Sarah 337
Newton Papers [Senate House Library, University of London] 334, 335, 336, 337, 362
Newtown (Cambridge), Massachusetts Bay 180
Nicholson, Francis, Lieutenant Governor of New York 260
Nickerie, Dutch Guiana [Surinam] 383
Nicholls, William Queen's purveyor 71, 126
   death sentence passed on 71, 126
Nicoll, John 203,
Nicolls, William 261
Noell, Martin, merchant 246, 253
   shipment of arms, horses, prisoners etc to Barbados 253
Norfolk xiii, 22-23, 109, 111, 113-117, 128, 144, 151, 375
Norfolk Hotel, London 304
Normandy, export of cloth to 82
Norreys family 29
Norreys, Sir William 27
Norrys, Clement 111
North Erpingham 115
North Sea 137
Northumberland, Henry Percy, Earl of, 83, 84, 85, 86, 89, 93, 96; see also Petworth
   arrest and interrogation of 85
   imprisonment of 96
   suicide of 96
Norwich 114, 181
Noy, Peter de la 261
Nurse, John Henry, Esq 326

Officers, Her Majesty's, 43, 133, 164, 166
Offley [Offeley], Henry 121, 122
Offley, Sir Thomas, Merchant Tailor and former Mayor of London 122, 133

'*Old Doll*', a slave 299
Old Fish Street, Parish of St Nicholas Cole Abbey; *see under* London
   Taverns in 48
Old Machar, Aberdeen 394
Oldmixon 255, 268
Oliver, Harriet Watkins, Mrs Henry Haynes 282, 356-357
   Antiguan marriage settlement of 356
Oliver, Vere Langford 234, 239, 251, 333
Orderson, Isaac Williamson 323
Orderson, Revd Dr Thomas Harrison 370
Orinoco (river), Venezuela 324
Osborne, Elizabeth 330, 331
Osburne, Peter 50
Osmotherley, Northallerton, Yorkshire 371, 372, 373, 374, 375, 377, 379, 396, 397
   St Peter's Church 373, 375, 377, 378
Oste [Ost, Oast or Oost] 136, 138, 139, 142, 149, 150
Oughterson, Anne Walker Glieison, daughter of Arthur 382
Oughterson, Arthur 382
Oughterson Plantation 312
Outlines of Equity 357
Ovey, John 108
Owen, Richard, purveyor of sheep and oxen 58
   dismissal of 58
   petition of 58
Owston [Aweston] Yorkshire, Rectory of 112, 126
Oxford, (16th) Countess of 127-128
   16th Earl of 119, 120, 127-128
      bigamous marriage of 127-128
      'unkynde dealing of' 127
   (17th) Countess of 201
      donation for the Hackney poor of 201
      disbursement of, by Richard Haynes and Walter Halliley 201, 213, 214
   17th Earl of, 85, 104, 121, 128
   livery of 38
Oxley, William 329

packhorse(s) 79, 138, 139, 151
packs, heavy 86, 87, 92, 93
Paddington, Peterborough Place 366, 384
Page, Edward, Esquire 58, 60, 63, 71
Page, Richard, of St Giles in the Fields 60
Page, William, gent., son of Edward 58, 60, 63, 71, 199
   action brought against John Haynes by in Court of Common Law 71, 199
Paget, Charles, fourth son of Baron, brother of Thomas 84, 85, 86, 90, 93, 96, 104
   involvement in Babington plot 104
   meeting of, with the Earl of Northumberland 84

INDEX

with Thomas Paget 84
with William Shelley 84
return to England through Arundel 84, 90
reversal of attainder 96
self-imposed exile in France of 84
stay at Conigar Lodge, Petworth
Walsingham's agent 104
Paget, Thomas, Lord 74, 84, 85, 86, 87, 88, 89, 91, 93, 94, 95, 96
attainder *in absentia* 96
confiscation of lands 96
flight (escape) to France 62, 85, 86, 87, 90, 94, 400
imprisonment for aiding Edmund Campion 84
Paget, William, Baron 84
Paleologus, Emperor Constantine 336
Paleologus, Prince Theodore, great, great, grandson of Thomas, husband of Mary Balls 336
children (princes/princesses): Dorothy 336
Ferdinand [d.1678, Barbados] 306, 336
famous tombstone of 306
John 336
Mary 336
Theodore 336
Paleologus, Prince Thomas, brother of Constantine 336
Pallingham Quay, Wisborough Green 108
Pallmer (Palmer), Mr (Robert) MP 314
Palmer, William 89
Pandora 382
Paradise Street Unitarian Chapel, Liverpool 347
Pari jugo dulcis tractus 64
Parker, John 53, 54
Parkins see Perkins
Parkinson, Robert, gunner 355
Parliament, Members of, claims [inappropriate] for expenses by 401
Parochial Medical Centre, St John, Barbados 363
Parre, Thomas 64
Parry, Governor David 298
gold coin value standardised by 298
Parvis, Edward, London merchant
Passage, St Michael 345
Pastures, royal 44, 58, 59; *see also* Creslow Pastures, Sayes Court and Tottenham Court
Patching 84, 85, 89, 93, 105
Patching Copse 85,
Patrick Brompton, North Yorkshire, The Hall at 372, 373, 376, 379
Paulet [Powlett], Sir Amyas 112, 125, 126
dislike of Queen Mary of Scotland 125,
jailer of Queen Mary of Scotland 125
Paulet, Sir Hugh, father of Amyas 125

Paulet, William, First Marquis of Winchester 164
Payne, Jane Alleyne, Mrs Richard Haynes, daughter of Joseph 296, 320, 345, 364, 387
residuary legatee, Richard Sharp of Philadelphia 364
Payne, Joseph Alleyne, husband of Jane Cragg 345, 364
Payne, Sarah Ann, Mrs Robert Haynes, daughter of Joseph 296, 349, 364, 365, 370, 374
children of: 349, 374
death of 349
memorial to 370
residuary legatee, Richard Sharp of Philadelphia 364
Pead, Lieutenant William
Peere, Robert 192
Peers, Richard 231
Peirse, James 371
Peirse, Richard William Christopher 371
Pellet, Thomas, of North Stoke, farmer and servant of William Shelley 88
conversation with William Dartnall implicating Haynes 88
examination of 88
Penruddock, Col. John 237, 247, 248
Pepys, Samuel 121, 223
Percival, Captain Thomas, RN 310, 336
Percy, Henry, Earl of Northumberland; *see* Northumberland
Perkin, William 287
Perkins, Dorothea, wife of Thomas 101
Perkins, Richard, son of Thomas 101
Perkins, Thomas, father of Richard 101
Perryman, Richard Jun, husband of Anne Haynes daughter of John 240, 242, 266, 267, 271, 387
children: Ann, supposed wife of Thomas Beachamp 241, 242
Ann, wife of James Critchlow 240, 241, 266, 267
children: James 266
Henry 266
James 266
Edmund 241, 266, 267
Elizabeth 241, 266
Elizabeth another daughter of Richard and Anne 241, 266
Frances 241, 266, 267
Richard, son of Richard and Anne 241, 266
Richard, another son of Richard and Anne 241, 266
Robert 266, 267
Samuel 241, 266, 267
Perryman (Berriman), Richard Snr 242
Perth Amboy, New Jersey 258, 259

439

Peter Bonaventure 249
Peterborough Cathedral 392
Pether, a man called 31
Pether, Bastian 37
   occupation of land at Hartley, Shinfield by 37
Pettiman, Joan 206
petty larceny, sentences for 55,
Petty School 74
Petworth 83, 84, 85, 86, 87, 91, 96, 99
Petworth House, seat of the Earl of Northumberland 83
Petworth Park, Conigar Lodge in 84
Philipse, Frederick 258, 260, 261, 262
Phillicott, Thomas 240, 257
Phillipot, John 122
Phillipott (Philpot), Sir George 122
Phillipott, Thomas 122
Phillips, Captain D 355
Phillips, Francis, a minor, husband of Hester Dartnall 192, 212
Pierrepont, Thomas 323
Pigeon, Robert esquire, escheator for Middlesex, 219
Pile, Charlotte, widow 365
Pilgrim Plantation 263
Pilkington, Annie May, wife of Henry Haynes 380
Pilkington, William, of Roby Hall, father of Annie May 380
Pinckard, Dr George 295, 297, 333
Pinder, Elizabeth (Senhouse) 385
Pinder, Francis Ford, bother of William
   residence: Gay St, Bath, England
Pinder, John Hothersal, husband of Thomasine Haynes 296 315, 316, 337, 385
   Major in the Barbados Militia 385
   memorial to, St Swithin's Parish Church, Walcot, Bath 385-386
   move to England 385
   Representative of the Assembly, St John 315, 316, 337
   sudden death of
   will of
Pinder, Reverend William Lake, brother of John Hothersal 385, 386
Pinder, Richard 385
Pinder, Thomasine (née Haynes) 386
Pinder, William, father of John Hothersal 385
   Ashford Plantation, owner of 385
Pinhorne, William 261
piracy 74, 81, 97, 103, 146, 195, 400
pirates 80, 81, 195
plague, bubonic 14, 60, 64, 106, 109, 121, 124, 135, 157, 158, 159, 167, 190, 206, 216, 218, 219, 223
   death from in the Haynes Family 14, 60, 64, 109, 124, 135, 159, 190, 206, 216, 218, 219
   England 158
   Hackney, 1593 124, 129
   1603 218,
   1607, 216, 219
   London 124, 158
   major epidemics of 158
   preventive ordinances 158
   treatments in Tudor times 158
plague of locusts 235
plantations; *see under individual names*
planters, calls for compensation of 299, 334
   departure from Barbados of 334, 344, 345, 371, 402
   indebtedness among 299, 334
plantocracy xvii, 272, 314, 349
plate (silver) 65, 262, 295, 305, 306, 323, 365, 381
   belonging to John Haynes of New York 262
   belonging to Lt Gen Robert Haynes 295, 365, 381
Plowman, Matthew, collector of revenues, New York 260
Pollard, Phillipa, wife of Sir Hugh Paulet 125
Pooler, Sarah 267
Pooler, Thomas 254, 267
Pope Leo XIII 8
Popham, Sir John 162, 169
Popham, RN, Admiral Sir Home 398
portage 78
Porter, Thomas 17, 169
Portsmouth 304
Poultry, London 122, 133
Poultry, the 43, 58, 163, 185
Powell, Captain Henry, 227
Powell, Captain John 227
pre-emption, purveyors' right of 47, 136, 138, 141, 165; see also prerogative
Prendergrass, Margaret 256
Prentis, Elizabeth, daughter of William 203
Prentis, Helen, daughter of William 203
Prentis, William 203
prerogative, royal 111, 136, 138, 141
Preswicke, Miles 202, 203
Price, Barbara, wife of Samuel Branch jnr 264-5
   married secondly John Miskett 265
Price, Elizabeth, mother of Barbara 264
   godmother of three grandchildren 264
Price, William 147
prison 57, 96, 105, 141, 143, 146, 149, 261, 283
   Bridewell 139, 149-150
   Gatehouse 96
   Marshalsea 90, 94, 146
   need for in Bridgetown 318
   Tower of London 93, 96, 181

INDEX

prisoners of war, escape of during hurricane 283
Prittlewell, Essex 394
privateers, 308, 390
Privy chamber, gentlemen of the 166
  groom of 66,
Privy Council 39, 50, 51, 57, 65, 81, 85, 114, 131, 141, 142, 145, 146, 195,
Privy Seal 133
prize fight (bare knuckle boxing, or pugilism) 285, 286
prizes 302, 346
Procter, Millegan, carpenter on the *Cockatrice* 355
Procter, William, Commander of the brig, *Marques Wellington* 382
Progresses, Royal 67, 110, 111, 112, 132, 158, 166
Prosser, Robert 7
Punch, Agnes Louise, Mrs George Barrow Haynes 393
Punch, Mrs Agnes S. 393
Punch, John J. 393
purveyance laws 57, 71, 114
purveyor[s] xv, 36, 39, 41, 45, 46, 47, 48, 49, 50, 51, 53, 57, 58, 66, 68, 74, 81, 82, 109, 112, 113, 114, 115, 116, 117, 126, 131, 132, 136, 137, 138, 139, 140, 141, 142, 143, 144, 145, 146, 147, 149, 151, 161, 163, 164, 165, 167, 171, 401
  abuses of 57, 58, 114, 115, 145
  of the Queen's mouth 113, 116
Pye, Robert 253
Pyott, Mary, of Tring 242, 252

Quakers, the 299, 300
Quarles, James, 43, 185
  Chief Clerk of the Kitchen 43, 185
  Keeper of Creslow pastures 43, 185
  Sergeant of the Poultry 43, 185
  Supervisor of the will of William Haynes of Much Hadham 185
  Victualler of the Queen's Ships 43, 185
Quarterly Review 315
Queenhithe Ward; *see* London
Queen's gunner 32, 161
Questhouse 219, 220
Quintyne, Ann, wife of Thomas 276
Quintyne, Edward 286
Quintyne, Major Thomas 276
  plantation of [Quintynes] 276

racial inequality 344-5
Radclyffe, Robert 56
Radipole, Weymouth 357
Rainscroft, Arthur 53, 54, 69
Ramsden, John, carrier 86, 87, 91, 93, 104
  injury to horse of 87
Randolph, Bernard 49
Ratcliff, Middlesex 246
Ravensbourne 42
Rawleigh, Rebecca 191
Raynebowe, Ralph, servant of Edward Caryll 89
Raynehurst (Rainhurst), Chalke, Kent 51, 52, 111, 112, 122
Reading [Berkshire],
  Abbey 4, 8, 12, 16, 17, 28,
    suppression of 4, 8, 17
  Abbot of 3, 8,
  Albion Place [Street], 347, 351
  Charter of, 1559 4, 5, 9
  Christchurch Road, 16
  cloth industry 23, 25, 36, 38, 74, 75, 161
  Dolphyn, the 8
  fulling at 75
  Highgrove Street, 16,
  John Hopton, Mayor of 8, 14
  Prosperity of 74
  Records of the Borough of 8
  St Giles, 3, 7, 8, 9, 11, 14, 16, 17, 18, 19, 26, 27, 29, 31, 36, 37, 38, 109, 125, 130, 347, 352
  St Lawrence Church Obituary 22
  Syvyorstrete [Siveor or Silver Street] 4
  tawny, yellow cloth (heraldic colour) 38
  Whitley, Fountain ale-house 5
    conduit 4, 5, 6, 16
    Conduit Close 3, 4, 5, 6, 7
    Crown-fields 6
    farmhouse 5
    fish ponds 6
    Gaunder's Coppice 6
    Manor and Park 3, 4, 6
    Spittlefields 6
    Street 4
    turnpike 4, 5
Redwar, Leah, widow 365
Reece, Bezsin K. 323
Reece, Catherine Garrett, wife of John Torrance Haynes 393
Reece, Elizabeth, Mrs Robert Haynes of Thimbleby 296, 306, 349, 370-374, 375
Reece, Robert, father of Elizabeth 306, 321, 337, 349, 350, 370, 375, 388
Reece, Mrs Sarah (née Knight) 370
Reece, William 321, 371
Reed, Ann Elvira, mother of Elizabeth Haffey, Jane Morris, Lucy Haynes and Frances Haynes 361
Reed, Elizabeth, sister of Lucretia 361
Reed, Elizabeth Haffey, wife of Henry Haynes Walton 358, 361, 362
Reed, Frances Haynes 361

441

Reed, George, of Barbados and Dockfour, Demerara, father of Lucretia 360, 361
Reed, Jane Morris 361
Reed, Lucretia [Lucy], Mrs Edmund Haynes 282, 360
Reed, Lucy Haynes 361
Reeves, John, plantation of 232
Reid, Archibald 243, 252
Relfe, William, of Ore 56
reprisals against America, Prince Regent's order of 308
residence, certificate of 112,
Restoration, The 181, 223, 248
Réunion, Ile de Bourbon 293
Reve, Christopher 56
Reve of Queenborough, Kent, merchant 106
Reynolds, T, pardon of 146
Richards, Catherine 276
Richards, Judith, Mrs Sampson Wood Sen. 358
Richards, Paulus 260
Richardson, Samuel, Coroner, plantation of 233
Richardson, William 262
Richbell, Leonard 108
Richings, Rebecca, wife of William Cheek Bousfield 269
Ricketts, Governor George Poyntz 300
Rickettsia, disease caused by 268
Rickettsia africae 268
Rio de Janeiro 393
Ripon, Yorkshire 379
Rippiers [rippers; ripiers] 138, 139, 150
Rist, Philip 273, 274
River Arun 81, 108
Rivers, Colonel Marsellus 248
Robert 382
Robert Haynes, schooner 382-383
Robert James Hayne, brig 383
Roberts, Clement 232
Robinson, Dr 394
Robinson, Dr Robert 395
Robinson, Haynes Sparrow 395
Robinson, Henry Higginson 395
Robinson, Rev John Matthews, husband of Margaret Sparrow 369
Robinson, Robert Alleyne 395
Robotham, servant of the Earl of Northumberland 91, 93, 106
Roby, St Bartholomew's Church 380
Roby Hall, Lancashire 380
Rockall [Rockell; Rokholl; Rock[h]all], William 15, 33
    marriage to Anne [Ford] Haynes 33
Rockell, Richard, wife of 38
Rockell, Joan, daughter of 38
Rockell, Robert, son of 38
Rockoll, widow of Wargrave 37
Rodney, Admiral Sir George Bridges 283, 284

Roger, Thomas, of St Andrew's Parish 252
Rogers, Hanna 243
Rombouts, Francis 261, 262
Rosalie Bay 305
Rose, Roger 202
Rosewell, Richard 263
Rous, Samuel, President of Barbados 281
rovers, *see* pirates
Rowe, Sir Henry of Hackney 219
Rowe, Lady Susanna, wife of Sir Henry 219
Royal African Company 257
    slave-trading monopoly 257
Royal Bank of Liverpool, insolvency of 368
Royal Naval Dockyard 42
Royal Stables 111, 213
Royalist[s] xv, 181, 196, 203, 218, 229, 237, 247, 248, 401
Ruby Plantation 312
    loss suffered during uprising 312
Rugby School 338
Rugemere, lands of Prebend of, 44
Rugg, John 8,
Rules, The Marquis of Queensbury 285
rum 297, 298, 303, 313, 383, 384
Rumboldswyke, Sussex 24, 26
Ruscombe, Berkshire 32, 170
Rush, Mary 348
Russel, Colonel Francis, Governor of Barbados 256
Rutland, Countess of 247
Ryan, Cheeseman Moe 355
Ryan, Mary wife of Michael 355
Ryan, Michael, father of Cheesman Moe Ryan 355
Rycroft, Hon. John 321
Rye 50, 51, 80, 91, 97, 104, 106, 130, 131, 132, 133, 134, 136, 137, 139, 140, 141, 142, 144, 145, 149, 150, 159, 165, 169
    Baddinges Warde 136
    House of William Haynes 136
    customer of 50, 80, 91, 97, 106, 137
    fish market 136
    fishermen; see under Fishermen of Rye
    jurat(s) of 50, 137, 138, 139, 140, 141, 149, 159-160
    survival of children of 159-160
    King's Shop 136
    Mayor of 138, 140, 141, 150, 165
    Middlestret Ward 141
    Strand 136
    Watchbell Street 136

Sadlair, Sir Rafe, Secretary 65
Sage, C.F. 323
Salisbury, Constable of the Castle of 162, 170
Salisbury, Sir William 215
Salisbury Uprising (Penruddock) 248

INDEX

Samuel of London 246
Sarah, the brig of Workington 305
Sares of Abinger, Surrey, yeoman 89, 93
Saunders, Nicholas, of Ewell, Surrey, husband of Elizabeth Haynes of London 177
Saunders, Thomas, husband of Elizabeth Haynes of London 177
Savage, Thomas, keeper of Marylebone Park 43, 66
Saxony, Duke of 191
Sayes Court, Deptford Strand, Kent 39, 41, 42, 44, 45, 59, 61, 66
   lands of
      Broade Marsh alias Mould Meade 66
      meadow land abutting the Thames 66
      Broomefield 66
      Great Crane Meadow, part of 66
      Neales Marsh 66
      Pott Mead alias Crabtree Mead 66
      Overseer of the Queen's meadow (John Haynes) 66
   Park of 42,
   pastures 42
   slaughterhouse of 42
Scarborough 137, 376, 377
   South Cliff
   St Martin's Church 376
   West Street 376
school(s)
   Grammar 12, 13, 74, 196, 201
   Monastic 12
   Petty 12, 74
   Reading Abbey 12, 74
Scott, Elizabeth [née (?) Blanchett, formerly Haynes] 256
Scott, James 241, 256
Scott, Peter, mayor of Albany 261
Scott, William, son of Elizabeth 256, 257
scullery, the 36, 59, 163
scurvy 166
Seacombe, Cheshire 348, 351, 381
Seaforth, Lord – see Windward Islands; Robert Humberstone Mackenzie
seal of gold, great 35, 63, 197
Seal of Hezekiah Haynes 180
Seal of Office xv, 63
Seal of Spalatro 63, 173
seal, ring 63, 197
   William Haynes's 73, 147, 151, 153
Sealey (Sealy), Joseph 328, 335
Sealey, Thomas 297
   loan to Robert Haynes 297
Sealey, William 303, 335
seam (of fish) 151
searcher 78, 80, 92, 94, 97, 106, 137, 141, 151
Seaward, Elizabeth, née Haynes, formerly Branch 241, 264, 265

Seaward, Elizabeth, wife of John Branch 265
Seaward, John 241, 264, 265
Seawell Estate 313, 316, 358
   destruction during the slave uprising 313
Senate House Library, University of London xiii
Senets (?),William [*See* Sennott]
Senhouse, Edward H. 337
Senhouse, Elizabeth, second wife of Francis Ford Pinder 385, 386
Senhouse, Mary Ward [Mrs Barrow] 304
Sennott (Senets), William, 233
Sentence on the Will of John Haynes 73, 193
Sentence on the Will of Christopher Haynes 100
Seringapatam, siege of 397
servants, white; see indentured servants
settlers, early 185, 231, 234, 306, 401
Shadwell 230, 234, 246
Sharp, Richard of Philadelphia 364
Sharpe, William 337,
Sharpe, William Thomas 362
Shaw, Samuel, Master of the brig *John and Mary* 382
Sheffield, Lord [the Earl of Oxford's brother-in-law] 128
Shelley, Jane, wife of William 96
   distress of 96
Shelley, John, recusant brother of William, of Clapham 90, 102
Shelley, Mary, wife of Sir George Cotton 105
Shelley, Richard, uncle of William 96
Shelley, William, of Michelgrove, Clapham, Sussex 74, 84, 85, 87, 88, 89, 90, 91, 92, 93, 94, 95, 96, 105, 107, 400
   arrest and interrogation of 85
   attainder for conspiracy with Charles Paget 96
   confession of 96
   death in Gatehouse prison 96
   lease of property in Climping to Christopher Haynes 90
   negotiation of the escape 90
   organisation of the escape of Paget and Arundell 87, 89
   property in Queenhithe Ward, London 105
   respite of death sentence 96
   sentence of 96, 107
   threatened with the rack 96
   trial of 96
Shellye (Shellie), John, the elder 200, 214
Shepherd, Anne Walker, widow of George Barrow Haynes 366, 385, 387
   burial at Clifton, Somerset
   guardian of great nephew, Richard Haynes of Clifton
   home in Carlisle
   home in Paddington

Shepherd, Edmund, burgess and customer of Arundel
Shepherd, George Frederick 384
Shepherd, Richard, controller of Arundel 102
Shepherd, William 293
Sheriffes and Barwick Hall, manors of 202
Shinfield, Berkshire 27, 29, 37
Shiplake 32
ships-of-the-line 283
ships, passenger lists of 228, 258
Shipton, Hampshire 132
Sibourne, John 99
Sidall, Robert 231
sidesmen 200
Sidney, Lady Mary 158
   facial disfigurement of 158
Simonds B.R. 192, 211
Simpson, Mary, wife of Benjamin Carleton Howell 378
Simpson, Thomas 232
Sindlesham, Berkshire 25
Sixteen Acres, land, part of The Hydes, Climping 105
Skeete, John Brathwaite, President in Council, Barbados 370, 384
Skidmore, Erasmus, purveyor of grain 113
Skinner(s) 50, 65, 183, 207, 208, 209
Slade, William, postmaster 382
Slade, Mary Ann (Haynes), wife of William 382
slave apprentices 343, 344
slave codes (laws) 311
slave compensation 299, 334, 345, 350, 368, 384, 388, 402
slave rebellion, Easter Day, 1816, *Bussa's Rebellion*; *see* Insurrection of 1816
slave registers 383
slave trade 299, 300, 307, 311, 334, 337, 383
   abolition of 289, 299, 307, 311, 315, 334, 383
slaves: African 227, 311
   African-European 322
   emancipation of; *see* Emancipation
   execution of 313, 314
   families, encouragement by planters of 311
   high price of 257, 307
   killed in the 1816 uprising 313
   looting by 283, 312
   manumission of; *see* manumission
   named: Abba [girl] 256
      Ambrabah [girl] 256
      Arran [man] 256
      Ashee [girl] 256
      Bussa [Busso or Bussoe] 311, 312, 336
      Cinaine Hagar [Mulatto girl] 256
      Crab 240
      Flora 256
      Hamlet 301, 301, 302, 382

Henry 388
James Cave 339
Jane 240
John Francis 388
John Haynes
King William 311
Kitty Thomas 309, 336 [mulatto]
Mary 242
Mary Thomas 309
Mary-Ann Saer [mulatto] 309
Minbah [girl] 256
Old Doll 299
Polly Kitty Williams [mulatto] 309
Robert James 344
Sandy 388
Sue 263
Will 240
religious instruction of 320, 321
wilful murder of 285
   capital offence for 285
female, punishment of by flogging 323-324
Sloughter, Colonel Henry, Governor of New England 261, 270
smallpox 158, 159, 204, 276, 278, 279, 294, 333; *see* inoculation
Smiles, John 263
Smitton, William, attorney to Christopher Haynes 77
Smith, Allderman 178
   house at London Stone 178
Smith, Bartholomew, of Hackney 214
Smith, John, [Earl of Oxford's servant] 128
   attack on Joan Jockey 128
Smith, John, merchant of Barbados 259
   business with Captain John Bowne 259
Smith, Sir Thomas, author 117, 118
Smith, William, former Governor of Tangier ('Tangier Smith') 261
Smyth, Simon, of Petworth, Secretary to the Earl of Northumberland 83, 86, 87, 91, 92, 93, 104
Smyth, Thomas, witness 32
Smythe, Thomas, customer of London 56, 71
Snap, Anthony, servant of William Shelley 85, 104
Society Estate 280-281
Society for Promoting Christian Knowledge 321
Society for the Propagation of the Gospel 280
Society of Friends [Quakers] 290, *see also* Quakers
soilmen 55
Solway Firth 8
Somer, Rosalie, second wife of John Higginson Haynes 370
Somerfyld, Margaret 7
Somers, Will, The, Old Fish Street 48,
Somerset, Edward, Duke of 3

INDEX

South Egney, Stepney 122
South Stockton, Yorkshire 377
Sowerby, Yorkshire 379
Spalatro 63
Sparrow, Margaret, wife of Revd John Matthews Robinson 369
Speaker of the Barbados House of Assembly; *see* Barbados, House of Assembly
Spensar, Richard 38
Spencer, Thomas C. 323
Spicery 59, 153
Spooner's Hill, St Michael, Barbados 393
sports, *see* entertainments
spotted fever 255, 268
spousal 156, 157
Springfield Estate 365
　mortgage on by Richard Haynes 365
Stanley, Sir William 146
　English Legion of 146
　treachery of 146
Stanley, Thomas 51
Stanner, Thomas, mariner 81
Star Chamber 96
Starton, Christopher 231
State Oath, Barbadian 370
Statute of Artificers 13
Stede, Edwin, Lieutenant Governor of Barbados 292
Stephen Jr, James 314, 337
Stepney, Manor of 122
Sterling Plantation 345
Stevenson, Louisa Georgiana [née Lardy], sister of Christian Frederick Lardy 387
Stevenson, Reverend Thomas, of Winchester 387
Stevenson, Thomas, surveyor 243
Steward, John 214
Stewart, Robert, husband of Anne Haynes 230, 235, 236, 238, 239, 242
Stirks 45, 164
St Andrew's Church, Holborn 285
St Anne's Parish 277
　gun implacements at 277
St Anne's Chapel, Winchelsea 132
St Augustine's Church, Hackney; *see* St John At Hackney
St Bartholomew's Hospital, Smithfield 214
St Clement Danes Church 120, 186, 192, 204, 210, 215
St Dunstan's (Dunstan and All Saints) Church, Stepney 121, 230, 234
St Gabriel Fenchurch, Parish of 40, 65
St George's Church, Barbados; *see under* Barbados Parishes
St George's Church, Bloomsbury 392
St George's Church, Everton 367
St George's Church, Guelph, Ontario 389

St Giles Reading; *see* Reading St Giles
St John's Church, Barbados; *see under* Barbados Parishes
St John's Wood 39, 54, 60,
　alleged illegal felling by John Haynes in 60
St John of Basing, William, Lord 164
St Katherine Crechurch 204
St Lucia 283
St Margaret of Scotland, church of, 376
St Margaret Pattens 65
St Mary Antill; *see* London Parish of
St Mary's Church Whitechapel 121
St Mary Magdalene; *see* London Parish of
St Mary Mounthaw 190
St Mary Somerset; *see* London Parish of
St Mary Woolchurch; *see* London Parish of
St Michael's Cathedral, Bridgetown 320, 327, 366, 290
St Nicholas Cole Abbey; *see* London Parish of
St Paul's Cathedral 44, 53, 54
St Peter ad Vincula, Wisborough Green 187, 188
St Peter Paul's Wharf 192, 211
St Vincent 283, 308, 324, 333
　earthquakes on 308
　La Soufriere, explosion of 308
　volcanic eruption on 308
　　climatic effects of 308
　　effects on Barbados of 308
　　effects of ash from 308
stockfish 41, 68, 149
Stocks Market, 49, 122, 130, 133-134, 140, 147
　lease by William Haynes 133
　cessation of 134, 147
　William Haynes's profit from 134
Stockton-on-Tees 377
Stonestrete, Robert 32
Stott, partner in Higginson, Deane and Stott 321, 335, 368
　purchased Mount Clapham estate 368
Strachan, Barbara, wife of William Clarke son of Henry Husbands Haynes 394
Straughan, Mary Harbourne, wife of Robert son of William Clarke Haynes 388
Stretten, Richard 38
Stuart, Abel 323
Stuart, Lady Arbella 146
Stubbs, Richard, notary public 64, 193
Study, Elizabeth 276
Sturminster Newton, Dorset 51, 69
　manor of 69
Styles, Robert, Mayor of Arundel, 1558 76
subsidy; *see* lay subsidy rolls
Suffren, Admiral Pierre André de 397
sugar, boiling house 254
　Bourbon cane; *see* Bourbon
　Brazilian cane 293, 295

445

extraction from beet 310
milling equipment 246, 365
muscovado 236, 240, 257
planter 217, 229, 230, 239, 243, 252, 256, 257, 274, 334, 344, 364, 371, 377, 378, 393, 401, 403
Summers, Ann, daughter of George and Ann, widow of Crossing 274
wife of George 274
Summers, George 274
Sunbury Plantation 303, 311, 312
Surrey, Mr, yeoman 89, 93
Surveyor of the Vestry 255
Sutton House, Sussex, home of Charles Arundel 85
Swan, The, Old Fish Street 48
Swellivant, Tobias, cooper of St Michaels, 273
godfather to Elizabeth Haynes 273
Swinnerton, Robert 287
Symons, lands called, Billingshurst 210
Syster, William, husbandman 60

tabling (gaming or gambling) house 86
Talbot County, Maryland 290
'Tangier Smith', *see* William Smith
Tarring Neville, Sussex 222
Tatershall, Earl of 336
Tavernor, John, Surveyor of Her Majesty's Woods South of the Trent 60,
Tawke [Taulke], John, recusant 94
Taylor, John, felon 203
sentenced to be whipped 203
Teach, Edward, pirate, known as Blackbeard 292
Tenham, Kent 7
Thacker, Christopher, of Hackney 203
Theft from the State 401
Theobalds 113
Theydon Bois 113
Thicket Plantation 292, 311
Thimbleby, Thimbleby 349, 353, 254, 371, 372, 373, 374, 375, 376, 377, 378, 379,380, 381, 396,
Third Mysore War 397
Thomas 382
Thomas, Commander Sir William Sidney, Bart RN, fifth Baronet of Yapton and Dale Park; 357; *see also* Thomasine Oliver Haynes
children: Frederick Louis Charles, George Sidney Meade, Isabella Montague Jane, Lucy Elizabeth 357
Thomas, Edward, of Brecknock, indentured servant 234, 250, 275
Thomas, Edward, possible father of Elvira Thomas 250, 275
Thomas, Elvira, wife of Richard Haynes 250, 273, 274, 275

Thomas Lee Haynes & Co, Liverpool 316, 382, 384
Thomas, Lord Buckhurst 56, 83, 104, 114
Thomas, Mary, a slave, mother of Polly Kitty Williams 309
Thompson, Adam, gent 232
Thompson and Rowlandson 285
Thompson, James F 383
Thomson, Agnes [Annis] 10, 15, 61, 157
Thorne, Ann, wife of Edmund Haynes 280
Thorne, Col. James 265
Thorne, Elvie 271, 275, 276, 280
Thorne, John 213
Thornhill, Mary, wife of Edward Bonnet 292
Thornton, Mary, first wife of John Haynes of Copford Hall, Essex 180, 181
Thorpe Estate 289
Thos. Lee and Co 384
Three leane sheaves 20, 21
Throckmorton conspiracy xvii, 84, 93
Throckmorton, Francis 83-84, 93, 96, 105
arrest of 84
execution of 84
plotting with expatriates 84
torture of 84
Throckmorton, Thomas, brother of Francis 84
Thurburn, Helen Elizabeth, wife of John Higginson 369
Thurloe, John 181
Thurston, Henry, Esquire, J P 201
Thwaites, Richard, carrier 86, 87, 91
Thyssen papers, Hackney Archives Department 172
Tiger 256
Timberlake, Henry, 190, 191, 193, 194, 211
bequests to Jasper, Rose and William Dartnall 191, 193
children:Sarah [Baleere] 193
Thomas 193
executor of Henry, William and Anne Haynes of Hackney and Chessington 190-191
friend of Haynes family of Hackney 190
Timberlake, Margaret, wife of Henry 194
Timbrell, William, Chief Justice of Barbados 243
Times, The 255, 335, 337, 338
Tippler (Tipler) 92, 120, 127
tobacco 227, 231, 232, 246, 249, 298
Tobago 283
toe (tow) 188
Toils, Toyles (nets) 145, 152
Yeoman of the 145, 152
Tooke, William, Auditor General of the Court of Wards and Liveries 56
Tottenham [Tottenhall] Court, London 39, 41, 43, 44, 52, 53, 54, 55, 66, 69, 169
Tower of London 32, 93, 96, 161, 181, 193
Townes, John, will of 276

446

INDEX

Trafalgar Square, Bridgetown 327
Trafalgar, victory at 335
Transatlantic crossing 257, 307, 322, 383, 384, 386
Tredcrafte, Thomas, mariner 81, 108
Tredcrofte, William, of Billingshurst 101, 108
Trend, Lawrence 243
Trevet, William 77
Trotman, Henry 362
Trotman, Robert 297, 362
Trotman, Thomas 362
Trown, Master, Clerk to Lord Burghley 59
tuberculosis 159, 279, 301
Tucker, William, of Shadwell, boatswain 246
Turner, Hercules, purveyor of the Queen's Mouth 113
Turner, Pattrick, pirate 146
Turriff, Aberdeenshire 394
Tuskar Rock, Wexford 305
Twyford, Berkshire 63, 208
Tyanson, Thomas, thief 55
Tyburn 84
Tyburn Manor 44
Tymberwood, Manor of 112
typhoid 159

United States xii, 279, 284, 308, 309, 346, 358, 359, 369
USS Constitution 390

vaccinia (cow-pox) 333
Vale, Giles 89, 93
Van Angelbeek, Johan Gerard, Dutch governor of Ceylon 397
Variola minor 294
variolation; *see* inoculation
Vaughan, N., Master of the brig *Amity* 382
Velis et Remis (with sails and oars) 63, 174, 181, 183, 354
Venerable HMS 309, 390, 398
venison from Royal Parks 152, 165
Venus 284, 285, 370
Vere; *see under* Oxford, 16th and 17th Earls of
Vestry, St John at Hackney 124, 200, 201, 202, 204, 213, 214, 219
  Long Memorandum Book
  meetings of 200, 201, 202, 204, 214
  members of 200, 201, 219
  minutes 200, 201, 202, 213, 214
  Richard Haynes, member of 200, 219
Vibert, Helear, Master of the brig *Highlander* 382
Vice-Admiral of Scotland 195
Villa Nova [formerly Haynesfield] 360, 362
Ville de Paris 284
Villeneuve, Vice-Admiral Pierre-Charles-Jean-Baptiste-Silvestre de 301

Vincent, John 53
vintry, the 56
violent distemper 255
Virginia 279, 283, 287, 391
visitation(s) of Berkshire, 1665 xvi, 9, 17, 28,
  of London, 1568, 1633, 1634, and 1635 10, 11, 14, 17, 19, 20, 24, 36, 49, 61, 66, 78, 120, 134, 153, 154, 155, 157, 172 178, 180, 184, 185, 186
  of Rutland 1618/19 63, 212
  of Surrey 28, 29
  of Wiltshire 28, 29

Wade, Michael (or Maurice), haberdasher 198, 213,
Walcote, The Hon. Daniel 294
Walcott [Woolcot; née Evans], Hannah, née Evans, widow of Nicholas 241, 258, 259
Walcott, Margaret, wife of Bishop Edward Beckles 394
Walcott, Nicholas 247, 268
Walcott, Robert J 317
Waleys, Henry le, mayor of London 68, 133
Walker, Anne, widow of Henry, and George Foster Clark; second wife of Richard Downes Haynes 278, 282, 284,
  allowing slaves to 'rob and plunder' 291
  indolence and extravagance of 291
  neglect of domestic concerns by 291
Walker, Henry 284, 290
Walker, Jane 373
Walker, Mr, the colonial agent 280
Wall, Agnes, wife of Thomas Haynes of Westbury on Trym 182
Wall, Jarret, of New Jersey 264
Waller, John, husband of Dorothy Dartnall 210
Walley, Master Thomas 222, 224
Wallis, Richard, will of 235
Walrond, Benjamin 323, 358
Walrond, Henry Jr 242
Walsh, J 383
Walsingham, Francis, the Queen's Secretary of State 62, 84, 85, 88, 92, 95, 96, 103, 146
  suspicion of Christopher Haynes 92,
Walter Hood, 394, 395
  death of Joshua Rowley Haynes 395
  memorial 395
  sinking of 395
Waltham St Lawrence, Berkshire 32
Walton, Ann Ellcock, daughter of Gen John Walton 358
Walton, Charlotte (Mrs Samuel Brown), daughter of Gen John 358
Walton, Dorothy, daughter of Gen John 358
  lived with her sister Haynes in London
Walton, Gen John, second husband of Ann Ellcock Wood, née Haynes 282, 358

447

Walton, Henry Haynes, son of Gen John 358
  children: Elizabeth (Bessie), George 358
  marriage to Elizabeth Haffey Reed 358, 361, 362
Walton, Haynes, daughter of Gen John 358
Walton, William 25
war with Spain 138, 300
Ward, Richard, cofferer 163, 165
Warde, Sir Henry, - see Barbados, Governor of
Warde, William, gentleman 40
Wardrober of the Bedds 166
Wargrave 31, 32, 33, 34, 35, 37, 38, 75, 98, 154, 157, 159, 168, 186, 205, 206, 207, 208, 209, 215, 399
Wargrave Church 32, 33, 34, 98, 157, 205, 206, 208, 209, 215
Warwick Court, Holborn 285
Warwick, Earl of 121
Washington, George 279, 280, 287
  Mount Vernon, inheritance of 379
  rented Bush Hill House 279
  smallpox 279
  visit to Barbados 279
Washington, Lawrence 279
  death from tuberculosis 279
Watson, Elizabeth Lee, wife of William Pilkington 380
Watson, Mrs Margaret [Hopkinson] 223
Watts, Thomas, Prebend of St Paul's Cathedral 53
weapons, as 'life-preservers' against vagabonds 98
Weaverham 375, 396
Wedgewood, John, Master of the brig *Betty* 382
Wedon, Andrew 43
Weeks, Sir Ronald, husband of Evylyne Elsie Haynes 381
Welch, Mary 276
Welche, Stephen, feter 143
Weldon, cofferer 170
Wellesley, Arthur [Duke of Wellington] 397
Wells, Robert 137
Welney, Norfolk 375, 376
Welshe, Ellin, widow, daughter of Hippolyte Lynnet 123, 195
Wenham, Thomas, merchant of New York 260, 261, 269
  assault of 261
  churchwarden of Trinity Church 269
  executor of John Haynes of New York 262
West Firle, Sussex 222
West India Estate 366
West India interests 320
West India Regiment, First 312
West India Regiment Second 379
West Indies xiv, 246, 248, 258, 284, 293, 294, 295, 299, 315, 320, 323, 333, 336, 346, 352, 386, 387, 395
Westbury Cemetery, Barbados 374
Westbury on Trym 182, 185, 386-387
Westdean, Ralph 77, 78
Westende Charity for almshouses, Wokingham 170, 173
Westende, John, clerk of Wokingham 170, 173
Westminster Hall 96
Weston, John 231, 249
Wharton, Isabella Carr, Mrs Nathaniel Cave 329, 350
wheat, spoilage of 126
Wheatley [Whetly, Whelly], Agnes, wife of Thomas Haynes 15, 20, 31, 36, 38, 82, 157,
  will of 38, 82
Wheatley, Robert, fourth Governor of Barbados 287
Wheatley, Solomon 36
Wheatley, William 36
Whistler, Henry 401
Whistley, manor of 29, 163
Whitborne; clerk of St Giles, Reading 109
Whitcombe, Thomas, artist 398
White Colne, Essex 123, 128, 202
  Church 128
  bigamous marriage of the Earl of Oxford at 128
  Cote Crofte [Colecroft] 202, 203
White Hart Inn, Southwark 86, 87, 93, 104
  in Shakespeare's *Henry VI* 104
White Horse, Arundel 99, 104
White Horse, emblem and pseudonym of the Earl of Arundel 104
White (Whyte, Wayght), Joan, widow of Raffe 133
White Raffe 133
White, Richard 293
Whitestaves 170
Whitley; *see under* Reading
Whittle, Caroline Haynes 338
Whittle, Eleanor Brenda Violet, gt-granddaughter of Robert Haynes of Thimbleby 338, 372, 373, 396
Wick [Wyke], Hackney, manor of 177
Wick, Gloucestershire 182
Wilberforce, William 311, 337
Wilkinson, Jonas 337
Willett, Thomas, of Long Island 261
William (of Orange) and Mary 223, 260, 261
William, King, a slave, friend of Bussa 311
William, Lord Cobham 56, 57, 121
Williams, Polly Kitty 309, 310
Willoughby, Edward, surgeon 203
Willoughby, Lord 239

INDEX

Willson, George, former churchwarden, Hackney 200
Wilson, Hannah 265
Wilson, Dr Thomas, Master of the Court of Requests 50
Wiltshire, Ernest M xii
Wimbledon 357
Winchelsea 132
   Black Friars; *see under* William Haynes, Purveyor of Sea-fish
   Grey Friars; *see under* William Haynes, Purveyor of Sea-fish
   Harbour, silting of 132
   Holy Cross Chapel; *see under* William Haynes, Purveyor of Sea-fish
   smuggling in 132
   St Anne's Chapel, *see under* William Haynes, Purveyor of Sea-fish
Winchester 65, 132, 162, 164, 170, 370, 387
Winchester, Bishop of 65, 132
Winchester, First Marquis of, 164
Windebank[e] xvi, 163, 170
Windebank, Francis [Sir] 170
Windebank, Sir Thomas, Clerk of the Signet to Queen Elizabeth 163, 170
Windsor, Henry G. 323
Windsor Royal Forest of 29
Windward Islands 279
   Lord Seaforth, Governor of 353
wine, belonging to Lt Gen Robert Haynes 365
Winnall Manor, Winchester, dispute over the ownership of 162
Wisborough Green 87, 91, 100, 101, 108, 187, 188, 189, 190, 194, 210
Withycombe, in Drayton, Oxfordshire 211
Wodecoke, William 27
Wokingham, Berkshire 25, 162, 170, 173; *see also* Westende Charity
Wolseley, Sir Garnet 379
Wolverhampton 10, 69, 134, 157, 169
Wood, Ann Ellcock, wife of Samson Wood 281, 282, 293, 298, 299, 309, 358
   advised appointment of brother as Attorney to John and Thomas Lane 358
   advised appointment of John Farrell as manager of Newton and Seawell 358
   marriage to General John Walton 358; *see under* Walton for children 358
Wood, Helen Haynes, daughter of Samson Wood 358
Wood, Henry Haynes, son of Samson Wood 358
Wood, John Jr 337
Wood, John, ship owner of Arundel 86
Wood, Richard Haynes, son of Samson Wood 358
Wood, Robert Haynes, son of Samson Wood 358

Wood, Sampson, husband of Ann Ellcock Haynes 282, 298, 309, 358
   death of 358
   Speaker of the House of Assembly, Barbados 358
Wood, Sampson Sen 358
Wood, Thomas, Sergeant of the Pantry 123, 162
Woodcock family of, Battle Abbey 29
Woodham Walter, Essex 18
Woodworth, Allegra 38, 41, 58,
Woodyard 163
Woolridge, Henry, marriage to Elena Haynes or Heynes, widow of Arundel 107
Woolridge, Henry, son of Richard 107
Woolridge, Richard, Servant of Christopher Haynes 99, 107
Woorte, (Wort) Agnes, vagrant and thief 121
Worth, Capt James Andrew, RN 390, 398
Worthing, Sussex 89, 222
Wreck Bay 395, 398
Wrenne, Charles, of Gray's Inn 56
Wrighte, William, clerk of Little Laver, Essex 47
   clerk of Colchester St Leonards 67
   monk of Hatfield Regis (Hatfield Broad Oak) 67
   will of 67
Wrightington, George, Commissioner of the Exchequer 53
Wrightson, Thomas, iron-founder, South Stockton 377
Wriothesley, Sir Thomas, Secretary 65
Wyatt, Elizabeth 243
Wyatt, Sir Henry 112
Wyatt, Sir Thomas, son of Sir Henry 51, 69, 112,
Wyther, Philip, Mayor of Arundel 102

Yapton 357
Yellow Fever (Barbados distemper) 185, 235, 264, 268, 270-271, 275, 276, 287
   relative immunity of Africans to 270-271
Yeomen; *see under* occupations
Yeoman, Anne (Annie) Elizabeth, of Osmotherley 379
Yeoman, Hannah, wife of William 379
Yeoman, John [1828-1890], memorial to 379
Yeoman, the Misses 379
Yeoman, William, of Osmotherley, father of Ann Elizabeth 379
Yonge, Elizabeth, wife of Robert Haynes Cave, 338
Yonge, John, 91, 92, 95, 97, 98, 106, 138
Yonge, Norman Bond 338
Yonge, Revd George Vernon 338
Yonger (Younger), Arms of, impaled Haynes of Hackney 203

449

possible wife of Richard Haynes of Hackney 203
York, 376,
　Marygate 373, 378
　Norwood House, Falsgrave 276

Yorkshire Gazette 397

Zippel, John Gottlieb, Moravian Minister 360
Zouche, Edward, Lord 121